Chemical Signatures of the Brain
Bridging Biochemistry, Neuroanatomy, and Brain Disorders

Chemical Biology

Editor-in-chief:
Kira J. Weissman, *University of Lorraine, France*

Series editors:
Konstantinos Beis, *Imperial College London, UK*
Lei Li, *Georgia State University, USA*

For a list of titles in this series see: rsc.li/chem-bio-series

How to obtain future titles on publication:
A standing order plan is available for this series. A standing order will bring delivery of each new volume immediately on publication.

For further information please contact:
Book Sales Department, Royal Society of Chemistry, Thomas Graham House, Science Park, Milton Road, Cambridge, CB4 0WF, UK
Telephone: +44 (0)1223 420066, Fax: +44 (0)1223 420247
Email: booksales@rsc.org
Visit our website at books.rsc.org

Chemical Signatures of the Brain

Bridging Biochemistry, Neuroanatomy, and Brain Disorders

Edited by

Sachchida Nand Rai
Saveetha University, India
Email: raibiochem@gmail.com

and

Abhishek Kumar Singh
Manipal Academy of Higher Education, India
Email: abhishek.singh@manipal.edu

ROYAL SOCIETY OF **CHEMISTRY**

Chemical Biology No. 26

Hardback ISBN: 978-1-83707-136-4
EPUB ISBN: 978-1-83707-137-1
PDF ISBN: 978-1-83707-139-5
Print ISSN: 2055-1975
Electronic ISSN: 2055-1983

A catalogue record for this book is available from the British Library

The Royal Society of Chemistry is a charity, registered in England and Wales, Number 207890, and a company incorporated in England by Royal Charter (Registered No. RC000524), registered office: Burlington House, Piccadilly, London W1J 0BA, UK, Telephone: +44 (0)20 7437 8656.

For further information see our website at www.rsc.org

For general enquiries, please contact books@rsc.org

For EU product safety enquiries, please email books@rsc.org or contact Royal Society of Chemistry Worldwide (Germany) GmbH, Römischer Hof, Unter den Linden 10, 10117 Berlin.

Printed in the United Kingdom by CPI Group (UK) Ltd, Croydon, CR0 4YY, UK

Preface

The human brain is an extraordinarily complex organ whose functions emerge from an intricate interplay between chemistry, structure, and dynamic cellular communication. Over the past few decades, remarkable progress in biochemistry, neuroanatomy, molecular biology, and systems neuroscience has transformed our understanding of how chemical signals shape brain development, function, and dysfunction. Yet, despite these advances, knowledge in these domains often remains fragmented across disciplines. **Chemical Signatures of the Brain: Bridging Biochemistry, Neuroanatomy, and Brain Disorders** has been conceived to address this gap by providing an integrated and coherent perspective on the chemical foundations of brain structure, signaling, and disease.

This book brings together fundamental principles of neurochemistry and neuroanatomy with contemporary insights into brain disorders, emphasizing how molecular and biochemical alterations translate into cellular, circuit-level, and clinical outcomes. Beginning with foundational concepts, the early chapters introduce essential aspects of brain chemistry, cellular architecture, and neurotransmitter systems that govern neuronal communication. Subsequent chapters expand this framework to include neuropeptides, energy metabolism, mitochondrial function, lipid signaling, oxidative stress, and neuroinflammation, processes that are increasingly recognized as central drivers of neurodegenerative and neuropsychiatric disorders.

A distinctive strength of this volume lies in its focus on emerging chemical signatures of the brain. Chapters dedicated to epigenetic and transcriptomic regulation, biomarker discovery, lipidomics, and metabolic profiling highlight how advanced technologies are reshaping our ability to diagnose, stratify, and treat neurological diseases. Environmental influences, including neurotoxicants, diet, and gut–brain interactions, are also explored to

Chemical Biology No. 26
Chemical Signatures of the Brain: Bridging Biochemistry, Neuroanatomy, and Brain Disorders
Edited by Sachchida Nand Rai and Abhishek Kumar Singh

Published by the Royal Society of Chemistry, www.rsc.org

provide a broader view of how external factors modulate neurochemical homeostasis. The concluding chapter looks toward the future, emphasizing interdisciplinary and systems-level approaches that bridge biochemistry, neuroscience, and precision medicine.

The development of this book has been a truly collaborative effort, bringing together contributions from accomplished researchers and clinicians across diverse geographical regions and areas of expertise. Each chapter has been authored by experts actively engaged in cutting-edge research, ensuring both scientific rigor and contemporary relevance. While the book is firmly grounded in basic science, a strong translational focus has been maintained throughout, with particular attention to therapeutic strategies, biomarker-based diagnostics, and clinical implications.

This volume is intended for a broad audience, including neuroscientists, biochemists, clinicians, pharmacologists, and advanced graduate students who seek a comprehensive and integrative understanding of brain disorders from a chemical and molecular perspective. We hope that this book will not only serve as a valuable reference but also stimulate interdisciplinary thinking and inspire new research directions aimed at improving brain health.

We would like to express our sincere gratitude to all the contributing authors for their enthusiasm, scholarly dedication, and timely efforts in shaping this book. We are also thankful to the editorial and production teams for their guidance and support throughout the development process. Finally, we acknowledge the broader scientific community whose relentless pursuit of knowledge continues to deepen our understanding of the brain and its disorders.

It is our hope that **Chemical Signatures of the Brain** will contribute meaningfully to this ongoing endeavor and help bridge molecular insights with clinical realities in neuroscience.

Sachchida Nand Rai and Abhishek Kumar Singh

Contents

Chapter 1 Foundations of Neurochemistry and Neuroanatomy 1
Falguni Goel, Vaishali Dobhal, Shipra and Pradeep Kumar

1.1 Introduction 1
1.2 Basic Structure of the Nervous System 3
1.3 Neuroanatomy of Key Brain Regions 5
1.4 Neuronal Communication and Synaptic Transmission 8
1.5 Neurochemistry of Neurotransmitters and Neuromodulators 11
1.6 Cellular and Molecular Aspects of Neurochemistry 15
1.7 Neurodevelopment and Neural Plasticity 17
1.8 Blood–Brain Barrier and Neuroprotection 21
1.9 Neurochemistry in Neurological and Psychiatric Disorders 22
1.10 Experimental Techniques in Neurochemistry and Neuroanatomy 26
1.11 Future Aspects 27
1.12 Conclusion 31
List of Abbreviations 31
References 31

Chapter 2 Neurotransmitter System in the Brain 47
Nilay Kumar Nandi, Priyanka Nayak, Pooja Yadav, Viney Lather and Prabhash Nath Tripathi

2.1 Introduction 47

Chemical Biology No. 26
Chemical Signatures of the Brain: Bridging Biochemistry, Neuroanatomy, and Brain Disorders
Edited by Sachchida Nand Rai and Abhishek Kumar Singh

Published by the Royal Society of Chemistry, www.rsc.org

2.2 Basics of Neurotransmission in the Brain 49
2.3 Major Neurotransmitter Systems in the Brain 52
2.4 The Glutamate Neurotransmitter System 52
2.5 The GABA Neurotransmitter System 56
2.6 Dopamine Neurotransmitter System 58
2.7 The Serotonin Neurotransmitter System 60
2.8 The Norepinephrine Neurotransmitter System 62
2.9 The Acetylcholine Neurotransmitter System 64
2.10 Other Neurotransmitters and Neuropeptides 65
2.11 Future Directions in Neurotransmitter System Research 67
2.12 Conclusion 68
References 69

Chapter 3 Neuropeptides and Their Role in Brain Function 82
Shivangi Gupta, Romil Sharma and Sandeep Singh

3.1 Introduction 82
3.2 Discovery of Neuropeptides 84
3.3 Neuropeptides and Their Structure 85
3.4 Biosynthesis of Neuropeptides 85
3.5 Classification of Neuropeptides 86
3.5.1 Hypothalamic Releasing Factors 86
3.5.2 Pituitary Neuropeptides 86
3.5.3 Gastrointestinal Peptides 88
3.6 Development of Neuropeptide Signaling Pathways 89
3.7 Neuropeptide Binding Receptors 91
3.8 Functional Aspects of Neuropeptides in the Brain 91
3.8.1 Regulation of Neuronal Development by Neuropeptides 92
3.8.2 Role of Neuropeptides in Maintaining Neurological Homeostasis 93
3.8.3 Role of Neuropeptides in Sensory Perception 94
3.8.4 Role of Neuropeptides During the Systemic Inflammatory Response 94
3.9 Conclusion 95
References 95

Chapter 4 Neurochemical Pathways in Energy Metabolism and Mitochondrial Function 102
Md Abubakar, Ritu Dahiya, Abhishek Ambawatiya, Krushna Chandra Maharana, Shahnaz Alom, Aruna Rawat, Amita Rai, Damini Yadav, Krishnendu Adhikary and Nitesh Kumar

4.1 Introduction 103
4.1.1 Brain Energy Demands and Mitochondrial Centrality 103
4.2 Mitochondrial Basics in Neurons 103
4.2.1 Structure, Dynamics, and ATP Production 103
4.2.2 Neuronal Mitochondrial Distribution and Its Trafficking 104
4.3 Primary Energy Substrates 104
4.3.1 Glucose and Glycolysis 104
4.3.2 Astrocyte–Neuron Lactate Shuttle (ANLS) 105
4.3.3 Ketone Bodies and Alternative Fuels 105
4.4 Key Regulatory Pathways 106
4.4.1 AMPK (AMP-activated Protein Kinase) and Energy Sensing 106
4.4.2 Sirtuins and NAD^+ Signaling 107
4.4.3 Mitochondrial Biogenesis and PGC-1α 107
4.4.4 mTOR and Anabolic Balance 108
4.5 Calcium (Ca^{2+}) Activity and Mitochondrial Function 108
4.5.1 Ca^{2+} Regulation of the TCA Cycle and Respiration 108
4.5.2 Risks of Ca^{2+} Overload 109
4.6 Redox Signaling and ROS 110
4.6.1 Sources and Physiological Roles 110
4.6.2 Antioxidant Defenses and Nrf2 Pathway 112
4.7 Neurotransmitter Metabolism and Energy Cost 113
4.7.1 Glutamate–Glutamine Cycle 113
4.7.2 GABA Shunt and Anaplerosis 114
4.7.3 Catecholamine Synthesis 115
4.8 Mitochondrial Dysfunction in Brain Disorders 115
4.8.1 NDDs 115
4.8.2 Mood Disorders and Epilepsy 119
4.9 Therapeutic Interventions 120
4.9.1 Mitochondrial Protectors and Antioxidants 120
4.9.2 Ketogenic Approaches and Emerging Therapies 122

4.10 Conclusion and Future Perspective 126
Author Contributions 127
Acknowledgements 128
References 128

Chapter 5 Oxidative Stress, Neuroinflammation, and Neuroprotection 137
Vishal Jain

5.1 Introduction 137
5.2 Molecular Mechanism of Oxidative Stress 140
5.2.1 Sources of Reactive Oxygen Species (ROS) 140
5.2.2 Antioxidant Defense Systems 143
5.2.3 Transcriptional Control *via* Nrf2 Pathway 144
5.2.4 Imbalance Between ROS/RNS and Antioxidant Defenses 146
5.3 Oxidative Stress and the Nervous System 146
5.3.1 Intrinsic Vulnerability of Neural Tissue 146
5.3.2 Mitochondria: The Core Source of ROS 147
5.3.3 Glial Involvement in Oxidative Homeostasis 148
5.3.4 Calcium, Excitotoxicity, and ROS 148
5.3.5 Targets of Oxidative Damage 148
5.3.6 Redox-sensitive Signaling and Gene Regulation 149
5.3.7 Disease-linked Manifestations of Oxidative Stress 149
5.3.8 Oxidative Stress and Aging 150
5.4 Neuroinflammation 151
5.4.1 The Physiological Role of Neuroinflammation 151
5.4.2 Microglia: The Primary Effectors of CNS Inflammation 151
5.4.3 Astrocytes and Their Regulatory Role 152
5.4.4 Cytokines and Inflammatory Mediators 153
5.4.5 Key Signaling Pathways in Neuroinflammation 153
5.4.6 The Interface Between Neuroinflammation and Oxidative Stress 154
5.4.7 Breakdown of the Blood–Brain Barrier 154
5.4.8 Neuroinflammation Across Neurological Disorders 154
5.4.9 Resolution and Regulation of Neuroinflammation 155

5.5 Therapeutic Approaches for Oxidative Stress Along with Neuroinflammation 156
5.5.1 Pharmacological Approaches 157
5.5.2 Non-pharmacological Approaches 158
5.5.3 Herbal and Nutraceutical Approaches 159
5.5.4 Behavioral and Environmental Approaches 160
5.5.5 Integrated Perspective on Neuroprotection 161
5.6 Summary 162
5.7 Limitations and Future Directions 163
5.8 Concluding Remarks 164
Acknowledgements 164
References 164

Chapter 6 Lipid Signaling and Brain Disorders **170**
Mukesh Kumar, Nitish Kumar Singh, Mayank Gangwar, Vivek Chandra Verma, Syed Mohammad Nasar Ata and Mudassir Jan Makhdoomi

6.1 Introduction to Lipid Signaling 170
6.2 Types of Lipids Involved in Signaling 172
6.3 Lipid Signaling Pathways in the Brain 172
6.3.1 G-protein-coupled Receptors (GPCRs) and Lipid Mediators in Neurology 175
6.3.2 Endocannabinoid Signaling *via* GPCRs 176
6.3.3 Lysophospholipid GPCRs: LPA and S1P Receptors 177
6.3.4 GPCR–Lipid Mediator Crosstalk and Neuroinflammation 177
6.4 Lipid–Protein Interactions in Neural Function 178
6.4.1 Intracellular Signaling Cascades Involving Lipids 179
6.4.2 Impact of Lipid Modifications on Synaptic Plasticity 180
6.5 Lipid Dysregulation in Brain Disorders 182
6.6 Lipid-based Biomarkers and Therapeutics 183
6.6.1 Targeting Lipid Pathways for Treatment 186
6.6.2 The Endocannabinoid System as a Potential Therapeutic Target 188
6.7 Advances in Lipidomics and Brain Research 189
6.7.1 Alzheimer's Disease 190
6.7.2 Parkinson's Disease 191
6.7.3 Huntington's Disease 192
6.7.4 Schizophrenia 192

6.8 Challenges in Lipid-based Drug Development 192
6.8.1 The Potential of Personalized Lipid Therapies in Neurology 193
6.9 Conclusion 194
Abbreviations 195
References 196

Chapter 7 Epigenetic and Transcriptomic Regulations in Brain Disorders 203
Ajay Pal and Jaishriram Rathored

7.1 Introduction 203
7.2 Overview of Epigenetic Mechanisms 205
7.2.1 DNA Methylation 205
7.2.2 Histone Modifications 206
7.2.3 Non-coding RNAs 207
7.2.4 Chromatin Remodelling 208
7.3 Transcriptomic Regulation in the Brain 209
7.3.1 Transcription Factors and Co-regulators 211
7.3.2 RNA Splicing and Isoform Diversity 211
7.3.3 Non-coding RNAs in Transcriptomic Control 211
7.4 Epigenetic Dysregulation in Brain Disorders 214
7.4.1 Neurodevelopmental Disorders 214
7.4.2 Neuropsychiatric Disorders 216
7.4.3 Neurodegenerative Diseases 217
7.5 Transcriptomic Alterations in Brain Disorders 218
7.5.1 Neurodevelopmental Disorders 219
7.5.2 Neuropsychiatric Disorders 220
7.5.3 Neurodegenerative Diseases 221
7.6 Epigenetic and Transcriptomic Crosstalk 223
7.6.1 Epigenetic Control of Transcription 223
7.6.2 Transcription Factors as Epigenetic Modulators 225
7.6.3 Non-coding RNAs as Integrators of Epigenetic and Transcriptomic Regulation 226
7.7 Technological Advances and Methodologies 227
7.7.1 Epigenomic and Transcriptomic Profiling 228
7.7.2 Multi-omics Integration 228
7.7.3 Functional Genomics and Epigenome Editing 228
7.7.4 Experimental Models and Computational Resources 229

7.8 Therapeutic Implications 230
7.8.1 Epigenetic Drugs 230
7.8.2 RNA-based Therapeutics 231
7.8.3 Epigenome Editing 231
7.8.4 Personalized Medicine and Biomarkers 231
7.8.5 Neuroprotective and Reprogramming Strategies 232
7.9 Challenges and Future Directions 232
7.9.1 Biological and Cellular Complexity 232
7.9.2 Temporal and Spatial Dynamics of Regulation 232
7.9.3 Annotation, Integration, and Interpretation 233
7.9.4 Translational and Ethical Barriers 233
7.9.5 Future Outlook 233
7.10 Conclusion 234
Abbreviations 234
References 236

Chapter 8 Biomarker and Precision Medicine in Neurology **258**
Syed Mohammad Nasar Ata, Nitish Kumar Singh, Mayank Gangwar, Jaikee Kumar Singh and Mukesh Kumar

8.1 Introduction 258
8.2 Evolution of Neurological Biomarkers in Clinical Practice 259
8.3 Types of Biomarkers in Neurology 261
8.3.1 Genetic Biomarkers 261
8.3.2 Proteomic Biomarkers 264
8.3.3 Metabolomic Biomarkers 267
8.3.4 Neuroimaging Biomarkers 269
8.4 Precision Medicine in Neurology 272
8.4.1 The Role of Precision Medicine in Neurological Diseases 273
8.4.2 Personalized Treatment Plans Based on Biomarkers 274
8.4.3 Genetic and Environmental Considerations in Precision Medicine 275
8.5 Applications for Biomarkers in Neurological Disorders 277
8.6 Challenges in Implementing Biomarkers and Precision Medicine 277
8.6.1 Ethical and Privacy Issues with Genetic Data 279

8.6.2 Cost and Accessibility of Precision Medicine in Neurology 280
8.6.3 Integrating Multi-omics Data for Comprehensive Disease Profiling 280
8.6.4 The Need for Global Cooperation in Neurological Research 280
8.7 Advances in Genomic Sequencing and CRISPR Technology 282
8.8 Future Directions and Innovations 284
8.9 Conclusion 286
Abbreviations 287
References 289

Chapter 9 Environmental Influence on Neurochemistry 297
Anushka Sharma and Daksh Kumar

9.1 Introduction 297
9.2 Heavy Metals, Pollutants and Neurotoxicity 301
9.2.1 Heavy Metals 301
9.2.2 Neurotoxicity and Contaminants 303
9.3 Diet, Gut-Microbiota, Neurochemical Interactions 307
9.3.1 Dietary Components and Neurochemical Modulation 307
9.4 Environmental Stressors and Neurochemical Consequences 312
9.5 Prevention Strategies for Environmental Neurotoxicity 315
9.6 Future Perspective 318
9.7 Conclusion 320
Abbreviations 321
References 321

Chapter 10 Future Directions in Neurochemistry and Brain Disorders 337
Prabhash Nath Tripathi, Mayank Gangwar, Nitish Kumar Singh, Vivek Chandra Verma, Jaikee Kumar Singh and Jaspreet Marwaha

10.1 Introduction 337
10.1.1 Advancing Neurochemical Research Tools 341
10.1.2 Biomarker Discovery and Precision Neurochemistry 353

10.1.3 The Neuroimmune Interface and Glial Modulation 356
10.1.4 Neuroregeneration, Synaptic Plasticity, and Chemical Repair 362
10.1.5 Environmental and Lifestyle Modulation of Neurochemistry 365
10.1.6 Future Outlook 368
10.2 Conclusion 369
Abbreviations 369
References 372

Subject Index **381**

CHAPTER 1

Foundations of Neurochemistry and Neuroanatomy

FALGUNI GOEL,[a] VAISHALI DOBHAL,[b] SHIPRA[c] AND PRADEEP KUMAR*[d]

[a] Department of Pharmaceutical Technology, Meerut Institute of Engineering & Technology (MIET), Meerut; [b] Gurukul Kangri Deemed to be University, Haridwar, India; [c] Motherhood University, Roorkee, India; [d] Department of Neurology, Faculty of Medicine, Hadassah Medical Center, Hebrew University of Jerusalem, Hadassah-9112001, Jerusalem, Israel
*Email: kumar.pradeep@mail.huji.ac.il

1.1 Introduction

The human nervous system is one of the most complex as well as essential systems in the body, regulating all aspects of our perception, movement, emotion, memory, and thought. The nervous system is comprised of a complex web of interconnected cells and chemical messengers facilitating communication and coordination across different areas of the body and brain.[1] Understanding the nervous system involves exploring a foundational understanding of two important areas: neuroanatomy and neurochemistry. Neuroanatomy and neurochemistry are complementary disciplines that describe the structure and function of the brain and the spinal cord.[2]

The nervous system has two major parts: the central nervous system (CNS), made up of the brain and spinal cord; and the peripheral nervous system (PNS), comprising all nerves that branch from the spinal cord to the limbs and organs. The CNS is the command center, while the PNS is the communication system linking information from the CNS back to the body.[3] The CNS is usually divided into more specific parts that carry out more

Chemical Biology No. 26
Chemical Signatures of the Brain: Bridging Biochemistry, Neuroanatomy, and Brain Disorders
Edited by Sachchida Nand Rai and Abhishek Kumar Singh

Published by the Royal Society of Chemistry, www.rsc.org

specific functions. In the CNS, certain functions are assigned to different parts of the brain, for example, the cerebral cortex is responsible for higher functioning cognitive reasoning (thinking), the cerebellum is the control center for balance and coordination, the brainstem is responsible for autonomic functions (respiration, cardiac, *etc.*) and the cerebellum is the control center for balance.[4] The spinal cord is a transmission conduit that allows information to pass from the brain to the rest of the body, and can also integrate simple reflexes without the brain's involvement.[5]

Neuroanatomy studies these properties at various specific levels, from broad anatomical features of various brain regions to ultrastructural features of dendritic spines on neurons. Neurons are specialized cells that can transmit chemical and electric signals and are the general elements of the nervous system.[6] Each neuron has a cell body or soma, which is the site that must receive incoming signals on the dendrites of that neuron before an action potential can travel down the axon to other neurons or effector cells. Other specialized cells, called glial cells, which consist of astrocytes, oligodendrocytes, and microglia, provide critical elements of support for neurons.[7] Glial cells serve many functions for neurons, including maintaining homeostasis, insulating myelin, and protecting neurons against disease and injury. The arrangement of these cells and the structure of the connections between these cells results in complex and large neural networks that support all brain functions.[8]

As neuroanatomy shows the map of the nervous system, neurochemistry equips us with the means to understand how information is communicated and processed according to this map. Neurochemistry examines the biochemical make-up of nervous tissue and the molecules responsible for the regulation of neuronal communication.[9] Central to communication is the synapse, which is a specialized connection point in which the axon terminal of one neuron physically interfaces with either the dendrite or soma of another neuron.[10] Information is transmitted across the synapse at the synaptic cleft using neurotransmitters which present as a chemical message released from the presynaptic neuron following an electrical impulse. The neurotransmitters pass across the synaptic cleft towards receptors on the postsynaptic membrane, triggering a series of active cellular circumstances within the target neuron that may subsequently excite or inhibit.[11]

There are a few major classes of neurotransmitters that serve particular functions in the brain and behavior. Glutamate is the most abundant excitatory neurotransmitter and is very important for learning and memory through the mechanisms of long-term potentiation. GABA (gamma-aminobutyric acid) is the most prevalent inhibitory transmitter.[12] GABA primarily provides stabilization and prevents over-excitation in the brain. The monoamines (dopamine, serotonin, norepinephrine) are critical for regulating mood, motivation, and arousal, so it is not surprising that dysregulation in their synapses can lead to a host of psychiatric and neurodegenerative diseases.[13] Neuropeptides and neuromodulators are used to adjust the degree of synaptic transmission. They are involved in a multitude of brain functions such as pain perception, appetite, and response to stress.[14]

The interplay between neuroanatomy and neurochemistry is not only additive but synergistic. Anatomical structures dictate where, how, and when chemical signals are generated, released and received.[15] Conversely, those chemical signals and interactions can dictate anatomical structures over time through processes such as neurogenesis, synaptic plasticity, and neuronal pruning.[16] To illustrate, one can take a single dopaminergic pathway consisting of a tract of neurons from substantia nigra to striatum called the nigrostriatal pathway. Anatomy can define the pathway as an anatomical structure. Function pathologically considers the pathway *via* the neurotransmitter purpose, and in the case of the nigrostriatal pathway it is a functional purpose of dopamine for movement control.[17] When dopaminergic neurons are lost in the nigrostriatal pathway, the resultant manifestations are signs and symptoms of Parkinson's disease. The hippocampus, a well-studied structure in memory formation, contains many relevant glutamatergic synapses. Deterioration in glutamate signaling through this pathway leads to cognitive dysfunction and subsequently may lead to Alzheimer's pathology.[18]

In addition, the establishment and operation of neural circuits are influenced by structural cues and chemical gradients that direct neuronal migration during development and synapse formation.[19] Neurochemical signals can also regulate the expression of particular genes and protein synthesis, which ultimately affects the growth, differentiation, and survival of neurons. In contrast, structural changes such as axonal injury or demyelination have the potential to alter the dynamics of neurotransmitter action and produce functional consequences with cascading impairments.[20]

Neurochemistry and neuroanatomy provide a two-sided approach to understanding the nervous system by marrying structural elements with molecular (chemical) basics. This representation is necessary to help scientists address complex behavioral phenomena and progress diagnostics and treatment of disabling neurological and mental disorders.[21] Moreover, with the continued development of neuroscience as a discipline, the alignment of anatomical and chemical perspectives as sources of information and knowledge will keep driving developments in basic knowledge and clinical applications in both functional and disordered states, providing strong structural foundation for modern neuroscience.[22]

1.2 Basic Structure of the Nervous System

The nervous system is an elaborate collection of cells and tissues that performs countless important functions, including receiving sensory information, coordinating motor activity, regulating organ systems, and facilitating high-level functions such as memory, emotion, and consciousness.[23] The nervous system is traditionally subdivided into a structural framework consisting of the CNS and PNS. The CNS and PNS serve two distinct roles that are complementary in support of the CNS–PNS duality of control and communication.[24]

The central nervous system consists of the brain and spinal cord, and is the central processing and integrating body. The brain is contained within the skull, is the center of consciousness, and is comprised of many specific parts that perform sensory, motor, language, reasoning, and autonomic regulation functions.[25] The cerebrum houses the cerebral cortex, which is responsible for the higher functioning of cognition as well as other deep subcortical structures that are responsible for more basic functioning (*e.g.*, basal ganglia, limbic system). The cerebellum has coordinating functions for voluntary movement as well as balance.[26] The brainstem consists of higher functioning control centers that regulate vital life-sustaining functions (*e.g.*, respiration, cardiovascular regulation). The spinal cord extends downward from the brainstem, providing a means of communication for sensory and motor information to/from the brain and life-sustaining functions, as well as mediating reflexes and local processing due to spinal cord segmentation.[27]

The peripheral nervous system, on the other hand, is the connection with the CNS to the limbs, organs, and sensory structures, and allows for communication between the external world and the internal physical state. One of the main sets of nerves, referred as cranial nerves, arises from the brain, and others are spinal nerves that arises from the spinal cord.[28] Functionally, the PNS can be divided into the somatic nervous system (which involves voluntary movement and sensory input from the skin and musculoskeletal system) and the autonomous nervous system (which controls involuntary activity, such as heart rate, digestion, glandular activity, *etc.*). The autonomous nervous system has sympathetic and parasympathetic branches, which have opposite effects to maintain a steady state or homeostasis.[29]

The nervous system at the cellular level consists of two major classes of cells: neurons and glial cells. Neurons are the major signaling blocks of the nervous system.[30] A neuron typically consists of a cell body, which contains the nucleus and some organelles; dendrites, which receive signals from other neurons; and a single axon, which communicates action potentials to distant targets.[31] The axon terminates in synaptic boutons, and once an action potential reaches a bouton, it releases a chemical known as a neurotransmitter that acts to communicate with adjacent neurons, glands, muscles and cells. Neurons are extremely specialized to communicate rapidly and in a targeted manner, and there is great diversity in the form and function of different types of neurons, including sensory neurons that detect environmental stimuli, motor neurons that elicit muscle contractions, interneurons that combine and modify ongoing central activity, and many others.[32]

Glial cells support neurons, and are far from passive participants. Glial cells are responsible for neural homeostasis, axonal insulation, metabolic support, neural modulation, and are the primary line of defense against damage and disease.[33] Astrocytes, the most prevalent type of glial cell in the CNS, regulate the cellular composition of the extracellular space, provide support for the blood–brain barrier (BBB), and uptake and recycle neurotransmitters.[34] Oligodendrocytes in the CNS produce myelin; a fatty layer that wraps around axons to help propagate electrical impulses, while Schwann cells provide this myelination in the PNS. Microglia provide resident immune functions in the CNS, monitoring the immune microenvironment, and activating and phagocytosing debris following injury or disease, as well as secreting cytokines. Glial cells compartmentalize neurons, modulate their function and ultimately determine what information is processed by neural circuits.[35]

The BBB is a unique and fundamental characteristic of the CNS, a selectively permeable barrier that separates the central nervous system from the environment, protecting the brain from potentially harmful compounds circulating in the bloodstream while allowing the transport of essential nutrients into the brain.[36] The BBB is employed by the endothelial cells that compose the cerebral capillaries, which, due to serous attractiveness and the presence of tight junctions between adjacent endothelial cells, strictly limit paracellular transport. The BBB is supported by astrocytic end-feet and a basal lamina that, together with the endothelium, form a tripartite structure, incorporated as part of the BBB, that facilitates the compartmentalization and regulation of other BBB functions.[37] The BBB is important because it provides the brain with the extracellular environment required for proper neuronal function and protects the brain from neuromolecular toxins, pathogens, and fluctuations in the composition of the bloodstream. However, this protective function simultaneously complicates drug delivery to the brain and has encouraged the development of advanced pharmacological strategies to avoid or manipulate the barrier for therapeutic reasons.[38]

The anatomical framework of the nervous system from gross anatomical division to cell-based division accommodates its broad and highly coordinated functions.[39] The central and peripheral elements collaboratively receive and process sensory information, initiate motor functions, and maintain the internal environment; the neuronal and glial components allow for intercellular communication with homeostasis; and the BBB preserves a protected environment for the neurosystem's complex activities. Together, they comprise the basis of neurobiology and the context for more detailed exploration of its physiological and pathological states.[40]

1.3 Neuroanatomy of Key Brain Regions

The human brain is a very organized structure with specialized areas that have distinct functions but share overlapping functions. The brain is organized neuroanatomically in a way that allows it to process complicated sensory inputs, initiate voluntary movements, regulate internal states, and perform high-order cognitive, emotional, and behavioral functions.[41] Understanding the neuroanatomical features of important areas in brain structure provides insight into how structural specificity supports functional specificity, while also linking different areas through complex networks to achieve integrated brain function.[42]

The cerebral cortex is the outermost layer of the cerebrum, and it is arguably the most evolutionarily advanced and functionally heterogeneous portion of the brain. It is comprised of gray matter, which consists of closely packed neuronal cell bodies, and is subdivided into four main lobes: the temporal, parietal, occipital and frontal lobes. Each lobe contains areas that serve distinct functions.[43] The frontal lobe is associated with executive functions, such as planning, reasoning, voluntary motor control and decision-making. The primary motor cortex is located in the precentral gyrus of the frontal lobe.[44] The parietal lobe is associated with somatosensory functions and proprioception through the postcentral gyrus. The temporal lobe is involved in auditory processing, language comprehension, and memory encoding.[45] The critical functions of the occipital lobe include the visual perception of received stimuli. The cortex can be organized into primary areas, secondary areas, and association areas that work together to interpret, comprehend, and integrate the multitude of sensory information received and coordinate complex behavior. The cortical surface consists of gyri and sulci that add surface area, and, along with the density of functional units (neurons), support the added computational capacity.[46]

The basal ganglia are located deep in the cerebral hemispheres and consist of different subcortical nuclei, including the putamen, caudate nucleus, globus pallidus, substantia nigra and subthalamic nucleus.[47] The basal ganglia have relevant connections with the cerebral cortex and the thalamus to form closed loops and are mainly involved in the regulation of procedural learning, voluntary motor movements and the formation of habits.[48] Activation of the direct and indirect pathways, respectively, promotes the initiation and inhibition of movement, which must be regulated with dopaminergic signals from the substantia nigra pars compacta.[49] The circuits connecting these nuclei are dysfunctional in motor disorders like Parkinson's disease, characterized by bradykinesia, rigidity, and tremor, whereas dysfunction of the circuits of the basal ganglia will lead to Huntington's disease which is a movement disorder characterized by involuntary, hyperkinetic movements.[50]

The limbic system is the brain region emphasized by emotions because it is recognized from interrelated structures of the hippocampus, amygdala, cingulate gyrus, and portions of the hypothalamus and thalamus. The limbic system takes a leading role in the management of emotions, motivations, and mechanisms for savings memories.[51] The hippocampus is highlighted for its role in blocking short-term memories from becoming long-term declarative memories and as it is also responsible for other tasks including spatial navigation skills and remembering the directions of previously traveled paths.[52] The amygdala is concerned with the emotional stimuli of the body, especially fear and aggression, and social behavior. The amygdala functions in proximity with the hippocampus in facilitating the storage of emotional memories. Dysfunctional limbic learning circuits often become established with several psychiatric diagnoses including anxiety, depression, and post-traumatic stress disorder.[53]

The thalamus is located near the middle of the brain, and is a fundamental relay station for sensory and motor signals to reach the cerebral cortex. The thalamus transmits all sensory modalities except smell. Each of the signals is processed within the thalamus to be filtered and integrated together before reaching the relevant cortical areas.[54] The thalamus is organized into multiple distinct nuclei with a unique pathway for each sensory or motor function. The thalamus serves the dual purpose of transmitting information and regulating attention, consciousness, and sleep by modulating cortical activation. Damage to the thalamus can lead to deficits in sensory stimulation, disturbances in motor function, and/or disturbances in consciousness.[55]

The region immediately below the thalamus is the hypothalamus, which is small but functionally critical because it is the body's primary homeostasis regulatory system. Functional homeostasis means that the body has a stable internal environment when faced with changing internal or external stimuli.[56] Therefore, the hypothalamus modulates autonomic, endocrine, and behavior responses to regulate homeostasis. The hypothalamus mediates body temperature, energy/hunger regulation, thirst regulation, circadian regulation, emotional expression and homeostatic mechanisms through its control of the pituitary gland to secrete hormones throughout the body.[57] The hypothalamus exists at the intersection of neural activity and hormonal function that is essential for survival and behavioral adaptation. Injury to the hypothalamus can disrupt metabolic function, endocrine functions, sleep–wake cycle homeostasis, and numerous other neurobiological activities and homeostatic regulation [58]

The brainstem is situated at the base of the brain and is directly continuous with the spinal cord. It consists of the midbrain, pons, and medulla oblongata. The brainstem acts as a conduit for ascending sensory fibers and descending motor fibers and contains the nuclei of most of the cranial nerves.[59] It performs many vital autonomic functions such as breathing, heart rate, blood pressure, and digestion and has a role in maintaining consciousness and arousal in the reticular activating system. Lesions in the brainstem can be catastrophic, as they may lead to respiratory failure, cardiovascular collapse, and/or coma.[60]

The cerebellum occupies the space posterior to the brainstem and inferior to the occipital lobes. It is critical for the coordination, precision, and timing necessary for movement. While it does not initiate voluntary movement, the cerebellum fine-tunes motor activity through the integration of sensory input with motor commands.[61] The cerebellum is divided into three functional divisions: the vestibulocerebellum for balance and eye movements, the spinocerebellum for muscle tone and posture, and the cerebrocerebellum for planning and coordination of complex movements.[62] While the mature cerebellum is thought to contain a full range of motor abilities, it also contributes to cognition and affective behaviors that involve language, attention, and emotionality. Damage to the cerebellum results in ataxia, dysmetria, and impaired coordination.[63]

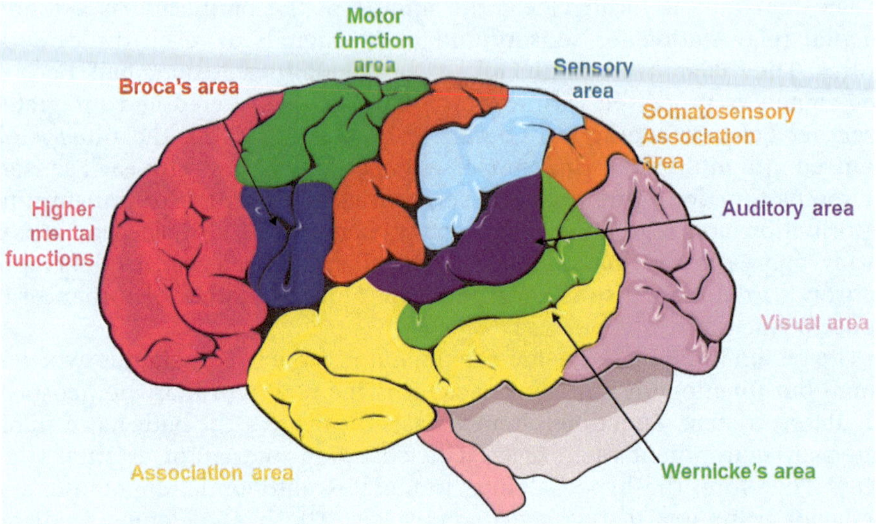

Figure 1.1 Structure of the nervous system. This diagram outlines the major anatomical subdivisions of the nervous system tissue including the CNS – the brain and spinal cord (containing signals traveling away from and towards, respectively), and the PNS which operates sensory (afferent) pathways and motor (efferent) pathways. The motor division, as noted above, is subdivided into the somatic nervous system (voluntary movements of skeletal muscles) and autonomic nervous system (regulates involuntary activities in the body). The autonomic nervous system is further subdivided into sympathetic (fight-or-flight response) and parasympathetic (calming effect) divisions. This structural overview illustrates the important points of relationship and coordination that allow for enabling of the neural signals for communication and control throughout the body.

These zones of the brain form a very specialized and interconnected system that encompasses most, if not all, of these human behaviors and physiological processes.[64] Getting to know the neuroanatomy of each structure, along with its functional contributions, is the first step in understanding the neural bases of cognition, emotion, movement, and autonomic regulation. The neuroanatomies of all of the components established above serve as a foundation for clinical neuroscience and the diagnosis and treatment of neurological and psychiatric disorders[10] (Figure 1.1).

1.4 Neuronal Communication and Synaptic Transmission

At the center of all functions of the nervous system – everything from simple reflexes to cognitive processing – is neuronal communication. Neuronal communication involves the fine balance of electrical and chemical signals, allowing for rapid and precise information transmission.[65] A basic understanding of the biological processes involved in neuronal excitability, synaptic transmission, and modulation of signals allows us to better understand how the brain processes information, adapts to experiences, and maintains functional integrity across the lifespan.[66]

Neurons communicate by generating an action potential, which is an electrical impulse that travels down a neuron's axon. The neuron maintains a resting membrane potential of approximately −70 mV through selective permeability of its membrane and the functioning of membrane ion pumps, specifically the sodium–potassium ATPase.[67] When the neuron receives enough stimulation, this will cause the voltage-gated sodium channels to open, allowing an influx of sodium ions to enter.[68] This leads to depolarization of the membrane. The sodium channels are now open, all at the same time as leading to a positive shift in membrane potential, which eventually leads to the opening of potassium channels to repolarize the membrane potential.[69] The entire process of depolarization and repolarization is extremely rapid, and the action potential propagates down the axon with no decrement in magnitude; this is because the positive change in membrane potential sequentially generates activation of adjacent ion channels.[70] In myelinated axons, the conduction of action potentials in any given direction is referred to as "saltatory conduction". The particular timing and eventual direction of this electrical signal give rise to information that will be transduced downstream.[71]

When the action potential arrives at the axon terminal, the next step in neuronal communication, synaptic transmission, takes place. The synapse is the specialized site of communication between two neurons, or between a neuron and an effector cell, that allows for the transfer of information.[72] The synaptic cleft is the small extracellular space that separates the presynaptic and postsynaptic membranes. When the action potential reaches the presynaptic terminal, it triggers the opening of voltage-gated calcium channels, which leads to an influx of calcium ions. The influx of calcium ions causes the synaptic vesicles to fuse with the presynaptic membrane, and the neurotransmitters are released into the synaptic cleft *via* exocytosis.[73]

Neurotransmitters then diffuse across the cleft and bind to specific receptors located on the postsynaptic membrane, activating signal transduction cascades to change the postsynaptic cell's membrane potential and its potential to fire an action potential.[74] Neurotransmitters are classified according to their chemical nature and functional roles. Classical neurotransmitters include amino acids like glutamate and GABA, biogenic amines such as dopamine, serotonin, and norepinephrine, in addition to acetylcholine.[69] Other peptides, or neuropeptides, including substance P, endorphins, and oxytocin, are modulating information carriers and are usually co-released with classical neurotransmitters. There are also unconventional transmitters like nitric oxide and endocannabinoids that act by non-vesicular release or retrograde signaling.[75]

The variety of neurotransmitter actions is largely determined by the receptors they bind to. Receptors can be classified into ionotropic and metabotropic receptors. Ionotropic receptors are actually ligand-gated ion channels that respond to the ligand with rapid and direct changes in membrane potential. Glutamate's NMDA and AMPA receptors utilize cations like sodium or calcium to mediate excitatory postsynaptic currents.[76] Metabotropic receptors produce slower but longer-lasting effects through intracellular signaling cascades. These signaling pathways can use second messengers to affect changes through modulation of ion channel activity, gene expression, or synaptic strength. Whether the effect is excitatory, inhibitory, or modulatory is dependent on the receptor subtype, distribution of receptors, and nature of the intracellular response.[77]

The nervous system has an astonishing ability to change with experience in addition to its basic function of signal transmission, which is called synaptic plasticity, or the ability of synapses to strengthen or weaken over time with increases or decreases in their activity.[78] LTP is one of the best-understood forms of synaptic plasticity and is often studied in the hippocampus, the part of the brain that is important for learning and memory.[79] LTP is produced by high-frequency stimulation of afferent fibers and is characterized by a prolonged increase in synaptic strength. The mechanism of LTP is defined at the molecular level as the activation of an NMDA receptor that allows calcium to enter the postsynaptic neuron, if the depolarization is strong enough.[78] The entry of calcium produces a cascade of intracellular signaling molecules and protein kinases such as CaMKII and PKA that insert additional AMPA receptors into the postsynaptic membrane and modify synaptic proteins in ways that should increase their efficacy.[80]

LTD represents another form of plasticity, but in essence it is the reverse, in which low-frequency stimulation induces synaptic weakening. Both LTP and LTD play crucial roles in synaptic reorganization, learning, memory consolidation, and fine-tuning of neural networks.[81] The physiological processes that support synaptic plasticity highlight the flexibility of synaptic function; specifically, previously established activity patterns may change how a neuron's synapse responds to a future stimulus, allowing the brain to encode and store information.[82]

Polarity, or the balance of excitatory and inhibitory transmission, is vital to enabling normal neural activity. The relationship between excitatory and inhibitory transmission is regulated through neurotransmitters, receptors, and intracellular signaling. Abnormalities that occur while regulating any one part of neuronal communication can lead to pathological states.[83] For example, excessive excitatory transmission can cause excitotoxicity and neurodegeneration, while deficits in inhibitory transmission have been observed in epilepsy, anxiety disorders, and schizophrenia. Abnormal dopamine signaling is implicated in both movement and psychiatric disorders, including Parkinson's disease, schizophrenia, and substance use disorders.[84]

The basis for all neural activity is neuronal communication and synaptic transmission. Neurons and the nervous system transfer information *via* combinations of electrical signals, chemical messages, receptors, and plasticity.[85] The nervous system is able to detect, record, respond and adapt to an

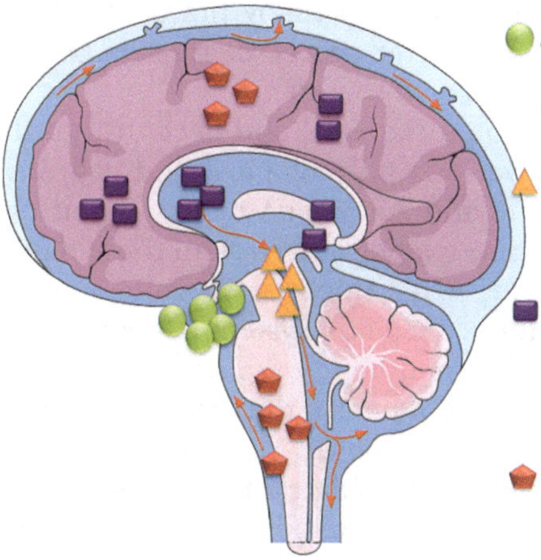

Figure 1.2 Neuronal communication: synaptic transmission. This figure illustrates the basic way in which neurons are able to communicate by synaptic transmission. An electrical impulse – or action potential – travels down the axon of the presynaptic neuron, ultimately leading to the release of neurotransmitter from the presynaptic neuron into the synaptic cleft. The neurotransmitter serves as a chemical messenger that binds to specific receptors located on the membrane of the postsynaptic neuron, triggering a new electrical signal in the postsynaptic neuron. In addition to synaptic transmission, this figure illustrates synaptic vesicles, voltage-gated calcium channels, neurotransmitter receptors and synaptic reuptake or enzymatic degradation that terminate the signal.

ever-changing internal and external environment in a precise manner. An understanding of neuronal communication and synaptic transmission mechanisms is important for understanding brain function in health and disease (Figure 1.2).[86]

1.5 Neurochemistry of Neurotransmitters and Neuromodulators

The human brain communicates using a complicated chemical language with the aid of neurotransmitters and neuromodulators as they are very important to many different types of processes that involve synaptic activity to modulating circuits of different neurons that influence behavior, cognition, and emotion. Neurotransmitters are fast-acting chemical messengers that are released at synapses that influence target neurons' excitability.[87] Neuromodulators do not act at synapses and, instead, have modes of action that are more overarching and slower to be released and diffused as they modulate germs in biological time frames and space, but do not directly generate a postsynaptic potential, but rather change the strength and efficacy of triggering synaptic transmission in time frames analogous with their half-life.[88] These chemicals mediate the process of many brain functions, and thus, are key to the neurophenomenology or processes that lie behind the pathology of neurological and psychological disorders.[89]

ACh was the first neurotransmitter to be discovered and is central to the cholinergic system, and is involved in various physiological and cognitive functions. ACh is produced from choline and acetyl-CoA by choline acetyltransferase and is primarily degraded by acetylcholinesterase in the synaptic cleft.[90] In the peripheral nervous system, Ach, which is basically a major neurotransmitter present in the autonomic nervous system as well as the neuromuscular junction, modulates parasympathetic output.[91] In the CNS, mostly cholinergic neurons arise from the basal part of the forebrain and are involved in attention, learning, memory, and arousal. Disruption of cholinergic signaling is best exemplified in neuropsychiatric disorders such as Alzheimer's disease. In this case, degeneration of cholinergic neurons originating from the basal forebrain is associated with cognitive dysfunction.[92]

Monoamine neurotransmitters such as DA, 5-HT, NE, and epinephrine are vital in mood, reward, arousal, and executive function. They are synthesized from amino acid precursors: dopamine and norepinephrine from tyrosine, and serotonin from tryptophan.[93] Dopaminergic systems include the nigrostriatal tract and mesolimbic and mesocortical pathways. The tuberoinfundibular pathway is involved in hormonal secretions.[94] The neuromodulatory actions of dopamine provide vital regulatory mechanisms for the brain's reward circuitry and play a role in addiction, schizophrenia, and Parkinson's disease. Serotonergic pathways are widespread throughout the brain and modulate mood, appetite, sleep, and emotional behavior.[95] SSRIs, which increase levels of serotonin in synaptic clefts, are common treatments for depression and anxiety disorders. Norepinephrine is mainly produced in the locus coeruleus and is involved in arousal and vigilance, as well as stress responses. Epinephrine plays a limited role in central nervous system activity, but is important in peripheral sympathetic responses.[96]

Glutamate and GABA are the primary excitatory and inhibitory neurotransmitters of the central nervous system, respectively, and maintain the excitatory–inhibitory balance within the brain, which is necessary for proper neural function.[97] Glutamate contributes to nearly all aspects of brain function, including learning and memory, through its action on various receptor types, including the AMPA, NMDA, and kainate receptors. Of these, the NMDA receptors are notable for their physiological importance in synaptic plasticity and LTP, which are well accepted as mechanisms for memory formation.[98] However, when released in too high quantities, or not cleared properly, excess glutamate can contribute to the phenomenon of excitotoxicity, which is an important consideration in neuronal injury for stroke and the regulation of seizures, particularly given the growing evidence for excitotoxicity in conditions associated with neurodegeneration. GABA is made from glutamate by glutamate decarboxylase and functionally acts on $GABA_A$ and $GABA_B$ receptors. As a significant player in neuronal inhibition, GABAergic inhibition is vital for regulating excessive excitation and stability within circuits. Most commonly, however, pharmacological agents that enhance GABA action are used to enhance GABA's anxiolytic, sedative, and anticonvulsant effects.[99]

Along with classical neurotransmitters, the brain utilizes a diverse family of neuropeptides as neuromodulators to fine-tune how neurons operate and fire. Biochemical assays of neuropeptides demonstrate some variability; however, in general, neuropeptides are synthesized in the soma and packaged for transport in the Golgi apparatus, before being sent to axon terminals in vesicles. Neuropeptides can generally be released from vesicles alongside classical neurotransmitters which can cause the effects of neuropeptides to last longer because they generally signal by binding to a G-protein coupled receptor along post-synaptic membranes.[100] Some well-known neuropeptides include: substance P, which is involved in pain transmission; endorphins and enkephalins, which are engaged in regulating pain and reward; oxytocin and vasopressin which are both responsible for social behavior and bonding; and neuropeptide Y which is involved in appetite and energy homeostasis. Consistent with these examples, the effects of neuropeptides often are context-dependent, as they contribute to the brain's ability to change synaptic responses to biological and environmental demands.

One of the newest and fastest-growing areas in neurochemistry is the endocannabinoid system, which largely orchestrates synaptic transmission and governs neural plasticity. Endocannabinoids, such as anandamide and 2-arachidonoylglycerol, are lipid-derived compounds synthesized on demand in postsynaptic neurons. Unlike classical neurotransmitters, endocannabinoids can act in a retrograde fashion as they diffuse back across the synapse to bind to cannabinoid receptors on presynaptic terminals and inhibit neurotransmitter release.[101] This mechanism generates a negative-feedback loop that can modulate excitatory or inhibitory signaling. Most of the activities like appetite, response to stress memory, mood and many physiological functions are regulated by the endocannabinoid system. There is an ongoing exploration of the pharmacological manipulation of this system for therapeutic reasons concerning pain, anxiety, neuroinflammation, and neurodegeneration.[102]

The neurochemistry of neurotransmitters and neuromodulators provides a molecular language for the brain. Each system adds a distinctive layer to the ways in which neural circuits and behavioral outcomes are regulated. These molecular messengers provide synaptic communication and plasticity based on their interaction with a particular receptor and downstream signaling pathway.[103] Disturbances in the synthesis, release, receptor activity and metabolism of these systems are core elements of a large number of neurological and psychiatric disorders. Understanding these systems is not only meaningful for enhancing our understanding of brain function, but also equally important in identifying possible targets for pharmacological drug discovery to treat mental health and neurodegenerative diseases (Table 1.1).[104]

Table 1.1 The neurochemistry of neurotransmitters and neuromodulators, including their types, synthesis, receptors, and functions.

Category	Name	Precursor	Receptors	Major functions	Ref.
Excitatory NT	Glutamate	Glutamine → glutamate (*via* glutaminase)	NMDA, AMPA, Kainate, mGluRs	Primary excitatory NT; involved in learning, memory, LTP	105
Inhibitory NT	GABA	Glutamate → GABA (*via* glutamate decarboxylase, GAD)	GABA-A (ionotropic), GABA-B (metabotropic)	Primary inhibitory NT; regulates anxiety, motor control	106
Biogenic amine	Dopamine (DA)	Tyrosine → L-DOPA → Dopamine	D1–D5 (metabotropic)	Reward, motivation, motor control (deficient in Parkinson's disease)	107
Biogenic amine	Norepinephrine (NE)	Dopamine → norepinephrine	α1, α2, β1–β3 adrenergic receptors	Arousal, attention, stress response, mood regulation	108
Biogenic amine	Histamine	Histidine → histamine (*via* histidine decarboxylase)	H1–H4 receptors	Wakefulness, appetite, immune response	109
Cholinergic NT	Acetylcholine (ACh)	Choline + acetyl-CoA (*via* choline acetyltransferase)	Nicotinic (ionotropic), muscarinic (M1–M5)	Muscle contraction, memory, attention, learning	110
Peptide NT	Substance P	Derived from preprotachykinin A	Neurokinin receptors (NK1–NK3)	Pain perception, inflammation	111
Peptide NT	Endorphins	Proopiomelanocortin-derived peptides	μ, κ, δ opioid receptors	Analgesia, stress relief, euphoria	112
Neuromodulator	Nitric oxide (NO)	L-Arginine → NO (*via* nitric oxide synthase, NOS)	Diffuses freely, acts *via* cGMP	Vasodilation, memory, synaptic plasticity	113
Neuromodulator	Endocannabinoids	Arachidonic acid derivatives (*e.g.*, anandamide)	CB1, CB2 receptors	Appetite, pain modulation, mood, retrograde inhibition of NT release	114
Neurohormone	Oxytocin	Produced in hypothalamus	Oxytocin receptor (GPCR)	Social bonding, childbirth, emotional regulation	115

1.6 Cellular and Molecular Aspects of Neurochemistry

Neurochemistry consists of the cellular and molecular machinery of neuronal activity. Neurotransmitters and receptors are the first messengers in neuronal communication; however, most of the communality consists of different regulatory intracellular components, including ion channels, second messengers, neurotrophins, and mitochondria.[104] These cellular and molecular components are not only supporting players in the transmission of signals; they determine synaptic strength, neuronal survival, and the development of the dynamic and active forms of plasticity that are essential in learning, memory, and adaptation.[116]

Ion channels are some of the most important molecular structures in neurons because they constitute the basis of electrical excitability and quick communication. Ion channels are proteins embedded in membranes that allow ions, such as Na^+, K^+, Ca^{2+}, and Cl^-, to move across the neuronal membrane in response to electrical or chemical signals. A type of ion channel called voltage-gated ion channels responds to changing membrane potentials to open or close. Generation of action potential is *via* sodium-gated channels.[117] For example, voltage-gated sodium channels mediate depolarization, whereas the voltage-gated potassium channels mediate the restoration of the resting membrane potential. Ligand-gated ion channels, which include ionotropic receptors for glutamate or GABA (*i.e.*, $GABA_A$), are activated by neurotransmitters binding and mediate fast synaptic transmission.[118] The spatial distribution/density, biophysical properties of channels, and other properties will influence the excitability of a neuron and integration of synaptic inputs. Dysfunction and/or aberrant expression/operation of ion channel function caused by genetic mutations or environmental causative factors may lead to various neuropsychiatric disorders collectively termed channelopathies such as epilepsy, ataxia, migraine, *etc.*[119]

Outside of requiring immediate fluxes of ions as a result of neurotransmitters, neurons can signal intracellular pathways that can modulate long-term effects. Neurons employ signaling pathways that often consist of second messenger systems that relay and amplify signals from membrane receptors to numerous intracellular targets. The second messengers are commonly cAMP, IP_3, DAG, or calcium ions. For example, neuromodulators such as dopamine or serotonin activate GPCRs that stimulate adenylyl cyclase to generate cAMP, which finally activates PKA.[120] The PKA phosphorylates ion channels, receptors and transcription factors to change the synaptic transmission, gene expression and cellular excitability. In a comparable pathway, the activation of phospholipase C produces IP_3 and DAG that access intracellular calcium stores and activate PKC, respectively. These cascades affect the short-term neuronal responses but they are the basis for lateral change also seen in areas like synaptic plasticity and memory formation. The amplification of effects from the second messengers means that low-level extracellular information can leave the cell with substantial intracellular effects. This enables adaptive and precise control over neuronal behavior.[121]

Along with these signaling avenues, neurotrophic factors are required for the development, maintenance, and plasticity of neurons. These are a range of proteins promoting neuronal survival, differentiation, and regeneration. Some of the best studied include NGF, BDNF, NT-3, and GDNF. Neurotrophic factors bind with specific receptors to signal.[122] The binding of neurotrophic factors to their respective receptors activates downstream signaling cascades intracellularly, such as the MAPK/ERK, PI3K/Akt, and PLCγ pathways that increase cell survival, promote axonal growth, and modulate synaptic strength and plasticity. BDNF is a significant neurotrophic factor to support plasticity, particularly activity-dependent synaptic modification, and learning and memory.[123] For example, BDNF is expressed more when neurons are active, and animals move to new and stimulating environments. The increase in expression following activity is relevant because it describes a type of molecular adaptation to developmental and life experiences. Therefore, neurotrophic signaling is supported by increased neuronal activity and experience that can promote synaptic strength and development.[124] Impaired neurotrophic signaling has been examined in a range of disorders, particularly many neurological and psychiatric conditions including Alzheimer's disease, depression, and schizophrenia. BDNF is mentioned with more frequency because of the evidence that shows circulating BDNF levels are associated with commonly observed synaptic losses associated with cognitive impairment, the exacerbation of depression and the onset and persistence of symptoms in schizophrenia.[125]

Mitochondria, the organelles responsible for energy production for the cell, are crucial to meet the metabolic requirements of a neuron, which is the most energy-demanding cell type in the body. Neuronal function requires continuous amounts of ATP that are made primarily by oxidative phosphorylation in mitochondria. ATP is required to support ion pumping, neurotransmitter synthesis, vesicle trafficking, and synaptic signaling processes.[125] Mitochondria also contribute to calcium buffering, redox signaling, and apoptotic pathway regulation. Mitochondria in neurons are very rapidly transported to the synapse, where energy demands are greatest based on increased neuronal activity. It has been shown that disruption of mitochondria can lead to unreliable synaptic transmission, reduced plasticity, and also can invoke cell death pathways. Mitochondrial dysfunction is indicative of a vast number of neuropsychiatric disorders such as Alzheimer's disease, Parkinson's disease and ALS.[125] In these neurodegenerative conditions, neuronal degeneration is attributed to impaired energy metabolism, oxidative stress, and genetic mutations to mitochondrial DNA. As such, studying mitochondria in the nervous system not only contributes to our overall understanding of basic neurochemistry, but also has a direct benefit towards the creation of specific therapeutic interventions.[126]

The cellular and molecular ecosystem of neurochemistry constitutes an elegantly orchestrated system in which ion channels provide the final layer of excitability control, second messengers control intracellular signaling, neurotrophic factors maintain neuronal health and plasticity, and mitochondria allow for any of these pathways to function. Together they facilitate the functional complexity and plasticity of the nervous system.[127] When this balance is disrupted, functional deficits and disease are bound to emerge. For this reason, actionable progress towards the basic science of neuroscience, and particularly towards the creation of new forms of treatment for neurological and psychiatric disorders, cannot occur without an integrated understanding of the different cellular and molecular processes involved.[128]

1.7 Neurodevelopment and Neural Plasticity

The capability of the neurological activity to alter in response to external or internal stimuli by rearranging its connections, functions, or structure is known as brain plasticity or neuroplasticity. Synaptic plasticity refers to the potency of neurons to alter the effectiveness and intensity of synaptic signaling through a variety of activity-dependent processes, and is a key characteristic of neurons. The activity of the nerve impulse and subsequent alterations in the nerve fibers at the cortical level are necessary for this plasticity hypothesis to function.[129] First, by performing excitability of the nerve fibers, a nerve impulse causes a functional change in the cortical network. Second, the neural system undergoes plastic alterations as a result of this functional transition, which leads to neuroplasticity. The potency of neurons that enables the creation of a neuro-connection and changing of the pre-existing strength of synaptic connections is termed neuroplasticity.[130] This statement indicates that the term neuroplasticity means the removal as well as creation of synapses in adult and growing brains, while a basic modification in the synaptic strength helps to mature synaptic connections. Similar processes may also take place at peripheral synapses, which are the sites of many of the early investigations into synaptic transmission. This covers a wide range of work, with the revival of synaptic transmission after destruction to the central nervous system being another aspect of neuroplasticity.[131]

In the brain, specialized connections known as synapses allow information to travel from one neuron to another. A synaptic cleft divides the presynaptic and postsynaptic terminals that make a synapse between two neurons. Small vesicles holding chemical neurotransmitters are found in the presynaptic terminal, whereas receptors unique to these neurochemicals are found in the postsynaptic terminal.[132] Neurons transmit information in the form of an electrical impulse, termed an action potential, that is triggered at the cell body and travels down the axon, effectively converting an electrical impulse into a chemical signal. Diffusing across the synaptic cleft, the release of neurotransmitter vesicles is voltage dependent, which is a result of action potential generation at the synaptic region and occurs after this conversion of an electrical signal to a chemical signal. After this diffusion within the synaptic cleft, neurotransmitters available at the postsynaptic region bind to the receptor to produce an electrical signal.[133] When the total number of synapses in the postsynaptic cell activates an electrical signal for firing, the postsynaptic cell will then fire an action potential. Each presynaptic cell may integrate information from several sources before transmitting it as an electrical code because a neuron can accept connections from a wide variety of presynaptic cells. The human brain is today recognized to be a highly dynamic and continually reorganizing mechanism capable of being formed and altered during the entire lifespan.[134] According to one theory, the structural organization of the brain is based on modification of experience. The property of remodulation is a key response of experience and learning in the brain and is generally described as neuroplasticity. The potency of neurons to generate new neurons as well as the generation of strong connections within the neurons over a lifetime is referred as to as neurogenesis.[135]

There are many phases in the formation of neurons. Neural proliferation, the initial phase of neuron development, starts in the neural tube's ventricular zone after the neural tube has closed. About four weeks after conception, during embryonic development, neurogenesis, the process of creating new neurons, begins. Significant cell growth in the neural tube, the precursor structure within the central nervous system, characterizes this crucial stage.[136] Prior to birth, the majority of neurons in the telencephalon of humans, which includes the cortex, basal ganglia and hippocampus, *etc.*, are created. In some parts of the brain, such as the cerebellum, extensive neurogenesis takes place after birth. The cells forming during brain growth are neural stem cells, with pluripotent and totipotent stem cells being the two main types. While multipotent cells are more restricted than pluripotent cells, the pluripotent cells have the ability to differentiate into the various cells that make up the body.[136] This ability of differentiation of pluripotent stem cells into any type of cell in the human body is why researchers are looking into using them in regenerative cell-based therapies for a number of illnesses, including diabetes, heart-related issues, and spinal cord injury. Neural stem cells are unique pluripotent cells that exclusively give rise to radial glial cells throughout CNS development.[137]

During the migration period, neurons remain immature and lack basic neuronal features like axons and dendrites. Neuronal migration is the movement of neurons from their original location to a new target location. For neurons of the central nervous system, neural migration is limited to the neural tube, while for neurons of the peripheral nervous system, neural migration may occur across different neural regions.[137] Migration follows an inside-out pattern, with neurons moving outward from the neural tube towards their destination. There are two types of this movement: tangential migration and radial migration. Long considered the main way that neurons travel in the cortex, radial migration happens when radial glial cells direct neurons to move towards the brain's surface in accordance with the neural tube's radial pattern, eventually forming the layered structure of the neocortex.[138] To form brain circuits, neurons must align and integrate with one another after migrating. We refer to this as aggregation. There are two main processes that facilitate aggregation. In order to facilitate cell-to-cell communication and tissue stabilization, cell-adhesion molecules on cell surfaces identify and attach to molecules on other cells. Furthermore, gap junctions provide channels of communication between neighboring cells, facilitating the passage of ions and metabolites like glucose and fostering metabolic coupling between the two cells.[139] The neural architecture of the human nervous system is ultimately the result of these processes, which promote neuronal integration and connections. The removal of neurons in the nervous system, known as neural cell death, takes place widely during development and aids in the growth of the brain. Usually, there are two types of neuron death: necrosis and apoptosis. Active, planned cell death to preserve proper growth is known as apoptosis, while passive, unintentional cell death brought on by environmental disruptions such as trauma, poisoning, or oxygen deprivation is known as necrosis.[140] Following neuronal growth, apoptosis takes place, removing surplus and immature neurons to allow for appropriate neural connection and the development of functional networks. The zones of dentate gyrus of the hippocampus are the lateral and subventricular zones, where neurogenesis has been most reliably detected. Furthermore, a large number of adult-generated neurons die and are unable to integrate into preexisting brain networks. Even though there is evidence of adult neurogenesis from several research works, this subject remains rather contentious. According to some research, the adult hippocampus produces 700 new neurons every day, however other research indicates that adult hippocampus neurogenesis is either invisible or non-existent.[141]

The molecular level is where synaptogenesis and pruning take place. Through long-term potentiation or long-term depression, excitatory transmission can either strengthen or weaken a synapse. Rapid activation of glutamate-using synapses results in long-term potentiation, while gradual, repeated stimulation of the same synapses results in long-term depression. *N*-methyl-D-aspartate-type glutamate receptors in the postsynaptic membrane open in response to rapid stimulation.[142] This process results in the insertion of alpha-amino-3-hydroxyl-5-methyl-4-isoxazole-propionic acid-type glutamate receptors in the postsynaptic membrane, the activation of signaling cascades (including calcium calmodulin II), and an influx of calcium. There appears to be a calcium threshold for both chronic depression and the change from chronic depression to chronic potentiation. Astrocytes' recognition of glutamate triggers the activity-dependent production of neurotrophins, such as tumor necrosis factor-α and brain-derived neurotrophic factor, which improve long-term potentiation and support neuronal survival.[143] The gradual, repeated activation of the synapse, on the other hand, causes a decrease in alpha-amino-3-hydroxyl-5-methyl-4-isoxazole-propionic acid-type glutamate receptors in the postsynaptic membrane, which leads to protracted depression. There is mounting evidence from animal research that environmental enrichment promotes synaptogenesis across the brain. The number of dendritic spines and dendritic branching in the hippocampus are increased in correlation with enrichment.[144]

The processes through which neurons and neural systems alter their input–output functions are the main focus of the broad study of neurobiology of learning and memory. How specific cell–cell communication may cause plasticity, such as long-lasting alterations in dendritic spines, receptor quantity or binding characteristics, or neurotransmitter release dynamics, is a problem at the cellular and molecular levels. These classes of neuronal plasticity are thought to describe the main underlying mechanisms of memory creation. However, overlying and interacting with these processes are layers of further richness.[145] First, it is evident that not all brain activity produces enduring alterations that may be indicative of memory. The majority of neuronal transmission is probably what allows coherent information to flow without requiring permanent alteration. Perceived but not memorized occurrences fall under this group. It is evident that an animal's internal condition may affect how it processes, learns, and remembers the same information. In this instance, the internal state may be seen as a background state rather than one that is brought on by the event.[146] Stress-related hormones, reproductive hormones, and motivational hormones seem to provide circumstances that favor learning some components of a scenario more than others (Figure 1.3).[147]

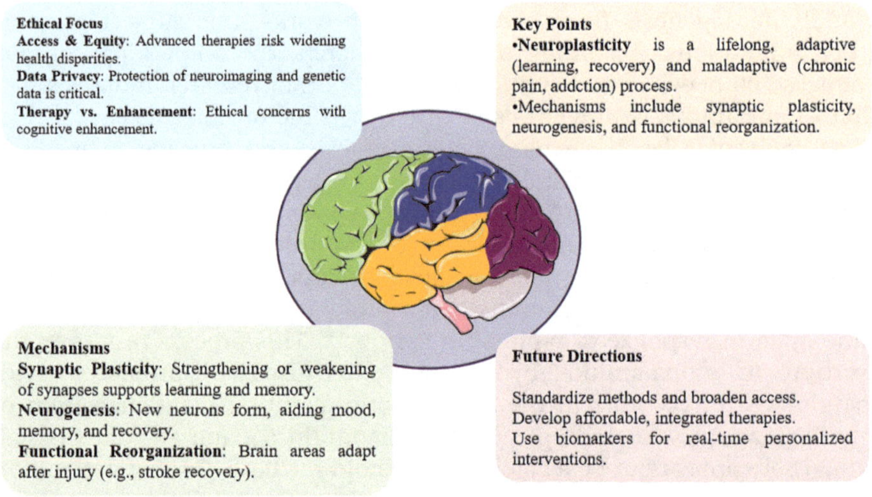

Figure 1.3 Neurodevelopment and neural plasticity: this figure depicts the critical stages in neurodevelopment. These stages including neuronal proliferation, neuronal migration, neuronal differentiation, synaptogenesis, and myelination. Neurodevelopmental plasticity is depicted in the diagram with respect to intrinsic and extrinsic factors that cause plastic changes. Two forms of plasticity are denoted: structural plasticity (the brain can alter its structure) and functional plasticity (the brain can alter its function). In the figure, structural plasticity of synaptic strength, connectivity, and circuitry are illustrated, as neural circuits in one sense can be modified over time, in addition to neuroplastic changes that take place throughout the lifespan during critical periods of development and learning, and in response to injury. This characteristic provides the foundation for cognitive development, memory, and recovery from neural damage, among others.

1.8 Blood–Brain Barrier and Neuroprotection

The CNS has specialized "barriers" that regulate neuronal transit at various surfaces. Specialized barriers play important roles in brain metabolism and fluid balance. These transport interfaces, which connect the blood to diverse parts of the CNS, are located either inside the endothelium or in the epithelia.[148] The spinal cord, brain and blood are surrounded by a solute traffic governed by an endothelial interface generally described as the blood–spinal cord barrier and BBB. In the cerebral microvessels of the majority of vertebrates, the BBB is selectively semipermeable and has a dynamic mechanism. It separates the extracellular fluid of the brain from the circulation. It is essential for controlling the movement of chemicals required for brain activity. By shielding the brain from potentially harmful compounds found in the blood, the BBB maintains a stable environment for the brain by controlling the movement of molecules into and out of the CNS. It also maintains homeostasis and keeps pathogens, blood cells, and plasma components out of the brain.[149] It does this by forming a neurovascular unit (NVU) including astrocytes, pericytes, and endothelial cells, all of which cooperate to maintain the chemical components of the neural environment and maintain normal brain function. The blood–brain barrier is formed by tightly joined endothelial cells connected through tight junctions, which restrict the paracellular movement of molecules and help maintain central nervous system homeostasis. In order to maintain homeostasis, the BBB, a highly specialized structure, regulates the admission of solutes into the brain. This is a problem in the treatment of CNS issues since it is a biological and enzymatic barrier that may restrict the brain distribution to less than a tenth of the intravenous dosage.[150] The BBB, the barrier separating the blood supply from the brain, contains a number of tightly regulated transport mechanisms that can be used to deliver medications and nanotherapeutics to the brain. In this regard, it is crucial to engineer medications with suitable physicochemical traits and ideal pharmacokinetic qualities, such as metabolic stability, particularly if passive diffusion is the goal. Endocytotic transport is the most effective method for delivering drugs to the brain, particularly antibodies and therapeutic macromolecules. Generally, drugs that are lipid soluble (basically less than 300 Da) follow transport routes such as receptor-mediated endocytosis, carrier-mediated transport, and absorptive endocytosis. Nanoparticulate systems have been effectively delivered utilizing both carrier-mediated and receptor-mediated endocytosis. However, receptor-mediated endocytosis using high-capacity carriers has been the most successful and is now in clinical development.[151]

Neuroprotective therapies that target the BBB are critical in illnesses such as traumatic brain injury, stroke, multiple sclerosis, and neurodegenerative disease, all of which decrease barrier integrity. They strive to preserve or restore the BBB's ability to prevent harmful chemicals, immune cells, and infections from entering the brain. Neuroprotective mechanisms include increased tight junction protein expression, decreased inflammatory processes, and stabilization of endothelial cell function. Furthermore, antioxidant therapies can reduce the oxidative stress that generally causes BBB breakdown.[151] Other pharmacological therapies known to reinforce the BBB include corticosteroids, matrix metalloproteinase inhibitors, and growth hormones. Novel approaches to provide targeted and long-term neuroprotection include gene therapy, stem cell treatment, and nanoparticle-based medication delivery.[152]

1.9 Neurochemistry in Neurological and Psychiatric Disorders

Changes in glutamate homeostasis, persistent inflammation, and excessive oxidative stress are associated with most neurodegenerative disorders. Chronic neuroinflammation and excessive oxidative stress are long-lasting, self-sustaining conditions that continue long after an injury. In addition to the long-term activation of astrocytes and microglia, chronic neuroinflammation also involves the prolonged production of inflammatory mediators that increases with oxidative stress.[153] The continuous production of cytokines and other inflammatory mediators, including proteases and adhesion molecules, contributes to the continuation of the inflammatory cycle by stimulating the growth of more astrocytes and microglia and causing them to produce more inflammatory chemicals. Redox status is changed with calcium homeostasis disruption, the production of excess glycerophospholipid-, sphingolipid-, and cholesterol-derived lipid mediators and their interactions with one another which support stress due to oxidation and inflammation in neuropsychiatric disorders.[154]

When sphingolipid-, glycerophospholipid- and cholesterol-derived lipid mediators interact more intensely, neuronal death results. Enzyme families linked to the production of the aforementioned lipid mediators include phospholipases A2, sphingomyelinases, and cholesterol hydroxylases. A class of serine/threonine protein kinases known as MAPKs is also essential for controlling the synthesis of inflammatory mediators and for the majority of cellular responses to cytokines and external stress signals.[155]

The development and classification of therapeutic medications for the treatment of various illnesses has mostly relied on how they affect the release of neurotransmitters and the subsequent activation of receptors. This viewpoint emphasizes the practical and theoretical ramifications of theories which deal with the fluctuating nature regarding the dysregulation of neurotransmitters, such as the various self-regulating/controlling synaptic processes, the tonic and phasic elements of neurotransmission, compartmentalized release, the distinction between disruption of activated and basal release and dysfunctional transmission by the neuronal site triggered by cognitive and behavioral activity.[156] In addition to refining neuropharmacological mechanisms to act as targets for novel treatment approaches, evidence from developing techniques for measuring the release of neurotransmitters at high temporal and spatial resolution is likely to advance theories explaining the significance of neurotransmitter dysregulation in the growth and evaluation of major neuropsychiatric disorders.[157]

It is a well-established notion that suggests oxidative damage from ROS is the main cause of aging. It is well known that ROS contributes to ND in conditions including Parkinson's disease, Huntington's disease, multiple sclerosis, Alzheimer's disease, amyotrophic lateral sclerosis, and Friedreich's ataxia. ROS can, in fact, change the genetic and epigenetic landscape by causing methylation of genes.[158] Environmental exposures like ROS can cause aberrant alterations in the epigenome and alter DNA methylation in chronic illnesses, which is why they are a trigger for many of these diseases. ROS are important in the pathophysiology of ND and are groupings of atoms with an odd, unpaired number of electrons that cause OS. Reduced expression of antioxidant genes or mitochondrial malfunction may make this worse. Oxygen, superoxide, alkoxy, hydroxyl, and peroxyl radicals, as well as nitric oxide and nitrogen dioxide, are a few types of free radicals. Hydrogen peroxide, hypochlorous acid, and other nitrogen compounds are examples of non-radical ROS. Tools for detecting OS biomarkers, such as immunofluorescence, are available to study ND.[159] One example of a detection technique is the use of nitro-tyrosine, 4-hydrononeal, and 8-hydroxyguanine in hippocampus slices. Protein carbonyls are another type of biomarker that might be helpful as diagnostic tools.[160]

The successful integration of new ideas characterizing the dynamic, compartmentalized, and activity-dependent features of dysregulated neurotransmitter systems is essential to the relevance and utility of hypotheses regarding the aberrant regulation of the release of extracellular concentrations of primary messengers. The effects of potential or actual therapeutic drugs on neurotransmitter release are a crucial part of drug development efforts.[161] Recent developments in new techniques for monitoring neurotransmitter release at high temporal and spatial resolution help to reveal the multiple and highly dynamic modes of neurotransmitter release. When used in conjunction with experiments intended to measure drug effects on activated release, these techniques are likely to uncover more complex and effective neuropharmacological strategies for treatment development. For instance, nicotine and other nAChR agonists appear to alleviate the cognitive symptoms of ADHD, schizophrenia, and other conditions.[162]

Various methods focus on minimizing the aggregation and production of amyloid-β (Aβ) oligomers and hyperphosphorylated tau tangles, which are the primary pathological hallmarks of neuropsychiatric diseases like AD. Another tactic involves altering neurotransmitter signaling, while yet another approach consists of clinically authorized *N*-methyl-D-aspartate receptor blockade therapies and cholinesterase inhibitors for the symptomatic management of neuropsychiatric disorders.[163] Further, the discussion of other targets that seek to stabilize neuron signaling by modifying receptors of neurotransmitters include another strategy consisting of a group of "sensitive targets" which generally affects the buildup of tau or Aβ indirectly. These targets are proteins whose function changes as a result of direct Aβ-target interaction or Aβ buildup in the brain (Table 1.2).

Table 1.2 Neurochemistry in neurological and psychiatric disorders, focusing on key neurotransmitter imbalances, affected brain regions, and associated conditions.

Disorder	Key neurochemical abnormalities	Affected neurotransmitters	Impacted brain regions	Clinical consequences	Ref.
Alzheimer's disease	↓ Acetylcholine, ↑ glutamate-induced excitotoxicity	Acetylcholine, glutamate	Hippocampus, cortex	Memory loss, cognitive decline, behavioral changes	164
Parkinson's disease	↓ Dopamine due to degeneration of substantia nigra	Dopamine	Substantia nigra, basal ganglia	Tremors, rigidity, bradykinesia, postural instability	165
Schizophrenia	↑ Dopamine (mesolimbic), ↓ dopamine (mesocortical), glutamate dysfunction	Dopamine, glutamate	Prefrontal cortex, limbic system	Delusions, hallucinations, cognitive and emotional disturbances	166
Depression (major depressive disorder)	↓ Serotonin, norepinephrine, dopamine	Serotonin, norepinephrine, dopamine	Limbic system, prefrontal cortex	Persistent low mood, anhedonia, sleep and appetite disturbances	167
Anxiety disorders	↓ GABAergic function, ↑ norepinephrine and serotonin dysregulation	GABA, norepinephrine, serotonin	Amygdala, hippocampus	Excessive worry, hypervigilance, panic, somatic symptoms	168
Bipolar disorder	Dysregulated monoamine levels (↑ in mania, ↓ in depression)	Dopamine, serotonin, norepinephrine	Prefrontal cortex, limbic regions	Mood swings, impulsivity, euphoria or depressive episodes	169
Huntington's disease	↓ GABA and acetylcholine; altered glutamate signaling	GABA, acetylcholine, glutamate	Basal ganglia, cortex	Involuntary movements, cognitive and psychiatric symptoms	170
Epilepsy	Imbalance between excitatory and inhibitory signals (↑ glutamate, ↓ GABA)	Glutamate, GABA	Cortex, hippocampus	Seizures, altered consciousness, behavioral changes	171
Addiction/substance use disorders	↑ Dopamine in reward pathways, altered GABA/glutamate homeostasis	Dopamine, GABA, glutamate	Ventral tegmental area (VTA), nucleus accumbens	Compulsive drug seeking, tolerance, withdrawal symptoms	172

Autism spectrum disorder (ASD)	Altered serotonin, dopamine, and glutamate transmission	Serotonin, dopamine, glutamate	Cortex, amygdala, cerebellum	Social deficits, repetitive behaviors, communication challenges	173
Obsessive-compulsive disorder (OCD)	↑ Serotonin transporter activity, dysfunction in glutamate system	Serotonin, glutamate	Orbitofrontal cortex, caudate nucleus	Obsessions, compulsions, anxiety	174
Attention deficit hyperactivity disorder (ADHD)	↓ Dopamine and norepinephrine activity in prefrontal cortex	Dopamine, norepinephrine	Prefrontal cortex, basal ganglia	Inattention, hyperactivity, impulsiveness	175

1.10 Experimental Techniques in Neurochemistry and Neuroanatomy

Within the field of imaging science, neuroimaging creates non-invasive pictures of the brain or other elements of the central nervous system using a variety of state-of-the-art technologies. For example, neuroimaging may quantitatively analyze the architecture, blood volume, blood flow, electrical, oxygen consumption, activity, metabolism receptor sites, and many physiological activities inside the CNS and offer a variety of directly or indirectly derived visual representations. Often referred to as brain scanning, neuroimaging falls into two main categories: structural neuroimaging and functional neuroimaging. The use of fMRI in conjunction with PET, CT, and SPECT has lately opened the door for multimodal neuroimaging, which integrates information from many imaging modalities to provide more in-depth insights into brain processes.[176]

The computerized X-ray imaging method known as "computed tomography," or CT, employs a narrow X-ray beam focused at the patient and rapidly rotating around the body to create signals that the machine's computer processes to create cross-sectional pictures, or "slices." Compared to traditional X-rays, these slices, known as tomographic images, can provide medical personnel with more comprehensive information.[177] Following the collection of many subsequent slices by the machine's computer, these slices may be digitally "stacked" to create a 3D image of the patient, which makes it simpler to identify both fundamental structures and potential abnormalities or tumors.[178]

A nuclear imaging scan (SPECT) is a technique that combines a radioactive tracer with CT. Doctors can examine how blood travels to tissues and organs thanks to the tracer. An injection of a tracer is administered prior to the SPECT scan. Due of its radiolabeled nature, the tracer releases gamma rays that the CT scanner can detect. The CT cross-sections show the data that the computer has gathered from the gamma rays.[179] To create a three-dimensional representation of the brain, these cross-sections may be re-assembled. SPECT often uses the following radioisotopes to identify tracers: fluorine-18, thallium-201, xenon-133, technetium-99m, and iodine-123. The scanner will pick up these radioactive versions of natural elements as they go through the body.[180] These isotopes are useful for labeling a variety of medications and other substances. The medical personnel will employ a tracer type based on what they wish to measure. For instance, the physician may employ radiolabeled glucose to examine a tumor and observe how the tumor metabolizes it. The tracer remains in the bloodstream instead of being absorbed by surrounding tissues, which limits the images to locations where blood circulates. This is how this test varies from a PET scan.[181]

Medical imaging comprises non-invasive techniques used to visualize the internal structures and functions of the body for clinical diagnosis, disease monitoring, and biomedical research. This technique using radio frequency in the presence of a well-regulated magnetic field recreates excellent fragmentary photographs of the body.[182] A well-organized MR image is obtained *via* a highly restrained magnetic field created by a large magnet inside the equipment. Proper patient alignment during imaging ensures accurate anatomical visualization, minimizes image distortion, and improves diagnostic quality. Upon cutting off the radio waves, the atoms revert to their initial positions and emit radio signals. A computer receives these signals and transforms them into a picture of the physical component under examination. This picture shows up on a viewing screen.[183]

In high-performance liquid chromatography (HPLC), the mobile phase consists of one or more suitable solvents that transport analytes through the stationary phase, enabling their separation based on physicochemical properties. Food analysis frequently uses HPLC to separate and purify macromolecules as well as to quantify ions and tiny molecules. A wide range of column packing materials have played a significant role in HPLC's widespread adoption. HPLC is an established chromatographic and analytical method for accurately quantifying neurotransmitters in biological materials such as brain tissue, cerebrospinal fluid, and plasma.[184] With its great sensitivity and specificity, HPLC can identify even the smallest amounts of neurotransmitters and their metabolites, such as dopamine, serotonin, and norepinephrine, in intricate biological materials. Analytical processes usually include sample preparation (such as deproteinization or derivatization) and product separation under controlled pressure in a chromatographic column. The analyte, typically a neurotransmitter or its metabolite, is separated and quantified using an appropriate analytical technique such as HPLC coupled with ultraviolet, fluorescence, or mass spectrometric detection. Crucially, HPLC is the best method for carrying out neurochemical analysis and researching the biochemical underpinnings of the consequences and outcomes of neurological illnesses because of its high resolving and quantitative capabilities.[185]

In neuroscience, electrophysiology is an important methodological technique that makes it easier to examine the electrical characteristics of neurons and other excitable cells. Scientists can investigate how neurons communicate and process information while interacting with other cells under regulated conditions within a certain timescale, and how they react to an external stimulus by detecting voltage changes across membranes. Microelectrodes are usually positioned within (intracellular recording) or next to (extracellular recording) the neurons of interest in order to capture neural activity.[186] When studying neurological illnesses, neural recordings can detect action potentials, synaptic events, network oscillation, and provide information about the normal and abnormal functioning of neuronal networks. Scientists are better able to record the high-resolution temporal characteristics of neural networks because of the development of new technologies such as multi-electrode arrays and simultaneous, optogenetically controlled recordings (Table 1.3).[187]

1.11 Future Aspects

The future of neurochemistry and neuroanatomy will be defined by innovative scientific developments and disciplines working together to change

Table 1.3 Experimental techniques in neurochemistry and neuroanatomy, summarizing the key methodologies, their principles, and applications in research.

Technique	Category	Principle	Applications	Ref.
Immunohistochemistry (IHC)	Neuroanatomy/ protein detection	Uses antibodies to detect specific proteins in tissue sections	Localizing neurotransmitter receptors, enzymes, or structural proteins in brain regions	188
In situ hybridization (ISH)	Neuroanatomy/ molecular biology	Detects mRNA expression in tissue using complementary nucleic acid probes	Gene expression mapping in neurons	189
Western blotting	Neurochemistry/ protein analysis	Separates proteins *via* electrophoresis; detects with antibodies	Quantification of neurotransmitter-related proteins or enzymes	190
ELISA (enzyme-linked immunosorbent assay)	Neurochemistry/ protein quantification	Antibody-based quantification of proteins in solution	Measuring levels of neurotransmitters, cytokines, or receptors in brain samples	191
HPLC (high-performance liquid chromatography)	Neurochemistry/ small-molecule analysis	Separates and quantifies small molecules	Detection of neurotransmitters (dopamine, serotonin, GABA, *etc.*) in tissues or fluids	192
Microdialysis	Neurochemistry/ *in vivo* sampling	Collects extracellular fluid from specific brain regions	Real-time measurement of neurotransmitter release and metabolism	193
Mass spectrometry (MS)	Neurochemistry/ metabolomics	Ionizes and detects molecular mass of compounds	Profiling neurotransmitters, neuromodulators, and metabolites	194
Autoradiography	Neuroanatomy/ receptor mapping	Uses radiolabeled ligands to visualize receptor distribution	Mapping neurotransmitter receptor density (*e.g.*, dopamine D2, GABA-A)	195
Confocal/fluorescence microscopy	Neuroanatomy/ imaging	Uses lasers to scan fluorescently labeled tissue in high resolution	Detailed imaging of neuronal structures and labeled neurotransmitters or receptors	196
Electron microscopy (EM)	Neuroanatomy/ ultrastructure	Uses electrons for high-resolution imaging of cells	Studying synaptic architecture, neuronal organelles, and ultrastructural changes	197
Patch clamp electrophysiology	Neurochemistry/ functional study	Measures ionic currents through individual ion channels	Analysis of synaptic transmission, action potentials, and drug effects on neuronal excitability	198

Electroencephalography (EEG)	Functional neuroanatomy	Records electrical activity of the brain *via* scalp electrodes	Studies of brain wave patterns in epilepsy, sleep, and cognitive functions	199
Functional MRI (fMRI)	Neuroanatomy/functional imaging	Detects blood oxygenation level-dependent (BOLD) signals	Mapping brain activity during sensory, motor, and cognitive tasks	200
Optogenetics	Neurochemistry/functional study	Uses light-sensitive proteins to control neuronal activity	Investigating neural circuit functions and behavioral responses	201
Brain slicing and staining	Neuroanatomy	Uses dyes (*e.g.*, Nissl, Golgi) to visualize neurons	Examining cytoarchitecture and neuroanatomical connections	202
qPCR (quantitative PCR)	Neurochemistry/gene expression	Quantifies gene expression levels	Monitoring changes in neurotransmitter biosynthetic enzymes, receptors, or inflammatory markers	203

how we think about the chemical and physical bases of brain structure and function.[156] With many new high-throughput molecular methods and integrated imaging methods becoming readily available, the lines separating neurochemistry from neuroanatomy are becoming rapidly blurred. Brain function is being viewed less as the result of one or the other chemical or physical state.[156] Rather, it is being considered as the output of various networks of molecular signals combined with spatial geometry, structure, and organization.[204]

With recent technologies such as single-cell transcriptomics, spatial proteomics, and real-time neuroimaging, we can now probe and acquire data about brain tissue as never before. These new technologies offer a level of detail in which to discern cell-type-specific neurochemical microenvironments of discrete anatomical regions, in a way that we could previously only hope to determine as to how molecules, neurotransmitters, neuromodulators, and structural proteins interact to modify cognition, emotion and behavior.[205] When discussing molecular analysis of specific brain regions, very small tissue samples can be isolated for biochemical, molecular, or histological examination to investigate localized cellular and metabolic changes.[206]

At the same time, the integration of computational neuroscience and artificial intelligence is creating the opportunity for more sophisticated, and more accurate, modeling of brain circuits and neurochemical dynamics.[207] These models not only refine theoretical understanding, but have also helped in predicting the outcomes of pharmacological treatments and the design neurosurgical approaches. As we continue to uncover the complexities of the brain, brain maps will likely play a central role in the diagnosis and treatment of neurological and psychiatric disorders as they become personalized representations of an individual's anatomical and neurochemical fingerprints.[208]

Studies of neurodevelopmental trajectories and neuroplasticity will continue to provide insight into the brain's capacity to respond to its environment over time, and how it responds to injury or aging.[209] Longitudinal imaging studies, coupled with longitudinal molecular studies, hold particular promise by providing an avenue for detecting neurodegenerative illness at an earlier stage, as well as designing interventions to restore or preserve the function of the brain, in targeted ways.[210]

These advancements are not without consequences as they will raise ethical, social, and clinical issues that require careful consideration. With increased abilities to alter the brain, chemically and structurally, in precise and specific ways, we will need to examine the concerns in relation to cognitive enhancement, neuroethics, and individual identities.[211] The pillars of neurochemistry and neuroanatomy will not only be the foundations of neuroanatomy and neuroscience training but will also represent the launching pads for new discoveries that will connect the molecular details to systems-level brain function.[212] The brain's complexity will be solved by drawing upon diverse approaches and perspectives while translating basic knowledge into clinical and technological opportunities.[213]

1.12 Conclusion

In conclusion, the foundational knowledge of neuroanatomy and neurochemistry is a fundamental tool for approaching the vast complexity of the human brain. In the past, neuroanatomy and neurochemistry have been approached as discrete disciplines, but today we appreciate that these different measures are connected, and understanding the relationships between the chemical and structural components in the brain provides greater insight into the chemical signaling processes that control cognition, behavior, and health. As our tools and techniques allow us to discover new levels of complexity and integration in the nervous system, the value of foundational information that comes from neurochemistry and neuroanatomy is reinforced. For both research and clinical innovation, using a foundational knowledge of neurochemistry and neuroanatomy provides clinicians and researchers with a framework to navigate complexity and to apply this knowledge to enormous challenges presented by the function and dysfunction of the brain, while offering potential future approaches to developing preventative, diagnostic, and therapeutic techniques to ameliorate the burden of neurological and psychiatric disorders.

List of Abbreviations

5-HT	Serotonin
Ach	Acetylcholine
BBB	Blood–brain barrier
BP	Blood pressure
cAMP	Cyclic adenosine monophosphate
CNS	Central nervous system
DA	Dopamine
DAG	Diacylglycerol
GABA	Gamma-aminobutyric acid
GPCR	G-protein-coupled receptors
IP_3	Inositol trisphosphate
LTD	Long-term depression
LTP	Long-term potentiation
NE	Norepinephrine
PK	Protein kinase
PNS	Peripheral nervous system
SSRIs	Selective serotonin reuptake inhibitors

References

1. S. Uysal, *Functional neuroanatomy and clinical neuroscience: foundations for understanding disorders of cognition and behavior*, Oxford University Press, 2023.

2. A. J. Steiner, L. Aguilar-Hernandez, R. Abdelsalam, K. Q. Mercado, A. M. Taran, L. E. Gelfond and W. W. IsHak, Neurobiological Sciences: Neuroanatomy, Neurophysiology, and Neurochemistry, in *Atlas of Psychiatry*, Springer, 2023, pp. 91–146.
3. M. Sabina, T. Zakhrokhon and D. B. Gulamovna, Nervous System And Its Main Functions, *Eurasian J. Med. Nat. Sci.*, 2024, **4**(1), 61–67.
4. N. H. Barmack and V. E. Pettorossi, Adaptive balance in posterior cerebellum, *Front. Neurol.*, 2021, **12**, 635259.
5. D. Banerjee, P. K. Das and J. Mukherjee, Nervous System, in *Textbook of Veterinary Physiology*, Springer, 2023, pp. 265–293.
6. M. J. Moore, E. Milosevich, J. B. Mattingley and N. Demeyere, The neuroanatomy of visuospatial neglect: A systematic review and analysis of lesion-mapping methodology, *Neuropsychologia*, 2023, **180**, 108470.
7. P. Saglam-Metiner, E. Duran, L. Sabour-Takanlou, C. Biray-Avci and O. Yesil-Celiktas, Differentiation of neurons, astrocytes, oligodendrocytes and microglia from human induced pluripotent stem cells to form neural tissue-on-chip: a neuroinflammation model to evaluate the therapeutic potential of extracellular vesicles derived from mesenchymal stem cells, *Stem Cell Rev. Rep.*, 2024, **20**(1), 413–436.
8. I. Kalafatakis and D. Karagogeos, Oligodendrocytes and microglia: key players in myelin development, damage and repair, *Biomolecules*, 2021, **11**(7), 1058.
9. R. Bhushan, F. Goel and S. Singh, Beyond dopamine: novel therapeutic pathways for Parkinson's disease through receptor signaling, *CNS Neurol. Disord.: Drug Targets*, 2025, **24**(6), 434–451.
10. M. T. Ullman, G. M. Clark, M. Y. Pullman, J. T. Lovelett, E. I. Pierpont, X. Jiang and P. E. Turkeltaub, The neuroanatomy of developmental language disorder: a systematic review and meta-analysis, *Nat. Hum. Behav.*, 2024, **8**(5), 962–975.
11. M. A. Patestas, A. J. Meyer and L. P. Gartner, *A textbook of neuroanatomy*, John Wiley & Sons, 2025.
12. J. V. Andersen, K. H. Markussen, E. Jakobsen, A. Schousboe, H. S. Waagepetersen, P. A. Rosenberg and B. I. Aldana, Glutamate metabolism and recycling at the excitatory synapse in health and neurodegeneration, *Neuropharmacology*, 2021, **196**, 108719.
13. R. I. Teleanu, A.-G. Niculescu, E. Roza, O. Vladâcenco, A. M. Grumezescu and D. M. Teleanu, Neurotransmitters—key factors in neurological and neurodegenerative disorders of the central nervous system, *Int. J. Mol. Sci.*, 2022, **23**(11), 5954.
14. F. Goel, Exploring the therapeutic role of Moringa oleifera in neurodegeneration: antioxidant, anti-inflammatory, and neuroprotective mechanisms: Exploring the therapeutic role of Moringa oleifera in neurodegeneration, *Inflammopharmacology*, 2025, 1–17.
15. F. Goel, D. Kumar and A. Sharma, Impact of corticoid receptors on Alzheimer's disease: a neuroendocrine perspective, *Inflammopharmacology*, 2025, 1–16.

16. J. He, T. L. Pham and H. E. Bazan, Neuroanatomy and neurochemistry of rat cornea: Changes with age, *Ocul. Surf.*, 2021, **20**, 86–94.
17. T. S. Moreira, A. C. Takakura, B. Falquetto, J.-M. Ramirez, L. M. Oliveira, P. E. Silva and E. V. Araujo, Neuroanatomical and neurochemical organization of brainstem and forebrain circuits involved in breathing regulation, *J. Neurophysiol.*, 2025, **133**(4), 1116–1137.
18. S. Booth and J. H. Ko, Radionuclide Imaging of the neuroanatomical and neurochemical substrate of Cognitive decline in Parkinson's Disease, *Nucl. Med. Mol. Imaging*, 2024, **58**(4), 213–226.
19. H. Wei, A. M. Frey and A. Jasanoff, Molecular fMRI of neurochemical signaling, *J. Neurosci. Methods*, 2021, **364**, 109372.
20. Y. Yang, B. Li and Y. Li, Genetically encoded sensors for the *in vivo* detection of neurochemical dynamics, *Annu. Rev. Anal. Chem.*, 2024, **17**(1), 367–392.
21. J. Leo, Chemical Neuroanatomy, in *Gross Anatomy, Neuroanatomy, and Embryology for Medical Students*, Springer, 2025, pp. 451–465.
22. B. M. Ascenzi, M. W. Myers and B. Buccilli, Neurochemistry of the central nervous system, in *From Anatomy to Function of the Central Nervous System*, Elsevier, 2025, pp. 59–107.
23. A. Pal, F. Goel and V. K. Garg, From Genetics to Function: the Role of ABCA12 in Autism Neurobiology, *J. Mol. Neurosci.*, 2025, **75**(2), 67.
24. B. K. Johnson, Physiology of the autonomic nervous system, in *Basic sciences in anesthesia*, Springer, 2025, pp. 377–386.
25. A. Pal, F. Goel, A. Sharma and V. K. Garg, Oxidative stress and antioxidant therapeutics in autism spectrum disorder: a biochemical and structure–activity relationship perspective, *Mol. Diversity*, 2025, 1–19.
26. A. Correia, A. Monteiro, R. Silva, J. Moreira, J. S. Lobo and A. Silva, Lipid nanoparticles strategies to modify pharmacokinetics of central nervous system targeting drugs: Crossing or circumventing the blood–brain barrier (BBB) to manage neurological disorders, *Adv. Drug Delivery Rev.*, 2022, **189**, 114485.
27. Y. Pei, L. Yan, Z. Wu, J. Lu, J. Zhao, J. Chen, Q. Liu and X. Yan, Artificial visual perception nervous system based on low-dimensional material photoelectric memristors, *ACS Nano*, 2021, **15**(11), 17319–17326.
28. F. Evans-Martin, *The nervous system*, Infobase Holdings, Inc, 2022.
29. T. Komori, The 2021 WHO classification of tumors, central nervous system tumors: the 10 basic principles, *Brain Tumor Pathol.*, 2022, **39**(2), 47–50.
30. A. Pal, F. Goel and V. K. Garg, Zebrafish as a tool for autism research: unraveling the roles of Shank3, Cntnap2, Neuroligin3, and Arid1b in synaptic and behavioral abnormalities, *Neurogenetics*, 2025, **26**(1), 48.
31. S. A. Korai, F. Ranieri, V. Di Lazzaro, M. Papa and G. Cirillo, Neurobiological after-effects of low intensity transcranial electric stimulation of the human nervous system: from basic mechanisms to metaplasticity, *Front. Neurol.*, 2021, **12**, 587771.
32. M. Yonesi, M. Garcia-Nieto, G. V. Guinea, F. Panetsos, J. Perez-Rigueiro and D. Gonzalez-Nieto, Silk fibroin: An ancient material for repairing the injured nervous system, *Pharmaceutics*, 2021, **13**(3), 429.

33. D. Kumar, F. Goel and S. N. Rai, *The role of ABCA12 in neurodegenerative diseases–a review of molecular mechanisms and potential therapeutic implications*, 2025.
34. T. Aksu, R. Gopinathannair, D. Gupta and D. H. Pauza, Intrinsic cardiac autonomic nervous system: what do clinical electrophysiologists need to know about the "heart brain"?, *J. Cardiovasc. Electrophysiol.*, 2021, **32**(6), 1737–1747.
35. T. V. Zhukova, M. A. Yurochkin, Y. G. Shanko, A. V. Beletsky, A. V. Borisov, S. M. Polyakova, Y. I. Rogov, K. A. Samochernykh, N. E. Ivanova and J. M. Zabrodskaya, Morphological features of diagnostics of repair and pathological processes of the central nervous system and their significance in clinical practice, *Russ. Neurosurg. J. (Prof. A. L. Polenov)*, 2025, 1263–1288.
36. D. Wu, Q. Chen, X. Chen, F. Han, Z. Chen and Y. Wang, The blood–brain barrier: Structure, regulation and drug delivery, *Signal Transduction Targeted Ther.*, 2023, **8**(1), 217.
37. E. G. Knox, M. R. Aburto, G. Clarke, J. F. Cryan and C. M. O'Driscoll, The blood–brain barrier in aging and neurodegeneration, *Mol. Psychiatry*, 2022, **27**(6), 2659–2673.
38. A. Alahmari, Blood–brain barrier overview: Structural and functional correlation, *Neural Plast.*, 2021, **2021**(1), 6564585.
39. M. Abubakar, A. Giri, F. Goel, M. Khan, J. Gupta, D. Kumar, M. Kaushik, S. N. Rai and N. Kumar, Diabetes, Alzheimer's Disease Risk Factors, and the Cafeteria Diet: A Comprehensive Review, *Curr. Neuropharmacol.*, 2025, **24**, 573–591.
40. L. Formery, F. Orange, A. Formery, S. Yaguchi, C. J. Lowe, M. Schubert and J. C. Croce, Neural anatomy of echinoid early juveniles and comparison of nervous system organization in echinoderms, *J. Comp. Neurol.*, 2021, **529**(6), 1135–1156.
41. K. Javed, V. Reddy and F. Lui, Neuroanatomy, cerebral cortex, in *StatPearls*, StatPearls Publishing, 2023.
42. T. H. Champney, *Essential clinical neuroanatomy*, John Wiley & Sons, 2023.
43. M. M. Mesulam, Temporopolar regions of the human brain, *Brain*, 2023, **146**(1), 20–41.
44. S. N. Rai, P. Singh, H. W. Steinbusch, E. Vamanu, G. Ashraf and M. P. Singh, The role of vitamins in neurodegenerative disease: An update, *Biomedicines*, 2021, **9**(10), 1284.
45. N. Aridan, M. Bernstein-Eliav, D. Gamzo, M. Schmeidler, N. Tik and I. Tavor, Neuroanatomy in virtual reality: Development and pedagogical evaluation of photogrammetry-based 3D brain models, *Anat. Sci. Educ.*, 2024, **17**(2), 239–248.
46. S. D. Lanooij, U. L. Eisel, W. H. Drinkenburg, E. A. van der Zee and M. J. Kas, Influencing cognitive performance *via* social interactions: a novel therapeutic approach for brain disorders based on neuroanatomical mapping?, *Mol. Psychiatry*, 2023, **28**(1), 28–33.

47. S. N. Rai, C. Singh, A. Singh, M. Singh and B. K. Singh, Mitochondrial dysfunction: a potential therapeutic target to treat Alzheimer's disease, *Mol. Neurobiol.*, 2020, **57**(7), 3075–3088.
48. A. Patel, G. M. N. R. Biso and J. B. Fowler, Neuroanatomy, temporal lobe, in *StatPearls*, StatPearls Publishing, 2023.
49. S. S. Singh, S. N. Rai, H. Birla, W. Zahra, A. S. Rathore, H. Dilnashin, R. Singh and S. P. Singh, *Research Article Neuroprotective Effect of Chlorogenic Acid on Mitochondrial Dysfunction-Mediated Apoptotic Death of DA Neurons in a Parkinsonian Mouse Model*, 2020.
50. N. Makris, R. Rushmore, J. Kaiser, M. Albaugh, M. Kubicki, Y. Rathi, F. Zhang, L. J. O'Donnell, E. Yeterian and V. S. Caviness, A proposed human structural brain connectivity matrix in the Center for Morphometric Analysis Harvard-Oxford Atlas Framework: a historical perspective and future direction for enhancing the precision of human structural connectivity with a novel neuroanatomical typology, *Dev. Neurosci.*, 2023, **45**(4), 161–180.
51. S. N. Rai, Effect of COVID19 Pandemic on the Vital Function of the Central Nervous System: A Literature-based Prospective, *EC Neurol.*, 2020, **12**, 163–168.
52. F. Sanaeifar, S. Pourranjbar, M. Pourranjbar, S. Ramezani, S. R. Mehr, A.-H. S. Wadan and F. Khazeifard, Beneficial effects of physical exercise on cognitive-behavioral impairments and brain-derived neurotrophic factor alteration in the limbic system induced by neurodegeneration, *Exp. Gerontol.*, 2024, **195**, 112539.
53. M. Connaughton, E. O'Hanlon, T. J. Silk, J. Paterson, A. O'Neill, V. Anderson, R. Whelan and J. McGrath, The limbic system in children and adolescents with attention-deficit/hyperactivity disorder: a longitudinal structural magnetic resonance imaging analysis, *Biol. Psychiatry: Glob. Open Sci.*, 2024, **4**(1), 385–393.
54. J. M. Shine, L. D. Lewis, D. D. Garrett and K. Hwang, The impact of the human thalamus on brain-wide information processing, *Nat. Rev. Neurosci.*, 2023, **24**(7), 416–430.
55. H. Arnts, S. E. Coolen, F. W. Fernandes, R. Schuurman, J. K. Krauss, H. J. Groenewegen and P. van den Munckhof, The intralaminar thalamus: a review of its role as a target in functional neurosurgery, *Brain Commun.*, 2023, **5**(3), fcad003.
56. C. Abagnale, A. Di Renzo, G. Giuliani, G. Sebastianelli, F. Casillo, L. Ziccardi, V. Parisi, C. Di Lorenzo, M. Serrao and F. Caramia, MRI-based analysis of the microstructure of the thalamus and hypothalamus and functional connectivity between cortical networks in episodic cluster headache, *J. Headache Pain*, 2025, **26**(1), 12.
57. C.-C. Chao, M.-T. Tseng, P.-C. Hsieh, C.-H. Lin, S.-L. Huang, S.-T. Hsieh and M.-C. Chiang, Brain mechanisms of pain and dysautonomia in diabetic neuropathy: connectivity changes in thalamus and hypothalamus, *J. Clin. Endocrinol. Metab.*, 2022, **107**(3), e1167–e1180.

58. M. H. Bear, V. Reddy and P. C. Bollu, Neuroanatomy, hypothalamus, in *StatPearls*, StatPearls Publishing, 2022.
59. R. Leiras, J. M. Cregg and O. Kiehn, Brainstem circuits for locomotion, *Annu. Rev. Neurosci.*, 2022, **45**(1), 63–85.
60. S. Arber and R. M. Costa, Networking brainstem and basal ganglia circuits for movement, *Nat. Rev. Neurosci.*, 2022, **23**(6), 342–360.
61. M. R. Carey, The cerebellum, *Curr. Biol.*, 2024, **34**(1), R7–R11.
62. J. Prakash, S. Chouhan, S. K. Yadav, S. Westfall, S. N. Rai and S. P. Singh, Withania somnifera alleviates parkinsonian phenotypes by inhibiting apoptotic pathways in dopaminergic neurons, *Neurochem. Res.*, 2014, **39**(12), 2527–2536.
63. S. Rudolph, A. Badura, S. Lutzu, S. S. Pathak, A. Thieme, J. L. Verpeut, M. J. Wagner, Y.-M. Yang and D. Fioravante, Cognitive-affective functions of the cerebellum, *J. Neurosci.*, 2023, **43**(45), 7554–7564.
64. S. K. Yadav, S. N. Rai and S. P. Singh, Mucuna pruriens reduces inducible nitric oxide synthase expression in Parkinsonian mice model, *J. Chem. Neuroanat.*, 2017, **80**, 1–10.
65. M. B. Kale, N. L. Wankhede, A. K. Bishoyi, S. Ballal, R. Kalia, R. Arya, S. Kumar, M. Khalid, M. Gulati and M. Umare, Emerging biophysical techniques for probing synaptic transmission in neurodegenerative disorders, *Neuroscience*, 2025, **565**, 63–79.
66. B. Wang and O. K. Dudko, A theory of synaptic transmission, *eLife*, 2021, **10**, e73585.
67. S. N. Rai, H. Birla, W. Zahra, S. Sen Singh and S. P. Singh, Commentary: Synaptic vesicle glycoprotein 2C (SV2C) modulates dopamine release and is disrupted in Parkinson disease, *Front. Synaptic Neurosci.*, 2018, **9**, 18.
68. S. Hunt, Y. Leibner, E. J. Mertens, N. Barros-Zulaica, L. Kanari, T. S. Heistek, M. M. Karnani, R. Aardse, R. Wilbers and D. B. Heyer, Strong and reliable synaptic communication between pyramidal neurons in adult human cerebral cortex, *Cereb. Cortex*, 2023, **33**(6), 2857–2878.
69. S. N. Rai, H. Birla, W. Zahra, S. Sen Singh and S. P. Singh, The role of Mucuna pruriens and Withania somnifera in the neuroprotection and treatment of Parkinson's disease, *SOJ Neurol.*, 2018, **5**, 1–7.
70. S. Li and Z.-H. Sheng, Energy matters: presynaptic metabolism and the maintenance of synaptic transmission, *Nat. Rev. Neurosci.*, 2022, **23**(1), 4–22.
71. F. Wang, X. Xie, X. Xing and X. Sun, Excitatory synaptic transmission in ischemic stroke: a new outlet for classical neuroprotective strategies, *Int. J. Mol. Sci.*, 2022, **23**(16), 9381.
72. B. Drukarch, M. M. Wilhelmus and S. Shrivastava, The thermodynamic theory of action potential propagation: a sound basis for unification of the physics of nerve impulses, *Rev. Neurosci.*, 2022, **33**(3), 285–302.
73. Q. Banks, H. Bibollet, M. Contreras, D. F. Bennett, R. A. Bannister, M. F. Schneider and E. O. Hernández-Ochoa, Voltage sensor movements of CaV1. 1 during an action potential in skeletal muscle fibers, *Proc. Natl. Acad. Sci. U. S. A.*, 2021, **118**(40), e2026116118.

74. L. M. Dicks, Gut bacteria and neurotransmitters, *Microorganisms*, 2022, **10**(9), 1838.
75. Z. Yang, Y. Zou and L. Wang, Neurotransmitters in prevention and treatment of Alzheimer's disease, *Int. J. Mol. Sci.*, 2023, **24**(4), 3841.
76. A. Barnfield, Role of Neurotransmitters, in *Encyclopedia of Personality and Individual Differences*, Springer, 2019, pp. 1–7.
77. J. Rizo, Mechanism of neurotransmitter release coming into focus, *Protein Sci.*, 2018, **27**(8), 1364–1391.
78. S. N. Rai, W. Zahra, S. S. Singh, H. Birla, C. Keswani, H. Dilnashin, A. S. Rathore, R. Singh, R. K. Singh and S. P. Singh, Anti-inflammatory activity of ursolic acid in MPTP-induced parkinsonian mouse model, *Neurotoxic. Res.*, 2019, **36**(3), 452–462.
79. A. Figurov, L. D. Pozzo-Miller, P. Olafsson, T. Wang and B. Lu, Regulation of synaptic responses to high-frequency stimulation and LTP by neurotrophins in the hippocampus, *Nature*, 1996, **381**(6584), 706–709.
80. F. Burette, T. M. Jay and S. Laroche, Reversal of LTP in the hippocampal afferent fiber system to the prefrontal cortex *in vivo* with low-frequency patterns of stimulation that do not produce LTD, *J. Neurophysiol.*, 1997, **78**(2), 1155–1160.
81. H. Birla, S. N. Rai, S. S. Singh, W. Zahra, A. Rawat, N. Tiwari, R. K. Singh, A. Pathak and S. P. Singh, Tinospora cordifolia suppresses neuroinflammation in parkinsonian mouse model, *NeuroMol. Med.*, 2019, **21**(1), 42–53.
82. X. G. Liu, C. Morton, J. Azkue, M. Zimmermann and J. Sandkühler, Long-term depression of C-fibre-evoked spinal field potentials by stimulation of primary afferent Aδ-fibres in the adult rat, *Eur. J. Neurosci.*, 1998, **10**(10), 3069–3075.
83. S. Kirischuk, Keeping excitation–inhibition ratio in balance, *Int. J. Mol. Sci.*, 2022, **23**(10), 5746.
84. G. Hennequin, E. J. Agnes and T. P. Vogels, Inhibitory plasticity: balance, control, and codependence, *Annu. Rev. Neurosci.*, 2017, **40**(1), 557–579.
85. P. Srivastava, P. N. Tripathi, P. Sharma, S. N. Rai, S. P. Singh, R. K. Srivastava, S. Shankar and S. K. Shrivastava, Design and development of some phenyl benzoxazole derivatives as a potent acetylcholinesterase inhibitor with antioxidant property to enhance learning and memory, *Eur. J. Med. Chem.*, 2019, **163**, 116–135.
86. L. Cathenaut, B. Leonardon, R. Kuster, P. Inquimbert, R. Schlichter and S. Hugel, Inhibitory interneurons with differential plasticities at their connections tune excitatory–inhibitory balance in the spinal nociceptive system, *Pain*, 2022, **163**(5), e675–e688.
87. H. Lagercrantz, E. Herlenius, M. Hanson, P. Evrard and C. Rodeck, Neurotransmitters and neuromodulators, *The Newborn Brain: Neuroscience and Clinical Applications*, ed. H. Lagercrantz, M. Hanson, P. Evrard and C. H. Rodeck, 2002, pp. 139–165.
88. E. Vamanu and S. N. Rai, The link between obesity, microbiota dysbiosis, and neurodegenerative pathogenesis, *Diseases*, 2021, **9**(3), 45.

89. E. Herlenius and H. Lagercrantz, Neurotransmitters and neuromodulators, *The Newborn Brain: Neuroscience and Clinical Applications*, Cambridge University Press, Cambridge, 2010, pp. 99–119.
90. K. Ramakrishna, M. K. Viswanadh, N. Dumala, G. Chakravarth, K. Venkateswarlu, G. Gutti, G. Yadagiri, S. Duguluri and S. N. Rai, Advanced biomaterials in neuroprotection: innovations and clinical applications, in *Biomaterials and neurodegenerative disorders*, Springer, 2025, pp. 69–92.
91. P. Świt, A. Pollap and J. Orzeł, Spectroscopic determination of acetylcholine (ACh): a representative review, *Top. Curr. Chem.*, 2023, **381**(4), 16.
92. C. A. Curcio, D. Kar, C. Owsley, K. R. Sloan and T. Ach, Age-related macular degeneration, a mathematically tractable disease, *Invest Ophthalmol. Visual Sci.*, 2024, **65**(3), 4.
93. N. Singh, S. N. Rai, D. Singh and S. P. Singh, Withania somnifera shows ability to counter Parkinson's Disease: An Update, *SOJ Neurol.*, 2015, **2**(2), 1–4.
94. C. Jiang, Q. Zhan and C. Zeng, The 5-HT-related gut-brain axis in obesity, *Life Sci.*, 2024, **358**, 123171.
95. R. Patil, S. N. Rai, J. Jadhav and S. P. Singh, Mucuna sanjappae shows promising anti-Parkinson's activity by reducing oxidative stress in mptp induced mouse model, *Eur. J. Pharm. Med. Res.*, 2016, **3**(11), 452–463.
96. V. Dionisie, G. A. Filip, M. C. Manea, M. Manea and S. Riga, The anti-inflammatory role of SSRI and SNRI in the treatment of depression: a review of human and rodent research studies, *Inflammopharmacology*, 2021, **29**(1), 75–90.
97. S. Duguluri and S. N. Rai, in Neuroprotection: Innovations and Clinical Applications, *Biomaterials and Neurodegenerative Disorders*, 2025, p. 69.
98. J. V. Andersen and A. Schousboe, Milestone Review: Metabolic dynamics of glutamate and GABA mediated neurotransmission—The essential roles of astrocytes, *J. Neurochem.*, 2023, **166**(2), 109–137.
99. S. M. Sears and S. J. Hewett, Influence of glutamate and GABA transport on brain excitatory/inhibitory balance, *Exp. Biol. Med.*, 2021, **246**(9), 1069–1083.
100. K. N. Fountoulakis, Neurotransmitters, in *The Human Connectome: How the Brain Works*, Springer, 2022, pp. 69–105.
101. M. Perisic, K. Woolcock, A. Hering, H. Mendel and M. Muttenthaler, Oxytocin and vasopressin signaling in health and disease, *Trends Biochem. Sci.*, 2024, **49**(4), 361–377.
102. V. B. Patel, V. R. Preedy and C. R. Martin, *Neurobiology and Physiology of the Endocannabinoid System*, Elsevier, 2023.
103. U. Chaudhary, Neuromodulation in Neural Circuits: Neurotransmitters and Synaptic Plasticity, in *Expanding Senses using Neurotechnology: Volume 1–Foundation of Brain-Computer Interface Technology*, Springer, 2025, pp. 225–245.

104. H. Samaripour, Neurotransmitters and mental health: understanding the chemical underpinnings of psychological disorders, *Int. J. New Chem.*, 2025, **12**(4), 825–841.
105. S. Murala, A. Boddu and P. C. Bollu, Glutamate, in *Neurochemistry in Clinical Practice*, Springer, 2022, pp. 91–107.
106. S. Murala, A. Yelam, M. M. Ismail and P. C. Bollu, Gaba, in *Neurochemistry in clinical practice*, Springer, 2022, pp. 73–89.
107. F. Bono, V. Mutti, Z. Tomasoni, G. Sbrini, C. Missale and C. Fiorentini, Recent advances in dopamine D3 receptor heterodimers: focus on dopamine D3 and D1 receptor–receptor interaction and striatal function, *Therapeutic Applications of Dopamine D3 Receptor Function: New Insight After 30 Years Of Research*, 2022, pp. 47–72.
108. J. Motiejunaite, L. Amar and E. Vidal-Petiot, in *Adrenergic receptors and cardiovascular effects of catecholamines*, Annales d'endocrinologie, Elsevier, 2021, pp. 193–197.
109. X. Zhang, G. Liu, Y.-N. Zhong, R. Zhang, C.-C. Yang, C. Niu, X. Pu, J. Sun, T. Zhang and L. Yang, Structural basis of ligand recognition and activation of the histamine receptor family, *Nat. Commun.*, 2024, **15**(1), 8296.
110. K. Nagori, M. Pradhan, M. Sharma, Ajazuddin, H. R. Badwaik and K. T. Nakhate, Current progress on central cholinergic receptors as therapeutic targets for Alzheimer's disease, *Curr. Alzheimer Res.*, 2024, **21**(1), 50–68.
111. C. Humes, A. Sic and N. N. Knezevic, Substance P's impact on chronic pain and psychiatric conditions—a narrative review, *Int. J. Mol. Sci.*, 2024, **25**(11), 5905.
112. L. Janicka, A. Dajnowska and M. B. Arciszewski, Role of proopiomelanocortin (POMC) derivatives in body weight regulation: A review of current knowledge and significance in the pathogenesis of obesity in mammals, *Med. Weter.*, 2025, **81**(5), 221–230.
113. S. Blanco, M. dM. Muñoz-Gallardo, R. Hernández and M. Á. Peinado, The Interplay Between Melatonin and Nitric Oxide: Mechanisms and Implications in Stroke Pathophysiology, *Antioxidants*, 2025, **14**(6), 724.
114. R. Greco, C. Demartini, A. M. Zanaboni, M. Francavilla, R. De Icco, L. Ahmad and C. Tassorelli, The endocannabinoid system and related lipids as potential targets for the treatment of migraine-related pain, *Headache*, 2022, **62**(3), 227–240.
115. Y. Jin, D. Song, Y. Yan, Z. Quan and H. Qing, The role of oxytocin in early-life-stress-related neuropsychiatric disorders, *Int. J. Mol. Sci.*, 2023, **24**(13), 10430.
116. J. Lee and H.-J. Kim, Normal aging induces changes in the brain and neurodegeneration progress: review of the structural, biochemical, metabolic, cellular, and molecular changes, *Front. Aging Neurosci.*, 2022, **14**, 931536.
117. W. A. Catterall, Structure and function of voltage-gated ion channels, *Annu. Rev. Biochem.*, 1995, **64**(1), 493–531.

118. C. M. Armstrong and B. Hille, Voltage-gated ion channels and electrical excitability, *Neuron*, 1998, **20**(3), 371–380.
119. A. Scholz, Mechanisms of (local) anaesthetics on voltage-gated sodium and other ion channels, *Br. J. Anaesth.*, 2002, **89**(1), 52–61.
120. M. Sobolczyk and T. Boczek, Astrocytic calcium and cAMP in neurodegenerative diseases, *Front. Cell. Neurosci.*, 2022, **16**, 889939.
121. H. Gakire Munyaneza, *The Role of Second Messengers in Neurodegenerative Disorders*, 2025.
122. E. Castrén and L. M. Monteggia, Brain-derived neurotrophic factor signaling in depression and antidepressant action, *Biol. Psychiatry*, 2021, **90**(2), 128–136.
123. Y. Li, F. Li, D. Qin, H. Chen, J. Wang, J. Wang, S. Song, C. Wang, Y. Wang and S. Liu, The role of brain derived neurotrophic factor in central nervous system, *Front. Aging Neurosci.*, 2022, **14**, 986443.
124. M. Albini, A. Krawczun-Rygmaczewska and F. Cesca, Astrocytes and brain-derived neurotrophic factor (BDNF), *Neurosci. Res. Lett.*, 2023, **197**, 42–51.
125. S. Jaberi and M. Fahnestock, Mechanisms of the beneficial effects of exercise on brain-derived neurotrophic factor expression in Alzheimer's disease, *Biomolecules*, 2023, **13**(11), 1577.
126. D. Trigo, C. Avelar, M. Fernandes, J. Sá and O. da Cruz e Silva, Mitochondria, energy, and metabolism in neuronal health and disease, *FEBS Lett.*, 2022, **596**(9), 1095–1110.
127. A. M. Antonio, G. G. Hernando, E. V. Cano, J. L. Rois, O. S. Moreira, P. Sanchez-Quinteiro and I. Ortiz-Leal, Neurochemical Features and Glycan Patterns in the Wapiti Vomeronasal Pathway, bioRxiv, 2025, preprint, DOI: 10.1101/2025.06. 25.661586.
128. K.-S. Hwang, Y. Son, S. S. Kim, D.-S. Shin, S. H. Lim, J. Y. Yang, H. N. Jeong, B. H. Lee and M. A. Bae, Size-dependent effects of polystyrene nanoparticles (PS-NPs) on behaviors and endogenous neurochemicals in zebrafish larvae, *Int. J. Mol. Sci.*, 2022, **23**(18), 10682.
129. L. G. Appelbaum, M. A. Shenasa, L. Stolz and Z. Daskalakis, Synaptic plasticity and mental health: methods, challenges and opportunities, *Neuropsychopharmacology*, 2023, **48**(1), 113–120.
130. R. Yasuda, Y. Hayashi and J. W. Hell, CaMKII: a central molecular organizer of synaptic plasticity, learning and memory, *Nat. Rev. Neurosci.*, 2022, **23**(11), 666–682.
131. J. Cornell, S. Salinas, H.-Y. Huang and M. Zhou, Microglia regulation of synaptic plasticity and learning and memory, *Neural Regen. Res.*, 2022, **17**(4), 705–716.
132. A. D. Madar, A. Jiang, C. Dong and M. E. Sheffield, Synaptic plasticity rules driving representational shifting in the hippocampus, *Nat. Neurosci.*, 2025, 1–13.
133. Y. Sun, H. Wang and D. Xie, Recent advance in synaptic plasticity modulation techniques for neuromorphic applications, *Nano-Micro Lett.*, 2024, **16**(1), 211.

134. A. Goto, Synaptic plasticity during systems memory consolidation, *Neurosci. Res. Lett.*, 2022, **183**, 1–6.
135. Y. K. Wu, C. Miehl and J. Gjorgjieva, Regulation of circuit organization and function through inhibitory synaptic plasticity, *Trends Neurosci.*, 2022, **45**(12), 884–898.
136. E. López-Merino, M. I. Cuartero, J. A. Esteban and V. Briz, Perinatal exposure to pesticides alters synaptic plasticity signaling and induces behavioral deficits associated with neurodevelopmental disorders, *Cell Biol. Toxicol.*, 2023, **39**(5), 2089–2111.
137. S. Zhu, X. Zhang, M. Zhou, K. M. Kendrick and W. Zhao, Therapeutic applications of transcutaneous auricular vagus nerve stimulation with potential for application in neurodevelopmental or other pediatric disorders, *Front. Endocrinol.*, 2022, **13**, 1000758.
138. J. Hao, H. Xie, K. Harp, Z. Chen and K.-C. Siu, Effects of virtual reality intervention on neural plasticity in stroke rehabilitation: a systematic review, *Arch. Phys. Med. Rehabil.*, 2022, **103**(3), 523–541.
139. S. F. Grieco, E. Castrén, G. M. Knudsen, A. C. Kwan, D. E. Olson, Y. Zuo, T. C. Holmes and X. Xu, Psychedelics and neural plasticity: therapeutic implications, *J. Neurosci.*, 2022, **42**(45), 8439–8449.
140. E. A. Fertuck, S. A. Fischer and R. D. Melara, Atypical neural plasticity and behavioral effects of trustworthiness learning in borderline personality disorder features, *J. Pers. Disord.*, 2023, **37**(5), 542–558.
141. D. Mercerón-Martínez, C. Ibaceta-González, C. Salazar, W. Almaguer-Melian, J. A. Bergado-Rosado and A. G. Palacios, Alzheimer's disease, neural plasticity, and functional recovery, *J. Alzheimers Dis.*, 2021, **82**(s1), S37–S50.
142. A. Tyrtyshnaia, A. Bondar, S. Konovalova and I. Manzhulo, Synaptamide improves cognitive functions and neuronal plasticity in neuropathic pain, *Int. J. Mol. Sci.*, 2021, **22**(23), 12779.
143. P. Escalada, A. Ezkurdia, M. J. Ramírez and M. Solas, Essential role of astrocytes in learning and memory, *Int. J. Mol. Sci.*, 2024, **25**(3), 1899.
144. D. G. Diniz, M. A. de Oliveira, C. M. de Lima, C. A. R. Fôro, M. C. K. Sosthenes, J. Bento-Torres, P. F. da Costa Vasconcelos, D. C. Anthony and C. W. P. Diniz, Age, environment, object recognition and morphological diversity of GFAP-immunolabeled astrocytes, *Behav. Brain Funct.*, 2016, **12**(1), 28.
145. J. Ridet, A. Privat, S. Malhotra and F. Gage, Reactive astrocytes: cellular and molecular cues to biological function, *Trends Neurosci.*, 1997, **20**(12), 570–577.
146. A. E. West, E. C. Griffith and M. E. Greenberg, Regulation of transcription factors by neuronal activity, *Nat. Rev. Neurosci.*, 2002, **3**(12), 921–931.
147. M. R. Alam, V. Dobhal and S. Singh, Neuroprotective potential of solanesol against tramadol induced zebrafish model of Parkinson's disease: insights from neurobehavioral, molecular, and neurochemical evidence, *Drug Chem. Toxicol.*, 2024, **47**(6), 1241–1256.

148. F. Goel, V. Dobhal, S. Singh, Protective effects of diacerein against quinolinic acid–induced Huntington's disease–like symptoms in adult zebrafish by targeting GSK-3β signalling. *Naunyn-Schmiedebergs Arch. Pharmacol.* 2025, 1–23.
149. J. D. Sánchez-Martínez, A. Valdés, R. Gallego, Z. J. Suárez-Montenegro, M. Alarcón, E. Ibañez, G. Alvarez-Rivera and A. Cifuentes, Blood–brain barrier permeability study of potential neuroprotective compounds recovered from plants and agri-food by-products, *Front. Nutr.*, 2022, **9**, 924596.
150. P. Guo, Z. Jin, J. Wang, A. Sang and H. Wu, Irisin Rescues Blood-Brain Barrier Permeability following Traumatic Brain Injury and Contributes to the Neuroprotection of Exercise in Traumatic Brain Injury, *Oxid. Med. Cell. Longevity*, 2021, **2021**(1), 1118981.
151. L. Subedi and B. P. Gaire, Neuroprotective effects of curcumin in cerebral ischemia: cellular and molecular mechanisms, *ACS Chem. Neurosci.*, 2021, **12**(14), 2562–2572.
152. D. Paramanick, V. D. Singh and V. K. Singh, Neuroprotective effect of phytoconstituents *via* nanotechnology for treatment of Alzheimer diseases, *J. Controlled Release*, 2022, **351**, 638–655.
153. Y. K. Ghallab and O. S. Elassal, Biochemical and neuropharmacology of psychiatric disorders, in *Nutrition and Psychiatric Disorders: An Evidence-Based Approach to Understanding the Diet-Brain Connection*, Springer, 2024, pp. 25–47.
154. M. J. Zigmond, C. A. Wiley and M.-F. Chesselet, *Neurobiology of brain disorders: biological basis of neurological and psychiatric disorders*, Academic Press, 2022.
155. A. A. Farooqui and T. Farooqui, Phospholipids, sphingolipids, and cholesterol-derived lipid mediators and their role in neurological disorders, *Int. J. Mol. Sci.*, 2024, **25**(19), 10672.
156. M. Haripriyaa and K. Suthindhiran, The important role of systems biology in neuroscience and neurology and its associated diseases, in *Systems Biology Approaches: Prevention, Diagnosis, and Understanding Mechanisms of Complex Diseases*, Springer, 2024, pp. 343–392.
157. S. Bano, M. A. Raza, S. Ghosh, N. S. Pandit, S. Srivastava, M. Azam, S. K. Dey, S. S. Han, J. K. Sinha and M. Ruwali, Emerging therapeutic targets in molecular neuropharmacology for Alzheimer's disease, *J. Biol. Regul. Homeost. Agents*, 2023, **37**(11), 5769–5784.
158. K. Jomova, R. Raptova, S. Y. Alomar, S. H. Alwasel, E. Nepovimova, K. Kuca and M. Valko, Reactive oxygen species, toxicity, oxidative stress, and antioxidants: chronic diseases and aging, *Arch. Toxicol.*, 2023, **97**(10), 2499–2574.
159. J. Yang, J. Luo, X. Tian, Y. Zhao, Y. Li and X. Wu, Progress in understanding oxidative stress, aging, and aging-related diseases, *Antioxidants*, 2024, **13**(4), 394.

160. V. N. Gladyshev, S. B. Kritchevsky, S. G. Clarke, A. M. Cuervo, O. Fiehn, J. P. de Magalhães, T. Mau, M. Maes, R. L. Moritz and L. J. Niedernhofer, Molecular damage in aging, *Nat. Aging*, 2021, **1**(12), 1096–1106.
161. A. Manolopoulos, P. J. Yao and D. Kapogiannis, Extracellular vesicles: translational research and applications in neurology, *Nat. Rev. Neurol.*, 2025, 1–18.
162. F. Di Domenico, C. Lanzillotta and M. Perluigi, Redox imbalance and metabolic defects in the context of Alzheimer disease, *FEBS Lett.*, 2024, **598**(17), 2047–2066.
163. D. Garbuz, O. Zatsepina and M. Evgen'ev, Beta amyloid, tau protein, and neuroinflammation: an attempt to integrate different hypotheses of Alzheimer's disease pathogenesis, *Mol. Biol.*, 2021, **55**(5), 670–682.
164. O. J. Olajide, I. T. Gbadamosi, E. O. Yawson, T. Arogundade, F. S. Lewu, K. Y. Ogunrinola, O. O. Adigun, O. Bamisi, E. Lambe and L. O. Arietarhire, Hippocampal degeneration and behavioral impairment during Alzheimer-like pathogenesis involves glutamate excitotoxicity, *J. Mol. Neurosci.*, 2021, **71**(6), 1205–1220.
165. S. Kumar, L. Goyal and S. Singh, Tremor and rigidity in patients with Parkinson's disease: emphasis on epidemiology, pathophysiology and contributing factors, *CNS Neurol. Disord.: Drug Targets*, 2022, **21**(7), 596–609.
166. A. O. Kruse and J. R. Bustillo, Glutamatergic dysfunction in Schizophrenia, *Transl. Psychiatry*, 2022, **12**(1), 500.
167. V. Singh, R. Singh and G. Singh, *Neuroanatomical basis of depression and bipolar disorder: insights into the brain's role in mood disorders*, Medknow, 2024, vol. 73, pp. 297–299.
168. M. Locatelli, I. Vanelli, L. Fregna, F. Martini, E. Manfredi, F. Pacchioni, G. Bigai, L. Tonet, C. Passani and C. Colombo, Anxiety-Related Disorders, in *Fundamentals of Psychiatry for Health Care Professionals*, Springer, 2022, pp. 121–179.
169. H. Raji, S. Dinesh and S. Sharma, Inside the impulsive brain: a narrative review on the role of neurobiological, hormonal and genetic factors influencing impulsivity in psychiatric disorders, *Egypt. J. Neurol. Psychiatry Neurosurg.*, 2025, **61**(1), 4.
170. N. Osafo and O. K. Yeboah, Therapeutic Targets and Neurochemical Signaling in Huntington's Disease, in *Neurochemical Systems and Signaling*, CRC Press, 2023, pp. 127–144.
171. G. L. Sarlo and K. F. Holton, Brain concentrations of glutamate and GABA in human epilepsy: A review, *Seizure*, 2021, **91**, 213–227.
172. A. Quadir, *Addiction: Critique Of The BDMA And Development Of a Transdisciplinary, Biopsychosocial-Ecological Model*, 2021.
173. V. Bamicha, P. Pergantis, C. Skianis and A. Drigas, Computational Neuroscience's Influence on Autism Neuro-Transmission Research: Mapping Serotonin, Dopamine, GABA, and Glutamate, *Biomedicines*, 2025, **13**(6), 1420.

174. K. Kalaiarasi and A. Augustine, Obsessive-Compulsive Disorder and Histopathology, in *Nutrition and Obsessive-Compulsive Disorder*, CRC Press, 2023, pp. 79–88.
175. D. Patel, K. Sharma, R. Tripathi, J. Shah and M. Shah, Attention deficit hyperactivity disorder (ADHD): A recent review, *Eur. J. Biomed. Pharm. Sci.*, 2021, **8**(4), 233–240.
176. M. Pereira, X. Chen, A. Paltarzhytskaya, Y. Pacheco, N. Muller, L. Bovy, X. Lei, W. Chen, H. Ren and C. Song, Sleep neuroimaging: Review and future directions, *J. Sleep Res.*, 2025, e14462.
177. S. Tabakov and P. Bregant, Introduction to Diagnostic Radiology (X-Ray and Computed Tomography Imaging), in *Introduction to Medical Physics*, CRC Press, 2022, pp. 95–142.
178. P. J. Withers, C. Bouman, S. Carmignato, V. Cnudde, D. Grimaldi, C. K. Hagen, E. Maire, M. Manley, A. Du Plessis and S. R. Stock, X-ray computed tomography, *Nat. Rev. Methods Primers*, 2021, **1**(1), 18.
179. Y. Bouchareb, A. AlSaadi, J. Zabah, A. Jain, A. Al-Jabri, P. Phiri, J. Q. Shi, G. Delanerolle and S. R. Sirasanagandla, Technological advances in SPECT and SPECT/CT imaging, *Diagnostics*, 2024, **14**(13), 1431.
180. P. Ritt, in *Recent developments in SPECT/CT*, Seminars in nuclear medicine, Elsevier, 2022, pp. 276–285.
181. S. Katal, L. S. Eibschutz, B. Saboury, A. Gholamrezanezhad and A. Alavi, Advantages and applications of total-body PET scanning, *Diagnostics*, 2022, **12**(2), 426.
182. A. Del Casale, S. Ferracuti, A. Alcibiade, S. Simone, M. N. Modesti and M. Pompili, Neuroanatomical correlates of autism spectrum disorders: A meta-analysis of structural magnetic resonance imaging (MRI) studies, *Psychiatry Res.: Neuroimaging*, 2022, **325**, 111516.
183. J. M. Slipsager, S. L. Glimberg, L. Højgaard, R. R. Paulsen, P. Wighton, M. D. Tisdall, C. Jaimes, B. A. Gagoski, P. E. Grant and A. van Der Kouwe, Comparison of prospective and retrospective motion correction in 3D-encoded neuroanatomical MRI, *Magn. Reson. Med.*, 2022, **87**(2), 629–645.
184. A. H. Ali, High-performance liquid chromatography (HPLC): A review, *Ann. Adv. Chem.*, 2022, **6**, 010–020.
185. B. P. Guiard and G. Gotti, The high-precision liquid chromatography with electrochemical detection (HPLC-ECD) for monoamines neurotransmitters and their metabolites: a review, *Molecules*, 2024, **29**(2), 496.
186. J. Mateus, M. Sousa, J. Burrone and P. Aguiar, Beyond a transmission cable—new technologies to reveal the richness in axonal electrophysiology, *J. Neurosci.*, 2024, **44**(11), e1446232023.
187. Y. Wang, S. Liu, H. Wang, Y. Zhao and X.-D. Zhang, Neuron devices: emerging prospects in neural interfaces and recognition, *Microsystems Nanoeng.*, 2022, **8**(1), 128.
188. E. Molnár, Localization of neurotransmitter receptor and ion channel proteins in unfixed brains using *in situ* immunoblotting, in *Receptor and Ion Channel Detection in the Brain*, Springer, 2021, pp. 175–190.

189. H. Y. Leung, M. H. Y. Yeung, W. T. Leung, K. H. Wong, W. Y. Tang, W. C. S. Cho, H. T. Wong, H. F. Tsang, Y. K. E. Wong and X. M. Pei, The current and future applications of *in situ* hybridization technologies in anatomical pathology, *Expert Rev. Mol. Diagn.*, 2022, **22**(1), 5–18.
190. Z. Xie, N. Cao, M. Li, H. Wang, H. Guan, X. Cheng and C. Wang, Exploration of the potential neurotransmitter or neuromodulator-like properties of harmine: evidence from synthesis to synaptic modulation, *Front. Pharmacol.*, 2025, **16**, 1588105.
191. F. Kobeissy, M. Goli, H. Yadikar, Z. Shakkour, M. Kurup, M. A. Haidar, S. Alroumi, S. Mondello, K. K. Wang and Y. Mechref, Advances in neuroproteomics for neurotrauma: unraveling insights for personalized medicine and future prospects, *Front. Neurol.*, 2023, **14**, 1288740.
192. A. A. Alhusban, A. M. Hammad, L. F. Alzaghari, A. I. Shallan and K. Shnewer, Rapid and sensitive HPLC–MS/MS method for the quantification of dopamine, GABA, serotonin, glutamine and glutamate in rat brain regions after exposure to tobacco cigarettes, *Biomed. Chromatogr.*, 2023, **37**(1), e5513.
193. C. Biesdorf and R. E. Stratford, Integrating Imaging and Microdialysis into Systems Neuropharmacology, *Front. Clin. Drug Res.: CNS Neurol. Disord.*, 2021, **9**, 1–53.
194. B. G. Anderson, P. Popov, A. R. Cicali, A. Nwamba, C. R. Evans and R. T. Kennedy, In-Depth Chemical Analysis of the Brain Extracellular Space Using *In Vivo* Microdialysis with Liquid Chromatography-Tandem Mass Spectrometry, *Anal. Chem.*, 2024, **96**(41), 16387–16396.
195. S. Vallabhajosula, Molecular Imaging in Neurology, in *Molecular Imaging and Targeted Therapy: Radiopharmaceuticals and Clinical Applications*, Springer, 2023, pp. 375–423.
196. D. Choquet, M. Sainlos and J.-B. Sibarita, Advanced imaging and labelling methods to decipher brain cell organization and function, *Nat. Rev. Neurosci.*, 2021, **22**(4), 237–255.
197. H. Tamada, Three-dimensional ultrastructure analysis of organelles in injured motor neuron, *Anat. Sci. Int.*, 2023, **98**(3), 360–369.
198. M. Forsberg, D. Zhou, S. Jalali, G. Faravelli, H. Seth, A. Björefeldt and E. Hanse, Evaluation of mechanisms involved in regulation of intrinsic excitability by extracellular calcium in CA1 pyramidal neurons of rat, *J. Neurochem.*, 2025, **169**(1), e16209.
199. P. Babaeeghazvini, L. M. Rueda-Delgado, J. Gooijers, S. P. Swinnen and A. Daffertshofer, Brain structural and functional connectivity: A review of combined works of diffusion magnetic resonance imaging and electro-encephalography, *Front. Hum. Neurosci.*, 2021, **15**, 721206.
200. E. S. Finn, R. A. Poldrack and J. M. Shine, Functional neuroimaging as a catalyst for integrated neuroscience, *Nature*, 2023, **623**(7986), 263–273.
201. S. Li, J. Zhang, J. Li, Y. Hu, M. Zhang and H. Wang, Optogenetics and chemogenetics: key tools for modulating neural circuits in rodent models of depression, *Front. Neural Circuits*, 2025, **19**, 1516839.

202. S. K. Ghosh and J. A. Walocha, Evolution of staining methods in neuroanatomy: Impetus for emanation of neuron doctrine during the turn of 20th century, *Anat. Rec.*, 2024, **307**(10), 3398–3412.
203. P. Durairaj and Z. L. Liu, Brain Cytochrome P450: Navigating Neurological Health and Metabolic Regulation, *J. Xenobiotics*, 2025, **15**(2), 44.
204. Y. Timsit and S.-P. Grégoire, Towards the idea of molecular brains, *Int. J. Mol. Sci.*, 2021, **22**(21), 11868.
205. J. Adams, S. Nishitani, J. Mun, M. Klinger, M. Landry and N. Komatsu, *Unraveling Neurochemistry*, American Chemical Society, 2025.
206. K. F. Ostrom, J. E. LaVigne, T. F. Brust, R. Seifert, C. W. Dessauer, V. J. Watts and R. S. Ostrom, Physiological roles of mammalian transmembrane adenylyl cyclase isoforms, *Physiol. Rev.*, 2022, **102**(2), 815–857.
207. M. Asif, D. Srivastava and P. Choudhary, Tools for Improving Synaptic Plasticity from Computational Neuroscience, in *Synaptic Plasticity in Neurodegenerative Disorders*, CRC Press, 2024, pp. 191–216.
208. W. Zhang, J. A. Sweeney, J. R. Bishop, Q. Gong and S. Lui, Biological subtyping of psychiatric syndromes as a pathway for advances in drug discovery and personalized medicine, *Nat. Ment. Health*, 2023, **1**(2), 88–99.
209. P. Marzola, T. Melzer, E. Pavesi, J. Gil-Mohapel and P. S. Brocardo, Exploring the role of neuroplasticity in development, aging, and neurodegeneration, *Brain Sci.*, 2023, **13**(12), 1610.
210. J. Das, U. Bhui, S. Chowdhary, S. Sarkar, I. K. Ghoshal, S. Nayak, R. Bishayee, N. Khurana, B. Kumar and N. Sharma, Biomarkers Unveiling Neurodegeneration: Keys to Progression and Therapeutic Insights, *Indian J. Pharm. Educ. Res.*, 2025, **59**(1s), s1–s15.
211. A. D. Mohamed, Neuroethical issues in pharmacological cognitive enhancement, *Wiley Interdiscip. Rev.: Cogn. Sci.*, 2014, **5**(5), 533–549.
212. L. Barker, *Human Brain*, Springer, 2024.
213. S. Datta Burton, T. Mahfoud, C. Aicardi and N. Rose, Clinical translation of computational brain models: Understanding the salience of trust in clinician–researcher relationships, *Interdiscip. Sci. Rev.*, 2021, **46**(1–2), 138–157.

CHAPTER 2

Neurotransmitter System in the Brain

NILAY KUMAR NANDI,[a,b] PRIYANKA NAYAK,[a] POOJA YADAV,[c] VINEY LATHER*[b] AND PRABHASH NATH TRIPATHI*[c,d]

[a] KSOP, KIET Group of Institutions, Delhi-NCR, Meerut Rd, Ghaziabad, Uttar Pradesh 201206, India; [b] Amity Institute of Pharmacy, Amity University Uttar Pradesh, Sector-125, Uttar Pradesh, India; [c] Department of Pharmaceutical Technology, Meerut Institute of Engineering and Technology, Meerut, Uttar Pradesh, India; [d] Department of Pharmaceutical Sciences, College of Pharmacy, University of Arkansas for Medical Sciences, Little Rock, AR, USA
*Emails: vlather@amity.edu; tripathiprabhash@gmail.com

2.1 Introduction

The human brain is an intricate biological network where communication between billions of neurons gives rise to thoughts, emotions, behavior, and consciousness itself.[1] At the heart of this communication lies the neurotransmitter system, a finely tuned chemical signaling framework that allows neurons to exchange information across tiny junctions known as synapses.[2] Unlike electrical wiring in machines, the brain's communication is both electrochemical and highly dynamic, shaped by molecules that act within milliseconds but can also influence long-term changes in mood, memory, and cognition.[3] Neurotransmitters are the key messengers of this system.[4] They are small chemical compounds, synthesized and stored within neurons, that depart in response to electrical activity and bind to receptors on neighboring cells, thus propagating or modulating neuronal signals.[5]

Chemical Biology No. 26
Chemical Signatures of the Brain: Bridging Biochemistry, Neuroanatomy, and Brain Disorders
Edited by Sachchida Nand Rai and Abhishek Kumar Singh

Published by the Royal Society of Chemistry, www.rsc.org

The diversity of neurotransmitters reflects the complexity of brain function.[6] Traditional neurotransmitters such as glutamate and gamma-amino butyric acid (GABA) govern excitatory and inhibitory balance, ensuring that neural circuits remain both active and stable.[7] Monoamines, including dopamine, serotonin, and norepinephrine, regulate processes as diverse as reward, emotion, arousal, and decision-making.[8] Meanwhile, neuropeptides, endocannabinoids, and other modulatory transmitters provide additional layers of control, often influencing the sensitivity of neurons or coordinating activity across larger brain regions.[9] This diversity ensures that the brain does not function as a simple relay of signals but rather as a dynamic, adaptive system capable of learning, flexibility, and resilience.

The neurotransmitter system is not uniform throughout the brain; rather, it is organized into networks with specialized roles.[10] For example, dopaminergic pathways such as the mesolimbic and mesocortical tracts are central to motivation and reward processing, while the serotonergic system originating from the raphe nuclei is essential for mood regulation and circadian rhythm.[11] Dysfunctions in these systems are strongly associated with neuropsychiatric and neurodegenerative disorders. Imbalances in dopamine signaling are implicated in Parkinson's disease and schizophrenia, while alterations in serotonin transmission are linked to depression and anxiety disorders.[12] Understanding these systems at a molecular and circuit level has therefore been a cornerstone of modern neuroscience and psychopharmacology.[13]

Neurotransmission is also highly plastic, meaning that synaptic strength and neurotransmitter release can be modified by experience, environmental inputs, and even internal states such as stress or hormonal changes.[14] This plasticity underlies the brain's ability to encode memories and adapt to new challenges. Moreover, neurotransmitter systems interact extensively with one another.[15] For instance, glutamatergic and dopaminergic signaling often converge in pathways involved in learning and reward, while GABAergic inhibition helps maintain the balance necessary for precise information flow.[7] Thus, the brain's chemical language is not linear but a constantly evolving conversation among multiple systems.

Studying neurotransmitter systems is not only of academic interest but also of profound clinical relevance. Many widely used medications, from antidepressants and antipsychotics to anxiolytics and stimulants, exert their effects by modulating neurotransmitter activity.[16] Advances in imaging, molecular biology, and computational modeling have further expanded our ability to map neurotransmitter dynamics in living brains, offering new hope for precision medicine in treating mental and neurological illnesses.[17] In essence, the neurotransmitter system represents the chemical foundation of human experience. It bridges the gap between electrical impulses and complex behaviors, linking molecular events to the emergence of thoughts and feelings.[18] By unraveling the intricacies of this system, neuroscience moves closer to understanding not just how the brain works, but also what makes us uniquely human.

2.2 Basics of Neurotransmission in the Brain

Neurotransmission is the crucial biological process through which neurons communicate with one another and with other cell types to coordinate perception, thought, emotion, and behavior.[19] At its core, neurotransmission is an elegant interplay of electrical excitability and chemical signaling, a system that converts electrical impulses into molecular messages.[20] This process underpins cognition, learning, and memory, while its dysregulation is central to many neurological and psychiatric disorders.[21] To appreciate the brain's complexity, one must first understand how neurons are able to transmit signals across microscopic spaces with remarkable precision and speed.

Approximately 86 billion neurons are present in a human brain, each specialized for signal processing and transmission.[22] A typical neuron consists of dendrites that receive incoming signals, a soma or cell body that integrates electrical inputs, and an axon that propagates action potentials toward its terminal endings.[22] At the terminal, the electrical impulse is converted into a chemical signal that crosses a specialized gap known as the synapse.[2] This chemical signal, mediated by neurotransmitters, then influences the excitability of the postsynaptic neuron.[23] Thus, neurons serve as the biological substrates of communication networks that parallel and surpass the most advanced artificial systems in efficiency and adaptability (Figure 2.1).

Neurons maintain a membrane potential (resting), typically about −70 mV, generated by selective permeability of the membrane and the activity of ion pumps such as the sodium–potassium ATPase.[24] The rapid, transient reversal of this potential – an action potential – is the starting point for neurotransmission.[25] When a neuron is sufficiently depolarized, voltage-gated Na^+ channels open, allowing Na^+ to rush inward and generate the rising phase of the action potential.[26] This is followed by inactivation of sodium channels and opening of potassium channels, which drive potassium efflux and repolarize the membrane.[27] A brief hyperpolarization may occur before the potential returns to baseline.[28] Because of its all-or-nothing nature, the action potential ensures fidelity, and its rapid propagation along axons, often accelerated by myelin sheaths and saltatory conduction, guarantees efficient communication across long distances.[29]

Neurotransmitters are diverse and can be broadly classified into categories. Amino acid transmitters such as glutamate, GABA, and glycine serve as the main excitatory and inhibitory signals.[30] Monoamines like dopamine, norepinephrine, serotonin, and histamine are modulatory and crucial for mood, motivation, and arousal.[8] Acetylcholine plays an essential role in neuromuscular transmission and cortical activation.[31] Neuropeptides such as substance P and endorphins mediate more specialized modulatory functions, while unconventional messengers such as nitric oxide and endocannabinoids defy the classic vesicular release paradigm.[32] Each neurotransmitter acts through specific receptors, and the postsynaptic response depends not only on the transmitter itself but also on the receptor subtype

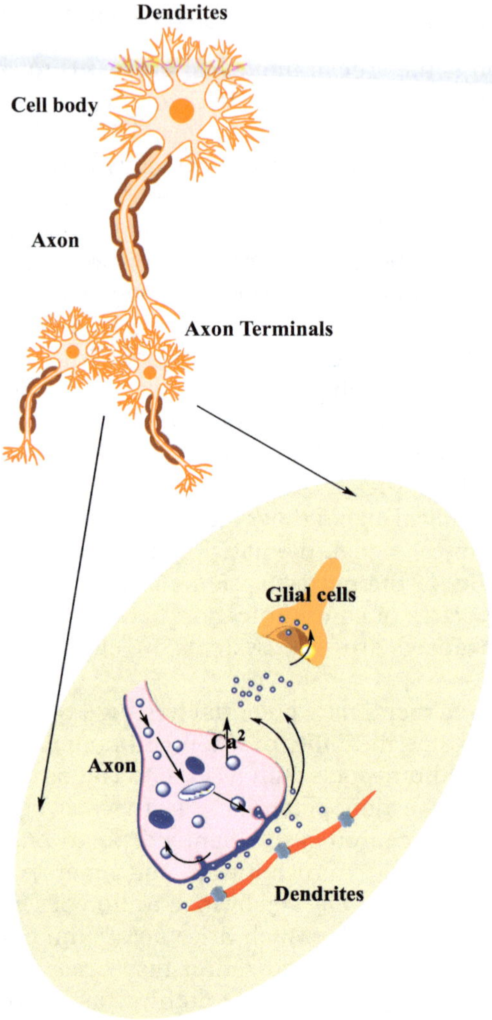

Figure 2.1 Potential site of action of neurotransmitter systems.

engaged. For example, glutamate acting on AMPA and NMDA receptors contributes to excitatory transmission and plasticity, while GABA binding to GABA_A receptors mediates rapid inhibition by allowing chloride influx.[33]

Receptors themselves are categorized into two functional types. Ionotropic receptors are ligand-gated ion channels that produce immediate changes in postsynaptic membrane potential within milliseconds.[34] Metabotropic receptors, in contrast, are G-protein-coupled receptors that activate intracellular second messenger cascades, producing slower but more versatile cellular responses.[35] The coexistence of these receptor systems allows the nervous system to combine speed with modulatory complexity, enabling both rapid information transfer and long-term regulation of neural networks.[36]

Because of synaptic plasticity, the capacity of synapses to become stronger or weaker in response to activity, neurotransmission is dynamically controlled rather than static.[37] Long-term potentiation (LTP), often dependent on NMDA receptor-mediated calcium entry, results in sustained enhancement of synaptic efficacy, whereas long-term depression (LTD) produces the opposite effect.[38] These processes are considered the cellular basis of learning and memory. Plasticity ensures that neuronal circuits can adapt to developing environments and experiences, which underlies the brain's remarkable capacity for learning and resilience.[39]

When neurotransmission is disrupted, the consequences manifest as disease. Dopamine deficits lead to the motor dysfunction of Parkinson's disease, while overactivity of dopaminergic pathways is linked to schizophrenia.[40] Imbalances in serotonin contribute to depression and anxiety disorders.[41] Dysregulation of glutamate signaling leads to excitotoxicity, a major factor in stroke and neurodegeneration, while inadequate GABAergic inhibition underlies epilepsy.[42] Pharmacological treatments, from SSRIs to benzodiazepines and antipsychotics, act by restoring balance within neurotransmitter systems, underscoring the clinical relevance of understanding these mechanisms[43] (Figure 2.2).

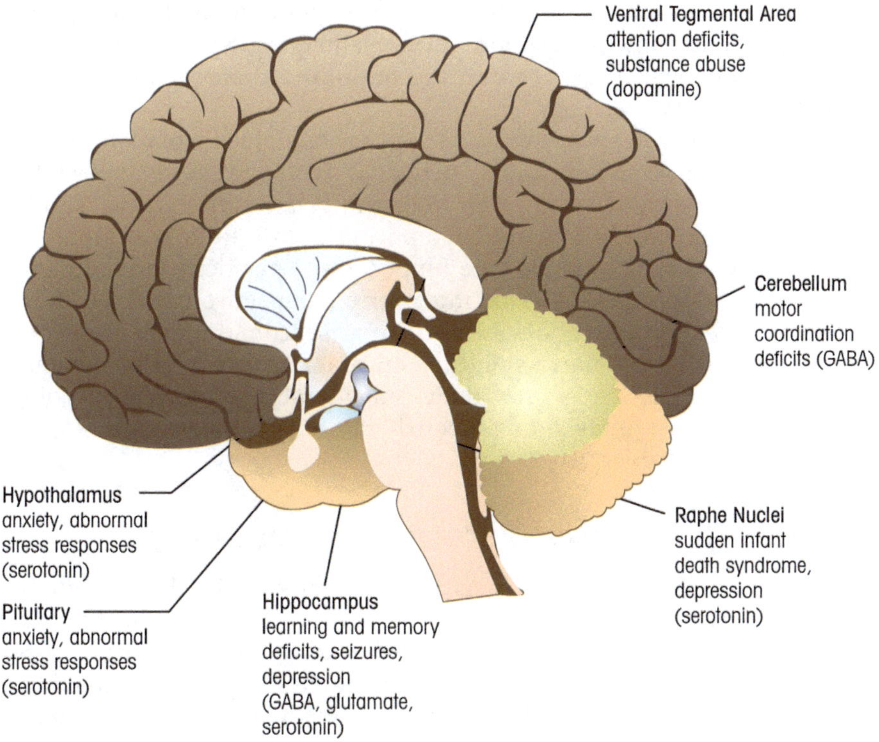

Figure 2.2 Schematic representation of the mature human brain.

Finally, neurotransmission represents one of the most intricate yet elegant processes in biology, integrating electrical excitability, chemical signaling, and synaptic adaptability. From the firing of an action potential to the binding of neurotransmitters and the modulation of synaptic strength, each step highlights the evolutionary refinement of neural systems. Just as importantly, disturbances in these processes give rise to human suffering, making their study both scientifically vital and clinically indispensable. Ultimately, neurotransmission is not only the machinery of the brain but also the foundation of thought, memory, behavior, and consciousness itself.

2.3 Major Neurotransmitter Systems in the Brain

This section provides a detailed exploration of the major neurotransmitter systems in the brain, emphasizing their neuroanatomical pathways, receptor subtypes, physiological roles, and clinical implications.

Beyond these classical systems, additional neurotransmitters and neuromodulators have gained recognition. Neuropeptides such as substance P, opioids, and orexins regulate pain, stress, and feeding. Adenosine, a purinergic neuromodulator, accumulates during wakefulness to promote sleep, while caffeine antagonizes adenosine receptors to counteract fatigue.[44,45] Endocannabinoids like anandamide and 2-arachidonoylglycerol act as retrograde messengers, binding to CB1 receptors to modulate synaptic plasticity, appetite, and nociception.[46] These emerging systems highlight the diversity of chemical messengers in the brain and their relevance to both physiology and pharmacology.

Importantly, neurotransmitter systems rarely operate in isolation. Instead, they function as interdependent networks with extensive cross-talk.[47] Dopamine and glutamate interact in basal ganglia circuits to regulate motor activity; serotonin and norepinephrine jointly influence mood and arousal; and cholinergic inputs gate hippocampal plasticity in conjunction with glutamatergic signaling. This integrative perspective underscores that psychiatric and neurological disorders often arise from dysregulation across multiple systems rather than a single chemical imbalance.[8,48] Consequently, future therapeutic strategies may increasingly embrace polypharmacology, targeting multiple systems simultaneously rather than pursuing narrow receptor selectivity (Tables 2.1–2.4).

2.4 The Glutamate Neurotransmitter System

The majority of fast excitatory synaptic transmission in the mammalian central nervous system (CNS) is attributed to glutamate, the primary excitatory neurotransmitter.[49] It is estimated that more than four-fifths of cortical synapses rely on glutamate as their signaling molecule, making it indispensable for cognition, learning, memory, and higher-order neural processing.[50] Unlike neuromodulators such as dopamine or serotonin that fine-tune neuronal activity, glutamate provides the fundamental excitatory

Table 2.1 Major classical neurotransmitters and their key features.

Neurotransmitter	Primary source in the brain	Main receptors	Principal functions	Clinical relevance
Glutamate	Pyramidal neurons (cortex, hippocampus)	NMDA, AMPA, kainate, mGluRs	Excitatory transmission, learning, memory	Stroke, epilepsy, Alzheimer's disease, schizophrenia
GABA	Interneurons (cortex, hippocampus, cerebellum)	GABAA_AA, GABAB_BB, GABAC_CC	Inhibitory control, oscillations, motor coordination	Epilepsy, anxiety, insomnia, schizophrenia
Dopamine	Substantia nigra, VTA, hypothalamus	D1–D5 (GPCRs)	Reward, motivation, motor control, cognition	Parkinson's disease, schizophrenia, addiction, ADHD
Serotonin (5-HT)	Raphe nuclei	5-HT1_11–5-HT7_77 (GPCRs, ionotropic 5-HT3_33)	Mood, sleep, appetite, pain modulation	Depression, anxiety, migraine, hallucinogen effects
Norepinephrine (NE)	Locus coeruleus	α1, α2, β1–3 (GPCRs)	Arousal, vigilance, stress response, attention	ADHD, depression, PTSD, anxiety
Acetylcholine (ACh)	Basal forebrain, brainstem nuclei	Nicotinic (ionotropic), muscarinic (M1–M5)	Memory, attention, REM sleep, motor control	Alzheimer's disease, myasthenia gravis, nicotine addiction
Histamine	Tuberomammillary nucleus (hypothalamus)	H1–H4 (GPCRs)	Wakefulness, appetite, thermoregulation	Narcolepsy, sedation (antihistamines), cognitive disorders

Table 2.2 Dopaminergic pathways and associated disorders.

Pathway	Origin → target	Primary function	Dysfunction
Nigrostriatal	Substantia nigra → striatum	Motor control	Parkinson's disease (degeneration)
Mesolimbic	VTA → nucleus accumbens, amygdala	Reward, reinforcement, motivation	Addiction, schizophrenia (positive symptoms)
Mesocortical	VTA → prefrontal cortex	Executive function, cognition	Schizophrenia (negative symptoms, cognitive deficits)
Tuberoinfundibular	Hypothalamus → pituitary	Prolactin regulation	Hyperprolactinemia with D2 antagonists

Table 2.3 Key receptor subtypes and their actions.

Neurotransmitter	Receptor subtype	Receptor type	Primary action	Clinical target
Glutamate	NMDA	Ionotropic	Ca^{2+} influx, plasticity	Alzheimer's (memantine)
GABA	GABAA_AA	Ionotropic	Fast inhibition (Cl^-)	Benzodiazepines, barbiturates
Dopamine	D2	GPCR (Gi)	Inhibition of cAMP	Antipsychotics, Parkinson's meds
Serotonin	5-HT1A_1A1A	GPCR (Gi)	Autoreceptor, anxiolytic	Buspirone
Serotonin	5-HT2A_2A2A	GPCR (Gq)	Hallucinogenic, cortical excitation	Atypical antipsychotics
Norepinephrine	α2	GPCR (Gi)	Presynaptic inhibition	Clonidine (antihypertensive, ADHD)
Acetylcholine	Nicotinic	Ionotropic	Fast excitation	Nicotine, varenicline
Histamine	H1	GPCR (Gq)	Wakefulness, alertness	Antihistamines (sedation)

Table 2.4 Emerging neurotransmitters.

Neurotransmitter class	Examples	Primary functions	Clinical/research relevance
Neuropeptides	Substance P, endorphins, orexins	Pain modulation, stress, feeding, reward	Pain therapy, obesity treatment, narcolepsy drugs
Purinergic	ATP, adenosine	Sleep–wake regulation, synaptic modulation	Caffeine (adenosine antagonist), stroke therapy
Endocannabinoids	Anandamide, 2-AG	Retrograde signaling, appetite, mood, pain	Cannabinoid-based therapies, epilepsy treatment

drive that enables circuits to function.[51] Yet, this central role comes with a paradox: the same neurotransmitter that sustains thought and memory also poses the risk of excitotoxic damage if its levels are not precisely regulated. This dual nature makes the glutamate system one of the most fascinating and critically balanced aspects of neuroscience.[52,53]

The synthesis of glutamate within the CNS is tightly coupled to cellular metabolism.[54] Although glutamate is a nonessential amino acid in the diet, neurons cannot depend on peripheral pools due to the impermeability of the blood–brain barrier.[55] Instead, neurons synthesize glutamate locally through transamination of α-keto-glutarate, an intermediate of the tricarboxylic acid (TCA) cycle.[56] A central mechanism that ensures neurotransmitter supply without excessive accumulation is the glutamate–glutamine cycle, which represents a cooperative exchange between neurons and astrocytes.[57] After glutamate is released into the synaptic cleft, astrocytes rapidly take it up through high-affinity excitatory amino-acid transporters (EAATs).[58] Glutamate is transformed into glutamine by glutamine synthetase, a reaction that effectively neutralizes its excitatory potential, inside astrocytes. Glutamine, being non-excitatory, is then transported back to neurons, where it is re-converted into glutamate by phosphate-activated glutaminase.[59] This shuttle system exemplifies the intimate metabolic partnership between neurons and glial cells and ensures both safety and efficiency of excitatory transmission.[60]

Once generated, glutamate is stored in synaptic vesicles by a specialized family of vesicular glutamate transporters (VGLUTs). Three major isoforms, VGLUT1, VGLUT2, and VGLUT3, are distributed in a region-specific manner, with VGLUT1 enriched in cortical and hippocampal neurons, VGLUT2 dominating in thalamic and subcortical regions, and VGLUT3 expressed more sparsely, often in interneurons and even in non-neuronal cells.[61] The proton electrochemical gradient generated by the vacuolar H^+-ATPase powers this vesicular loading.[62] Voltage-gated calcium channels activate when the presynaptic junction depolarizes, causing vesicles to fuse with the plasma membrane and glutamate to be quantally released into the synaptic cleft.[63] The precision of this machinery allows neurons to regulate the magnitude and timing of excitatory signals with exquisite fidelity.

The postsynaptic effects of glutamate are mediated through two broad classes of receptors: ionotropic and metabotropic.[64] Ionotropic glutamate receptors (iGluRs) are ligand-gated ion channels that generate rapid post-synaptic currents.[65] They consist of kainate, AMPA, and NMDA receptors, each of which has unique structural and functional characteristics. Most fast excitatory transmission is mediated by AMPA receptors, and the GluA2 subunit's RNA editing status has a significant impact on their ion permeability.[66] NMDA receptors, by contrast, possess the unique property of being both ligand- and voltage-gated, as their pore is blocked by Mg^{2+} ions at resting membrane potential.[67] Relief of this blockage during depolarization, coupled with glutamate binding, allows Ca^{2+} influx that initiates signaling cascades critical for synaptic plasticity.[68] This coincidence detection property makes NMDA receptors central to mechanisms such as long-term potentiation (LTP) and long-term depression (LTD), processes widely regarded as cellular correlates of learning and memory. Kainate receptors, although less prevalent, contribute to synaptic integration and presynaptic regulation of neurotransmitter release.[69]

Complementing the ionotropic receptors are metabotropic (mGluRs) glutamate-receptors, which belong to the G-protein-coupled receptor superfamily.[64] They are subdivided into three groups based on their signaling mechanisms. Group I receptors, comprising mGluR1 and mGluR5, couple to Gq proteins and activate phospholipase C, thereby generating second messengers that elevate intracellular calcium and activate protein kinase C.[70] Groups II and III, including mGluR2/3 and mGluR4/6/7/8, respectively, couple to Gi/o proteins, inhibit adenylyl cyclase, and decrease cyclic AMP production.[71] These latter groups often function as autoreceptors at presynaptic terminals, modulating the probability of glutamate release. Collectively, mGluRs act as modulators of excitatory tone and are crucial for fine-tuning synaptic responses over longer timescales than ionotropic receptors.[72]

The termination of glutamatergic signaling is achieved not through enzymatic degradation, as is the case for acetylcholine, but through highly efficient uptake mechanisms.[73] EAATs located primarily on astrocytic membranes clear glutamate from the extracellular space within milliseconds. Of the five known isoforms, EAAT1 and EAAT2 are glial dominant, while EAAT3 is primarily neuronal.[74] Their failure or dysfunction leads to persistent elevation of extracellular glutamate, which drives overactivation of NMDA receptors and initiates excitotoxic cascades.[75] This excitotoxicity is characterized by calcium overload, activation of proteases and lipases, mitochondrial dysfunction, and ultimately cell death.[76] Such mechanisms underlie the pathogenesis of several neurological conditions, including ischemic stroke, amyotrophic lateral sclerosis (ALS), Huntington's disease, and traumatic brain injury.

2.5 The GABA Neurotransmitter System

Among the many neurotransmitters that govern brain function, γ-aminobutyric acid (GABA) occupies a uniquely important role as the principal inhibitory messenger of the central nervous system.[77] While excitatory transmitters such as glutamate promote neuronal firing and drive activity through the brain's vast networks, GABA ensures balance by restraining excessive excitation and maintaining the delicate equilibrium between inhibition and excitation that is fundamental to healthy cognition, emotion, and movement.[78] Without this regulatory influence, neural activity would easily spiral into uncontrolled states of hyperexcitability, and indeed, disruption of GABAergic signaling underlies a wide range of clinical conditions, from epilepsy and anxiety to schizophrenia and neurodegenerative disorders.[43] Thus, GABA is not simply a chemical brake; it is a cornerstone of the brain's ability to function within safe and meaningful physiological boundaries.

The molecular story of GABA begins with its synthesis from glutamate, the brain's main excitatory transmitter, through the enzymatic action of glutamic acid decarboxylase (GAD). Two isoforms of this enzyme, GAD65 and GAD67, maintain distinct yet complementary roles. GAD65, concentrated in presynaptic terminals, supports activity-dependent production of GABA for rapid inhibitory signaling, while GAD67 is more evenly distributed throughout the neuronal cytoplasm, providing basal levels of the transmitter required for continuous inhibitory tone.[79,80] This process depends on pyridoxal phosphate, a derivative of vitamin B6, reminding us that even micronutrients have profound consequences for brain function. Once released, GABA does not remain indefinitely in the synaptic cleft;[81] specialized transporters GAT1 through GAT4 rapidly remove it from the extracellular space, recycling it into neurons or transporting it into astrocytes.[82] In astrocytes, GABA is metabolized by GABA transaminase into succinic semialdehyde, which then feeds into the tricarboxylic acid cycle.[83] This metabolic shunt not only prevents toxic accumulation but also links neurotransmission with cellular energy metabolism, creating a seamless loop between signaling and bioenergetics[84] (Figure 2.3).

The influence of GABA is executed through receptor systems that can be broadly divided into ionotropic and metabotropic types.[85] Ionotropic GABAA_AA receptors are ligand-gated chloride channels composed of various subunits whose arrangement dictates their pharmacological sensitivity and kinetic behavior.[86] When GABA binds, the channel opens, allowing chloride ions to flow across the membrane. Depending on the chloride gradient, this typically results in hyperpolarization, effectively dampening the excitability of the postsynaptic neuron.[87] These receptors are the targets of many clinically important drugs, including benzodiazepines, barbiturates, and anesthetics, which act as positive allosteric modulators and enhance the inhibitory effect of GABA.[88] In this way, manipulation of the GABA system has been central to the treatment of epilepsy, anxiety, and sleep disorders for decades. By contrast, GABAB_BB receptors are metabotropic, belonging to the large family of G-protein-coupled receptors.[89] These exist as

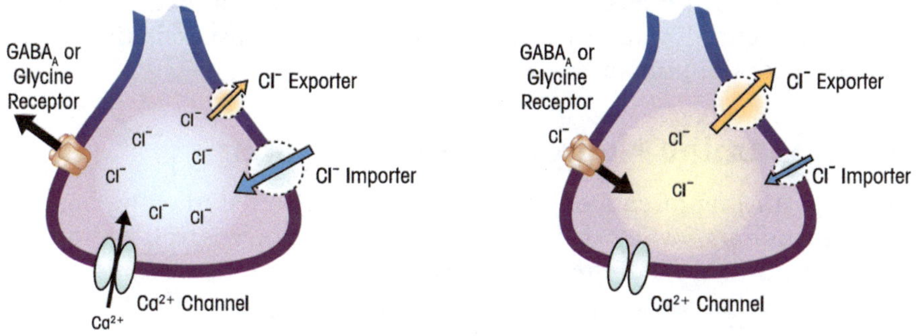

Figure 2.3 γ-Aminobutyric acid (GABA) receptors stimulate immature neurons and inhibit mature neurons.

obligatory heterodimers and, when activated, trigger intracellular cascades that inhibit adenylyl cyclase, reduce calcium entry through voltage-gated channels, and increase potassium conductance.[90] The result is a slower but longer-lasting inhibition compared with the rapid currents mediated by GABAA_AA. Presynaptic GABAB_BB receptors, in particular, provide feedback regulation of neurotransmitter release, ensuring that neural networks do not slip into runaway excitation.[91] Clinically, the drug baclofen, a GABAB_BB agonist, exemplifies the therapeutic potential of this receptor system, especially in managing muscle spasticity.[92]

GABA's role also evolves with development, adding another layer of complexity. In the immature brain, GABA is paradoxically excitatory due to the dominance of the chloride importer NKCC1, which keeps intracellular chloride concentrations high.[93,94] Under these conditions, GABA receptor activation drives chloride efflux and depolarization, promoting neuronal activity. This excitatory action is essential for early brain development, facilitating synapse formation and network wiring. As the brain matures, expression of the chloride exporter KCC2 increases, lowering intracellular chloride levels and switching GABA's action to the inhibitory mode characteristic of adulthood.[95] Disruptions in this developmental transition are implicated in neurodevelopmental disorders such as autism and epilepsy. Moreover, GABAergic synapses, like their glutamatergic counterparts, are capable of long-term plasticity, shaping learning and memory by modulating inhibitory tone in concert with excitatory plasticity.[96]

The clinical and translational implications of GABA research are profound. Drugs targeting GABAA_AA receptors have transformed the management of epilepsy and anxiety, although issues of tolerance and dependence remain challenges.[97] Meanwhile, GABAB_BB receptor modulators are being investigated for use in addiction, pain, and psychiatric disorders.[98]

By restraining excitatory drive, GABA creates the conditions under which complex thought, coordinated movement, and emotional regulation can emerge. Far from being a passive brake, it is a dynamic and adaptive system, deeply interwoven with metabolism, development, and plasticity. Its study continues to bridge molecular neuroscience with clinical medicine, underscoring the central truth that inhibition is as vital as excitation in the architecture of the human brain.

2.6 Dopamine Neurotransmitter System

Dopamine (DA) represents one of the most extensively studied monoamine neurotransmitters, owing to its central role in motor control, reward processing, cognition, and the pathophysiology of several neuropsychiatric disorders.[99] Biochemically, dopamine is synthesized from the amino acid tyrosine through a tightly regulated enzymatic cascade.[100] The first rate-limiting step involves the hydroxylation of tyrosine by tyrosine hydroxylase (TH), generating L-3,4-dihydroxyphenylalanine (L-DOPA). Subsequently, aromatic L-amino acid decarboxylase (AADC) converts L-DOPA into dopamine. This synthesis occurs predominantly in dopaminergic neurons, where dopamine is packaged into synaptic vesicles *via* the vesicular monoamine transporter 2 (VMAT2).[101,102] The exquisite regulation of these steps ensures that dopamine release is spatially restricted and temporally precise, a requirement for the fidelity of neuronal signaling.

Once released into the synaptic cleft through calcium-dependent exocytosis, dopamine exerts its effects by binding to a family of G-protein-coupled receptors (GPCRs).[103] These receptors are broadly categorized into D1-like (D1 and D5) and D2-like (D2, D3, and D4) subfamilies, distinguished by their intracellular signaling pathways.[104] D1-like receptors couple to Gs proteins, stimulating adenylyl cyclase and increasing cyclic AMP (cAMP) production, thereby enhancing neuronal excitability.[105] Conversely, D2-like receptors couple to Gi/o proteins, inhibiting adenylyl cyclase and reducing cAMP levels, leading to diminished neuronal activity.[106] This bidirectional modulation allows dopamine to fine-tune neuronal circuits, orchestrating complex functions that range from movement to motivation.

The anatomical distribution of dopaminergic neurons is relatively sparse compared to other neurotransmitter systems, yet their influence is disproportionately profound. Dopamine neurons are concentrated in midbrain nuclei, most notably the substantia nigra pars compacta (SNc) and the ventral tegmental area (VTA).[107] From these origins, distinct dopaminergic pathways project to various brain regions, forming four canonical systems: the nigrostriatal, mesolimbic, mesocortical, and tuberoinfundibular pathways.[108] The nigrostriatal pathway, projecting from the SNc to the dorsal striatum, is primarily responsible for motor control, and its degeneration underlies the classical motor symptoms of Parkinson's disease.[109] The mesolimbic pathway, originating in the VTA and projecting to the nucleus accumbens, amygdala, and hippocampus, is central to reward processing and reinforcement learning.[110] The mesocortical pathway extends from the VTA to the prefrontal cortex, governing executive functions, decision-making, and working memory.[111] Lastly, the tuberoinfundibular pathway regulates neuroendocrine functions by modulating prolactin secretion from the anterior pituitary.[112]

Functionally, dopamine is unique in its ability to bridge the domains of movement and motivation. The basal ganglia, a set of interconnected subcortical nuclei, depend critically on dopaminergic input to regulate motor output.[113] Dopamine here acts as a gatekeeper, facilitating the initiation of desired motor programs while suppressing competing actions. Beyond movement, dopamine's role in reinforcement learning has been demonstrated through the concept of reward prediction error (RPE).[114] Electrophysiological recordings from VTA neurons reveal that dopamine activity increases in response to unexpected rewards and decreases when anticipated rewards fail to occur.[115] This phasic firing encodes a teaching signal that allows organisms to adapt behavior in a dynamic environment. In the prefrontal cortex, dopamine fine-tunes cognitive control, enabling flexible decision-making, goal setting, and sustained attention.

Clinically, disturbances in the dopamine system manifest across a spectrum of neurological and psychiatric conditions. The most archetypal example is Parkinson's disease, where selective degeneration of SNc dopaminergic neurons leads to profound striatal dopamine deficiency, resulting in bradykinesia, rigidity, and tremor.[40] Conversely, hyperdopaminergic states are implicated in schizophrenia, particularly in the mesolimbic pathway, where excessive dopamine transmission is thought to contribute to positive symptoms such as hallucinations and delusions.[116] The dopamine hypothesis of schizophrenia, although refined over decades, continues to provide a framework for understanding the efficacy of antipsychotics, most of which act as D2 receptor antagonists.[117] Furthermore, dopamine dysregulation plays a central role in addiction, where drugs of abuse such as cocaine, amphetamine, and opioids hijack dopaminergic reward circuits, producing maladaptive reinforcement and compulsive drug seeking.[118]

From a therapeutic standpoint, manipulation of the dopamine system has yielded some of the most successful pharmacological interventions in medicine. L-DOPA replacement therapy, often combined with peripheral decarboxylase inhibitors, remains the gold standard for Parkinson's disease.[119] In contrast, dopamine receptor antagonists serve as the cornerstone of antipsychotic therapy, while dopamine agonists and reuptake inhibitors are employed in disorders ranging from depression to attention-deficit hyperactivity disorder (ADHD).[120] However, the clinical use of dopaminergic drugs is frequently limited by side effects, reflecting the intricate balance of dopamine's actions across different circuits. For example, while enhancing dopamine transmission may alleviate motor symptoms in Parkinson's disease, it may also induce dyskinesias or impulse-control disorders.[121]

2.7 The Serotonin Neurotransmitter System

Serotonin, or 5-hydroxytryptamine (5-HT), represents one of the most evolutionarily conserved and biologically influential neurotransmitter systems in the brain.[122] First isolated in the mid-20th century from blood serum, serotonin has since been recognized as a central regulator of mood, cognition, appetite, circadian rhythm, and numerous homeostatic processes.[123] Unlike dopamine or norepinephrine, which are more tightly associated with reward and arousal, serotonin is best conceptualized as a neuromodulator that broadly tunes brain circuits toward adaptive behavioral responses.[123] Its complexity is reflected not only in the chemical's widespread distribution but also in the remarkable diversity of receptor subtypes and downstream signaling cascades through which it exerts its influence.

At the biochemical level, serotonin is synthesized from the essential amino acid tryptophan through a tightly regulated enzymatic pathway.[8] The first and rate-limiting step involves hydroxylation of tryptophan by the enzyme tryptophan hydroxylase (TPH), which exists in two isoforms: TPH1, largely peripheral, and TPH2, expressed predominantly in neurons of the central nervous system.[124] This yields 5-hydroxytryptophan (5-HTP), which is subsequently decarboxylated by aromatic L-amino acid decarboxylase to form serotonin. The availability of tryptophan across the blood–brain barrier, influenced by diet and competition with other large neutral amino acids, critically shapes central serotonin synthesis.[125] Once produced, serotonin is stored in synaptic vesicles by the vesicular monoamine transporter (VMAT2), poised for activity-dependent release.[126]

The anatomical foundation of the serotonergic system lies in the brainstem, particularly within the raphe nuclei.[127] These nuclei, distributed along the midline of the medulla, pons, and midbrain, house the vast majority of serotonin-producing neurons. Despite their relatively modest numbers, estimated to be around 300 000 in the human brain, these neurons project extensively, forming a dense and diffuse network that innervates nearly every region of the central nervous system, from the spinal cord to the prefrontal cortex. This architecture highlights serotonin's function as a modulatory system, capable of orchestrating widespread effects rather than localized, discrete synaptic transmission alone.[128]

Serotonin signaling is mediated through a large family of receptors, grouped into seven major classes (5-HT1 through 5-HT7) with multiple subtypes within each.[129] These receptors encompass a broad spectrum of mechanisms, including G-protein-coupled receptors (GPCRs) and ligand-gated ion channels.[130] For example, the 5-HT1A receptor acts predominantly through inhibitory Gi/o proteins, reducing cyclic AMP levels and hyperpolarizing neurons *via* potassium channel activation. Conversely, 5-HT2 receptors couple to Gq proteins, activating phospholipase C and mobilizing intracellular calcium, thereby enhancing excitability.[131] The 5-HT3 receptor is unique as a ligand-gated cation channel, directly mediating rapid depolarization.[132] This receptor diversity explains the multifaceted roles of serotonin across physiological and psychological domains, from regulating anxiety and sleep to shaping sensory perception and gastrointestinal motility.

Functionally, serotonin has been implicated in an astonishing array of behavioral and physiological processes. In mood regulation, serotonin modulates limbic-prefrontal circuits, promoting emotional resilience and adaptive stress responses. Dysregulation of serotonergic signaling is strongly associated with major depressive disorder, anxiety, and obsessive compulsive disorder, though the relationship is far more complex than the outdated "serotonin deficiency hypothesis."[133] In cognition, serotonin influences learning, memory, and decision-making through its modulation of hippocampal plasticity and cortical information processing. Furthermore, serotonin plays a central role in regulating sleep–wake cycles by interacting with melatonin synthesis pathways in the pineal gland.[134]

Beyond its neural actions, serotonin exerts systemic effects that underscore its role as a true integrative signal. Approximately 90–95% of the body's serotonin is synthesized in the enterochromaffin cells of the gastrointestinal tract, where it modulates motility, secretion, and gut–brain communication.[135] Peripheral serotonin also participates in platelet aggregation, vascular tone, and metabolic regulation, linking it to cardiovascular and metabolic disorders.[136] This dual central and peripheral presence highlights serotonin as both a neurotransmitter and a hormone-like signaling molecule with far-reaching implications.

Recent scientific advances have further emphasized the importance of serotonergic system diversity and plasticity. Genetic polymorphisms in SERT (such as 5-HTTLPR), receptor variants, and enzymes involved in serotonin metabolism contribute to individual differences in mood vulnerability, stress reactivity, and antidepressant responsiveness.[137] Cutting-edge imaging studies have revealed dynamic changes in serotonin release during affective and cognitive tasks, while optogenetic experiments in animal models have demonstrated that distinct subsets of serotonergic neurons mediate opposing behavioral outcomes, from promoting patience to driving impulsivity.[138] These findings underscore the system's remarkable heterogeneity and challenge simplistic models of its function.

2.8 The Norepinephrine Neurotransmitter System

The norepinephrine (NE) neurotransmitter system represents one of the most influential neuromodulatory networks in the mammalian brain.[139] Derived from the catecholamine family, norepinephrine is synthesized through a well-defined biosynthetic pathway beginning with the amino acid tyrosine.[140] Tyrosine is hydroxylated by tyrosine hydroxylase to form L-DOPA, which is subsequently decarboxylated to dopamine. Within noradrenergic neurons, dopamine β-hydroxylase catalyzes the conversion of dopamine into norepinephrine, a step localized primarily to synaptic vesicles.[100] This compartmentalized synthesis ensures that norepinephrine is both readily available for release and tightly regulated to prevent aberrant signaling. Unlike classical excitatory or inhibitory transmitters such as glutamate or γ-aminobutyric acid (GABA), norepinephrine functions primarily as a neuromodulator, exerting widespread influence on neuronal circuits through metabotropic receptor signaling.[141]

The anatomical substrate of the norepinephrine system is remarkably compact yet widely projecting. The locus coeruleus (LC), a small nucleus located bilaterally in the dorsal pons, serves as the principal source of noradrenergic innervation to nearly the entire neuraxis.[142] From the LC, highly branched axons ascend to the cerebral cortex, hippocampus, amygdala, thalamus, and cerebellum, while descending projections target the spinal cord and brainstem autonomic nuclei. This architecture allows the NE system to broadcast signals across large cortical territories, synchronizing activity in distant brain regions. Electrophysiological recordings demonstrate that LC neurons fire in tonic and phasic modes, corresponding respectively to baseline arousal regulation and transient responses to salient stimuli.[143] Thus, norepinephrine integrates environmental demands with internal physiological states to modulate adaptive behavior.

At the receptor level, norepinephrine acts through a family of G-protein-coupled receptors (GPCRs) broadly divided into α- and β-adrenergic receptors, each with multiple subtypes (α1A, α1B, α1D; α2A, α2B, α2C; β1, β2, β3).[144] These receptors differ in their distribution, intracellular signaling cascades, and functional outcomes. α1-Adrenergic receptors, generally coupled to the Gq/11 family of G proteins, activate phospholipase C pathways, leading to inositol trisphosphate (IP3) production, calcium mobilization, and increased neuronal excitability.[145] In contrast, α2 receptors are typically inhibitory, coupling to Gi/o proteins to reduce cyclic adenosine monophosphate (cAMP) levels and inhibit presynaptic neurotransmitter release.[146] This autoreceptor function of α2A-adrenergic receptors is critical for limiting norepinephrine output and maintaining homeostasis.[146] β-Adrenergic receptors, linked to Gs proteins, stimulate adenylyl cyclase activity, elevate cAMP, and activate protein kinase A, thereby facilitating synaptic plasticity and long-term potentiation in cortical and hippocampal circuits.[147] The diversity of receptor subtypes underlies the paradoxical ability of norepinephrine to simultaneously enhance attention and suppress noise, sharpen memory encoding while reducing irrelevant processing, and promote vigilance without provoking hyperexcitability.[148]

The functional implications of norepinephrine signaling are vast. In the domain of arousal and vigilance, norepinephrine is indispensable. Experimental lesion studies in animals reveal that loss of LC neurons results in profound deficits in wakefulness and responsiveness to environmental stimuli.[149] Human neuroimaging has demonstrated increased LC activity during tasks requiring sustained attention, while pharmacological blockade of adrenergic receptors impairs performance on such tasks.[150] Importantly, norepinephrine also shapes emotional regulation, particularly through dense projections to the amygdala and prefrontal cortex. Activation of β-adrenergic receptors within the basolateral amygdala enhances memory consolidation for emotionally salient events, a phenomenon supported by both rodent models and human psychophysiological studies.[151] Conversely, dysregulated norepinephrine signaling is implicated in pathological anxiety, post-traumatic stress disorder, and panic disorders, where hyperactivation of the system biases perception toward threats.[152]

Beyond cognition and emotion, norepinephrine plays a pivotal role in stress physiology. The LC-NE system acts in concert with the hypothalamic–pituitary–adrenal (HPA) axis to coordinate adaptive responses to stressors.[149] Acute stress triggers heightened LC firing, elevating norepinephrine release throughout the cortex, thereby promoting vigilance and mobilizing metabolic resources.[149] However, chronic overactivation of this system contributes to maladaptive outcomes, including impaired prefrontal regulation, hippocampal atrophy, and vulnerability to mood disorders.[153] Thus, the balance of norepinephrine signaling is essential: sufficient activation supports resilience, while dysregulation fosters psychopathology.[154]

On a cellular and molecular scale, norepinephrine exerts profound influence on synaptic plasticity and network dynamics.[155] β-Adrenergic receptor activation enhances long-term potentiation (LTP) by facilitating NMDA receptor phosphorylation and increasing AMPA receptor trafficking, processes essential for memory formation.[147] Norepinephrine also modulates astrocytic calcium signaling, influencing glutamate uptake and metabolic support to neurons.[156] Furthermore, noradrenergic modulation of cortical interneurons regulates oscillatory rhythms in the gamma and theta range, thereby coordinating information flow between brain regions.[157] This multilevel regulation from molecular signaling to systems-level network dynamics illustrates why the norepinephrine system is considered a master regulator of cognitive flexibility and adaptive behavior.[48]

2.9 The Acetylcholine Neurotransmitter System

Acetylcholine (ACh) is historically the first identified neurotransmitter and continues to occupy a central place in neuroscience because of its ubiquitous presence and functional versatility.[158] Synthesized within presynaptic terminals from choline and acetyl coenzyme A by the enzyme choline acetyltransferase (ChAT), acetylcholine functions as a chemical messenger that bridges electrical activity in neurons with physiological responses in both the central and peripheral nervous systems.[159] The storage of ACh within synaptic vesicles, followed by its calcium-dependent exocytotic release, exemplifies the fundamental principles of chemical neurotransmission.[160] Once released into the synaptic cleft, ACh interacts with specific cholinergic receptors to initiate diverse effects, ranging from muscle contraction to modulation of higher-order cognitive processes.[161] Termination of signaling is achieved rapidly through enzymatic hydrolysis by acetylcholinesterase (AChE), underscoring the necessity of precise temporal regulation in cholinergic communication.[162]

The cholinergic system encompasses two broad classes of receptors, nicotinic and muscarinic, each with distinct structural and functional attributes. Nicotinic acetylcholine receptors (nAChRs) are pentameric ligand-gated ion channels that mediate fast synaptic transmission.[163] Upon binding of acetylcholine, conformational changes in nAChRs open a central pore, allowing influx of cations such as sodium and calcium, leading to rapid depolarization. These receptors are prominently expressed at neuromuscular junctions, autonomic ganglia, and selected regions of the brain.[164] Muscarinic acetylcholine receptors (mAChRs), in contrast, belong to the G-protein-coupled receptor (GPCR) superfamily and produce slower but more prolonged modulatory responses.[165] The five known subtypes (M1–M5) display diverse coupling preferences to G proteins, which in turn regulate ion channels, second messengers, and intracellular signaling cascades.[166] This dual-receptor architecture allows acetylcholine to exert both rapid and modulatory influences depending on the neural context.

Within the central nervous system (CNS), cholinergic neurons are relatively sparse compared to glutamatergic or GABAergic populations, yet their projections are extensive and functionally indispensable.[167] The basal forebrain cholinergic system, originating primarily from the nucleus basalis of Meynert and medial septal nuclei, innervates wide cortical and hippocampal territories.[168] Experimental evidence from lesion studies, pharmacological manipulations, and *in vivo* imaging converges on the conclusion that cortical cholinergic tone enhances signal-to-noise ratios, thus optimizing sensory processing and cognitive performance. The brainstem cholinergic nuclei, including the pedunculopontine and laterodorsal tegmental nuclei, project to thalamic and midbrain structures, where they regulate arousal states, REM sleep, and locomotor activity.[169]

The acetylcholine system is tightly linked to neuropathological conditions. In Alzheimer's disease, degeneration of basal forebrain cholinergic neurons is a hallmark finding, correlating strongly with the severity of cognitive decline.[170] Pharmacological strategies that inhibit acetylcholinesterase, thereby prolonging ACh action at synapses, form the mainstay of symptomatic treatment, though they do not halt disease progression.[171]

Modern research continues to unravel the subtleties of cholinergic signaling. Modern optogenetic and chemogenetic methods enable precise control of cholinergic neurons in animals, allowing researchers to directly assess their causal involvement in cognition and behavior. Structural studies of nAChRs and mAChRs at atomic resolution have revealed conformational states that underlie ligand specificity and receptor desensitization, opening avenues for rational drug design.[172] Moreover, the recognition that acetylcholine transmission is not restricted to synaptic contacts but also operates through volume transmission where the transmitter diffuses to extrasynaptic receptors has broadened our understanding of how cholinergic tone shapes network dynamics.[173]

2.10 Other Neurotransmitters and Neuropeptides

While much of the neuroscientific discourse has historically centered on the so-called "classical" neurotransmitters such as glutamate, γ-aminobutyric acid (GABA), acetylcholine, dopamine, norepinephrine, and serotonin, the nervous system in reality relies on a far broader repertoire of signaling molecules.[174] Among these are biogenic amines, peptides, and even lipid mediators that exert profound effects on physiology and behavior.[175] Four systems, histamine, endorphins, endocannabinoids, and glycine, illustrate the remarkable diversity of chemical communication in the brain and spinal cord.[141]

Histamine is perhaps best known in the context of allergy and immune responses, but within the brain it functions as a key neuromodulator.[176] Histaminergic neurons are clustered primarily in the tuberomammillary nucleus of the posterior hypothalamus, from which they send widespread projections across the cortex, hippocampus, basal forebrain, and brainstem.[177] This architecture allows histamine to exert global influences on arousal, attention, and circadian rhythms.[178] The release of histamine during wakefulness promotes cortical activation, whereas diminished histaminergic activity contributes to the initiation of sleep.[179] Clinically, this explains why first-generation antihistamines that readily cross the blood–brain barrier often induce drowsiness: they block H1 histamine receptors involved in wake maintenance.[180] Beyond arousal, histamine modulates learning, appetite regulation, vestibular function, and neuroendocrine secretion, making it a multifaceted neuromodulator rather than a peripheral immune signal alone.[176]

Endorphins represent the prototypical neuropeptides with opioid-like activity.[181] Derived from larger precursor proteins such as proopiomelanocortin, endorphins bind to μ-, δ-, and κ-opioid receptors distributed throughout the central and peripheral nervous systems.[182] Their most recognized function is analgesia: by dampening nociceptive signaling in the spinal cord and periaqueductal gray, endorphins serve as the brain's intrinsic pain-relief system.[183] Yet their influence extends far beyond simple analgesia. Endorphins are released in response to stress, exercise, and emotional stimuli, often contributing to states of euphoria or "runner's high." Evolutionarily, this system likely provided a survival advantage by enabling organisms to endure pain or stress in life-threatening circumstances. At the cellular level, endorphin receptor activation leads to G-protein-coupled inhibition of adenylyl cyclase, reduced cAMP production, and opening of potassium channels mechanisms that hyperpolarize neurons and suppress excitability.[184] Endorphins also engage in reciprocal regulation with dopamine in mesolimbic circuits, thereby influencing reward, addiction, and mood.[185]

Another strikingly different category of neuromodulators is the endocannabinoids, lipid-derived messengers synthesized "on demand" rather than stored in vesicles.[186] The two best characterized are anandamide (N-arachidonoylethanolamide) and 2-arachidonoylglycerol (2-AG).[187] Unlike conventional neurotransmitters that act in a forward direction, endocannabinoids often function as retrograde messengers: they are released from postsynaptic neurons and travel backward across the synapse to bind presynaptic cannabinoid CB1 receptors.[188] This retrograde signaling decreases the probability of neurotransmitter release, effectively tuning synaptic strength. Physiologically, the endocannabinoid system regulates appetite, stress reactivity, emotional processing, and memory consolidation.[189] For example, CB1 receptor activation in the hypothalamus enhances feeding behavior, while signaling in the amygdala and prefrontal cortex influences fear extinction and resilience to stress.[190] Importantly, endocannabinoids are deeply implicated in synaptic plasticity phenomena such as depolarization-induced suppression of inhibition (DSI) and long-term depression (LTD).[191] Dysregulation of this system has been linked to psychiatric disorders, chronic pain syndromes, and metabolic dysfunctions.[192] Unlike plant-derived cannabinoids such as Δ9-tetrahydrocannabinol (THC), endogenous cannabinoids act with exquisite temporal precision, revealing the nervous system's reliance on lipid chemistry for fine-tuned regulation.[193]

Glycine illustrates how small amino acids can exert disproportionately large effects on neuronal excitability.[194] Concentrated primarily in the spinal cord, brainstem, and retina, glycinergic neurons provide inhibitory control essential for motor coordination, reflexes, and sensory processing.[195] Glycine receptors are ligand-gated chloride channels; upon activation, they hyperpolarize postsynaptic neurons and suppress action potential firing; this inhibitory balance prevents excessive excitation, thereby ensuring smooth motor output and controlled reflex arcs.[196] Glycine also functions as a co-agonist at the *N*-methyl-D-aspartate (NMDA) subtype of glutamate receptors, where its binding is required for proper receptor activation and excitatory transmission.[197] Thus, glycine sits at the intersection of inhibition and excitation, contributing both to spinal inhibitory tone and cortical synaptic plasticity.[198]

Together, these systems remind us that brain communication extends far beyond a handful of canonical neurotransmitters. Histamine modulates arousal, endorphins orchestrate endogenous pain relief and reward, endocannabinoids provide lipid-based synaptic regulation, and glycine ensures inhibitory balance within motor and sensory circuits.[199,200] Their diversity underscores the evolutionary creativity of the nervous system: proteins, peptides, and lipids all serve as signals, each adapted to specific physiological demands. A deeper understanding of these molecules not only enriches our view of brain function but also provides therapeutic entry points for disorders of sleep, pain, mood, appetite, and motor control.

2.11 Future Directions in Neurotransmitter System Research

The study of neurotransmitter systems is rapidly entering a new era shaped by technological convergence, integrative approaches, and translational imperatives. While classical neurotransmitters such as glutamate, γ-aminobutyric acid (GABA), dopamine, serotonin, and acetylcholine have been extensively characterized in terms of synaptic transmission, receptor pharmacology, and second messenger coupling, the next frontier lies in unraveling their dynamic interplay across scales of organization. Modern neuroscience increasingly views neurotransmission not as a set of isolated pathways, but as a polyphonic network where chemical messengers, neuromodulators, and co-transmitters interact to sculpt brain function. This necessitates the integration of systems biology, high-resolution connectomics, and multi-omics profiling to decipher the spatiotemporal logic of neurotransmitter release, receptor cross-talk, and downstream signaling cascades.

A major future direction involves the development of ultra-sensitive molecular imaging platforms capable of tracking neurotransmitter flux in real time within living brains. Advances in genetically encoded biosensors, combined with multiphoton microscopy and optogenetics, now permit visualization of transmitter release with unprecedented precision. Extending these tools to deep-brain structures and across longer temporal windows will be crucial for mapping neurotransmitter signatures of cognition, emotion, and disease. Equally important is the need to refine single-cell transcriptomics and spatial proteomics, which can reveal how receptor subtypes and transporter distributions vary across neuronal subpopulations and evolve with development, aging, or pathological insult.

On the translational horizon, therapeutic strategies are moving beyond "one-transmitter, one-target" models toward polypharmacology and circuit-level modulation. For instance, emerging antidepressants exploit simultaneous actions on monoamine transporters, glutamatergic plasticity, and neurotrophic signaling, reflecting a shift toward network stabilization rather than receptor selectivity. Similarly, precision neuromodulation approaches, such as chemogenetics and closed-loop deep brain stimulation, aim to correct dysfunctional neurotransmitter dynamics with cell-type and circuit specificity. Another compelling direction is the integration of artificial intelligence and computational modeling, which can simulate neurotransmitter system perturbations and predict emergent network behaviors under pharmacological or genetic manipulations.

Finally, the future of neurotransmitter system research is inseparable from human-centric approaches. Large-scale neuroimaging consortia, longitudinal cohort studies, and brain–gut axis research are highlighting how neurotransmitters link molecular biology to behavior, environment, and systemic physiology. Understanding these connections will be vital for developing next-generation therapeutics for psychiatric, neurodegenerative, and metabolic disorders. In sum, the future trajectory of neurotransmitter system science rests on converging molecular precision with holistic network-level insights, ensuring that discoveries not only expand our mechanistic understanding but also translate into tangible clinical benefit.

2.12 Conclusion

The neurotransmitter system represents the cornerstone of neural communication, orchestrating an intricate balance between excitation and inhibition that underpins all aspects of brain function. Through the precise release, receptor binding, and reuptake of chemical messengers such as glutamate, GABA, dopamine, serotonin, and acetylcholine, the nervous system maintains homeostasis while enabling complex behaviors, cognition, and emotional regulation. Advances in molecular neuroscience have elucidated that neurotransmitters do not act in isolation; rather, they engage in dynamic cross-talk, modulating synaptic plasticity and network oscillations critical for learning and memory. Dysregulation of these systems has been directly linked to a spectrum of neuropsychiatric and neurodegenerative disorders, highlighting the translational importance of understanding neurotransmitter pathways for therapeutic development. Cutting-edge imaging and electrophysiological techniques have further revealed the spatiotemporal precision of neurotransmitter action, demonstrating that both regional specificity and receptor subtype diversity are vital for functional outcomes. Collectively, the study of neurotransmitter systems underscores the brain's remarkable complexity and adaptability, revealing a delicate chemical symphony where even subtle perturbations can manifest as profound behavioral or cognitive deficits. Continued exploration of these pathways promises not only deeper mechanistic insights but also the potential for highly targeted interventions in brain disorders.

References

1. M. Saleem and M. Hamid, Consciousness, cognition and neural networks in the brain: Advances and perspectives in neuroscience, *Int. J. Comput. Sci. Netw. Secur.*, 2023, **23**(2), 47–54.
2. A. E. Pereda, Electrical synapses and their functional interactions with chemical synapses, *Nat. Rev. Neurosci.*, 2014, **15**(4), 250–263.
3. M. Veletić and I. Balasingham, Synaptic communication engineering for future cognitive brain–machine interfaces, *Proc. IEEE*, 2019, **107**(7), 1425–1441.
4. R. I. Teleanu, A.-G. Niculescu, E. Roza, O. Vladâcenco, A. M. Grumezescu and D. M. Teleanu, Neurotransmitters—key factors in neurological and neurodegenerative disorders of the central nervous system, *Int. J. Mol. Sci.*, 2022, **23**(11), 5954.
5. D. M. Lovinger, Communication networks in the brain: neurons, receptors, neurotransmitters, and alcohol, *Alcohol Res. Health*, 2008, **31**(3), 196.
6. F. Bloom, The functional significance of neurotransmitter diversity, *Am. J. Physiol.: Cell Physiol.*, 1984, **246**(3), C184–C194.
7. S. M. Sears and S. J. Hewett, Influence of glutamate and GABA transport on brain excitatory/inhibitory balance, *Exp. Biol. Med.*, 2021, **246**(9), 1069–1083.
8. S. A. Azizi, Monoamines: dopamine, norepinephrine, and serotonin, beyond modulation,“switches” that alter the state of target networks, *Neuroscientist*, 2022, **28**(2), 121–143.
9. A. N. Van Den Pol, Neuropeptide transmission in brain circuits, *Neuron*, 2012, **76**(1), 98–115.
10. O. Sporns, Structure and function of complex brain networks, *Dialogues Clin. Neurosci.*, 2013, **15**(3), 247–262.
11. B. Radwan, H. Liu and D. Chaudhury, The role of dopamine in mood disorders and the associated changes in circadian rhythms and sleep-wake cycle, *Brain Res.*, 2019, **1713**, 42–51.
12. A. S. Correia, A. Cardoso and N. Vale, Highlighting immune system and stress in major depressive disorder, Parkinson's, and Alzheimer's diseases, with a connection with serotonin, *Int. J. Mol. Sci.*, 2021, **22**(16), 8525.
13. R. Vidyalakshmi, K. Rajaganapathy and S. Punithavalli, Psychopharmacology: Investigating the Link Between Mind and Body, *J. Pharma Insights Res.*, 2024, **2**(2), 021–025.

14. R. Carr, Neurotransmitters, neuromodulators and hormones: Putting it all together, *Art therapy and clinical neuroscience*, 2008, pp. 76–91.
15. J. D. Sweatt, Neural plasticity and behavior–sixty years of conceptual advances, *J. Neurochem.*, 2016, **139**, 179–199.
16. O. J. Fasipe, Neuropharmacological classification of antidepressant agents based on their mechanisms of action, *Arch. Med. Health Sci.*, 2018, **6**(1), 81–94.
17. A. T. Mohammadi, A. F. Marvast, Y. Pishkari, F. Aghaei, A. Janbozorgi, A. J. Bozorgi, *et al.*, *Neuroscience in the 21st Century: New Tools and Techniques Driving Exciting Discoveries*, Nobel Sciences, 2024.
18. H. Samaripour, Neurotransmitters and mental health: understanding the chemical underpinnings of psychological disorders, *Int. J. New Chem.*, 2025, **12**(4), 825–841.
19. D. R. Patel and C. Feucht, Basic concepts of neurotransmission, *Pediatr. Clin.*, 2011, **58**(1), 21–31.
20. G. M. Ashraf, *Neurochemical Systems and Signaling: From Molecules to Networks*, CRC Press, 2023.
21. M. J. Millan, Y. Agid, M. Brüne, E. T. Bullmore, C. S. Carter and N. S. Clayton, *et al.*, Cognitive dysfunction in psychiatric disorders: characteristics, causes and the quest for improved therapy, *Nat. Rev. Drug Discovery*, 2012, **11**(2), 141–168.
22. J. Zhang, Basic neural units of the brain: neurons, synapses and action potential, *arXiv*, 2019, preprint, arXiv:190601703.
23. R. W. Holz and S. K. Fisher, Synaptic transmission and cellular signaling: an overview, *Basic Neurochemistry*, 2012, 235–257.
24. V. Egger and D. Feldmeyer, Electrical activity in neurons, *Neurosciences-From Molecule to Behavior: a university textbook*, Springer, 2013, pp. 113–143.
25. S. Jana, Action potential, *Textbook of Veterinary Physiology*, Springer, 2023, pp. 37–44.
26. C. Hammond, The voltage-gated channels of Na+ action potentials, *Cellular and molecular neurophysiology*, 2015, 74–77.
27. C. M. Armstrong and S. Hollingworth, A perspective on Na and K channel inactivation, *J. Gen. Physiol.*, 2018, **150**(1), 7–18.
28. L. A. Tremere, R. Pinaud, R. P. Irwin and C. N. Allen, Postinhibitory rebound spikes are modulated by the history of membrane hyperpolarization in the SCN, *Eur. J. Neurosci.*, 2008, **28**(6), 1127–1135.
29. J. Jacak and W. Jacak, New wave-type mechanism of saltatory conduction in myelinated axons and micro-saltatory conduction in C fibres, *Eur. Biophys. J.*, 2020, **49**(5), 343–360.
30. M. Patri, Synaptic transmission and amino acid neurotransmitters, *Neurochemical basis of brain function and dysfunction*, IntechOpen, 2019.
31. M. Fagerlund and L. Eriksson, Current concepts in neuromuscular transmission, *Br. J. Anaesth.*, 2009, **103**(1), 108–114.
32. M. Gabrielli, N. Battista, L. Riganti, I. Prada, F. Antonucci and L. Cantone, *et al.*, Active endocannabinoids are secreted on the surface of microglial microvesicles, *SpringerPlus*, 2015, **4**(Suppl 1), L29.

33. F. Knoflach, M.-C. Hernandez and D. Bertrand, GABAA receptor-mediated neurotransmission: Not so simple after all, *Biochem. Pharmacol.*, 2016, **115**, 10–17.
34. C. J. Weir, Ion channels, receptors, agonists and antagonists, *Anaesth. Intensive Care Med.*, 2020, **21**(1), 62–68.
35. K. L. Pierce and R. J. Lefkowitz, Classical and new roles of β-arrestins in the regulation of G-protein-coupled receptors, *Nat. Rev. Neurosci.*, 2001, **2**(10), 727–733.
36. A. Roy, S. Kumari, A. K. Singh, R. K. Gupta and A. Kumar, The Nervous System Decoded: Structural Dynamics, Functional Integration, and Emerging Neuroscientific Frontiers, *Int. J. Sci. Res. Technol.*, 2025, **2**(12), 28–52.
37. R. C. Malenka and S. A. Siegelbaum, Synaptic plasticity, *Synapse*, 2001, 393–453.
38. C. Lüscher and R. C. Malenka, NMDA receptor-dependent long-term potentiation and long-term depression (LTP/LTD), *Cold Spring Harbor Perspect. Biol.*, 2012, **4**(6), a005710.
39. E. J. Hermans, T. Hendler and R. Kalisch, Building resilience: The stress response as a driving force for neuroplasticity and adaptation, *Biol. Psychiatry*, 2025, **97**(4), 330–338.
40. C. Mehler-Wex, P. Riederer and M. Gerlach, Dopaminergic dysbalance in distinct basal ganglia neurocircuits: implications for the pathophysiology of Parkinson's disease, schizophrenia and attention deficit hyperactivity disorder, *Neurotoxic. Res.*, 2006, **10**(3), 167–179.
41. K. J. Ressler and C. B. Nemeroff, Role of serotonergic and noradrenergic systems in the pathophysiology of depression and anxiety disorders, *Depress. Anxiety*, 2000, **12**(S1), 2–19.
42. A. Sood, K. Preeti, V. Fernandes, D. K. Khatri and S. B. Singh, Glia: a major player in glutamate–GABA dysregulation-mediated neurodegeneration, *J. Neurosci. Res.*, 2021, **99**(12), 3148–3189.
43. Y. K. Ghallab and O. S. Elassal, Biochemical and neuropharmacology of psychiatric disorders, *Nutrition and Psychiatric Disorders: An Evidence-Based Approach to Understanding the Diet-Brain Connection*, Springer, 2024, pp. 25–47.
44. V. Casadó Anguera, *Allosteric interactions between catecholamine receptors and other G protein-coupled receptors: Pharmacological and functional characterization*, 2018.
45. Y. Mrad, *The role of tryptophan dysregulation in a mouse model of migraine*, Université Clermont Auvergne, 2023.
46. D. D. Pete and S. N. Narouze, Endocannabinoids: Anandamide and 2-arachidonoylglycerol (2-AG), *Cannabinoids and Pain*, Springer, 2021, pp. 63–69.
47. S. Rani, S. Yadav and S. M. Ghorai, Cross-talk between Peptide Neurotransmitters and their Role in Homeostasis of Brain, Behavior, and Immunity, *Curr. Psychiatry Res. Rev.*, 2025, **21**(2), 123–144.
48. X. Qi, Neuropharmacology of Neurotransmitter Systems: Current Drugs and Their Effects on Neural and Neuroendocrine Pathways, *Asia Pac. J. Clin. Med. Res.*, 2025, **1**(1), 336.

49. M. J. Niciu, B. Kelmendi and G. Sanacora, Overview of glutamatergic neurotransmission in the nervous system, *Pharmacol., Biochem. Behav.*, 2012, **100**(4), 656–664.
50. J. W. Crabtree, Functional diversity of thalamic reticular subnetworks, *Front. Syst. Neurosci.*, 2018, **12**, 41.
51. H. E. Marei, Neural Circuit Mapping and Neurotherapy-Based Strategies, *Cell. Mol. Neurobiol.*, 2025, **45**(1), 1–31.
52. H. M. Balkhi, T. Gul, M. Z. Banday and E. Haq, Glutamate excitotoxicity: an insight into the mechanism, *Int. J. Adv. Res.*, 2014, **2**(7), 361–373.
53. M. P. Mattson, *Sculptor and Destroyer: Tales of Glutamatethe Brain's Most Important Neurotransmitter*, MIT Press, 2023.
54. P. Newsholme, J. Procopio, M. M. R. Lima, T. C. Pithon-Curi and R. Curi, Glutamine and glutamate—their central role in cell metabolism and function, *Cell Biochem. Funct.*, 2003, **21**(1), 1–9.
55. N. Ruderisch, *Amino acid transport across the murine blood-brain barrier*, University of Zurich, 2010.
56. L. Peng, L. Hertz, R. Huang, U. Sonnewald, S. B. Petersen and N. Westergaard, *et al.*, Utilization of glutamine and of TCA cycle constituents as precursors for transmitter glutamate and GABA, *Dev. Neurosci.*, 1993, **15**(3–5), 367–377.
57. J. V. Andersen, The Glutamate/GABA-Glutamine Cycle: Insights, Updates, and Advances, *J. Neurochem.*, 2025, **169**(3), e70029.
58. S. Magi, S. Piccirillo, S. Amoroso and V. Lariccia, Excitatory amino acid transporters (EAATs): glutamate transport and beyond, *Int. J. Mol. Sci.*, 2019, **20**(22), 5674.
59. S. Armano, S. Coco, A. Bacci, E. Pravettoni, U. Schenk and C. Verderio, *et al.*, Localization and functional relevance of system a neutral amino acid transporters in cultured hippocampal neurons, *J. Biol. Chem.*, 2002, **277**(12), 10467–10473.
60. C. L. Powell, A. R. Davidson and A. M. Brown, Universal glia to neurone lactate transfer in the nervous system: physiological functions and pathological consequences, *Biosensors*, 2020, **10**(11), 183.
61. W. Xin, *Functional heterogeneity in the macroglial network*, Johns Hopkins University, 2018.
62. Z. Farsi, R. Jahn and A. Woehler, Proton electrochemical gradient: driving and regulating neurotransmitter uptake, *BioEssays*, 2017, **39**(5), 1600240.
63. A. C. Dolphin, Functions of presynaptic voltage-gated calcium channels, *Function*, 2021, **2**(1), zqaa027.
64. A. Reiner and J. Levitz, Glutamatergic signaling in the central nervous system: ionotropic and metabotropic receptors in concert, *Neuron*, 2018, **98**(6), 1080–1098.
65. D. R. Madden, The structure and function of glutamate receptor ion channels, *Nat. Rev. Neurosci.*, 2002, **3**(2), 91–101.
66. V. Zinchenko, L. Dolgacheva and S. Tuleukhanov, Calcium-permeable AMPA and kainate receptors of GABAergic neurons, *Biophys. Rev.*, 2024, **16**(2), 165–171.

67. A. Fernandes, *Impact of open channel blockers on the surface dynamics and organization of NMDA receptors*, Université de Bordeaux, 2020.
68. M. Cavazzini, T. Bliss and N. Emptage, Ca^{2+} and synaptic plasticity, *Cell Calcium*, 2005, **38**(3–4), 355–367.
69. P. Pinheiro and C. Mulle, Kainate receptors, *Cell Tissue Res.*, 2006, **326**(2), 457–482.
70. M. F. Ribeiro, M. Paquet, P. S. Cregan and S. G. S. Ferguson, Group I metabotropic glutamate receptor signalling and its implication in neurological disease, *CNS Neurol. Disord.: Drug Targets*, 2010, **9**(5), 574–595.
71. D. Olivares-Berjaga, A. Martínez-Pinteño, N. Rodríguez, S. Mas, C. Morén and E. Parellada, *et al.*, Effectiveness of positive allosteric modulators of metabotropic glutamate receptor 2/3 (mGluR2/3) in animal models of schizophrenia, *Transl. Psychiatry*, 2025, **15**(1), 11.
72. M. Kalinowska and A. Francesconi, Group I metabotropic glutamate receptor interacting proteins: fine-tuning receptor functions in health and disease, *Curr. Neuropharmacol.*, 2016, **14**(5), 494–503.
73. A. Gasiorowska, M. Wydrych, P. Drapich, M. Zadrozny, M. Steczkowska and W. Niewiadomski, *et al.*, The biology and pathobiology of glutamatergic, cholinergic, and dopaminergic signaling in the aging brain, *Front. Aging Neurosci.*, 2021, **13**, 654931.
74. A. R. Malik and T. E. Willnow, Excitatory amino acid transporters in physiology and disorders of the central nervous system, *Int. J. Mol. Sci.*, 2019, **20**(22), 5671.
75. F. J. Carvajal, H. A. Mattison and W. Cerpa, Role of NMDA receptor-mediated glutamatergic signaling in chronic and acute Neuropathologies, *Neural Plast.*, 2016, **2016**, 12701526.
76. K. Szydlowska and M. Tymianski, Calcium, ischemia and excitotoxicity, *Cell Calcium*, 2010, **47**(2), 122–129.
77. Q. Zhang, L. Zhu, H. Li, Q. Chen, N. Li and J. Li, *et al.*, Insights and progress on the biosynthesis, metabolism, and physiological functions of gamma-aminobutyric acid (GABA): a review, *PeerJ*, 2024, **12**, e18712.
78. A. Sideraki and A. Drigas, GABA and Executive Functions in ASD, *Sci. Electron. Arch.*, 2024, **17**(3), 1–5.
79. C. G. Lau and V. N. Murthy, Activity-dependent regulation of inhibition *via* GAD67, *J. Neurosci.*, 2012, **32**(25), 8521–8531.
80. S.-E. Lee, Y. Lee and G. H. Lee, The regulation of glutamic acid decarboxylases in GABA neurotransmission in the brain, *Arch. Pharmacal Res.*, 2019, **42**(12), 1031–1039.
81. S. Dakshinamurti and K. Dakshinamurti, 11 Multifaceted Therapeutic Potential of Vitamin B6, *Vitamin-Binding Proteins: Functional Consequences*, 2013, p. 183.
82. Y. Zhou and N. C. Danbolt, GABA and glutamate transporters in brain, *Front. Endocrinol.*, 2013, **4**, 165.
83. J. V. Andersen and A. Schousboe, Milestone Review: Metabolic dynamics of glutamate and GABA mediated neurotransmission—The essential roles of astrocytes, *J. Neurochem.*, 2023, **166**(2), 109–137.

84. Y. He, K. Xie, K. Yang, N. Wang and L. Zhang, Unraveling the Interplay Between Metabolism and Neurodevelopment in Health and Disease, *CNS Neurosci. Ther.*, 2025, **31**(5), e70427.
85. J. N. Kew and J. A. Kemp, Ionotropic and metabotropic glutamate receptor structure and pharmacology, *Psychopharmacology*, 2005, **179**(1), 4–29.
86. A. Ghit, D. Assal, A. S. Al-Shami and D. E. E. Hussein, GABAA receptors: structure, function, pharmacology, and related disorders, *J. Genet. Eng. Biotechnol.*, 2021, **19**(1), 123.
87. A. Cupello, Neuronal transmembrane chloride electrochemical gradient: A key player in GABAA receptor activation physiological effect, *Amino Acids*, 2003, **24**(4), 335–346.
88. A. B. Philip, J. Brohan and B. Goudra, The role of GABA receptors in anesthesia and sedation: an updated review, *CNS Drugs*, 2025, **39**(1), 39–54.
89. W. Xie, Y. Li, X. Wang, E. Blokhina, E. Krupitsky and M. Vetrova, *et al.*, GABAB receptor: Structure, biological functions, and therapy for diseases, *MedComm*, 2025, **6**(5), e70163.
90. A. S. Kshatri, A. Gonzalez-Hernandez and T. Giraldez, Physiological roles and therapeutic potential of Ca^{2+} activated potassium channels in the nervous system, *Front. Mol. Neurosci.*, 2018, **11**, 258.
91. Y. Ben-Ari, J.-L. Gaiarsa, R. Tyzio and R. Khazipov, GABA: a pioneer transmitter that excites immature neurons and generates primitive oscillations, *Physiol. Rev.*, 2007, **87**(4), 1215–1284.
92. N. G. Bowery, Baclofen: therapeutic use and potential of the prototypic GABAB receptor agonist, *GABAB receptor*, Springer, 2016, pp. 337–356.
93. E. Cherubini, G. Di Cristo and M. Avoli, Dysregulation of GABAergic signaling in neurodevelomental disorders: targeting cation-chloride cotransporters to re-establish a proper E/I balance, *Front. Cell. Neurosci.*, 2022, **15**, 813441.
94. A. Nascimento, D. Pereira-Figueiredo, V. Borges-Martins, R. Kubrusly and K. Calaza, GABAergic system and chloride cotransporters as potential therapeutic targets to mitigate cell death in ischemia, *J. Neurosci. Res.*, 2024, **102**(5), e25355.
95. M. Watanabe and A. Fukuda, Development and regulation of chloride homeostasis in the central nervous system, *Front. Cell. Neurosci.*, 2015, **9**, 371.
96. A. H. Bazzari and H. R. Parri, Neuromodulators and long-term synaptic plasticity in learning and memory: A steered-glutamatergic perspective, *Brain Sci.*, 2019, **9**(11), 300.
97. S. Kaur, S. Singh, A. Arora, P. Ram, S. Kumar, P. Kumar, *et al.*, Pharmacology of GABA and its receptors, *Frontiers in pharmacology of neurotransmitters*, Springer, 2020, pp. 241–292.
98. J. Ong and D. I. Kerr, Clinical potential of GABAB receptor modulators, *CNS Drug Rev.*, 2005, **11**(3), 317–334.
99. M. O. Klein, D. S. Battagello, A. R. Cardoso, D. N. Hauser, J. C. Bittencourt and R. G. Correa, Dopamine: functions, signaling, and

association with neurological diseases, *Cell. Mol. Neurobiol.*, 2019, **39**(1), 31–59.
100. S. C. Daubner, T. Le and S. Wang, Tyrosine hydroxylase and regulation of dopamine synthesis, *Arch. Biochem. Biophys.*, 2011, **508**(1), 1–12.
101. J. M. King, G. Muthian, V. Mackey, M. Smith and C. Charlton, L-Dihydroxyphenylalanine modulates the steady-state expression of mouse striatal tyrosine hydroxylase, aromatic L-amino acid decarboxylase, dopamine and its metabolites in an MPTP mouse model of Parkinson's disease, *Life Sci.*, 2011, **89**(17–18), 638–643.
102. M. D. S. Tai, G. Gamiz-Arco and A. Martinez, Dopamine synthesis and transport: current and novel therapeutics for parkinsonisms, *Biochem. Soc. Trans.*, 2024, **52**(3), 1275–1291.
103. J. M. Beaulieu, S. Espinoza and R. R. Gainetdinov, Dopamine receptors–IUPHAR R eview 13, *Br. J. Pharmacol.*, 2015, **172**(1), 1–23.
104. G. Ayano, Dopamine: receptors, functions, synthesis, pathways, locations and mental disorders: review of literatures, *J. Ment. Disord. Treat.*, 2016, **2**(120), 2.
105. M. Glass and C. C. Felder, Concurrent stimulation of cannabinoid CB1 and dopamine D2 receptors augments cAMP accumulation in striatal neurons: evidence for a Gs linkage to the CB1 receptor, *J. Neurosci.*, 1997, **17**(14), 5327–5333.
106. M. C. Clark and D. J. Baro, Arthropod D2 receptors positively couple with cAMP through the Gi/o protein family, *Comp. Biochem. Physiol., Part B*, 2007, **146**(1), 9–19.
107. S. C. Gantz, C. P. Ford, H. Morikawa and J. T. Williams, The evolving understanding of dopamine neurons in the substantia nigra and ventral tegmental area, *Annu. Rev. Physiol.*, 2018, **80**, 219–241.
108. L. Annunziato, Regulation of the tuberoinfundibular and nigrostriatal systems: evidence for different kinds of dopaminergic neurons in the brain, *Neuroendocrinology*, 1979, **29**(1), 66–76.
109. J. Blesa, G. Foffani, B. Dehay, E. Bezard and J. A. Obeso, Motor and non-motor circuit disturbances in early Parkinson disease: which happens first?, *Nat. Rev. Neurosci.*, 2022, **23**(2), 115–128.
110. S. Ikemoto, Brain reward circuitry beyond the mesolimbic dopamine system: a neurobiological theory, *Neurosci. Biobehav. Rev.*, 2010, **35**(2), 129–150.
111. S. B. Floresco and O. Magyar, Mesocortical dopamine modulation of executive functions: beyond working memory, *Psychopharmacology*, 2006, **188**(4), 567–585.
112. G. A. Gudelsky, Tuberoinfundibular dopamine neurons and the regulation of prolactin secretion, *Psychoneuroendocrinology*, 1981, **6**(1), 3–16.
113. S. N. Haber, The place of dopamine in the cortico-basal ganglia circuit, *Neuroscience*, 2014, **282**, 248–257.
114. P. W. Glimcher, Understanding dopamine and reinforcement learning: the dopamine reward prediction error hypothesis, *Proc. Natl. Acad. Sci. U. S. A.*, 2011, **108**(supplement_3), 15647–15654.

115. W. Schultz, Electrophysiological correlates of reward processing in dopamine neurons, *Decision neuroscience*, Elsevier, 2017, pp. 21–31.
116. Dopamine, psychosis, and symptom fluctuation: a narrative review, ed. G. Novak and M. V. Seeman, in *Healthcare*, MDPI, 2022.
117. B. K. Madras, History of the discovery of the antipsychotic dopamine D2 receptor: a basis for the dopamine hypothesis of schizophrenia, *J. Hist. Neurosci.*, 2013, **22**(1), 62–78.
118. C. Gu, Y.-C. Geng and L.-N. Zhu, Dysregulation of dopamine neurotransmission in drug addicts: implications for criminal behavior and corrective interventions, *Front. Psychiatry*, 2024, **15**, 1434083.
119. A. Demailly, C. Moreau and D. Devos, Effectiveness of continuous dopaminergic therapies in Parkinson's disease: A review of L-DOPA pharmacokinetics/pharmacodynamics, *J. Parkinsons Dis.*, 2024, **14**(5), 925–939.
120. R. P. K. Babu and S. N. Maiti, Norepinephrine reuptake inhibitors for depression, ADHD and other neuropsychiatric disorders, *Heterocycles*, 2006, **69**(1), 539.
121. V. Voon, P.-O. Fernagut, J. Wickens, C. Baunez, M. Rodriguez and N. Pavon, *et al.*, Chronic dopaminergic stimulation in Parkinson's disease: from dyskinesias to impulse control disorders, *Lancet Neurol.*, 2009, **8**(12), 1140–1149.
122. S. Russo, I. P. Kema, F. Bosker, J. Haavik and J. Korf, Tryptophan as an evolutionarily conserved signal to brain serotonin: molecular evidence and psychiatric implications, *World J. Biol. Psychiatry*, 2009, **10**(4), 258–268.
123. J. Wiederhofer, *Psychoneuroendocrinology in Psychosocial and Psychotherapeutic Practice*, 2024.
124. D. J. Walther and M. Bader, A unique central tryptophan hydroxylase isoform, *Biochem. Pharmacol.*, 2003, **66**(9), 1673–1680.
125. W. Roth, K. Zadeh, R. Vekariya, Y. Ge and M. Mohamadzadeh, Tryptophan metabolism and gut-brain homeostasis, *Int. J. Mol. Sci.*, 2021, **22**(6), 2973.
126. H. F. D. Krantz, 7 Vesicular Neurotransmitter Transporters, *Handbook of Neurochemistry and Molecular Neurobiology*, 2009, p. 687.
127. J.-P. Hornung, The human raphe nuclei and the serotonergic system, *J. Chem. Neuroanat.*, 2003, **26**(4), 331–343.
128. Ö. D. Özçete, A. Banerjee and P. S. Kaeser, Mechanisms of neuromodulatory volume transmission, *Mol. Psychiatry*, 2024, **29**(11), 3680–3693.
129. J. L. Cortes-Altamirano, A. Olmos-Hernandez, H. B. Jaime, P. Carrillo-Mora, C. Bandala and S. Reyes-Long, *et al.*, 5-HT1, 5-HT2, 5-HT3 and 5-HT7 receptors and their role in the modulation of pain response in the central nervous system, *Curr. Neuropharmacol.*, 2018, **16**(2), 210–221.
130. M. M. Rahman, M. R. Islam, S. A. Mim, N. Sultana, D. K. Chellappan and K. Dua, *et al.*, Insights into the Promising Prospect of G Protein and GPCR-Mediated Signaling in Neuropathophysiology and Its Therapeutic Regulation, *Oxid. Med. Cell. Longevity*, 2022, **2022**(1), 8425640.

131. N. Kushwaha, *Characterization of the G-protein Coupling Domains of the 5-HT1A Receptor*, University of Ottawa, Canada, 2007.
132. D. C. Reeves and S. C. Lummis, The molecular basis of the structure and function of the 5-HT 3 receptor: a model ligand-gated ion channel, *Mol. Membr. Biol.*, 2002, **19**(1), 11–26.
133. M. Fakhoury, Revisiting the serotonin hypothesis: implications for major depressive disorders, *Mol. Neurobiol.*, 2016, **53**(5), 2778–2786.
134. S. Vaseghi, S. Arjmandi-Rad, M. Eskandari, M. Ebrahimnejad, G. Kholghi and M.-R. Zarrindast, Modulating role of serotonergic signaling in sleep and memory, *Pharmacol. Rep.*, 2022, **74**(1), 1–26.
135. Z. Cingelová, Important functions of serotonin in the gastrointestinal tract, *Int. J. Health New Technol. Soc. Work*, 2023, **18**(2), 73–83.
136. M. Rieder, N. Gauchel, C. Bode and D. Duerschmied, Serotonin: a platelet hormone modulating cardiovascular disease, *J. Thromb. Thrombolysis*, 2021, **52**(1), 42–47.
137. K. Andre, O. Kampman, A. Illi, M. Viikki, E. Setälä-Soikkeli and N. Mononen, *et al.*, SERT and NET polymorphisms, temperament and antidepressant response, *Nord. J. Psychiatry*, 2015, **69**(7), 531–538.
138. S. S. Desrochers, M. G. Spring and K. M. Nautiyal, A role for serotonin in modulating opposing drive and brake circuits of impulsivity, *Front. Behav. Neurosci.*, 2022, **16**, 791749.
139. C. Slater, Y. Liu, E. Weiss, K. Yu and Q. Wang, The neuromodulatory role of the noradrenergic and cholinergic systems and their interplay in cognitive functions: A focused review, *Brain Sci.*, 2022, **12**(7), 890.
140. M. Cosentino and F. Marino, Nerve Driven immunity: noradrenaline and adrenaline, *Nerve-Driven Immunity: Neurotransmitters and Neuropeptides in the Immune System*, Springer, 2012, pp. 47–96.
141. P. Kumar, S. N. Abed, Y. A. Bataineh and M. S. Salem, Neurotransmitters and their receptors—State of the art, *Front. Pharmacol.*, 2020, 1–29.
142. L. A. Schwarz and L. Luo, Organization of the locus coeruleus-norepinephrine system, *Curr. Biol.*, 2015, **25**(21), R1051–R1056.
143. D. M. Devilbiss, Consequences of tuning network function by tonic and phasic locus coeruleus output and stress: regulating detection and discrimination of peripheral stimuli, *Brain Res.*, 2019, **1709**, 16–27.
144. B. Gorain, S. Dutta, U. Nandy, P. Sengupta and H. Choudhury, Pharmacology of adrenaline, noradrenaline, and their receptors, *Frontiers in Pharmacology of Neurotransmitters*, Springer, 2020, pp. 107–142.
145. J. Chen and H. A. Rockman, The GPCRs that are particularly crucial for cardiovascular function include the adrenergic, angiotensin II, endothelin, and mus-carinic cholinergic receptors. 1 Upon extracellular ligand binding, GPCRs undergo conformational change and coupling of their intracellular domain with a heterotrimeric guanine-nucleotide-binding regulatory protein (G protein). The coupling allows for the exchange of guanosine diphosphate, *Acute Heart Failure*, 2009, p. 89.

146. Y. Yudin and T. Rohacs, Inhibitory Gi/O-coupled receptors in somatosensory neurons: potential therapeutic targets for novel analgesics, *Mol. Pain*, 2018, **14**, 1744806918763646.
147. T. J. O'Dell, S. A. Connor, R. Guglietta and P. V. Nguyen, β-Adrenergic receptor signaling and modulation of long-term potentiation in the mammalian hippocampus, *Learn. Mem.*, 2015, **22**(9), 461–471.
148. O. Borodovitsyna, M. Flamini and D. Chandler, Noradrenergic modulation of cognition in health and disease, *Neural Plast.*, 2017, **2017**(1), 6031478.
149. J. A. Ross and E. J. Van Bockstaele, The locus coeruleus-norepinephrine system in stress and arousal: unraveling historical, current, and future perspectives, *Front. Psychiatry*, 2021, **11**, 601519.
150. B. P. Ramos and A. F. Arnsten, Adrenergic pharmacology and cognition: focus on the prefrontal cortex, *Pharmacol. Ther.*, 2007, **113**(3), 523–536.
151. K. Tully and V. Y. Bolshakov, Emotional enhancement of memory: how norepinephrine enables synaptic plasticity, *Mol. Brain*, 2010, **3**(1), 15.
152. O. M. Oleshko, K. V. Berladir, T. B. Oleshko, V. V. Hlushchenko, O. S. Korol and V. V. Bilokonskyi, *et al.*, Neurobiological aspects of pathogenetic mechanisms in the development of post-traumatic stress disorder (literature review), *East. Ukr. Med. J.*, 2025, **13**(1), 39–54.
153. H. Krasner, C. V. Ong, P. Hewitt and T. A. Vida, From stress to synapse: the neuronal atrophy pathway to mood dysregulation, *Int. J. Mol. Sci.*, 2025, **26**(7), 3219.
154. G. Goseph, M. N. A. Halawa and S. Ali, Neurobiology of Resilience: Insights from Psychiatric Research and Implications for Treatment Strategies, *Arch. Med. Rep.*, 2024, **1**(3), 13–18.
155. S. Maity, R. Abbaspour, D. Nahabedian and S. A. Connor, Norepinephrine, beyond the synapse: coordinating epigenetic codes for memory, *Int. J. Mol. Sci.*, 2022, **23**(17), 9916.
156. C. Schnell, M. Negm, J. Driehaus, A. Scheller and S. Hülsmann, Norepinephrine-induced calcium signaling in astrocytes in the respiratory network of the ventrolateral medulla, *Respir. Physiol. Neurobiol.*, 2016, **226**, 18–23.
157. M. J. Dahl, M. Mather and M. Werkle-Bergner, Noradrenergic modulation of rhythmic neural activity shapes selective attention, *Trends Cogn. Sci.*, 2022, **26**(1), 38–52.
158. D. A. Brown, Acetylcholine and cholinergic receptors, *Brain Neurosci. Adv.*, 2019, **3**, 2398212818820506.
159. Y. Oda, Choline acetyltransferase: the structure, distribution and pathologic changes in the central nervous system, *Pathol. Int.*, 1999, **49**(11), 921–937.
160. E. Neher, Vesicle pools and Ca^{2+} microdomains: new tools for understanding their roles in neurotransmitter release, *Neuron*, 1998, **20**(3), 389–399.
161. J. S. Rosenthal and Q. Yuan, Constructing and tuning excitatory cholinergic synapses: the multifaceted functions of nicotinic acetylcholine

receptors in Drosophila neural development and physiology, *Front. Cell. Neurosci.*, 2021, **15**, 720560.
162. G. Zimmerman and H. Soreq, Termination and beyond: acetylcholinesterase as a modulator of synaptic transmission, *Cell Tissue Res.*, 2006, **326**(2), 655–669.
163. E. X. Albuquerque, E. F. Pereira, M. Alkondon and S. W. Rogers, Mammalian nicotinic acetylcholine receptors: from structure to function, *Physiol. Rev.*, 2009, **89**, 73–120.
164. J. A. Dani, Neuronal nicotinic acetylcholine receptor structure and function and response to nicotine, *Int. Rev. Neurobiol.*, 2015, **124**, 3–19.
165. M. G. Kaoullas, D. M. Thal, A. Christopoulos and C. Valant, Ligand bias at the muscarinic acetylcholine receptor family: Opportunities and challenges, *Neuropharmacology*, 2024, **258**, 110092.
166. S. K.-F. Wong, G protein selectivity is regulated by multiple intracellular regions of GPCRs, *Neurosignals*, 2003, **12**(1), 1–12.
167. G. He, Y. Li, H. Deng and H. Zuo, Advances in the study of cholinergic circuits in the central nervous system, *Ann. Clin. Transl. Neurol.*, 2023, **10**(12), 2179–2191.
168. M. Blake and M. Boccia, Basal forebrain cholinergic system and memory, *Curr. Top. Behav. Neurosci.*, 2017, **37**, 253–273.
169. K. Semba, The mesopontine cholinergic system: a dual role in REM sleep and wakefulness, *Handbook of behavioral state control*, CRC Press, 2019, pp. 161–180.
170. X.-Q. Chen and W. C. Mobley, Exploring the pathogenesis of Alzheimer disease in basal forebrain cholinergic neurons: converging insights from alternative hypotheses, *Front. Neurosci.*, 2019, **13**, 446.
171. ŁJ. Walczak-Nowicka and M. Herbet, Acetylcholinesterase inhibitors in the treatment of neurodegenerative diseases and the role of acetylcholinesterase in their pathogenesis, *Int. J. Mol. Sci.*, 2021, **22**(17), 9290.
172. C. R. Johnson, B. D. Kangas, E. M. Jutkiewicz, J. Bergman and A. Coop, Drug design targeting the muscarinic receptors and the implications in central nervous system disorders, *Biomedicines*, 2022, **10**(2), 398.
173. B. Lendvai and E. S. Vizi, Nonsynaptic chemical transmission through nicotinic acetylcholine receptors, *Physiol. Rev.*, 2008, **88**(2), 333–349.
174. R. Traub and A. Draguhn, *Brain leitmotifs*, Springer, Cham, 2024, vol. 10, pp. 978-3.
175. E. Pessione and S. Cirrincione, Bioactive molecules released in food by lactic acid bacteria: Encrypted peptides and biogenic amines, *Front. Microbiol.*, 2016, **7**, 876.
176. E. Carthy and T. Ellender, Histamine, neuroinflammation and neurodevelopment: a review, *Front. Neurosci.*, 2021, **15**, 680214.
177. H. Haas and P. Panula, The role of histamine and the tuberomamillary nucleus in the nervous system, *Nat. Rev. Neurosci.*, 2003, **4**(2), 121–130.
178. L. Tuomisto, V. Lozeva, A. Valjakka and A. Lecklin, Modifying effects of histamine on circadian rhythms and neuronal excitability, *Behav. Brain Res.*, 2001, **124**(2), 129–135.

179. M. B. Passani, P. Giannoni, C. Bucherelli, E. Baldi and P. Blandina, Histamine in the brain: beyond sleep and memory, *Biochem. Pharmacol.*, 2007, **73**(8), 1113–1122.
180. F. E. R. Simon and K. J. Simons, H1 antihistamines: current status and future directions, *World Allergy Organ. J.*, 2008, **1**(9), 145–155.
181. L. Terenius, From opiate pharmacology to opioid peptide physiology, *Upsala J. Med. Sci.*, 2000, **105**(1), 1–15.
182. E. E. Benarroch, Endogenous opioid systems: current concepts and clinical correlations, *Neurology*, 2012, **79**(8), 807–814.
183. A. S. Sprouse-Blum, G. Smith, D. Sugai and F. D. Parsa, Understanding endorphins and their importance in pain management, *Hawaii Med. J.*, 2010, **69**(3), 70.
184. M. J. Kelly, J. Qiu and O. K. Rønnekleiv, Estrogen modulation of G-protein-coupled receptor activation of potassium channels in the central nervous system, *Ann. N. Y. Acad. Sci.*, 2003, **1007**(1), 6–9.
185. Z. Peng, Q. Jia, J. Mao, X. Luo, A. Huang and H. Zheng, *et al.*, Neurotransmitters crosstalk and regulation in the reward circuit of subjects with behavioral addiction, *Front. Psychiatry*, 2025, **15**, 1439727.
186. C. A. Gallelli, S. Calcagnini, A. Romano, J. B. Koczwara, M. De Ceglia and D. Dante, *et al.*, Modulation of the oxidative stress and lipid peroxidation by endocannabinoids and their lipid analogues, *Antioxidants*, 2018, **7**(7), 93.
187. J. Barnett-Norris, F. Guarnieri, D. P. Hurst and P. H. Reggio, Exploration of biologically relevant conformations of anandamide, 2-arachidonylglycerol, and their analogues using conformational memories, *J. Med. Chem.*, 1998, **41**(24), 4861–4872.
188. T. F. Freund, I. Katona and D. Piomelli, Role of endogenous cannabinoids in synaptic signaling, *Physiol. Rev.*, 2003, 1017–1066.
189. A. D. Grzesińska, The Involvement of the Endocannabinoid, Glutamatergic, and GABAergic Systems in PTSD, *Int. J. Mol. Sci.*, 2025, **26**(13), 5929.
190. B. Lutz, G. Marsicano, R. Maldonado and C. J. Hillard, The endocannabinoid system in guarding against fear, anxiety and stress, *Nat. Rev. Neurosci.*, 2015, **16**(12), 705–718.
191. M. A. Diana and A. Marty, Endocannabinoid-mediated short-term synaptic plasticity: depolarization-induced suppression of inhibition (DSI) and depolarization-induced suppression of excitation (DSE), *Br. J. Pharmacol.*, 2004, **142**(1), 9–19.
192. B. W. Penninx and S. M. Lange, Metabolic syndrome in psychiatric patients: overview, mechanisms, and implications, *Dialogues Clin. Neurosci.*, 2018, **20**(1), 63–73.
193. R. S. Rodrigues, D. M. Lourenço, S. L. Paulo, J. M. Mateus, M. F. Ferreira and F. M. Mouro, *et al.*, Cannabinoid actions on neural stem cells: implications for pathophysiology, *Molecules*, 2019, **24**(7), 1350.
194. M. S. Hernandes and L. R. Troncone, Glycine as a neurotransmitter in the forebrain: a short review,*J. Neural Transm.*, 2009, **116**(12), 1551–1560.

195. P. Legendre, The glycinergic inhibitory synapse, *Cell. Mol. Life Sci.*, 2001, **58**(5), 760–793.
196. J. W. Lynch, Molecular structure and function of the glycine receptor chloride channel, *Physiol. Rev.*, 2004, **84**(4), 1051–1095.
197. F. Zafra, I. Ibáñez, D. Bartolomé-Martín, D. Piniella, M. Arribas-Blázquez and C. Giménez, Glycine transporters and its coupling with NMDA receptors, *Glial Amino Acid Transporters*, 2017, pp. 55–83.
198. C. L. Cioffi, Modulation of glycine-mediated spinal neurotransmission for the treatment of chronic pain, *J. Med. Chem.*, 2017, **61**(7), 2652–2679.
199. Z. Liu and S. Song, Neurotransmitters and Receptors in Anesthesia and Pain Modulation, *Neuroscience of Anesthesia: From Cellular Mechanisms to Clinical Applications*, Springer, 2025, pp. 33–44.
200. D. M. Flannery, *The Endocannabinoid System's Influence and Modulation of Nociception*, 2025.

CHAPTER 3

Neuropeptides and Their Role in Brain Function

SHIVANGI GUPTA,[a,†] ROMIL SHARMA[b,†] AND SANDEEP SINGH*[c]

[a] Neuroscience and Ageing Biology Division, CSIR-Central Drug Research Institute, Lucknow, Uttar Pradesh, 226031, India; [b] Molecular Microbiology and Immunology Division, CSIR-Central Drug Research Institute, Lucknow, Uttar Pradesh, 226031, India; [c] The Faculty of Medicine, The Hebrew University of Jerusalem, Israel
*Email: singh.sandeep@mail.huji.ac.il

3.1 Introduction

In order to convey our thoughts, feelings, emotions and other perceptions to someone, we need to talk or communicate with them. Likewise, the human brain is a master organ that communicates with the whole body through a range of chemical messengers and/or signaling molecules using an elaborate network of neurons and supporting cells. Among various chemical messengers and/or signaling molecules, neuropeptides are tiny molecules that resemble proteins and are used by neurons to exchange messages. These small, controlled-release proteinaceous substances are produced by neurons and act on either neural substrates, such as neurons or glial cells, or non-neuronal target cells, such as muscles or glands. By acting as neuromodulators and, occasionally, neurotransmitters, they have a variety of physiological and behavioral impacts on the body and brain.

† These authors contributed equally.

Chemical Biology No. 26
Chemical Signatures of the Brain: Bridging Biochemistry, Neuroanatomy, and Brain Disorders
Edited by Sachchida Nand Rai and Abhishek Kumar Singh

Published by the Royal Society of Chemistry, www.rsc.org

Since their discovery at the beginning of the 20th century, neuropeptides have been found in more than 100 distinct forms in mammals, including humans. These substances are produced in the cell body of neurons and then carried by the axon to be stored in large dense-core vesicles. They attach to G-protein-coupled receptors (GPCRs) of the target cell after being released into the extracellular space or synaptic cleft in response to stimulation. Because of their diffusion characteristics and longer-lasting effects, neuropeptides frequently work more slowly than classical neurotransmitters, which typically act quickly and locally.[1] This enables them to alter more extensive brain circuits and impact a variety of processes, including mood, learning, memory, appetite, sleep, stress response, and pain perception.[2] The importance of neuropeptides in preserving homeostasis and responding to environmental cues is highlighted by the variety of neuropeptides and their receptors as well as their extensive distribution throughout the brain. They exhibit several functions such as social behavior and emotional regulation, cognition, stress resilience, anxiety modulation, circadian rhythms and sleep–wake cycles, nociceptive signaling, migraine, and others.

New advances in neuropharmacology, molecular biology, and imaging technologies have significantly increased our comprehension of the role that neuropeptides play in the brain. Neuropeptide-expressing neurons can now be precisely manipulated thanks to the advent of optogenetic and chemogenetic technologies, which provides information about the precise roles these neurons play in behavior and brain function. Additionally, a variety of neurological and behavioral disorders, including depression, obesity, neurodegenerative illnesses, and chronic pain, are being investigated for therapy with neuropeptide-based therapies.

Despite these developments, there are still obstacles in the way of completely understanding how neuropeptide signaling works. Because of their complicated receptor systems, low concentrations, and quick breakdown, neuropeptides are more challenging to quantify than small-molecule neurotransmitters. Furthermore, it can be difficult to pinpoint their precise functions in particular brain processes due to their wide-ranging and occasionally overlapping effects. However, new studies keep revealing the amazing adaptability and significance of neuropeptides in the central nervous system (CNS).

In summary, neuropeptides modulate a vast range of physiological and behavioral processes, making them an essential part of the brain communication system. They are essential for regulating brain activity and preserving homeostasis because of their special qualities, which include co-transmission with traditional neurotransmitters and long-lasting, long-range effects. Gaining insight into the role of neuropeptides advances our understanding of brain biology and creates new opportunities for the development of focused treatments for mental and neurological illnesses. The importance of neuropeptides in brain function is becoming increasingly clearer as research advances, making them key players in the fields of neuroscience and medicine going forward.

3.2 Discovery of Neuropeptides

The first finding of mammalian neuropeptides was made by isolating them from hundreds of kilograms of brain or intestinal tissue, or from thousands of pieces of the hypothalamus. New peptides were simultaneously extracted from other sources, such as amphibian skin.[3] Others have used the C-terminal amidation as a chemical identifier to find more peptides.[4] In a stunning illustration of tissue-specific RNA processing, Rosenfeld and colleagues in 1983 looked at the gene for calcitonin, which is only expressed in the thyroid, and discovered that it encodes calcitonin-gene-related peptide, which is only expressed in the nervous system.[5] The pro-opiomelanocortin gene, which produces endorphin in the brain and adrenocorticotropic hormone in the pituitary, has comparable tissue specificity.[6,7] Further, subtraction analysis was used to identify a number of distinct transcripts that are only expressed in the hypothalamus;[8] the most well-known of these are the hypocretins,[9] also known as orexins.[10] OrphaninFQ, also known as nociceptin, is a ligand for the opioid-like receptor and the first of numerous novel peptides discovered as a result of the hunt for the endogenous ligands of G-protein-coupled receptors (GPCRs).[11,12] Additionally, a plethora of neuropeptides has been discovered as these researches progress.[13]

Substance P was identified as the first neuropeptide in 1931 by von Euler and Gaddum. However, it was not until 1971 that Susan Leeman and her associates demonstrated that substance P is a C-terminal amidated, 11-amino-acid residue peptide, describing its chemical makeup. Following the identification of substance P, more related peptides, including neurokinins A and B, were discovered. Multiple receptors (neurokinins 1, 2, and 3) have been characterized and cloned, and the distributions of these peptides have been mapped thanks to radioimmunoassay, immunohistochemistry, and *in situ* hybridization. After the development of pepididic (such as spantide) and nonpeptide receptor antagonists, the Food and Drug Administration has now approved the neurokinin 1 antagonist aprepitant (MK 869) for the treatment of chemotherapy-induced nausea and has demonstrated clinical efficacy in the treatment of major depression.[14]

After that, the hypothalamic neuropeptides vasopressin and oxytocin, which function as hormones systemically (*e.g.*, controlling breastfeeding and diuresis) and inside the brain to affect social behavior, were among the first neuropeptides to be chemically identified in mammals.[15]

Finally, Arvid Carlsson, Paul Greengard, and Eric Kandel were granted the 2000 Nobel Prize in Physiology or Medicine for their studies of neurotransmitters, their transduction patterns, and the ways in which certain neurotransmission systems can affect behavior.[16–18] These honors emphasized 50 years of research that had demonstrated that chemical mediation underlies intercellular communication in the CNS, a finding that had been made approximately 50 years prior for the peripheral nervous system.

3.3 Neuropeptides and Their Structure

A class of endogenous peptides known as neuropeptides function in the nervous system and other target organs as persistent neurotransmitters. The neuropeptides are smaller than normal proteins, have a less complex three-dimensional structure, and range in length from 3 to 100 amino acid residues. They are also up to 50 times larger than traditional neurotransmitters. Compared to smaller neurotransmitter molecules, they have multiple additional receptor-binding sites and more chemical information. As a result, neuropeptides and classic transmitters have 1000 times different binding affinities (nmol L^{-1} *vs.* μmol L^{-1}), higher selectivity, and a requirement for fewer copies of a neuropeptide to be released than a small transmitter (reducing the likelihood of side effects from pharmacological intervention with antagonists). Compared to traditional neurotransmitters, the bigger neuropeptides diffuse and bind more slowly but firmly, and the receptor–neuropeptide complex can be internalized to result in receptor desensitization. Nevertheless, compared to other neurotransmitters, neuropeptide concentrations in the brain are substantially lower. A neuron contains one or perhaps two tiny neurotransmitters in addition to nearly all neuropeptides.[19] Both slow (100–500 ms) and quick (2–5 ms) synaptic transmission are made possible by these combinations. Coexistence has occasionally been interpreted as a violation of Sir Henry Dale's "one neuron, one transmitter principle," however this theory only asserts that a neuron uses the same chemical transmission mechanism at each synaptic connection.[20]

However, in the mollusk *Aplysia*, the precursor of the hormone that lays eggs produces distinct peptides that are grouped into distinct granules and sent to various neuronal branches.[21] Peptides differ from other messenger molecules in a number of key ways, including how they are synthesized, stored, and used for signaling.[22,23] Peptide release, for instance, necessitates greater discharge frequencies, and once released, they may reach receptors located distant from the release site – a process known as volume transmission. This highlights the capacity of messenger molecules, particularly peptides, to affect wide target areas and broadens the traditional idea of synaptic transmission based on point-to-point contact.

3.4 Biosynthesis of Neuropeptides

Neuropeptide synthesis is a complicated process that differs greatly from that of traditional transmitters.[24] The precursor of almost all bioactive neuropeptides is a bigger, inactive molecule that is synthesized in the endoplasmic reticulum, then moved to the Golgi apparatus for packing, centrifugal transport, and exocytotic release.

The precursor proteins and processing enzymes, also known as convertases, which separate the bioactive peptides, are kept in what are known as big dense core vesicles or secretory granules. A number of these enzymes have been identified.[25,26] The secretory granules are where most endo-proteolytic cleavages take place, primarily at pairings of basic amino acid residues. Significant posttranslational enzymatic changes of the final peptide, including glycosylation, C-terminal amidation, acetylation, phosphorylation, and sulfation, may also take place.

Both distinct peptide types and multiple copies of the same peptide may be present in the precursor molecules. For instance, there are five copies of the tripeptide TRH in the precursor for TRH, and the proopiomelanocortin (POMC) precursor produces the opioid peptide β-endorphin, melanotropins, adrenocorticotropin (ACTH; corticotropin), and other substances with unique biological effects.[27] In theory, peptidases, of which several have been identified, degrade the neuropeptides in the extracellular space.[28]

3.5 Classification of Neuropeptides

Depending upon the history of evolution, location of synthesis, and target, neuropeptides are classified into various types. These include those described in the following subsections.

3.5.1 Hypothalamic Releasing Factors

An area of the brain called the hypothalamus is home to a variety of neuronal types that secrete various hormones. Just above the brainstem but beneath the thalamus is the hypothalamus. The hypothalamus, which is about the size of an almond in humans, is found in the brains of all vertebrates. In addition to synthesizing and secreting neurohormones or neuropeptides, the hypothalamus is in charge of certain metabolic functions and related autonomic nervous system functions. These peptides, which can promote or inhibit pituitary hormone production, are frequently referred to as releasing hormones or hypothalamus hormones. Body temperature, appetite, thirst, exhaustion, sleep, the circadian rhythm, and significant parenting and attachment behaviors are all regulated by the hypothalamus. All of the secreted peptides are released into the blood in the capillaries and travel immediately in portal veins to a second capillary bed in the anterior lobe of the pituitary, where they exert their effects (see Table 3.1).

3.5.2 Pituitary Neuropeptides

The midline pocket or fossa, a little hollow or dip in the sphenoid bone, is where the pituitary gland is situated beneath the brain. In front of the temporal bone and the basilar portion of the occipital bone, in the center of the skull, is the sphenoid bone, an unpaired cranial bone. Located in the rear and base of the skull, the occipital bone is a membranous bone in the shape of a saucer. The sella turcica is another name for this depression. The sella turcica, commonly known as the "Turkish Chair," is a saddle-shaped depression in the sphenoid bone of the human skull that is also present in

Table 3.1 Hypothalamic releasing factors neuropeptides.

S. no.	Neuropeptide	Target	Physiological effects
1.	Corticotrophin releasing hormone (CRH) is a peptide of 41 amino acids	Anterior lobe of the pituitary	CRH is produced by the placenta and acts on cells in the anterior lobe of the pituitary to release adrenocorticotropic hormone (ACTH). It also seems to control how long a pregnancy lasts. It is also believed to help prevent the mother's T cells from attacking the fetus with an immunological response
2.	Growth hormone releasing hormone (GHRH), mixture of two peptides, one containing 40 amino acids, the other 44 amino acids	Anterior lobe of the pituitary	Growth hormone (GH) is secreted by cells in the pituitary's anterior lobe in response to stimulation by GHRH
3.	Luteinizing hormone releasing hormone (LHRH). Reproductive hormone, also called gonadotropin-releasing hormone	Anterior pituitary gland	The anterior pituitary gland starts secreting follicle-stimulating hormone and luteinizing hormone in response to a hormone generated by the hypothalamus. A hormone made of glycoprotein that controls reproductive activities
4.	Gonadotropin-releasing hormone also known as luteinizing-hormone-releasing hormone (GnRH) and luliberin.	Anterior pituitary gland	The anterior pituitary releases luteinizing hormone (LH) and follicle-stimulating hormone (FSH). During childhood, GnRH activity is quite low; it becomes active during puberty or adolescence
5.	Somatostatin, or growth hormone release inhibiting hormone having two active forms of 14 amino acids, and 28 amino acids	Heart, stomach, acid-producing parietal cells	Controls the endocrine system, inhibits the release of many secondary hormones, and interacts with G-protein-coupled somatostatin receptors to influence neurotransmission and cell proliferation. Somatostatin reduces secretion in the stomach by acting on the parietal cells that produce acid through a G-coupled receptor. Additionally, by inhibiting the release of other hormones such as histamine, secretin, and gastrin, somatostatin indirectly reduces the formation of stomach acid
6.	Thyrotropin-releasing hormone (TRH), or thyrotropin-releasing factor (TRF), or thyroliberin or protirelin		Promotes the anterior pituitary to release prolactin and thyroid-stimulating hormone (TSH)

Table 3.2 Pituitary neuropeptides.

S. no.	Gland	Neuropeptide	Target	Physiological effects
1.	Anterior pituitary	Growth hormone	Liver, adipose tissue	Indirectly stimulates development and regulates the metabolism of proteins, fats, and carbohydrates
2.	Anterior pituitary	Thyroid-stimulating hormone	Thyroid gland	Promotes secretion of thyroid hormones
3.	Anterior pituitary	Adrenocortico-tropic hormone	Adrenal gland (cortex)	Enhances secretion of glucocorticoids
4.	Anterior pituitary	Prolactin	Mammary gland	Milk production
5.	Anterior pituitary	Luteinizing hormone	Ovary and testis	Regulates reproductive function
6.	Anterior pituitary	Follicle-stimulating hormone	Ovary and testis	Regulates reproductive function
7.	Posterior pituitary	Antidiuretic hormone	Kidney	Maintenance of body water
8.	Posterior pituitary	Oxytocin	Ovary and testis	Promotes milk ejection and uterine contractions

the skulls of other primates in the Hominidae family, which includes gorillas, chimpanzees, and orangutans. The anterior lobe makes up two-thirds of the human gland's volume, while the posterior lobe makes up the remaining one-third.

At the base of the brain, at the base of the hypothalamus, is a protrusion known as the posterior section of the pituitary gland. The hypophyseal nerve tract is made up of about 100 000 axons, and neurons in the hypothalamus reach the posterior pituitary gland directly. The axons and nerve terminals of hypothalamic neurons make up the posterior pituitary gland. Hormones held in the terminals are released upon electrical excitation. Furthermore, modified astrocytes called pituicytes encircle nerve terminals. Hormones produced by various pituitary cell types are released into the bloodstream and have an impact on other body organs. Because it regulates the actions of numerous other systems, the pituitary gland is sometimes referred to as the master gland. It secretes a variety of peptide hormones (see Table 3.2).

3.5.3 Gastrointestinal Peptides

The majority of gastrointestinal (GI) peptides are polypeptides that are made and released by specific gut endocrine cells. Numerous chemical transmitters that are involved in GI motility, secretion, absorption, growth, development, and food intake regulation are produced by these cells. The enteric and central nervous systems contain a large number of the same peptides as the GI tract (see Table 3.3).[29]

Table 3.3 Gastrointestinal peptides.

S. no.	Neuropeptide	Target	Physiological effects
1.	Gastrin	G-cells in the stomach (antrum)	Increases the development of the stomach mucosa and release of gastric acid (HCl)
2.	Cholecystokinin (CCK)	I-cells in the duodenum and jejunum	Slows stomach emptying, increases satiety, and stimulates the contraction of gallbladder and release of pancreatic enzymes
3.	Secretin	S-cells in the duodenum	Suppresses the production of stomach acid and increases the pancreatic release of bicarbonate
4.	Gastric inhibitory peptide (GIP)	K-cells in the duodenum and jejunum	Restricts the production of gastric acid and promotes the release of insulin (incretin effect)
5.	Motilin	M-cells in the small intestine	During fasting, it increases intestinal and stomach motility (migrating motor complex)
6.	Somatostatin	D-cells in the stomach, intestine, pancreas	Decreases the formation of stomach acid and inhibits the release of numerous other hormones, including glucagon, insulin, and gastrin
7.	Vasoactive intestinal peptide (VIP)	Enteric nervous system	Increases the secretion of intestinal water and electrolytes while relaxing smooth muscle
8.	Neuropeptide Y (NPY)	Enteric neurons	Increases appetite and controls blood flow and intestinal motility
9.	Ghrelin	P/D1 cells in the stomach	Increases hunger and encourages the release of growth hormone (GH)
10.	Enteroglucagon/ GLP-1	L-cells in ileum and colon	Delays stomach emptying, suppresses glucagon, and increases insulin secretion

3.6 Development of Neuropeptide Signaling Pathways

The evolution of neuropeptide signaling systems in the animal kingdom has been easier to reconstruct thanks to the availability of genome/ transcriptome sequence data from an increasing variety of invertebrate and vertebrate species. Thus, despite several instances of lineage-specific loss, the evolutionary genesis of at least 30 neuropeptide signaling systems may be linked to the common ancestor of protostomes and deuterostomes. It can be challenging to determine neuropeptide relationships only by comparing neuropeptide sequences because neuropeptide orthologous in various phyla are frequently very diverse.

Determining the orthology of representatives of neuropeptide families in other phyla has therefore depended heavily on the finding of receptors that mediate the effects of neuropeptides. This method has allowed for the full reconstruction of the evolutionary history of neuropeptide signaling pathways (such as the neuropeptide-S, NG peptide, and CCAP neuropeptide family) by revealing unexpected linkages.[24] We still don't fully understand the diversity and development of neuropeptide signaling in the bilateria, though. For instance, leucokinin-type receptor ligands have not yet been found in deuterostomes, while pedal peptide/orcokinin-type peptide receptors and likely peptide ligands for GPR19 and GPR150 have not yet been found in any phylum. Additionally, transcriptomic/genomic-level analysis of neuropeptide signaling is either still in its early stages of exploration in many bilaterian phyla (*e.g.*, Onychophora, Priapulida, and Tardigrada) or has not yet been investigated (*e.g.*, Xenacoelomorpha, Nemertea, Brachiopoda, Phoronida, Entoprocta, Rotifera, and others).[25]

Additionally unclear are the bilaterian neuropeptide signaling pathway's pre-bilaterian origins. To the best of our knowledge, none of the four non-bilaterian phyla—Porifera, Plasmodium, Ctenophora, and Cnidaria—have neuropeptides that function as GPCR ligands. Novel insights into the evolutionary origins of the bilaterian neuropeptide signaling systems may be gained if GPCRs that mediate the effects of neuropeptides or neuropeptide-like compounds in non-bilaterian phyla are found.[30,31]

Many pieces of the "jigsaw puzzle" of neuropeptide evolution are still lacking. Nonetheless, there has been a notable advancement in the last ten or so years. It didn't seem long ago that the animal kingdom's neuropeptide variety was portrayed as an unorganized group of mostly unrelated chemicals. Additionally, profiling of neuropeptide expression in recognized populations of neurons across many species is now possible thanks to developments in transcriptomics and peptidomics.[32] Opportunities to investigate the evolution of neuropeptide function at various levels, from identified neurons to networks of synaptically (and non-synaptically) connected populations of neurons to entire animal physiology and behavior, are growing as a result of the availability of such high-resolution data.[33]

The online neuropeptide database, www.neuropeptides.nl, contains information on around 90 genes that encode both classical and candidate neuropeptides in mammalian genomes.[1] The website includes links to bioinformatics databases on genomes, transcripts, brain expression, and homologies to other species in addition to genes, precursors, and the names of processed active peptides. The primary purpose of the NeuroPedia database was to identify neuropeptides using mass spectrometry data.[34] It obtained 3401 recognized spectra from NIST spectrum libraries and their own in-house spectral datasets, and it retrieved 847 neuropeptide sequences from eight animals that belong to the phylum Chordata.[35]

Many invertebrate-specific neuropeptides, including allatostatins, pyrokinins, crustacean cardioactive peptides, pigment-dispersing factors, *etc.*, were not taken into consideration because the two databases were tailored for the author's research, and neither provided thorough annotations for every neuropeptide sequence.[36–38] The development of new technology in recent years has led to the identification of numerous new invertebrate neuropeptides, despite the fact that their characterization lags behind that of vertebrate neuropeptides.

3.7 Neuropeptide Binding Receptors

Neuropeptides may be secreted from various locations not limited to the synaptic region and act differently depending on the receptor to which it binds. For example, gonadotropin-releasing hormone (GnRH)-like peptide present in frogs secreted from pre-ganglion axons and impacted cells a few microns from the discharge location.[39] A neuropeptide may still have an effect on cells postsynaptic to the axon releasing it, even in the event of nonsynaptic release. For example, the arcuate nucleus's GABAergic Neuropeptide Y (NPY) cells form synaptic connections with neighboring arcuate nucleus neurons that produce proopiomelanocortins (POMCs); by hyperpolarizing the POMC neurons,[40] NPY can inhibit the cell postsynaptic to its parent axon even though it may not be released synaptically.

Apart from this, neuropeptides may act distant from the release site by diffusing over vast distances, which is a second possibility that has drawn a lot of attention. It has been discovered that certain neuroactive peptides and proteins have very long-distance signaling. Leptin from adipose tissue, ghrelin from the stomach, and insulin from the pancreas, for example, are released far from the brain but function as signals of energy homeostasis on receptors within the CNS. Numerous bloodborne peptides may be prevented from entering the brain by the blood–brain barrier (BBB); nevertheless, certain brain regions, such as the median eminence/arcuate nucleus, may retain a weak BBB that allows signaling from the blood to the brain.[2]

In 1987, the first receptor for substance K was cloned.[41] For a certain peptide ligand, there are typically many receptor subtypes. The snail *Helix aspersa* has an ionotropic receptor for the non-vertebrate tetrapeptide FMRFamide, and sortilin, a neurotensin receptor, is an exception to the rule.[42,43] Sortilin is a large protein (833 amino acid residues) with a long luminal domain, a single transmembrane sequence, and a short cytoplasmic tail.[43] Almost all GPCRs have seven transmembrane regions.[44] However, GPCRs, which have been the focus of medication therapy for many years, are used by neuropeptides and traditional transmitters. Actually, almost half of all medications on the market now work by binding to receptors. The human genome has roughly 550 GPCR genes, of which 20% are ligands for neuropeptides, 55% have traditional transmitters or substances like prostaglandin E2 and anandamide as ligands, and 25% have no known ligands. Similar to peptide ligands, GPCRs are extensively dispersed throughout numerous tissues in addition to the nervous system.

3.8 Functional Aspects of Neuropeptides in the Brain

Over the past 20 years, many exceptional achievements have been made possible by the rapid advancement of modern medicine and scientific understanding. For instance, new treatment approaches have decreased cardiovascular disease mortality by 40%,[45] and using biologics or small molecule medications to target particular biochemical pathways has changed some malignancies from being fatal to chronic.[46,47] Additionally, the global COVID-19 pandemic crisis is now easier to control because of advancements in mRNA vaccine technology.[48] Nonetheless, there has been a comparatively low advancement in neurological problem therapies and remedies, and late-stage clinical trials for brain disorders have a remarkably high failure rate.[49] Conventional small-molecule drugs are dreadfully ineffective at treating the underlying causes of neurological conditions, frequently having unintended neurological effects.[50–52] This is due to the complexity of neurological processes, the irreversible nature of neurological decline, and the presence of a highly selective BBB. Naturally, medications with observable effects in palliative care are unlikely to significantly ameliorate neurological conditions.

An alternative pathway for the creation of innovative treatments for neurological disorders is offered by small-peptide medications. Natural peptides or their mimetics, which are intrinsic signaling molecules in normal cellular function, mimic physiological modulation of organ or cellular processes and offer specificity that small-molecule medications created or repurposed for a particular function cannot match.[53] Additionally, during medication manufacturing, the brief, well-defined sequences of neuropeptides can be utilized or altered immediately, negating the need for additional structural optimization for functioning. The production of insulin and ACTH for the treatment of type I diabetes and endocrine diseases in humans during the 1980s and 1990s served as the basis for peptide therapy techniques.[54] More than 80 peptide medications are now on the market, and numerous others are being developed in preclinical and clinical trials.[55] The expanding peptide drug market is reflected in the fact that 65% of these medications were approved in the early 21st century.

3.8.1 Regulation of Neuronal Development by Neuropeptides

The hippocampus dynorphin has yielded one of the earliest indications of a peptide physiological role in CNS processes. Granule cells and associated mossy fibers express this opioid peptide, and it has been demonstrated that endogenous dynorphin can inhibit long-term potentiation (LTP) through kappa opioid receptors.[56] Further, understanding chemical neurotransmission, which was established when Sir Henry Dale and Otto Loewi demonstrated acetylcholine function in neural communication, upending the notion of merely electrical transmission, is essential to neurological study and treatment.[57,58] As noted by Arvid Carlsson, Paul Greengard, and Eric Kandel, further research connected brain chemicals to behavior and verified that neurotransmitters control physiology, cognition, and behavior. These systems are modulated by a number of neurological medications, including fluoxetine for depression and galantamine for Alzheimer's disease (AD).[59,60] Crucially, in addition to traditional neurotransmitters, the nervous system also uses neuropeptides, which are bigger signaling molecules with more extensive and long-lasting effects.

From birth and throughout life, the brain develops through strictly controlled neuronal creation, migration, and communication. During early embryogenesis, vasoactive intestinal peptide (VIP) promotes neuronal differentiation and survival; abnormalities of this process are associated with microcephaly and an elevated risk of autism spectrum disorder (ASD).[61] Additionally, VIP uses neurons in the supra-chiasmatic nucleus (SCN) to control circadian cycles. The experience-dependent maturation of brain circuits is also influenced by oxytocin and vasopressin. Oxytocin, in particular, has an impact on early sensory processing, and genetic abnormalities in its receptor (OXTR) are linked to autism spectrum disorders.[62] Social behaviors are further linked to synaptic plasticity, long-term potentiation (LTP) and long-term depression (LTD), when oxytocin levels are altered. Other peptides, including NPY, pituitary adenylate cyclase-activating polypeptide (PACAP), and TLQP-62, support neurite outgrowth, neuron differentiation, and hippocampus neurogenesis. Neurogenesis continues throughout adulthood in specific areas like the dentate gyrus, where NPY and its receptor Y1R control the division and proliferation of precursor cells.[63] It is interesting to note that while NPY shortage decreases neurogenesis; memory learning is unaffected, underscoring the intricate functions that peptides play in brain growth and function.[64]

3.8.2 Role of Neuropeptides in Maintaining Neurological Homeostasis

Beyond development, neuropeptides remain essential for the nervous system, assisting in the preservation of homeostasis and averting behavioral instability or excessive neuronal activity. They shape neuronal excitability by controlling neurotransmitter release, synaptic scaling, and ion channel activation. For instance, nocistatin,[65] large dynorphin,[66] and RFamide[67] control acid-sensing ion channels that are important to stroke pathology, whereas FMRFamide was the first neuropeptide to be demonstrated to directly activate Na^+ channels. Neuropeptide S affects pain responses by modifying HCN channels in the amygdala. Additionally, neuropeptides regulate synaptic transmission. For example, dynorphin retrogradely inhibits the release of neurotransmitters, and amyloid-beta, which is typically associated with AD, functions similarly to a neuropeptide in controlling synaptic activity at low physiological levels. Furthermore, neuropeptides affect synaptic plasticity.[68,69] For example, gastrin-releasing peptide (GRP), NPY, corticotropin-releasing factor (CRF),[70] and substance P aid in balancing excitatory/inhibitory networks in areas such as the neocortex[71] and amygdala, while VIP, oxytocin, and neuromedin modify associative learning *via* LTP/LTD. All things considered, neuropeptides are essential for preserving the equilibrium of brain networks and controlling behavior; and neurological diseases are associated with their dysfunction.

3.8.3 Role of Neuropeptides in Sensory Perception

Neuropeptides have a crucial role in controlling sensory perception and behavior, especially in areas like pain perception and nociception. Excitatory and inhibitory neuropeptide signaling are both involved in pain responses inside the amygdala.[72] While neuropeptide S, somatostatin, and oxytocin decrease excitatory drive and lower nociceptive tone, peptides including calcitonin gene-related peptide (CGRP),[73] substance P, and vasopressin improve pain-related plasticity.[74] Opioid peptides like β-endorphin and enkephalins also contribute to analgesia by suppressing substance P and altering neuronal activity. CGRP and substance P also play important roles in migraine and headache pathophysiology because inflammation and pain are strongly related. Additionally, CGRP links brain and cardiovascular control.[75]

These peptides have an impact on smell, taste, vision, and hearing in addition to pain, indicating a wide range of sensory modulation functions. Neuropeptides also have a role in multisensory integration, which is essential for coherent perception.[76] For example, NPF in *Drosophila* and urocortin 3 (UCN3) and corticotropin-releasing hormone receptor 2 (CRFR2) signaling in mammals facilitate the connection between sensory input and behavioral responses.[77] Neuropeptides play a crucial role as mediators in the coordination of sensory processing, pain management, and behavioral adaptability; and neurological and psychiatric disorders are exacerbated by their dysregulation.

3.8.4 Role of Neuropeptides During the Systemic Inflammatory Response

Neuropeptides are essential for controlling the activity of immune cells and connecting neuroinflammation to neurological disorders. Direct neuropeptide modulation is made possible by the expression of NPY, opioid, VIP, and PACAP receptors by microglia.[78] By activating mast cells and microglia, pro-inflammatory peptides such as substance P, CRH, and CGRP cause the production of cytokines (such as TNF-α, IL-6, IL-1β, and IL-33) that increase the permeability of the BBB and encourage neuroinflammation.[79,80] These processes are involved in the pathophysiology of neuropathic pain, Huntington's disease (HD), Parkinson's disease (PD), AD, stroke, obesity, and aging,[81–85] all of which exhibit altered neuropeptide signaling. VIP specifically controls peripheral inflammation, such as the circadian-dependent modulation of IL-22 production in the gut, demonstrating the integration of immunological and metabolic regulation by neuropeptides.

Widespread cytokine release during systemic inflammation destroys neurons, interferes with the blood–brain barrier, and encourages toxic tryptophan metabolism, all of which exacerbate neurodegeneration. It is interesting to note that neuropeptides like oxytocin, orexin, enkephalins, and β-endorphin are increased early in systemic inflammation, even before cytokine production, indicating that they function as immune signaling first responders.[86,87] While some neuropeptides cause inflammation, others, such as kyotorphin-amide, somatostatin, VIP, NPY, and α-melanocyte stimulating hormone (α-MSH), have potent anti-inflammatory properties that prevent leukocyte and microglial activation.[88,89] As a result, neuropeptides have two opposing effects on immunity: they can exacerbate neuronal damage or, if their anti-inflammatory pathways are activated, they may have a therapeutic benefits.[90]

3.9 Conclusion

Neuropeptides are the essential communicator as well as the modulator of both the peripheral and central nervous systems. From the control of pain, behavior, sensory processing, and immunological responses to early brain development and synaptic plasticity, they maintain the delicate balance between excitement and inhibition necessary for healthy brain function. Their dual function as pro-inflammatory or anti-inflammatory mediators emphasizes how crucial they are in connecting systemic physiology and disease with brain activity. Dysfunction of neuropeptide signaling is linked to a variety of neurological, psychiatric, and metabolic conditions, making them both potential therapeutic targets and disease biomarkers. They are ultimately important in both health and disease because they are adaptable molecular links between the immunological, endocrine, and neurological systems.

References

1. J. P. Burbach, Neuropeptides from concept to online database www.neuropeptides.nl, *Eur. J. Pharmacol.*, 2010, **626**(1), 27–48.
2. A. N. van den Pol, Neuropeptide transmission in brain circuits, *Neuron*, 2012, **76**(1), 98–115.
3. V. Erspamer, P. Melchiorri, M. Broccardo, G. F. Erspamer, P. Falaschi, G. Improota, L. Negri and T. Renda, The brain-gut-skin triangle: new peptides, *Peptides*, 1981, **2**(Suppl 2), 7–16.
4. V. Mutt, Chemical properties of some polypeptides affecting vascular function, in *Neuronal messengers in vascular function*, ed. A. Nobin, C. Owman and B. Arneklo-Nobin, Elsevier, Amsterdam, 1987, pp. 125–198.
5. M. G. Rosenfeld, J. J. Mermod, S. G. Amara, L. W. Swanson, P. E. Sawchenko, J. Rivier, W. W. Vale and R. M. Evans, Production of a novel neuropeptide encoded by the calcitonin gene *via* tissue-specific RNA processing, *Nature*, 1983, **304**(5922), 129–135.

6. J. L. Roberts and E. Herbert, Characterization of a common precursor to corticotropin and beta-lipotropin: identification of beta-lipotropin peptides and their arrangement relative to corticotropin in the precursor synthesized in a cell-free system, *Proc. Natl. Acad. Sci. U. S. A.*, 1977, **74**(12), 5300–5304.
7. S. Nakanishi, Y. Teranishi, M. Noda, M. Notake, Y. Watanabe, H. Kakidani, H. Jingami and S. Numa, The protein-coding sequence of the bovine ACTH-beta-LPH precursor gene is split near the signal peptide region, *Nature*, 1980, **287**(5784), 752–755.
8. K. M. Gautvik, L. de Lecea, V. T. Gautvik, P. E. Danielson, P. Tranque, A. Dopazo, F. E. Bloom and J. G. Sutcliffe, Overview of the most prevalent hypothalamus-specific mRNAs, as identified by directional tag PCR subtraction, *Proc. Natl. Acad. Sci. U. S. A.*, 1996, **93**(16), 8733–8738.
9. L. de Lecea, T. S. Kilduff, C. Peyron, X. Gao, P. E. Foye, P. E. Danielson, C. Fukuhara, E. L. Battenberg, V. T. Gautvik, F. S. Bartlett 2nd, W. N. Frankel, A. N. van den Pol, F. E. Bloom, K. M. Gautvik and J. G. Sutcliffe, The hypocretins: hypothalamus-specific peptides with neuroexcitatory activity, *Proc. Natl. Acad. Sci. U. S. A.*, 1998, **95**(1), 322–327.
10. T. Sakurai, A. Amemiya, M. Ishii, I. Matsuzaki, R. M. Chemelli, H. Tanaka, S. C. Williams, J. A. Richardson, G. P. Kozlowski, S. Wilson, J. R. Arch, R. E. Buckingham, A. C. Haynes, S. A. Carr, R. S. Annan, D. E. McNulty, W. S. Liu, J. A. Terrett, N. A. Elshourbagy, D. J. Bergsma and M. Yanagisawa, Orexins and orexin receptors: a family of hypothalamic neuropeptides and G protein-coupled receptors that regulate feeding behavior, *Cell*, 1998, **92**(4), 573–585.
11. J. C. Meunier, C. Mollereau, L. Toll, C. Suaudeau, C. Moisand, P. Alvinerie, J. L. Butour, J. C. Guillemot, P. Ferrara and B. Monsarrat, *et al.*, Isolation and structure of the endogenous agonist of opioid receptor-like ORL1 receptor, *Nature*, 1995, **377**(6549), 532–535.
12. R. K. Reinscheid, H. P. Nothacker, A. Bourson, A. Ardati, R. A. Henningsen, J. R. Bunzow, D. K. Grandy, H. Langen, F. J. Monsma Jr and O. Civelli, Orphanin FQ: a neuropeptide that activates an opioidlike G protein-coupled receptor, *Science*, 1995, **270**(5237), 792–794.
13. O. Civelli, H. P. Nothacker, Y. Saito, Z. Wang, S. H. Lin and R. K. Reinscheid, Novel neurotransmitters as natural ligands of orphan G-protein-coupled receptors, *Trends Neurosci.*, 2001, **24**(4), 230–237.
14. T. Hökfelt, T. Bartfai and F. Bloom, Neuropeptides: opportunities for drug discovery, *Lancet Neurol.*, 2003, **2**(8), 463–472.
15. Z. R. Donaldson and L. J. Young, Oxytocin, vasopressin, and the neurogenetics of sociality, *Science*, 2008, **322**(5903), 900–904.
16. A. Carlsson, A paradigm shift in brain research, *Science*, 2001, **294**(5544), 1021–1024.
17. P. Greengard, The neurobiology of slow synaptic transmission, *Science*, 2001, **294**(5544), 1024–1030.
18. T. Hökfelt, O. Johansson, A. Ljungdahl, J. M. Lundberg and M. Schultzberg, Peptidergic neurons, *Nature*, 1980, **284**(5756), 515–521.

19. J. C. Eccles, Chemical transmission and Dale's principle, *Prog. Brain Res.*, 1986, **68**, 3–13.
20. W. S. Sossin, A. Sweet-Cordero and R. H. Scheller, Dale's hypothesis revisited: different neuropeptides derived from a common prohormone are targeted to different processes, *Proc. Natl. Acad. Sci. U. S. A.*, 1990, **87**(12), 4845–4848.
21. M. Ludwig and Q. J. Pittman, Talking back: dendritic neurotransmitter release, *Trends Neurosci.*, 2003, **26**(5), 255–261.
22. L. Bao, S. X. Jin, C. Zhang, L. H. Wang, Z. Z. Xu, F. X. Zhang, L. C. Wang, F. S. Ning, H. J. Cai, J. S. Guan, H. S. Xiao, Z. Q. Xu, C. He, T. Hökfelt, Z. Zhou and X. Zhang, Activation of delta opioid receptors induces receptor insertion and neuropeptide secretion, *Neuron*, 2003, **37**(1), 121–133.
23. E. R. Kandel, The molecular biology of memory storage: a dialogue between genes and synapses, *Science*, 2001, **294**(5544), 1030–1038.
24. B. A. Eipper and R. E. Mains, Neuropeptide precursors, in *Encyclopedia of Neuroscience*, ed. G. Adelman and B. H. Smith, Elsevier, Amsterdam, 1999, p. 422.
25. N. G. Seidah and M. Chrétien, Eukaryotic protein processing: endoproteolysis of precursor proteins, *Curr. Opin. Biotechnol*, 1997, **8**(5), 602–607.
26. D. F. Steiner, The proprotein convertases, *Curr. Opin. Chem. Biol.*, 1998, **2**(1), 31–39.
27. D. de Wied, Behavioral pharmacology of neuropeptides related to melanocortins and the neurohypophyseal hormones, *Eur. J. Pharmacol.*, 1999, **375**(1–3), 1–11.
28. B. P. Roques and F. Noble, Dual inhibitors of enkephalin-degrading enzymes (neutral endopeptidase 24.11 and aminopeptidase N) as potential new medications in the management of pain and opioid addiction, *NIDA Res. Monogr.*, 1995, **147**, 104–145.
29. H. A. Al-hussaniy and H. A. AL-Biati, The Role of Leptin Hormone, Neuropeptide Y, Ghrelin and Leptin/Ghrelin ratio in Obesogenesis, *Med. Pharm. J.*, 2022, **1**(2), 12–23.
30. D. C. Semmens, I. Beets, M. L. Rowe, L. M. Blowes, P. Oliveri and M. R. Elphick, Discovery of sea urchin NGFFFamide receptor unites a bilaterian neuropeptide family, *Open Biol.*, 2015, **5**(4), 150030.
31. A. E. Christie, D. H. Nolan, Z. A. Garcia, M. D. McCoole, S. M. Harmon, B. Congdon-Jones, P. Ohno, N. Hartline, C. B. Congdon, K. N. Baer and P. H. Lenz, Bioinformatic prediction of arthropod/nematode-like peptides in non-arthropod, non-nematode members of the Ecdysozoa, *Gen. Comp. Endocrinol.*, 2011, **170**(3), 480–486.
32. K. DeLaney, A. R. Buchberger, L. Atkinson, S. Gründer, A. Mousley and L. Li, New techniques, applications and perspectives in neuropeptide research, *J. Exp. Biol.*, 2018, **221**(Pt 3), jeb151167.
33. M. R. Elphick, O. Mirabeau and D. Larhammar, Correction: Evolution of neuropeptide signalling systems (doi:10.1242/jeb.151092), *J. Exp. Biol.*, 2018, **221**(Pt 19), jeb193342.

34. Y. Kim, S. Bark, V. Hook and N. Bandeira, NeuroPedia: neuropeptide database and spectral library, *Bioinformatics*, 2011, **27**(19), 2772–2773.
35. Y. Wang, M. Wang, S. Yin, R. Jang, J. Wang, Z. Xue and T. Xu, NeuroPep: a comprehensive resource of neuropeptides, *Database*, 2015, **2015**, bav038.
36. D. R. Nässel, Neuropeptides in the nervous system of Drosophila and other insects: multiple roles as neuromodulators and neurohormones, *Prog. Neurobiol.*, 2002, **68**(1), 1–84.
37. M. O'Shea and M. Schaffer, Neuropeptide function: the invertebrate contribution, *Annu. Rev. Neurosci.*, 1985, **8**, 171–198.
38. L. Y. Jan and Y. N. Jan, Peptidergic transmission in sympathetic ganglia of the frog, *J. Physiol.*, 1982, **327**, 219–246.
39. M. A. Cowley, J. L. Smart, M. Rubinstein, M. G. Cerdán, S. Diano, T. L. Horvath, R. D. Cone and M. J. Low, Leptin activates anorexigenic POMC neurons through a neural network in the arcuate nucleus, *Nature*, 2001, **411**(6836), 480–484.
40. Y. Masu, K. Nakayama, H. Tamaki, Y. Harada, M. Kuno and S. Nakanishi, cDNA cloning of bovine substance-K receptor through oocyte expression system, *Nature*, 1987, **329**(6142), 836–838.
41. G. A. Cottrell, The first peptide-gated ion channel, *J. Exp. Biol.*, 1997, **200**(Pt 18), 2377–2386.
42. E. Lingueglia, G. Champigny, M. Lazdunski and P. Barbry, Cloning of the amiloride-sensitive FMRFamide peptide-gated sodium channel, *Nature*, 1995, **378**(6558), 730–733.
43. J. P. Vincent, J. Mazella and P. Kitabgi, Neurotensin and neurotensin receptors, *Trends Pharmacol. Sci.*, 1999, **20**(7), 302–309.
44. J. Drews, Drug discovery: a historical perspective, *Science*, 2000, **287**(5460), 1960–1964.
45. G. A. Mensah, G. S. Wei, P. D. Sorlie, L. J. Fine, Y. Rosenberg, P. G. Kaufmann, M. E. Mussolino, L. L. Hsu, E. Addou, M. M. Engelgau and D. Gordon, Decline in Cardiovascular Mortality: Possible Causes and Implications, *Circ. Res.*, 2017, **120**(2), 366–380.
46. S. Lv, Y. Wang, T. Sun, D. Wan, L. Sheng, W. Li, H. Zhu, Y. Li, J. Lu Overall Survival Benefit from Trastuzumab-Based Treatment in HER2-Positive Metastatic Breast Cancer: A Retrospective Analysis. *Oncol. Res. Treat.* . Erratum in: *Oncol. Res. Treat.*, 2018, **41**(10), 655.
47. S. L. Maude, N. Frey, P. A. Shaw, R. Aplenc, D. M. Barrett, N. J. Bunin, A. Chew, V. E. Gonzalez, Z. Zheng, S. F. Lacey, Y. D. Mahnke, J. J. Melenhorst, S. R. Rheingold, A. Shen, D. T. Teachey, B. L. Levine, C. H. June, D. L. Porter, S. A. Grupp Chimeric antigen receptor T cells for sustained remissions in leukemia. *N. Engl. J. Med.* . Erratum in: *N. Engl. J. Med.*, 2016, 374(10), 998.
48. J. Abbasi, COVID-19 mRNA Vaccines Blunt Breakthrough Infection Severity, *JAMA*, 2021, **326**(6), 473.
49. A. Bespalov, T. Steckler, B. Altevogt, E. Koustova, P. Skolnick, D. Deaver, M. J. Millan, J. F. Bastlund, D. Doller, J. Witkin, P. Moser, P. O'Donnell,

U. Ebert, M. A. Geyer, E. Prinssen, T. Ballard and M. Macleod, Failed trials for central nervous system disorders do not necessarily invalidate preclinical models and drug targets, *Nat. Rev. Drug Discovery*, 2016, **15**(7), 516.

50. A. M. Clarfield, The decreasing prevalence of reversible dementias: an updated meta-analysis, *Arch. Intern. Med.*, 2003, **163**(18), 2219–2229.
51. R. Dutta and B. D. Trapp, Mechanisms of neuronal dysfunction and degeneration in multiple sclerosis, *Prog. Neurobiol.*, 2011, **93**(1), 1–12.
52. W. M. Pardridge, Blood-Brain Barrier and Delivery of Protein and Gene Therapeutics to Brain, *Front. Aging Neurosci.*, 2020, **11**, 373.
53. A. Groß, C. Hashimoto, H. Sticht and J. Eichler, Synthetic Peptides as Protein Mimics, *Front. Bioeng. Biotechnol.*, 2016, **3**, 211.
54. J. L. Lau and M. K. Dunn, Therapeutic peptides: Historical perspectives, current development trends, and future directions, *Bioorg. Med. Chem.*, 2018, **26**(10), 2700–2707.
55. M. Muttenthaler, G. F. King, D. J. Adams and P. F. Alewood, Trends in peptide drug discovery, *Nat. Rev. Drug Discovery*, 2021, **20**(4), 309–325.
56. T. Hökfelt, C. Broberger, Z. Q. Xu, V. Sergeyev, R. Ubink and M. Diez, Neuropeptides–an overview, *Neuropharmacology*, 2000, **39**(8), 1337–1356.
57. E. S. Valenstein, The discovery of chemical neurotransmitters, *Brain Cogn.*, 2002, **49**(1), 73–95.
58. D. Krehbiel, B. Bartel, M. Dirks and W. Wiens, Behavior and brain neurotransmitters: correlations in different strains of mice, *Behav. Neural Biol.*, 1986, **46**(1), 30–45.
59. J. Coyle and P. Kershaw, Galantamine, a cholinesterase inhibitor that allosterically modulates nicotinic receptors: effects on the course of Alzheimer's disease, *Biol. Psychiatry*, 2001, **49**(3), 289–299.
60. W. F. Byerley, F. W. Reimherr, D. R. Wood and B. I. Grosser, Fluoxetine, a selective serotonin uptake inhibitor, for the treatment of outpatients with major depression, *J. Clin. Psychopharmacol.*, 1988, **8**(2), 112–115.
61. J. M. Hill, Vasoactive intestinal peptide in neurodevelopmental disorders: therapeutic potential, *Curr. Pharm. Des.*, 2007, **13**(11), 1079–1089.
62. T. V. Miller and H. K. Caldwell, Oxytocin during Development: Possible Organizational Effects on Behavior, *Front. Endocrinol.*, 2015, **6**, 76.
63. O. W. Howell, K. Doyle, J. H. Goodman, H. E. Scharfman, H. Herzog, A. Pringle, A. G. Beck-Sickinger and W. P. Gray, Neuropeptide Y stimulates neuronal precursor proliferation in the post-natal and adult dentate gyrus, *J. Neurochem.*, 2005, **93**(3), 560–570.
64. M. C. Geloso, V. Corvino, V. Di Maria, E. Marchese and F. Michetti, Cellular targets for neuropeptide Y-mediated control of adult neurogenesis, *Front. Cell. Neurosci.*, 2015, **9**, 85.
65. D. I. Osmakov, S. G. Koshelev, I. A. Ivanov, Y. A. Andreev and S. A. Kozlov, Endogenous Neuropeptide Nocistatin Is a Direct Agonist of Acid-Sensing Ion Channels (ASIC1, ASIC2 and ASIC3), *Biomolecules*, 2019, **9**(9), 401.
66. L. Leisle, M. Margreiter, A. Ortega-Ramírez, E. Cleuvers, M. Bachmann, G. Rossetti and S. Gründer, Dynorphin Neuropeptides Decrease

Apparent Proton Affinity of ASIC1a by Occluding the Acidic Pocket, *J. Med. Chem.*, 2021, **64**(18), 13299–13311.

67. M. Reiners, M. A. Margreiter, A. Oslender-Bujotzek, G. Rossetti, S. Gründer and A. Schmidt, The Conorfamide RPRFa Stabilizes the Open Conformation of Acid-Sensing Ion Channel 3 *via* the Nonproton Ligand-Sensing Domain, *Mol. Pharmacol.*, 2018, **94**(4), 1114–1124.
68. J. Busciglio, D. H. Gabuzda, P. Matsudaira and B. A. Yankner, Generation of beta-amyloid in the secretory pathway in neuronal and non-neuronal cells, *Proc. Natl. Acad. Sci. U. S. A.*, 1993, **90**(5), 2092–2096.
69. T. Toneff, L. Funkelstein, C. Mosier, A. Abagyan, M. Ziegler and V. Hook, Beta-amyloid peptides undergo regulated co-secretion with neuropeptide and catecholamine neurotransmitters, *Peptides*, 2013, **46**, 126–135.
70. T. L. Kash and D. G. Winder, Neuropeptide Y and corticotropin-releasing factor bi-directionally modulate inhibitory synaptic transmission in the bed nucleus of the stria terminalis, *Neuropharmacology*, 2006, **51**(5), 1013–1022.
71. X. F. Wang, T. T. Ge, J. Fan, W. Yang and R. J. Cui, The role of substance P in epilepsy and seizure disorders, *Oncotarget*, 2017, **8**(44), 78225–78233 . Erratum in: Oncotarget, 2019, **10**(37), 3576.
72. V. Neugebauer, M. Mazzitelli, B. Cragg, G. Ji, E. Navratilova and F. Porreca, Amygdala, neuropeptides, and chronic pain-related affective behaviors, *Neuropharmacology*, 2020, **170**, 108052.
73. K. Shinohara, A. M. Watabe, M. Nagase, Y. Okutsu, Y. Takahashi, H. Kurihara and F. Kato, Essential role of endogenous calcitonin gene-related peptide in pain-associated plasticity in the central amygdala, *Eur. J. Neurosci.*, 2017, **46**(6), 2149–2160.
74. B. Cragg, G. Ji and V. Neugebauer, Differential contributions of vasopressin V1A and oxytocin receptors in the amygdala to pain-related behaviors in rats, *Mol. Pain*, 2016, **12**, 1744806916676491.
75. Z. Kee, X. Kodji and S. D. Brain, The Role of Calcitonin Gene Related Peptide (CGRP) in Neurogenic Vasodilation and Its Cardioprotective Effects, *Front. Physiol.*, 2018, **9**, 1249.
76. R. A. Stevenson, S. Park, C. Cochran, L. G. McIntosh, J. P. Noel, M. D. Barense, S. Ferber and M. T. Wallace, The associations between multisensory temporal processing and symptoms of schizophrenia, *Schizophr. Res.*, 2017, **179**, 97–103.
77. S. Pagella, J. M. Deussing and C. Kopp-Scheinpflug, Expression Patterns of the Neuropeptide Urocortin 3 and Its Receptor CRFR2 in the Mouse Central Auditory System, *Front. Neural Circuits*, 2021, **15**, 747472.
78. L. Carniglia, D. Ramírez, D. Durand, J. Saba, J. Turati, C. Caruso, T. N. Scimonelli and M. Lasaga, Neuropeptides and Microglial Activation in Inflammation, Pain, and Neurodegenerative Diseases, *Mediators Inflammation*, 2017, **2017**, 5048616.
79. R. Cocchiara, N. Lampiasi, G. Albeggiani, A. Bongiovanni, A. Azzolina and D. Geraci, Mast cell production of TNF-alpha induced by substance P

evidence for a modulatory role of substance P-antagonists, *J. Neuroimmunol.*, 1999, **101**(2), 128–136.
80. A. C. Voirin, N. Perek and F. Roche, Inflammatory stress induced by a combination of cytokines (IL-6, IL-17, TNF-α) leads to a loss of integrity on bEnd.3 endothelial cells *in vitro* BBB model, *Brain Res.*, 2020, **1730**, 146647.
81. P. A. C. Valadão, K. B. S. Santos, T. H. Ferreira E Vieira, T. Macedo E Cordeiro, A. L. Teixeira, C. Guatimosim and A. S. de Miranda, Inflammation in Huntington's disease: A few new twists on an old tale, *J. Neuroimmunol.*, 2020, **348**, 577380.
82. Z. Fan, Y. T. Pan, Z. Y. Zhang, H. Yang, S. Y. Yu, Y. Zheng, J. H. Ma and X. M. Wang, Systemic activation of NLRP3 inflammasome and plasma α-synuclein levels are correlated with motor severity and progression in Parkinson's disease, *J. Neuroinflammation*, 2020, **17**(1), 11.
83. K. A. Walker, B. N. Ficek and R. Westbrook, Understanding the Role of Systemic Inflammation in Alzheimer's Disease, *ACS Chem. Neurosci.*, 2019, **10**(8), 3340–3342.
84. G. Escobedo, E. López-Ortiz and I. Torres-Castro, Gut microbiota as a key player in triggering obesity, systemic inflammation and insulin resistance, *Rev. Invest. Clin.*, 2014, **66**(5), 450–459.
85. T. Lin, G. A. Liu, E. Perez, R. D. Rainer, M. Febo, Y. Cruz-Almeida and N. C. Ebner, Systemic Inflammation Mediates Age-Related Cognitive Deficits, *Front. Aging Neurosci.*, 2018, **10**, 236.
86. M. Y. Hamasaki, H. V. Barbeiro, D. F. Barbeiro, D. M. Cunha, M. K. Koike, M. C. Machado and F. Pinheiro da Silva, Neuropeptides in the brain defense against distant organ damage, *J. Neuroimmunol.*, 2016, **290**, 33–35.
87. D. E. Van Epps and S. L. Kutvirt, Modulation of human neutrophil adherence by beta-endorphin and met-enkephalin, *J. Neuroimmunol.*, 1987, **15**(3), 219–228.
88. J. Perazzo, C. Lima, M. Heras, E. Bardají, M. Lopes-Ferreira and M. Castanho, Neuropeptide Kyotorphin Impacts on Lipopolysaccharide-Induced Glucocorticoid-Mediated Inflammatory Response. A Molecular Link to Nociception, Neuroprotection, and Anti-Inflammatory Action, *ACS Chem. Neurosci.*, 2017, **8**(8), 1663–1667.
89. G. Colombo, A. Sordi, C. Lonati, A. Carlin, F. Turcatti, P. Leonardi, S. Gatti and A. Catania, Treatment with alpha-melanocyte stimulating hormone preserves calcium regulatory proteins in rat heart allografts, *Brain, Behav., Immun.*, 2008, **22**(6), 817–823.
90. E. Reinke and Z. Fabry, Breaking or making immunological privilege in the central nervous system: the regulation of immunity by neuropeptides, *Immunol. Lett.*, 2006, **104**(1–2), 102–109.

CHAPTER 4

Neurochemical Pathways in Energy Metabolism and Mitochondrial Function

MD ABUBAKAR (0000-0001-8467-0498),*[a,†] RITU DAHIYA,[b,†] ABHISHEK AMBAWATIYA,[a,†] KRUSHNA CHANDRA MAHARANA,[a,‡] SHAHNAZ ALOM,[c,‡] ARUNA RAWAT,[b,‡] AMITA RAI,[d,‡] DAMINI YADAV,[e,‡] KRISHNENDU ADHIKARY[f,‡] AND NITESH KUMAR (0000-0002-4929-3954)*[a]

[a] Department of Pharmacology and Toxicology, National Institute of Pharmaceutical Education and Research, Hajipur, Vaishali-844102, Bihar, India; [b] School of Pharmacy, Lingaya's Vidyapeeth (Deemed to be university), Faridabad, Haryana-121002, India; [c] Girijananda Chowdhury Institute of Pharmaceutical Science, Girijananda Chowdhury, University-Tezpur campus, Assam, 784501, India; [d] Department of Pharmaceutical Analysis, National Institute of Pharmaceutical Education and Research, Hajipur, Vaishali – 844102, Bihar, India; [e] Department of Pharmaceutics, Amity Institute of Pharmacy, Amity University, Gurugram, Haryana 122 413, India; [f] Department of Medical Laboratory Technology, Paramedical College Durgapur, West Bengal 713212, India

*Emails: mdabuabakr731@gmail.com; niteshkumar43@gmail.com; nitesh.kumar04@niperhajipur.ac.in

† These authors contributed equally to this work.
‡ Second position in contribution.

Chemical Biology No. 26
Chemical Signatures of the Brain: Bridging Biochemistry, Neuroanatomy, and Brain Disorders
Edited by Sachchida Nand Rai and Abhishek Kumar Singh

Published by the Royal Society of Chemistry, www.rsc.org

4.1 Introduction

4.1.1 Brain Energy Demands and Mitochondrial Centrality

The human brain is a metabolically privileged organ: although it constitutes only about 2% of the overall body mass, it accounts for nearly one-fifth of the body's basal energy expenditure. This extraordinary demand reflects the brain's continual engagement in electrochemical signaling, maintenance of ion gradients across expansive neuronal membranes, turnover of neurotransmitters, and the dynamic remodeling of synapses involved in learning and memory.[1] Unlike many peripheral tissues that can transiently rely on anaerobic pathways, neurons have a narrow metabolic safety margin—glycolysis alone cannot sustain their high-frequency firing or synaptic integration. Consequently, neurons depend overwhelmingly on mitochondrial oxidative phosphorylation (OXPHOS) as their principal ATP source.[2]

Beyond ATP synthesis, neuronal mitochondria act as multifunctional metabolic regulators. They fine-tune redox homeostasis, modulate intracellular Ca^{2+} waves that shape synaptic responsiveness, and relay metabolic cues to nuclear transcription machinery.[3] Their spatial organization within neurons is equally strategic: mitochondria accumulate at synaptic boutons to support vesicle cycling, localize near nodes of Ranvier to sustain action potential propagation, and populate dendritic spines where plasticity-associated signaling is initiated.[4] This anatomically precise distribution underscores their role as integrators of bioenergetic supply and neuronal performance. Disruptions in mitochondrial positioning, turnover, or metabolic responsiveness can therefore destabilize neural circuits—an early event observed across diverse neurological disorders. In this sense, mitochondrial health is central not only to sustaining moment-to-moment neuronal activity but also to preserving long-term cognitive capacity.[5]

4.2 Mitochondrial Basics in Neurons

4.2.1 Structure, Dynamics, and ATP Production

Neuronal mitochondria exhibit structural refinements that align with the energetic demands of the nervous system. Their inner membrane is densely folded into cristae, which increases the surface area available for the electron transport chain (ETC) complexes and ATP synthase, allowing for rapid ATP generation even during bursts of synaptic activity.[6] Beyond their morphology, neuronal mitochondria are highly dynamic entities that constantly undergo fusion and fission. Fusion events, regulated by mitofusins (MFN1/2) and OPA1, help sustain mitochondrial genome stability and maintain metabolic efficiency by blending the contents of partially damaged organelles. Conversely, fission—mediated primarily by DRP1—segregates dysfunctional regions of the mitochondrial network, allocating them for removal *via* mitophagy.[6] These dynamic cycles function as a quality-control system, ensuring that neurons, which are long-lived and rarely replaced, maintain a metabolically competent mitochondrial population. ATP production in these organelles is tightly coupled to the demands of neuronal signaling. During intense synaptic transmission, mitochondrial respiration accelerates to match the rapid consumption of ATP by ion pumps, vesicular transport systems, and cytoskeletal remodeling. The ability of mitochondria to rapidly adjust OXPHOS underscores their centrality in maintaining neuronal vigor and circuit reliability.[7]

4.2.2 Neuronal Mitochondrial Distribution and Its Trafficking

A hallmark of neuronal physiology is the exquisite spatial distribution of mitochondria along axons, dendrites, and synaptic terminals. Because energy requirements differ across these domains, mitochondria must be transported with precision. Axonal transport relies on microtubule-associated motor proteins: kinesin propels mitochondria toward distal axonal regions, while dynein mediates their retrograde movement toward the soma.[8] This bidirectional movement is further modulated by Ca^{2+}-sensitive adaptors such as Miro and Milton, which act as "traffic regulators." When synaptic Ca^{2+} rises—signaling increased local activity—these adaptors temporarily halt mitochondrial movement, anchoring them where ATP and Ca^{2+} buffering is urgently needed. This trafficking system ensures that high-energy microdomains—active synapses, nodes of Ranvier, and dendritic spines—receive prompt metabolic support. Disruptions in mitochondrial motility, docking, or turnover compromise the neuron's ability to sustain prolonged firing or maintain synaptic plasticity.[9] Such defects represent an early pathological signature in multiple neurological disorders, including Alzheimer's disease (AD), Parkinson's disease (PD), and inherited neuropathies. Thus, mitochondrial trafficking is not simply a logistical process but a critical determinant of neuronal survival and functional adaptability.

4.3 Primary Energy Substrates

4.3.1 Glucose and Glycolysis

Glucose is the principal fuel that sustains most neuronal activity. The brain secures a continuous supply through specialized GLUT transporters—GLUT1 at the blood–brain barrier and GLUT3 on neuronal membranes—which maintain efficient glucose uptake even when peripheral levels fluctuate.[10] While neurons do utilize glycolysis, its role is largely supportive rather than primary. Glycolysis generates pyruvate and small amounts of ATP, but the majority of pyruvate is rapidly shuttled into mitochondria for oxidative metabolism, a process that yields far more ATP and minimizes intracellular lactate buildup.[11]

This dependence on oxidative metabolism provides neurons with a significant energetic advantage, enabling them to produce the energy necessary for their functions efficiently. However, this process also renders them susceptible to any interruptions. Suppose there is a disruption in mitochondrial function or a decrease in glucose delivery. In that case, the impact on synaptic signaling can be profound and almost instantaneous, resulting in compromised communication between neurons.[12] Therefore, the steady operation of glycolysis, pyruvate oxidation, and OXPHOS represents a tightly coordinated metabolic axis essential for maintaining neuronal performance.

4.3.2 Astrocyte–Neuron Lactate Shuttle (ANLS)

Astrocytes provide an indispensable metabolic partnership that enhances neuronal energy security. Equipped with a robust glycolytic capacity, astrocytes convert glucose into lactate, which is then exported through monocarboxylate transporters (MCT1/MCT4). Neurons, expressing the high-affinity MCT2, readily import lactate and oxidize it within mitochondria to generate ATP.[13] This cooperative system, known as the ANLS, offers several physiological advantages. It safeguards neurons from the metabolic burden of high glycolytic rates, conserves glucose for rapid signaling needs, and stabilizes brain energy metabolism during hypoglycemia or heightened activity.[14] Lactate itself is emerging as more than just a fuel; it modulates NMDA receptor signaling, supports long-term memory formation, and protects neurons from excitotoxic stress. Thus, the ANLS represents a vital neuroenergetic strategy that couples glial metabolism with neuronal function.[15]

4.3.3 Ketone Bodies and Alternative Fuels

Under specific conditions, such as fasting, prolonged periods of physical activity, following a ketogenic diet, or during the early stages of development, the brain adapts to a different source of fuel by shifting its metabolism to ketone bodies. The primary players in this metabolic transition are β-hydroxybutyrate and acetoacetate. These molecules possess the remarkable ability to cross the blood–brain barrier (BBB) with ease, enabling them to be efficiently utilized by the brain. Once inside, they are readily oxidized by the mitochondria of neurons, resulting in the production of ATP, which serves as the energy currency of the cell. This mode of energy generation not only yields a high amount of ATP but also minimizes the production of reactive oxygen species (ROS), making it a safer and more efficient energy source for the brain under these specific conditions.[16]

Ketone metabolism enhances mitochondrial efficiency, improves redox balance, and may even activate signaling pathways associated with stress resilience and synaptic plasticity. Beyond ketones, neurons can utilize acetate, specific amino acids, and short-chain fatty acids under specialized physiological conditions, each of which contributes to metabolic flexibility.

This adaptability ensures that the brain maintains energetic continuity even when glucose availability is compromised.[17] Grasping the intricacies of these alternative substrates is essential for developing effective therapeutic strategies that address neurological disorders, metabolic dysfunctions, and cognitive decline associated with aging. By examining the distinct characteristics and roles of these substrates, researchers can develop targeted interventions that may enhance outcomes for individuals affected by these complex health issues.

4.4 Key Regulatory Pathways

4.4.1 AMPK (AMP-activated Protein Kinase) and Energy Sensing

AMPK functions as a crucial metabolic sentinel within neurons, vigilantly tracking variations in cellular energy status. When the levels of ATP—a primary energy currency—begin to decline, and the ratio of AMP to ADP increases, this typically occurs during periods of heightened synaptic activity, OS, or a scarcity of nutrients. In these scenarios, AMPK is activated by the action of upstream kinases, such as LKB1 and CaMKKβ. Once activated, AMPK initiates a carefully orchestrated transition from energy-consuming processes to those that conserve it, effectively helping the neuron adapt to energy deficits and maintain its functions.[18] In neurons, AMPK enhances glucose uptake, promotes fatty acid oxidation, and suppresses energy-intensive biosynthetic pathways (Table 4.1). Notably, AMPK's role extends beyond metabolism; it is also involved in remodeling dendritic spines, regulating the cycling of neurotransmitter vesicles, and adjusting synaptic activity during stress. While AMPK's activation can be beneficial in acute situations, chronic or excessive activation may disrupt synaptic plasticity and contribute to neurodegeneration. Thus, AMPK serves as a finely tuned regulator, balancing metabolic survival with functional flexibility.[19]

Table 4.1 Key neurochemical regulators of mitochondrial function in neurons.

Pathway/regulator	Primary role	Impact on neuronal energy metabolism	Ref.
AMPK	Energy sensor activated by low ATP	Enhances glucose uptake, promotes autophagy, stabilizes energy balance	25
SIRT (1 and 3)	NAD^+ (nicotinamide adenine dinucleotide)-dependent deacetylases	Improve mitochondrial efficiency, regulate oxidative stress (OS), support longevity pathways	26
PGC-1α	Master regulator of mitochondrial biogenesis	Increases mitochondrial number, enhances ATP production and antioxidant capacity	27
mTOR	Nutrient-sensitive anabolic regulator	Promotes protein synthesis; excessive activation suppresses autophagy and mitochondrial quality control	28

4.4.2 Sirtuins and NAD$^+$ Signaling

Sirtuins are a fascinating family of enzymes known as NAD$^+$-dependent deacetylases, playing a crucial role in cellular health and longevity (Table 4.1). These proteins serve as intricate molecular bridges that connect the energy status of neurons to the long-term resilience of cells. Their enzymatic activity is intricately linked to the availability of intracellular NAD$^+$, meaning that fluctuations in this critical metabolite directly impact the effectiveness of sirtuins. As a result, sirtuins find themselves at a vital intersection, influencing essential processes such as metabolism, the balance of oxidative and reductive states within cells, and the overall stability of the genome. This interplay emphasizes their importance in maintaining cellular integrity and responding to various stressors.[20]

SIRT1, mainly found in the nucleus, regulates transcriptional programs that promote mitochondrial biogenesis, stress tolerance, and metabolic adaptation. SIRT3, a mitochondrial sirtuin, directly deacetylates and activates essential metabolic enzymes, improving respiratory efficiency and reducing the accumulation of ROS. Additionally, other sirtuins—such as SIRT (2, 6, and 7)—play essential roles in maintaining the cytoskeleton, repairing DNA, and ensuring chromatin stability.[21] Declines in NAD$^+$ levels associated with aging weaken the function of sirtuins, reducing the ability of neurons to cope with OS and metabolic challenges. Restoring NAD$^+$ levels or activating sirtuins is emerging as a promising therapeutic approach for cognitive decline related to aging, mitochondrial dysfunction, and neurodegenerative diseases (NDDs).[22]

4.4.3 Mitochondrial Biogenesis and PGC-1α

Peroxisome proliferator-activated receptor gamma coactivator-1α (PGC-1α) plays a crucial role in maintaining and regenerating mitochondrial networks in neurons (Table 4.1). This transcriptional coactivator interacts with essential proteins such as NRF (1 and 2) and TFAM. Through these interactions, PGC-1α promotes the replication of mitochondrial DNA, which is vital for the production of new mitochondria (Table 4.1). Additionally, it enhances the expression of key components involved in the complex process of the ETC, ultimately supporting energy metabolism and the overall health of neuronal cells.[23]

Beyond fostering mitochondrial biogenesis, PGC-1α enhances antioxidant capacity by regulating the expression of enzymes, including superoxide dismutase 2 (SOD2) and catalase (CAT). In neuronal populations that rely heavily on sustained energy, such as hippocampal and DAergic (c) neurons, PGC-1α supports synaptic integrity, learning-associated plasticity, and resilience to metabolic stress. Deficits in PGC-1α expression or signaling contribute to impaired energy metabolism and are associated with various disorders. Enhancing PGC-1α activity is being explored as a strategy to restore bioenergetic stability in vulnerable neuronal circuits.[24]

4.4.4 mTOR and Anabolic Balance

The mammalian target of rapamycin (mTOR) integrates signals from nutrients, growth factors, and intracellular energy stores to regulate anabolic growth and protein synthesis (Table 4.1). In neurons, mTOR plays a fundamental role in dendritic development, synaptic maturation, and long-term potentiation—processes that require high levels of biosynthetic activity (Table 4.1). However, mTOR also intersects with mitochondrial maintenance pathways. When excessively activated, mTOR suppresses both mitophagy and autophagy, which causes the accumulation of dysfunctional mitochondria and impaired metabolic flexibility. This imbalance is associated with cognitive decline and several neurodegenerative conditions. Optimal neuronal function therefore requires a finely balanced mTOR tone—not too low to hamper synaptic development, and not so excessively high as to suppress mitochondrial quality control. Pharmacological and nutritional approaches targeting mTOR signaling are being extensively investigated for disorders characterized by metabolic dysregulation, including AD, epilepsy, and metabolic encephalopathies.[29]

4.5 Calcium (Ca^{2+}) Activity and Mitochondrial Function

Mitochondrial calcium ions (mCa^{2+}) serve as a rapid and versatile integrator, tightly coupling neuronal activity to ATP production in the mitochondria. Neurons possess a unique metabolic architecture that demands rapid ATP production, which is facilitated through Ca^{2+}-sensitive OXPHOS. During neuronal stimulation, Ca^{2+} is released from the endoplasmic reticulum (ER) primarily through IP3 receptors, generating cytosolic Ca^{2+} at mitochondria-associated membranes (MAMs). These ER-mitochondrial association sites serve as an effective hotspot for Ca^{2+} entry into the mitochondrial matrix through the mCa^{2+} uniporter (MCU) complex, which is regulated by the gatekeeping proteins MICU (1 and 2), as well as a strong cytosolic negative potential.[30,31] The excessive gain of Ca^{2+} in the mitochondrial matrix is controlled by the efflux mechanism, notably the H^+/Ca^{2+} exchanger (HCX) and Na^+/Ca^{2+} exchanger (NCX). This mechanism helps in achieving the optimal energy required for neuronal activity.[32]

4.5.1 Ca^{2+} Regulation of the TCA Cycle and Respiration

Before the cytosolic entry into mitochondria, Ca^{2+} stimulates the key enzymes of the tricarboxylic acid (TCA) cycle. It further enhances the enzymatic activity, thereby accelerating the production of reduced electron carriers like nicotinamide adenine dinucleotide (NADH) and flavin adenine dinucleotide ($FADH_2$), which are indispensable for the ETC and OXPHOS. The mCa^{2+} influx activates pyruvate dehydrogenase (PDH) through pyruvate dehydrogenase phosphatase (PDP)-mediated dephosphorylation, thereby increasing the

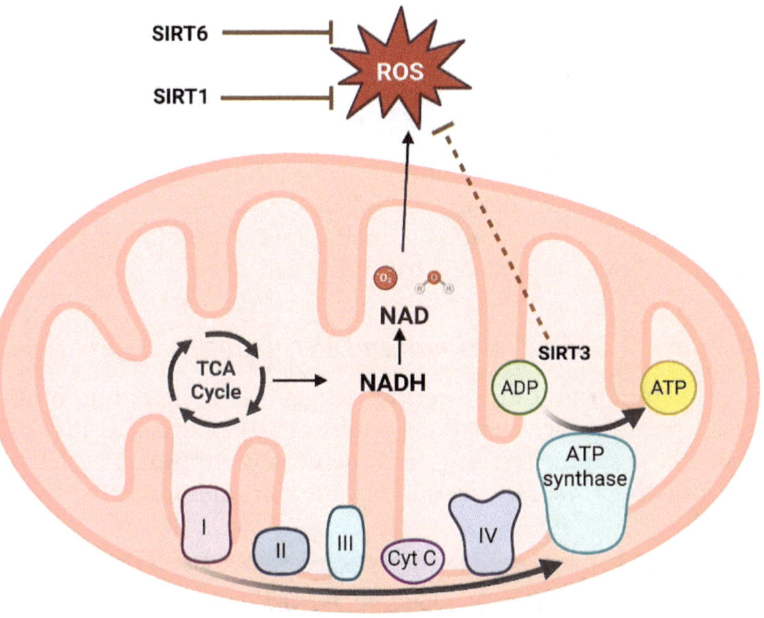

Figure 4.1 Neurochemical integration of energy metabolism and mitochondrial function.

supply of acetyl-CoA. Subsequently, the enzyme isocitrate dehydrogenase (ICDH) catalyses the conversion of isocitrate to α-ketoglutarate dehydrogenase, which further escalates production of NADH and $FADH_2$ and subsequent flux through the ETC. The production of NADH is also accompanied by α-ketoglutarate dehydrogenase, which stimulates α-ketoglutarate conversion to succinyl CoA (Figure 4.1).[33] Neuronal mitochondria are exclusively positioned to respond to these Ca^{2+} signals. The ER-mitochondria tethering proteins mitofusin-2 (MFN2) and GRP75 help sustain optimal spacing, which supports Ca^{2+} transport to the matrix, enabling ATP production to rise proportionately with neuronal load. Disruption of ER-mitochondria Ca^{2+} coupling resulted in a reduction in mitochondrial respiration and ATP production, underscoring the reliance of OXPHOS on tightly controlled Ca^{2+} transfer.[30]

The TCA cycle integrates substrates from amino acid catabolism, glycolysis and fatty acid oxidation to maintain production of ATP in high-energy states. Ca^{2+}, therefore, plays a vital role both as a metabolic regulator and as a synchronizer of neuronal activity with mitochondrial bioenergetics.[31]

4.5.2 Risks of Ca^{2+} Overload

While physiological mitochondrial Ca^{2+} entry supports efficient ATP production, excessive Ca^{2+} accumulation leads to metabolic dysfunction and structural damage. Ca^{2+} overload can occur during prolonged neuronal depolarisation, ER stress or impaired Ca^{2+} extrusion pathways.[30]

4.5.2.1 *Moderate Mitochondrial Ca^{2+} Overload*

At intermediate levels of Ca^{2+} loading, ATP synthesis becomes progressively inhibited. Carvalho *et al.* demonstrated that moderate Ca^{2+} overload suppresses mitochondrial respiration, causing reversible alterations to cristae morphology without immediately triggering cell death.[30]

4.5.2.2 *Severe Ca^{2+} Overload and Permeability Transition*

Critical mitochondrial Ca^{2+} overload triggers the opening of the mitochondrial permeability transition pore (mPTP), leading to loss of membrane potential, osmotic swelling, and membrane rupture. This leads to the release of pro-apoptotic factors, such as cytochrome, initiating irreversible cell death.[30]

4.5.2.3 *Ca^{2+} Overload in Disease Pathology*

Disturbances in Ca^{2+} homeostasis play a central role in the pathology of AD and PD, driven by ER stress, impaired mitophagy, mitochondrial fragmentation and chronic OS, which altogether contribute to persistent Ca^{2+} overload and progressive mitochondrial failure.[30]

4.5.2.4 *Protective Mechanisms*

Cells employ countermeasures to avoid Ca^{2+} toxicity:

- MICU1/2 regulation to prevent MCU opening at low cytosolic Ca^{2+}
- NCLX-mediated Ca^{2+} extrusion
- Reduction of ER-mitochondria coupling under chronic stress
- Mitophagy activation to eliminate damaged mitochondria
- Antioxidant responses to mitigate Ca^{2+}-induced ROS

Failure of these mechanisms shifts Ca^{2+} signaling from physiological regulation to pathological overload (Figure 4.2).

4.6 Redox Signaling and ROS

4.6.1 Sources and Physiological Roles

The ROS and reactive nitrogen species (RNS) are incessantly generated in aerobic organisms as byproducts of metabolism. Mitochondria are the primary endogenous source of cellular ROS. During oxidative energy production *via* the ETC, the incomplete oxygen reduction inevitably produces

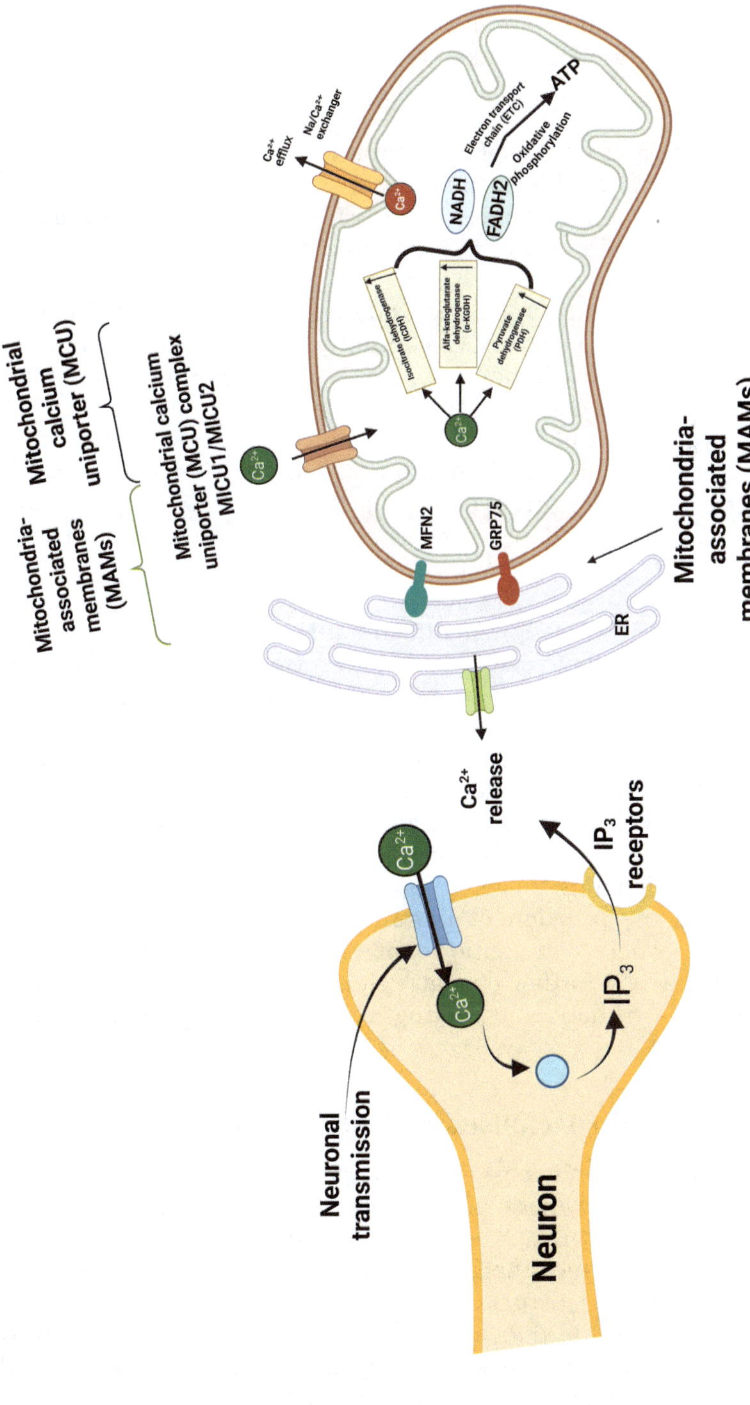

Figure 4.2 Illustrative representation of the role of mitochondrial Ca^{2+} in coupling neurological activity to bioenergetics.

key species like the superoxide anion (O_2^-), which is dismutated to a more stable species, *i.e.*, hydrogen peroxide (H_2O_2).[34,35]

Besides mitochondria, various enzymatic sources equally contribute towards the generation of regulated ROS for signaling and microbicidal functions.

- NADPH oxidases (NOX isoforms) produce localized H_2O_2 in response to growth factors, cytokines and receptor stimulation.
- Xanthine oxidase, uncoupled nitric oxide synthases, cytochrome P450S and peroxisomal oxidases also contribute context-dependent ROS/RNS production.
- The ER generates H_2O_2 during oxidative protein folding (Ero1/PDI activity) and, under ER stress, can amplify cellular ROS and communicate this stress to mitochondria *via* Ca^{2+} and ROS transfer at ER-mitochondria contact sites.
- Phagocytes, a type of immune cell, produce substantial bursts of ROS as part of their function in eliminating pathogens. This process highlights the crucial physiological role that ROS play in the body's innate immune response, helping to neutralize and destroy harmful invaders effectively.[34,35]

Physiologically, ROS act as signaling modulators. H_2O_2 reversibly oxidizes cysteine thiols to sulfenic acid or disulfides, transiently modifying the activity of protein tyrosine phosphatases and kinases and thereby shifting MAPK, PI3K/Akt, JAK/STAT and NF-κB pathways. This redox signaling regulates proliferation, differentiation, autophagy, mitochondrial biogenesis and immune cell activation. Reactive species play a dual role in human physiology, acting both as harmful agents and as essential regulatory mediators.[7] At basal and controlled concentrations, ROS and RNS play a crucial role in redox signaling-triggered processes, including gene regulation, apoptosis, and the function of immune cells. However, reactive species generation overwhelms the antioxidant capacity, and oxidative/nitrosative stress ensues, which further escalates lipid peroxidation, protein carbonylation, and DNA oxidation, triggering apoptosis, ferroptosis, and chronic inflammation.[34,36]

4.6.2 Antioxidant Defenses and Nrf2 Pathway

Cells maintain redox homeostasis through an organized antioxidant network composed of enzymatic and non-enzymatic defenses. Key enzymatic antioxidants include SOD (1, 2, and 3), which convert $O_2^{\bullet-}$ to H_2O_2, and CAT and peroxisomal enzymes that decompose H_2O_2. Additionally, glutathione peroxidases (GPx) and peroxiredoxins (Prx) reduce peroxides using trace elements (Se, Zn, and Mn) as essential cofactors.[35,36] Glutathione (GSH) is the most abundant and critical non-enzymatic determinant of the cellular redox state, typically quantified by the GSH/GSSG ratio.[37]

The nuclear factor erythroid 2-related factor 2 (Nrf2) pathway is the cell's master regulator and primary transcriptional defense mechanism against oxidative and electrophilic stress.[7] Under basal conditions, the Nrf2 protein is sequestered in the cytosol by its inhibitory partner, KEAP1. Oxidative or electrophilic stress, often signaled by H_2O_2 or similar molecules, modifies KEAP1, triggering the release of Nrf2. The released Nrf2 rapidly translocates to the nucleus, where it binds to the Antioxidant Response Element (ARE) in the promoter regions of target genes.[36] This binding activates the transcription of numerous protective genes, including those encoding core antioxidant and detoxification enzymes, such as γ-glutamylcysteine ligase (GCL), NAD(P)H: quinone oxidoreductase 1 (NQO1), and Heme Oxygenase-1 (HO-1). This activation significantly increases the cell's antioxidant capacity, notably elevating the GSH/GSSG ratio, thereby re-establishing redox homeostasis.[37]

4.7 Neurotransmitter Metabolism and Energy Cost

The brain exhibits exceptional metabolic demands relative to its mass, consuming a substantial proportion of the organism's total energy expenditure. The ATP derived from OXPHOS sustains ionic gradients and neurotransmitter biosynthesis, release, and reuptake. Consequently, impairments in cellular energy production result in neuronal dysfunction.[38,39] Here, we characterize the neurochemical mechanisms underlying this bioenergetic dependence, quantifying the metabolic requirements of discrete neurotransmitter pathways.

4.7.1 Glutamate–Glutamine Cycle

Excitatory neurotransmission is maintained chiefly by the glutamate–glutamine cycle, a pathway characterized by high metabolic turnover. Vesicular loading relies on VGLUT systems driven by ATP-dependent proton gradients, which concentrate glutamate to a concentration of 60–120 mM. Post-release, the re-establishment of ionic homeostasis *via* Na^+/K^+-ATPase activity dominates the energy budget of glutamatergic transmission.[39–41]

4.7.1.1 Astrocytic Uptake and Metabolism

Astrocytes take up 80–90% of the released glutamate through GLT-1 and GLAST transporters, which cotransport Na^+ and H^+ and countertransport K^+, imposing a significant ATP demand to restore ion homeostasis through Na^+/K^+-ATPase cycles. Within astrocytes, glutamate is amidated to glutamine *via* an astrocyte-specific glutamine synthetase (GS), an ATP-dependent enzyme that consumes one ATP per molecule of glutamine formed.[40,42]

4.7.1.2 Neuronal Replenishment

Glutamine is subsequently translocated to neurons *via* SNAT transporters. Within the neuronal compartment, phosphate-activated glutaminase (PAG) hydrolyzes glutamine to regenerate glutamate. This reaction yields ammonium (NH_4^+) as a byproduct, which must be exported back to astrocytes for detoxification through ammonia transport mechanisms or amino-acid shuttles to prevent neurotoxicity.[40,42]

4.7.1.3 Energetic and Mitochondrial Integration

The glutamate–glutamine cycle functions as an open metabolic system, as a fraction of glutamate undergoes oxidative catabolism within the tricarboxylic acid (TCA) cycle in both neurons and astrocytes. This oxidative loss necessitates continuous anaplerotic compensation. Due to the exclusive expression of pyruvate carboxylase in astrocytes, these cells serve a critical role in replenishing α-ketoglutarate (α-KG) depleted during glutamine synthesis. Consequently, this pathway establishes a direct functional coupling between neurotransmitter turnover and mitochondrial oxidative metabolism.[40]

This cycle supports not only neurotransmitter pools but also:

- TCA cycle intermediates (α-KG, oxaloacetate),
- Amino-acid metabolism
- Redox balance through NADH-dependent pathways.

Disruption of glutamate clearance leads to excitotoxicity, mitochondrial dysfunction, and neurodegeneration, as highlighted in disorders such as AD and HD.[39]

4.7.2 GABA Shunt and Anaplerosis

The GABA shunt provides an energetically efficient bypass of two steps of the TCA cycle and is essential in inhibitory neurons for maintaining GABA levels while supporting mitochondrial metabolism. GABA is synthesized from glutamate through the action of glutamate decarboxylase (GAD). Upon release, GABA is taken up by both neurons and astrocytes through GAT transporters.

Astrocytes metabolize GABA through GABA transaminase (GABA-T), which gets converted to succinic semialdehyde following the production of succinate *via* succinic semialdehyde dehydrogenase (SSADH). This succinate enters the TCA cycle, generating α-KG and ultimately regenerating glutamate and glutamine, integrating the GABA shunt directly with the glutamate–glutamine cycle.[42] Because the GABA shunt supplies succinate directly to the TCA cycle, it avoids α-KG dehydrogenase, a rate-limiting and ROS-sensitive step. It also supports mitochondrial ATP synthesis under high GABAergic activity, enabling efficient anaplerosis during inhibitory neurotransmission.

The GABA shunt is significant during high-frequency inhibitory signaling, development, and synapse maturation, as well as in metabolic stress, where it supports ATP generation while maintaining stable redox ratios. Disruption of GABA metabolism contributes to epilepsy, neurodevelopmental disorders, and altered energy pathways in ASD and other conditions.[12]

4.7.3 Catecholamine Synthesis

Catecholamine neurotransmitters, including DA, norepinephrine, and epinephrine, are metabolically synthesized from tyrosine, with each enzymatic step linked to mitochondrial bioenergetics and redox metabolism. A highly mitochondrial ROS-sensitive enzyme, tyrosine hydroxylase (TH), requires O_2 and tetrahydrobiopterin (BH_4) to convert tyrosine to L-DOPA. Aromatic L-amino acid decarboxylase (AADC) requires PLP vitamin B6 for the production of DA. Synaptic vesicles contain the enzyme DA β-hydroxylase (DBH), which requires O_2 and ascorbate and is linked to vesicular ATP-dependent processes. Phenylethanolamine *N*-methyltransferase (PNMT) converts norepinephrine to epinephrine, consuming ATP-derived *S*-adenosyl-methionine.

4.7.3.1 Energy and Mitochondrial Requirements

Catecholaminergic neurons, especially DAergic neurons, have a high metabolic demand due to pacemaking activity. They generate high mitochondrial ROS during DA turnover and metabolically rely on OXPHOS to sustain TH activity and vesicular monoamine transporter-2 (VMAT2) loading.[12]

4.8 Mitochondrial Dysfunction in Brain Disorders

4.8.1 NDDs

4.8.1.1 AD

Mitochondrial dysfunction is a crucial and early pathological feature of AD, as well as a primary cause of OS and bioenergetic failure.[43] Dysfunctional mitochondria produce an excess of ROS during electron transport through respiratory chain complexes I–IV, causing oxidative damage to mitochondrial DNA, proteins, and membrane lipids.[44] This oxidative injury increases mitochondrial degeneration, creating a vicious cycle of increased ROS production and diminishing mitochondrial efficiency. Reduced expression of OXPHOS genes in the neocortex is associated with brain hypometabolism identified by positron emission tomography, emphasizing the role of mitochondrial failure in cerebral homeostasis.[45] Dysregulation of complex I is associated with tau-mediated toxicity, while impairment of complex IV is significantly linked to Aβ pathology, resulting in decreased ATP production (Figure 4.3).[46] Changes in essential enzymes of the TCA cycle, such as pyruvate dehydrogenase and α-ketoglutarate dehydrogenase, might affect

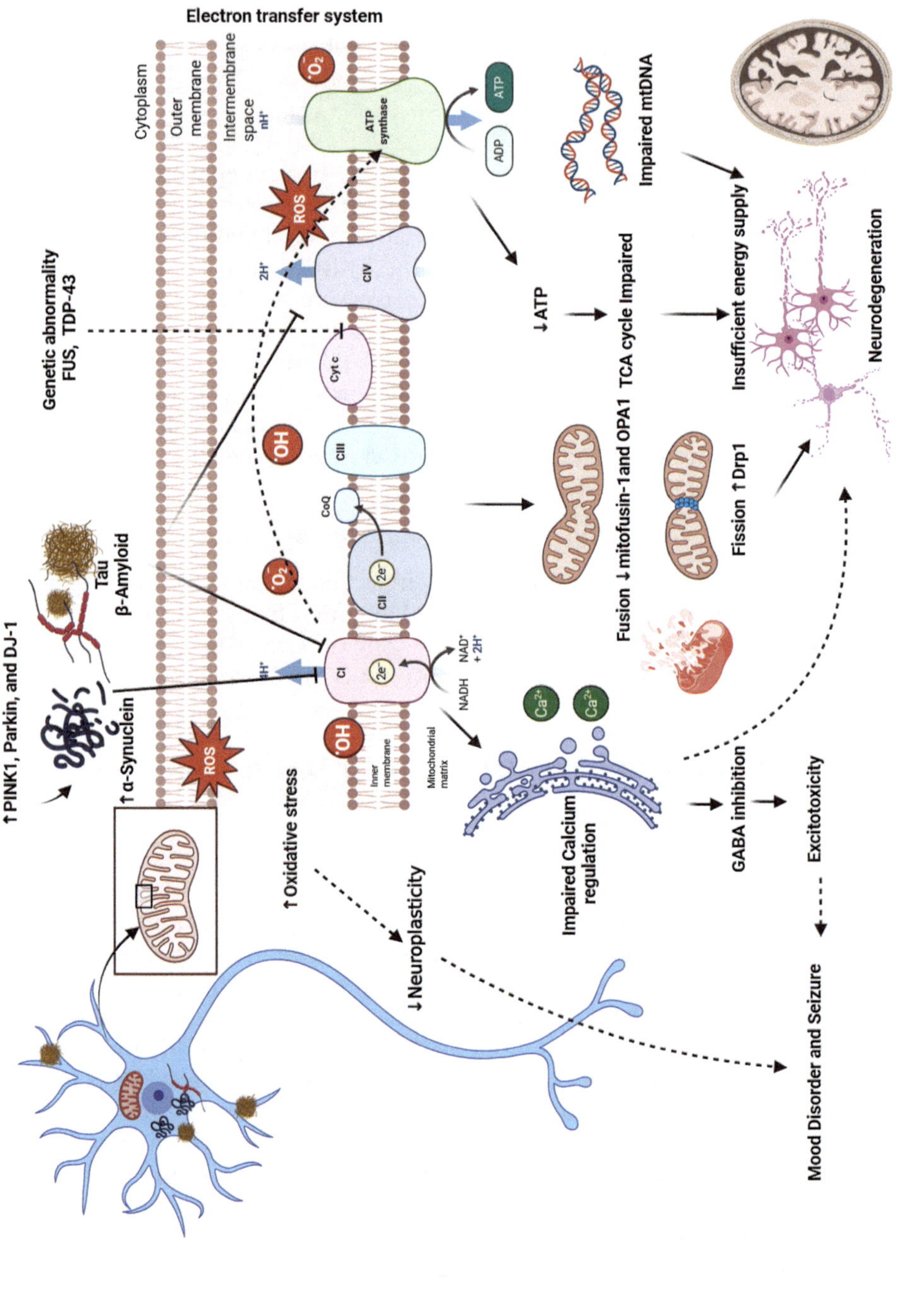
Electron transfer system
Cytoplasm
Outer membrane
Intermembrane space
nH⁺
Inner membrane
Mitochondrial matrix
ROS
·OH
·O₂⁻
CI
CII
CoQ
CIII
Cyt c
CIV
ATP synthase
4H⁺
2H⁺
2e⁻
NADH
NAD⁺ + 2H⁺
ADP
ATP
↑PINK1, Parkin, and DJ-1
↑α-Synuclein
Tau
β-Amyloid
Genetic abnormality
FUS, TDP-43
↑Oxidative stress
↓Neuroplasticity
Ca²⁺
Impaired Calcium regulation
GABA inhibition
Excitotoxicity
Mood Disorder and Seizure
↓ATP
TCA cycle Impaired
Insufficient energy supply
Impaired mtDNA
Fusion ↓mitofusin-1and OPA1
Fission ↑Drp1
Neurodegeneration

acetyl-CoA synthesis, leading to cholinergic dysfunction and cognitive decline.[47] Products of lipid peroxidation, such as hydroxynonenal, promote tau aggregation, confirming the relationship between OS and neurodegeneration. Importantly, mitochondrial deficiencies have been detected in peripheral tissues, implying that energy metabolism changes in AD may be a systemic disorder rather than a brain-restricted pathology.[48]

Mitochondria are extremely active organelles that constantly divide and fuse to preserve cellular integrity and metabolic efficiency. The disease severity disrupts this equilibrium, shifting it toward excessive fission, which results in mitochondrial fragmentation, impaired mitophagy, and the accumulation of defective mitochondria in neurons.[49] Drp1 and its receptors on the outer mitochondrial membrane predominantly govern mitochondrial fission, whereas mitofusins (MFN1 and 2) and OPA1 facilitate mitochondrial fusion.[50] In AD, the expression of fusion proteins is reduced, while the expression of fission-related proteins is dysregulated, leading to structural instability.[51] In addition to structural changes, mitochondrial transport along axons is significantly hindered, resulting in insufficient energy supply at synapses, which are areas of high metabolic demand. Aβ affects mitochondrial trafficking and interacts with the fission machinery, resulting in synaptic dysfunction.[52] Abnormal mitochondrial distribution, fragmentation, and reduced mobility all lead to poor Ca^{2+} buffering, increased OS, and activation of apoptotic pathways, which accelerate neuronal degeneration.

4.8.1.2 PD

Mitochondrial dysfunction is increasingly recognized as a significant cause of neurodegeneration in PD, specifically impacting DAergic neurons in the substantia nigra pars compacta.[53] In familial types of PD, mutations in PARK genes, including PINK1, Parkin (PRKN), and DJ-1, suggest that mitochondrial failure is the primary pathogenic event. Idiopathic PD is characterized by the abnormal accumulation of α-synuclein (α-syn), a key factor in mitochondrial toxicity (Figure 4.3).[54] Under normal conditions, α-syn regulates

Figure 4.3 Illustration of how mitochondrial dysfunction links NDDs and neuropsychiatric disorders. Disease-specific protein aggregates Aβ and tau in AD; α-synuclein in PD; TDP in ALS, and mutations in mitochondrial quality-control proteins (PINK1, Parkin, DJ-1) disrupt the mitochondrial electron transport chain (complex I–V), resulting in excessive ROS generation, impaired OXPHOS, and reduced ATP production. Oxidative stress damages mitochondrial DNA (mtDNA) and TCA cycle enzymes, further aggravating bioenergetic failure. Abnormal mitochondrial dynamics, characterized by increased fission (Drp1) and reduced fusion (OPA1, Mitofusin-1), lead to mitochondrial fragmentation and loss of functional integrity. Defective Ca^{2+} handling promotes excitotoxicity and GABAergic imbalance, contributing to seizures and mood dysregulation, while progressive energy failure and ROS-mediated injury drive synaptic dysfunction, reduced neuroplasticity, and neuronal death, ultimately culminating in neurodegeneration.

the function of ATP synthase and mitochondrial efficiency. However, pathological α-syn species reduce complex I activity, membrane potential, and OXPHOS and generate excessive ROS.[55] Oligomeric α-syn disrupts mitochondrial function by interacting with key proteins like ATP synthase subunit α and TOM20. This results in the impaired translocation of nuclear-encoded proteins into mitochondria and a collapse of respiratory chain activity.[56] Additionally, α-syn opens the mitochondrial permeability transition pore, causing Ca^{2+} dyshomeostasis and the release of proapoptotic proteins.

Pathological α-syn affects ER-mitochondrial contact sites, altering intracellular Ca^{2+} signaling and impairing mitochondrial respiration. Sustained mitochondrial OS degrades lipids, proteins, and mtDNA, triggering mitophagy pathways and gradually depleting healthy mitochondrial reserves.[57] In substantia nigra neurons, enhanced mitochondrial oxidant stress overwhelms autophagic ability, resulting in the buildup of defective mitochondria. Cytosolic ROS further harms mitophagy regulators, exacerbating mitochondrial failure and affecting lysosomal and proteasomal clearance mechanisms. This leads to a cycle of protein misfolding, diminished α-syn degradation, and increased cellular toxicity. Concurrently, neuroinflammation and cell death processes are directly impacted by mitochondrial malfunction. Increased mitochondrial antigen presentation, resulting from a lack of PINK1-Parkin activity, stimulates cytotoxic T-cell responses in the brain, whereas the release of mtDNA into the cytosol triggers inflammatory signaling pathways.[58] Additionally, apoptosis and parthanatos, a caspase-independent form of programmed neuronal death, are triggered by mitochondrial failure. By decreasing the effectiveness of energy-dependent processes, such as proteostasis, synaptic transmission, and membrane repair, compromised ATP generation exacerbates the pathogenesis of PD. By integrating various pathological mechanisms in PD, such as OS, α-syn toxicity, Ca^{2+} dysregulation, immune activation, and malfunctioning quality control systems, mitochondrial impairment establishes mitochondria as a central hub in PD pathogenesis and a crucial therapeutic target.[59]

4.8.1.3 ALS

One of the main pathogenic mechanisms in ALS is mitochondrial malfunction, which directly leads to the degeneration of motor neurons. Neuronal survival is significantly impacted by mitochondrial dysfunction, as mitochondria play a crucial role in regulating cellular energy production, Ca^{2+} homeostasis, redox balance, and apoptosis. In ALS, OXPHOS is disrupted by damage to the ETC, especially complex I (Figure 4.3).[60] This leads to decreased ATP production, loss of mitochondrial membrane potential, and excessive production of ROS, all of which exacerbate OS and mitochondrial damage.[61] Neuronal mitochondria buffer intracellular Ca^{2+}, and disrupting mitochondrial–endoplasmic reticulum interaction can cause Ca^{2+} dysregulation and excitotoxicity. Genetic abnormalities in critical mitochondrial-associated genes, such as SOD1, FUS, C9orf72, TDP-43, and CHCHD10, contribute to disease development by impairing ETC function, promoting protein aggregation, and triggering apoptotic pathways.[62] Dysfunctional mitochondrial dynamics, caused by altered expression of Drp1, Fis1, Mfn1, and Opa1, result in mitochondrial fragmentation and bioenergetic failure.[63] Simultaneously, the failure of mitochondrial quality control systems, including poor PINK1-Parkin-OPTN-mediated mitophagy and reduced autophagy-lysosomal degradation caused by mutations in TBK1 and SQSTM1, leads to the accumulation of damaged mitochondria within neurons.[64] Structural problems, such as decreased mitochondrial membrane potential and altered shape, can lead to poor ATP generation, Ca^{2+} imbalance, and ROS overproduction. In ALS, mitochondrial dysfunction combines genetic, metabolic, and oxidative problems into a single route that causes synaptic failure, axonal degradation, and progressive motor neuron death.

4.8.2 Mood Disorders and Epilepsy

4.8.2.1 Mood Disorders

Mitochondrial dysfunction is a fundamental biological disorder found in mood disorders, including major depressive disorder and bipolar disorder, and is strongly associated with poor energy metabolism, OS, and neural vulnerability.[65] At the bioenergetic level, ETC failure (particularly complexes I, II, and IV) results in decreased OXPHOS and ATP synthesis, followed by a compensatory metabolic shift toward glycolysis and lactate buildup.[66] These metabolic alterations are followed by decreased phosphocreatine, creatine kinase activity, and *N*-acetylaspartate levels, indicating impaired neural mitochondrial capability.[67] The excessive generation of reactive oxygen and nitrogen species leads to lipid peroxidation, mitochondrial DNA damage, and protein oxidation. In contrast, antioxidant mechanisms, such as GSH, SOD, and CAT, are commonly disrupted. Mitochondrial Ca^{2+} buffering is compromised, leading to intracellular Ca^{2+} overload, increased excitability, and disruption of Ca^{2}-dependent signaling pathways for synaptic plasticity and gene regulation.[68] Structural problems also contribute, as altered mitochondrial shape and uneven fusion–fission dynamics promote organelle disintegration and malfunction.

Further, mitochondrial quality control is compromised by impaired mitophagy and enhanced apoptotic signaling. Downregulation of Parkin, p62, and LC3 hinders mitochondrial removal, but stimulation of cytochrome c release and caspase cascades promotes cell death.[69] Inflammatory signaling pathways, including NF-κB and proinflammatory cytokines, can exacerbate mitochondrial injury by suppressing biogenesis and amplifying OS. These interrelated mechanisms confirm mitochondrial dysfunction as a primary cause of metabolic instability, neuroprogression, and poor neuroplasticity in mood disorders.[70]

4.8.2.2 Epilepsy

Through reduced energy metabolism, OS, and Ca^{2+} dysregulation, mitochondrial dysfunction plays a role in the onset and development of epilepsy. Defective OXPHOS leads to ATP depletion, which in turn reduces the activity of energy-dependent ion pumps, such as Ca^{2+}-ATPase and Na^+/K^+-ATPase.[71] This results in hyperexcitability and destabilization of neuronal membrane potentials. Mitochondrial ROS generation is increased by ETC failure, which further impairs mitochondrial function by causing oxidative damage to lipids, proteins, and mtDNA.[72] Increased ROS disrupts intracellular Ca^{2+} homeostasis, leading to mitochondrial Ca^{2+} overload, the opening of the permeability transition pore, and the activation of apoptotic signaling.[73] This pathway stimulates glutamate release while decreasing GABAergic inhibition, resulting in excitotoxicity and spontaneous seizure activity.[74] Recurrent seizures exacerbate mitochondrial injury, resulting in a self-sustaining loop of metabolic failure, oxidative damage, and neuronal excitability that promotes epileptogenesis.

4.9 Therapeutic Interventions

4.9.1 Mitochondrial Protectors and Antioxidants

All eukaryotic cells have mitochondria, which are responsible for converting around 98% of oxygen into usable energy.[75] The energetic process of OXPHOS is made possible by the ETC protein complexes, which are found in mitochondria.[76] Under typical physiological circumstances, about 3–5% of the oxygen in mitochondrial ETC complexes (I and III) can leak out, leading to the creation of superoxide radicals ($O_2^{\bullet -}$). These unstable and highly reactive molecules can amplify the production of additional ROS.[77,78] To deal with increased amounts of free radicals, mitochondria utilize antioxidant defense systems that are crucial for cellular repair. Enzyme systems, such as peroxiredoxin and thioredoxin, neutralize or break down superoxide ($O_2^{\bullet -}$) into water or the less reactive H_2O_2. The enzyme SOD accelerates this process. Using the Fenton reaction, superoxide ($O_2^{\bullet -}$) may create hydroxyl ($^{\bullet}OH$) radicals. These radicals can then be used to make peroxynitrite ($ONOO^-$) and other RNS. Mitochondria typically maintain redox balance by reducing RNS and ROS; however, when both levels are excessive, they can cause oxidative damage to proteins, lipids, nucleic acids, and the extracellular matrix.[79] Oxidized protein adducts (such as 3-nitrotyrosine), protein carbonylation, and 4-hydroxynonenal, a byproduct of lipid peroxidation, are essential indicators of OS caused by peroxynitrite. Damage to nuclear and mitochondrial DNA by ROS and RNS may impact gene expression and cellular function. Subunits of the ETC may be damaged by an excess of both, which can perpetuate a dysfunctional cycle in the mitochondria.[80,81] As shown in Table 4.1, antioxidants help mitigate cellular oxidative damage by scavenging or neutralizing free radicals. Since OS plays a significant role in the development and progression of NDDs, antioxidant treatment has recently garnered interest as a novel approach to treating these conditions. Potential treatment methods to reduce OS and free radical generation include restoring antioxidant levels in the brain *via* two main processes. Following brain-related diseases, this repair improves redox homeostasis. Detoxifying or scavenging excessive ROS and RNS through natural or synthetic methods, inhibiting free radical overproduction, and affecting signaling pathways that promote endogenous antioxidant synthesis are all ways in which antioxidants can modulate redox processes in pathological neural tissues. This, in turn, preserves the redox equilibrium.[82] The antioxidants in Table 4.2 are classified as either natural or synthetic ROS–RNS scavengers and detoxifiers. There are three types of synthetic antioxidants: those that target the mitochondria, those that do not, and those that are precursors of GSH. Enzymatic and non-enzymatic antioxidants are the two main categories of natural antioxidants. Instead of nontargeted antioxidants (NAs), researchers are focusing on developing targeted mitochondrial antioxidants (TMAs) (Table 4.2). Antioxidants may limit ROS formation if used excessively or incorrectly, which can lead to the activation of compensatory mechanisms in MAPK pathways. These mechanisms, in turn, can compromise the body's natural antioxidant system and disrupt normal cellular growth.[83] Nontargeted antioxidant uptake and metabolism across different organs is another essential concern. To enhance therapeutic effectiveness and facilitate targeted delivery to mitochondria, antioxidants can be coupled with carriers such as lipophilic cations, liposomes, or peptides. Through various pathways, this approach enables substantial accumulation within cells and mitochondria, thereby protecting against oxidative damage. Effective antioxidants must be readily absorbed into the bloodstream, either through the intestines or *via* intravenous injection, and they should exhibit high bioavailability. Muscles, the brain, the liver, the kidneys, the heart, and even the ears represent just a few of the vital tissues that TMAs focus on to combat OS. Over the past decade, researchers have made significant strides in the development of TMAs, yielding promising

Table 4.2 Antioxidant classifications based on pharmacological activities and subcellular receptors. The information provided has been sourced from ref. 86.

Natural ROS–RNS scavengers		Synthetic ROS–RNS scavengers and detoxifiers		
Non-enzymatic antioxidants	Enzymatic antioxidants	Mitochondrial-targeted antioxidants	Cytosolic antioxidants	GSH-precursors
Flavonoids	NADPH	Edaravone	Tiron	SAMe
Vitamins (A, C and E)	CAT	Carbon-PTIO	Allicin	NAC
Glutathione	SOD	Nano-CeO2	Cysteine	D-NAC
Minerals (Zn, Cu, Se and Mg)	GPx	MitoVit-E	Allin	NACA
Polyphenols	Trx	MitoQ	MnTBAP	
NADPH	Prx	Metallopolypherons		
CoQ10	TrxR	PBN		
CytC		SKQ1		
		Mito TEMPOL		
		Elamipretide		

outcomes that warrant a thorough examination. These advancements hold great potential for enhancing cellular health and mitigating oxidative damage across various organ systems.[83–85]

4.9.2 Ketogenic Approaches and Emerging Therapies

4.9.2.1 AD

4.9.2.1.1 The Possible Role of a Ketogenic Diet in the Etiopathogenesis of AD. Dementia is a general term for a loss of cognitive ability to the point that carrying out everyday tasks becomes difficult. Nearly two-thirds of dementia cases in persons over the age of 65 are caused by AD, making it the most frequent type of dementia, particularly among the elderly. The patient's risk of developing this condition increases with age. The number of reported instances of AD is expected to rise over the next several years, according to experts. It is quite challenging to pinpoint precisely when these alterations began to manifest as the illness progresses. Full impairment of cognitive functioning is a result of this illness. The capacity to concentrate, language skills, memory, and comprehension of simple problems are all significantly impacted. Most people experience moderate amnesia, characterized by a loss of recent memories, as their initial symptom. Identifying the causes of this illness is crucial to grasping its nature. Factors that may accelerate the disease's progression include advancing age, severe head trauma, cerebral vascular diseases, nicotinism, and depression.[87–89] Additionally, it appears that hereditary factors play a significant role in the development of the illness. Plaques and tangles formed by aberrant neuronal cells are the primary pathogenic process in AD.[90] Neuronal plaques are small-scale alterations characterized by an Aβ-core encased in clusters of larger axons. The amyloid precursor protein (APP) is the physiological scaffold from which Aβ is generated. APP is primarily divided by beta and alpha secretases. Small, inert Aβ fragments form as a byproduct of this reaction. The β- and γ-secretases separate APP when abnormalities occur. This procedure leads to the formation of Aβ (42 peptides). The brain is the final destination of its buildup and subsequent aggregation. An essential connection in the familial etiology of AD is the gene encoding the aforementioned mechanism of APP breakdown that is situated on chromosome 21, which plays a significant role in the stated process. Microbleeding occurs throughout the brain as a result of Aβ aggregates deposited in the brain's blood vessels, leading to angiopathies. The current consensus is that amyloid plaque deposition starts 20 years before the onset of clinical symptoms. The second primary process in the development of AD is the buildup of tau protein-based neurofibrillary tangles. When Aβ aggregates excessively, it causes this structure to hyperphosphorylate, which in turn leads to the aggregation of larger, abnormal conglomerates. These structures are present in the hippocampus from the beginning of the illness, according to the research. Neurons across the entire cerebral cortex may become infected as the disease progresses.[87,91]

On the other hand, this action stops new plaque from forming. As a consequence, a ketogenic diet (KD) has the potential to be an intriguing adjuvant treatment, which might halt the advancement of the illness and the corresponding decline in cognitive function. The spread of the disease to other parts of the brain may be influenced by inflammatory processes initiated by clusters of Aβ and tau proteins.[92–98] Inadequate clearance of neurotoxic 4-hydroxynonenal and disturbance of the hypothalamic–pituitary–adrenal (HPA) axis may be caused by OS and environmental variables, which may lead to onset of the illness.[99] In addition to having a role in late-onset AD, the ApoEε4 allele causes neuroinflammation, OS, and accelerated cellular aging.[100]

4.9.2.1.2 Medicinal Foods. Currently, there is no effective treatment for AD. Existing strategies primarily aim to delay the onset of severe symptoms for as long as possible. These strategies can be pharmacological or non-pharmacological in nature. Non-pharmacological approaches often emphasize diet, exercise, and cognitive training. One notable example of a beneficial diet is the Mediterranean diet, which includes the use of extra-virgin olive oil (EVOO). Research suggests that oleuropein, a secoiridoid found in EVOO, has neuroprotective properties, indicating its potential role in preventing NDDs, including AD. Studies conducted using mouse models have demonstrated that the active components of EVOO can enhance cognitive abilities in mice by reducing the accumulation of Aβ aggregates and promoting synaptic activity in the hippocampus. Moreover, EVOO has the potential to mitigate the cytotoxic and neuroinflammatory effects associated with Aβ aggregation. Long-term supplementation with EVOO significantly lowers tau protein phosphorylation, which is essential regardless of its impact on Aβ metabolism. Several studies have reported that a diet high in EVOO does not produce any harmful side effects, such as cell death or neurodegeneration.[101,102] Walnuts (*Juglans regia* L.) are another staple of the Mediterranean diet. They contribute to the diet's anti-neuroinflammatory effects because of their high antioxidant content, which includes *n*-3 α-linolenic acid, juglone, and tocopherol (vitamin E). When added to the diet of mice in a controlled environment, walnuts enhanced their memory and learning capacity.[103]

4.9.2.1.3 Clinical Relevance. A large body of research suggests that insulin resistance could have a role in the onset of NDDs.[103] Memory loss occurs as a result of brain alterations brought on by concurrent hyperglycemia. A decrease in hippocampus volume and an increase in gray matter volume have been reported in young individuals with hyperglycemia, respectively, by Kerti *et al.* and Weinstein *et al.*[104,105] Impaired insulin signaling has been linked to protein Aβ and, consequently, AD in several studies. Research involving AD patients indicates that a KD can improve insulin sensitivity, reduce insulin levels, and restore normal brain glucose metabolism. Patients have also shown enhanced performance on cognitive function tests, suggesting the potential effectiveness of the diet in NDDs. Table 4.3 summarizes

Table 4.3 Findings from randomized controlled studies and clinical trials that looked at the effects of nutritional ketosis on AD patients during the last seven years.

Diet	Group	Duration	Findings	Ref.
MCT-based ketogenic formula	20	12 weeks	Patients showed considerable improvement in both their immediate and delayed logical memory tests after 8 weeks, as well as in the digit-symbol coding test and immediate logical memory test after 12 weeks, as compared to their baseline scores	106
KD	26	Two 12-week treatment durations	Enhancement of everyday functioning and overall well-being	107
The administration of caprylidene, a ketogenic agent	16	45 days	Patients without the ApoEε4 gene have increased blood flow in specific brain areas	108
MCT ketogenic drink	39	6 months	The blood indicators revealed no clinically significant side effects. Substantial increases in plasma IL-8 levels have been observed following the intervention	109
MCT ketogenic drink	83	6 months	The kMCT group outperformed the placebo group on many tests, including the Trail-Making Test, the Boston Naming Test, and free and cued recall verbal fluency	110

the results of clinical trials, indicating that the KD may benefit AD patients in terms of cognitive function. Researchers continue to investigate the precise role of risk factors in neurodegeneration. However, it is clear that effective treatment and risk reduction heavily rely on appropriate prevention strategies and maintaining a healthy lifestyle. Considering that current pharmacotherapy primarily focuses on alleviating symptoms rather than addressing the underlying causes of the illness, this aspect is of utmost importance.[87]

4.9.2.2 PD

4.9.2.2.1 The Possible Role of a KD in the Etiopathogenesis of PD. PD is increasingly prevalent among advanced civilizations, where lifestyle factors and environmental conditions may play a significant role. Various elements contribute to the onset of this neurological disorder, including exposure to environmental pollutants, which can disrupt normal brain function. Additionally, certain medications may have side effects that impact neurological health. Genetic abnormalities can also predispose individuals to develop PD, highlighting the complex interplay between genetics and disease. Furthermore, injuries to the brain's blood vessels may lead to impairments that trigger the symptoms associated with PD. Together, these factors paint a comprehensive picture of the potential causes behind this condition.[111,112] A patient's chance of developing PD increases with age. People in the 85–89 age group had a 3.5% chance of getting PD. When compared to those over 60, the risk of developing PD in those under 60 is 1%.[111,113] Muscle stiffness, resting tremors, and motor slowness are the first psychomotor signs used to diagnose PD. It is believed that bradykinesia is the main indicator of the condition.[114,115] Constipation, salivation, dysgraphia, melancholy, sensory abnormalities, sleep problems, dementia, hallucinations, and highly critical cognitive and behavioural impairments are all indications of PD.[116] These symptoms accompany the classic motor symptoms. Postural abnormalities and gait problems are hallmarks of the early stages of the illness. The involuntary cessation of foot movement, known as a "freezing of gait",[117,118] is a common symptom. The degeneration of neurons in the basal ganglia, specifically those involved in DA transmission in the striatum and nucleus basalis, is the primary cause of the disease's symptoms.[119] Movement disorders arise from the malfunction of neural circuits that involve key areas of the brain, namely the basal ganglia and the motor cortex. These disorders are often caused by damage to the neurons within these regions, leading to a cascade of effects that ultimately disrupt the transport and regulation of DA, a crucial neurotransmitter for coordinating movement. This dysfunction can result in a range of motor symptoms, making it difficult for individuals to control their movements effectively.[120] Symptoms of PD will only manifest when DA levels in the brain's basal nuclei and substantia nigra drop below 20% of their maximum value.[121] Furthermore, Lewy bodies and Lewy neurites, which are composed of misfolded α-syn, are a neuropathological feature seen in PD.[122] These abnormalities could play a significant role in the degeneration of DAergic neurons, which are vital for motor control and coordination. Currently, the primary treatment for PD is levodopa (L-DOPA). While L-DOPA is effective in alleviating the symptoms of PD, it unfortunately lacks any neuroprotective properties that might halt or slow down the progression of the disease. Concerningly, the metabolite known as 5-*S*-cysteineldopamine has been identified as a contributor to OS within the body. This OS not only exacerbates DA depletion but may also facilitate the increased aggregation of α-syn, a protein associated with the development of NDDs.[123] There is evidence that a KD increases the bioavailability of L-DOPA, which is associated with a reduction in the amount of protein that the body derives from food. Furthermore, some studies suggest that KD, when combined with pharmaceutical treatments for symptoms, may slow the progression of the disease. Studies have demonstrated that the heroin analogue known as 1-methyl-4-phenylpyridinium (MPP^+) possesses the ability to induce cell death specifically in DAergic neurons located in the substantia nigra, a region of the brain crucial for movement and coordination. This compound functions by inhibiting the multi-enzyme complex of mitochondrial NADH dehydrogenase, thereby disrupting cellular energy production. The effect generates a pathological condition that mirrors the symptoms of PD when observed in cultures of midbrain neurons. A single study has supported the hypothesis that β-hydroxybutyrate (β-HB) protects DAergic neurons from damage.[87] Enhanced ATP generation and mitochondrial respiration are associated with this. By lowering levels of oxygen-free radicals, KDs also improve mitochondrial ETC function.[124] Recently, targeting K^+-ATP channels has emerged as a possible treatment for PD. Research has shown that activating these channels has neuroprotective benefits and reduces neuronal excitability. It is believed, however, that activation of K^+-ATP channels on GABAergic neurons, which block GABA-B receptors and activate glutamatergic terminals to produce α-synuclein, may contribute to the development of PD. Because K^+-ATP channels are so abundant in the brain, it is challenging to predict their effects on these channels based on the results of agonist and antagonist tests conducted in animal models.[125,126] The KD may be more effective as a treatment since it targets multiple diseased processes simultaneously, rather than just one, unlike particular drugs that impact K^+-ATP channels.

4.9.2.2.2 Clinical Trials. There has been mounting evidence linking changes in the gut microbiota to PD in the last few years.[87] The effects of food, probiotic, prebiotic, and synbiotic supplementation on pathogenic processes occurring in PD were studied by Alfonsetti *et al.* in conjunction with the microbiome's composition.[127] There is a notable difference between the microbiota of people with PD and healthy individuals, according to studies. Specifically, there are more *Enterobacteriaceae* and fewer *Prevotellaceae*.[128] Changes in the microbiota's quality cause impaired intestinal permeability (also known as "leaky gut"). This, in turn, promotes α-syn aggregation by inducing inflammatory processes and OS *via* bacterial LPS.[129] It is well-documented that the KD alters this ratio, leading to an upregulation of *Prevotella* and a downregulation of *Enterobacteriaceae*.[130] Improvements in gut sealing, anti-inflammatory effects, OS relief, and BDNF upregulation can be achieved through dietary changes, such as increasing consumption of omega-3 polyunsaturated fatty acids, probiotics, prebiotics, and synbiotics.[127] Phillips *et al.* conducted an 8-week randomized controlled trial in 2018 with 47 patients, and found that both high-fat and low-fat diets improved motor and non-motor symptoms. When it came to non-motor elements of everyday activity performance, such as pain, cognitive impairment, urine problems, and weariness, the KD outperformed the low-fat diet.[131]

4.10 Conclusion and Future Perspective

Several serious diseases are associated with mitochondria, which are essential organelles that produce energy, respond to infections, synthesize neurotransmitters, and regulate Ca^{2+}. These diseases range from NDDs like AD and PD to metabolic and cardiovascular disorders like diabetes. To develop effective treatments that address the underlying causes of these disorders, it is essential to comprehend mitochondrial biology and its applications in medicine. To fully appreciate the consequences of mitochondrial variability among tissues for health and disease, it is necessary to approach improving mitochondrial health in a context-specific manner, considering both disease and tissue uniqueness. It is problematic to rely solely on antioxidant drugs to reduce ROS, as their effectiveness in clinical studies has been limited. Therefore, the efficacy and safety of mitochondrial-based treatments must be confirmed through more thorough preclinical investigations, which should include human cell models as well as larger animal models, before proceeding to human trials. Consequently, NDDs treatment and advancement in mitochondrial biology depend on sustained research and multidisciplinary cooperation.[86,132]

Neuronal activity relies on a finely coordinated interaction between Ca^{2+} dynamics, redox signaling and neurotransmitter metabolism, each of which critically depends on mitochondrial function. Controlled mCa^{2+} uptake stimulates key TCA cycle enzymes to match ATP production with rapid synaptic demands, while physiological ROS support essential signaling functions. Disruptions in these pathways, such as Ca^{2+} overload, impaired antioxidant systems or altered neurotransmitter cycling, compromise mitochondrial integrity, increase OS and contribute to excitotoxicity and neurodegeneration. The glutamate–glutamine cycle, GABA shunt, and catecholamine biosynthesis exemplify the substantial metabolic investment required for neurotransmission and highlight the central role of astrocyte–neuron metabolic coupling in sustaining brain energy homeostasis.

Future work should prioritize strategies that stabilize mitochondrial Ca^{2+} flux, including modulation of the MCU, MICU (1/2), and NCLX activities, to protect neurons from Ca^{2}-induced mitochondrial failure. Targeting cellular redox systems, particularly through Nrf2 activation or mitochondrial-targeted antioxidants, offers the potential to restore redox balance without suppressing physiological signaling. Optimizing neurotransmitter metabolic pathways by enhancing glutamate clearance, supporting GABA shunt anaplerosis and improving catecholaminergic resilience may help alleviate metabolic stress in neurological and neurodegenerative diseases. Emerging multi-omics tools, metabolic imaging, and mitochondrial biomarkers will enable precise characterization of cell-specific energy pathways, paving the way for targeted therapeutic interventions that preserve mitochondrial health and maintain neuronal bioenergetic stability.

Author Contributions

The idea and design of this chapter wase a collaborative effort by all authors. Md Abubakar, Ritu Dahiya, and Abhishek Ambawatiya were involved in the conceptualization, writing of the original draft, mechanistic insight, input on critical points, and editing. Krushna Chandra Maharana, Shahnaz Alom, Aruna Rawat, Amita Rai, Damini Yadav, and Krishnendu Adhikary contributed to the manuscript writing. Nitesh Kumar was involved in the conceptualization, manuscript writing, reviewing, conceptualization of figures and approved the version to be published.

Acknowledgements

The images were created using www.Biorander.com and www.Grammarly.com was used to correct the grammar of this manuscript.

References

1. V. Balasubramanian, Brain power, *Proc. Natl. Acad. Sci. U. S. A.*, 2021, **118**, DOI: 10.1073/pnas.2107022118.
2. E. A. Melkonian and M. P. Schury, Biochemistry, Anaerobic Glycolysis, *StatPearls*, StatPearls Publishing LLC, Treasure Island (FL) ineligible companies, Disclosure: Mark Schury declares no relevant financial relationships with ineligible companies, 2025.
3. R. P. Chakrabarty and N. S. Chandel, Beyond ATP, new roles of mitochondria, *Biochemist*, 2022, **44**, 2–8.
4. F. Sayehmiri, F. Motamedi and Z. Batool, *et al.*, Mitochondrial plasticity and synaptic plasticity crosstalk; in health and Alzheimer's disease, *CNS Neurosci. Ther.*, 2024, **30**, e14897.
5. D. Valenti and R. A. Vacca, Brain Mitochondrial Bioenergetics in Genetic Neurodevelopmental Disorders: Focus on Down, Rett and Fragile X Syndromes, *Int. J. Mol. Sci.*, 2023, **24**, DOI: 10.3390/ijms241512488.
6. Z. Qin, X. Zhu and Y. Shen, *et al.*, An emerging role of mitochondrial quality control in bone metabolism: from molecular mechanisms to targeted therapeutic interventions, *Cell. Mol. Life Sci.*, 2025, **82**, 291.
7. V. J. Clemente-Suárez, L. Redondo-Flórez and A. I. Beltrán-Velasco, *et al.*, Mitochondria and Brain Disease: A Comprehensive Review of Pathological Mechanisms and Therapeutic Opportunities, *Biomedicines*, 2023, **11**, DOI: 10.3390/biomedicines11092488.
8. D. Lu, Y. Feng and G. Liu, *et al.*, Mitochondrial transport in neurons and evidence for its involvement in acute neurological disorders, *Front. Neurosci.*, 2023, **17**, 1268883.
9. München MATU, *Exploring the lifecycle of synaptic mitochondria in the healthy and diseased nervous system*, 2024.
10. N. Madrer, N. D. Perera and N. A. Uccelli, *et al.*, Neural Metabolic Networks: Key Elements of Healthy Brain Function, *J. Neurochem.*, 2025, **169**, e70084.
11. H. Li, C. Guglielmetti and Y. J. Sei, *et al.*, Neurons require glucose uptake and glycolysis *in vivo*, *Cell Rep.*, 2023, **42**, 112335.
12. Y. He, K. Xie and K. Yang, *et al.*, Unraveling the Interplay Between Metabolism and Neurodevelopment in Health and Disease, *CNS Neurosci. Ther.*, 2025, **31**, e70427.

13. Z. Liu, Y. Guo and Y. Zhang, *et al.*, Metabolic reprogramming of astrocytes: Emerging roles of lactate, *Neural Regen. Res.*, 2026, **21**, 421–432.
14. S. Zhang, J. Xia and W. He, *et al.*, From energy metabolism to mood regulation: The rise of lactate as a therapeutic target, *J. Adv. Res.*, 2025, DOI: 10.1016/j.jare.2025.04.018.
15. B. Vaccari-Cardoso, M. Antipina and A. G. Teschemacher, *et al.*, Lactate-Mediated Signaling in the Brain-An Update, *Brain Sci.*, 2022, **13**, DOI: 10.3390/brainsci13010049.
16. Y. Li, Q. Liu and Z. Jia, *et al.*, Ketone Bodies in Exercise, Health and Disease: Metabolic Mechanisms, Pathophysiology, and Therapeutic Implications, *Adv. Exerc. Health Sci.*, 2025, **2**, DOI: 10.1016/j.aehs.2025.05.004.
17. Y. Arima, The Impact of Ketone Body Metabolism on Mitochondrial Function and Cardiovascular Diseases, *J. Atheroscler. Thromb.*, 2023, **30**, 1751–1758.
18. A. Gupta and R. Singh, AMP-Activated Protein Kinase (AMPK): A Cellular Energy Sensor and the Guardian of Metabolism, *Int. J. Biochem. Res. Rev.*, 2025, **34**, 147–153.
19. C. L. Lyons and H. M. Roche, Nutritional Modulation of AMPK-Impact upon Metabolic-Inflammation, *Int. J. Mol. Sci.*, 2018, **19**, DOI: 10.3390/ijms19103092.
20. V. Voicu, C. Toader and M. Şerban, *et al.*, Systemic Neurodegeneration and Brain Aging: Multi-Omics Disintegration, Proteostatic Collapse, and Network Failure Across the CNS, *Biomedicines*, 2025, **13**, DOI: 10.3390/biomedicines13082025.
21. J. Zhang, H. Xiang and J. Liu, *et al.*, Mitochondrial Sirtuin 3: New emerging biological function and therapeutic target, *Theranostics*, 2020, **10**, 8315–8342.
22. P. Sah, A. K. Rai and D. Syiem, Sirtuin activators as an anti-aging intervention for longevity, *Explor. Drug Sci.*, 2025, **3**, 100881.
23. L. Qian, Y. Zhu and C. Deng, *et al.*, Peroxisome proliferator-activated receptor gamma coactivator-1 (PGC-1) family in physiological and pathophysiological process and diseases, *Signal Transduct. Target. Ther.*, 2024, **9**, 50.
24. Y. Yuan, Y. Tian and H. Jiang, *et al.*, Mechanism of PGC-1α-mediated mitochondrial biogenesis in cerebral ischemia-reperfusion injury, *Front. Mol. Neurosci.*, 2023, **16**, 1224964.
25. J. M. Park, D. H. Lee and D. H. Kim, Redefining the role of AMPK in autophagy and the energy stress response, *Nat. Commun.*, 2023, **14**, 2994.
26. X. Peng, H. Ni and B. Kuang, *et al.*, Sirtuin 3 in renal diseases and aging: From mechanisms to potential therapies, *Pharmacol. Res.*, 2024, **206**, 107261.
27. S. Kong, B. Cai and Q. Nie, PGC-1α affects skeletal muscle and adipose tissue development by regulating mitochondrial biogenesis, *Mol. Genet. Genomics*, 2022, **297**, 621–633.

28. N. Deleyto-Seldas and A. Efeyan, The mTOR-Autophagy Axis and the Control of Metabolism, *Front. Cell Dev. Biol.*, 2021, **9**, 655731.
29. H. Querfurth and H. K. Lee, Mammalian/mechanistic target of rapamycin (mTOR) complexes in neurodegeneration, *Mol. Neurodegener.*, 2021, **16**, 44.
30. E. J. Carvalho, P. B. Stathopulos and M. Madesh, Regulation of Ca(2+) exchanges and signaling in mitochondria, *Curr. Opin. Physiol.*, 2020, **17**, 197–206.
31. F. Borbolis, C. Ploumi and K. Palikaras, Calcium-mediated regulation of mitophagy: implications in neurodegenerative diseases, *npj Metab. Health Dis.*, 2025, **3**, 4.
32. S. Rumpf, N. Sanal and M. Marzano, Energy metabolic pathways in neuronal development and function, *Oxf. Open Neurosci.*, 2023, **2**, kvad004.
33. S. H. Lee, H. E. Duron and D. Chaudhuri, Beyond the TCA cycle: new insights into mitochondrial calcium regulation of oxidative phosphorylation, *Biochem. Soc. Trans.*, 2023, **51**, 1661–1673.
34. S. Di Meo, T. T. Reed and P. Venditti, *et al.*, Role of ROS and RNS Sources in Physiological and Pathological Conditions, *Oxid. Med. Cell. Longev.*, 2016, **2016**, 1245049.
35. C. F. Manful, E. Fordjour and D. Subramaniam, *et al.*, Antioxidants and Reactive Oxygen Species: Shaping Human Health and Disease Outcomes, *Int. J. Mol. Sci.*, 2025, **26**(15), 7520.
36. Y. Hong, A. Boiti and D. Vallone, *et al.*, Reactive Oxygen Species Signaling and Oxidative Stress: Transcriptional Regulation and Evolution, *Antioxidants*, 2024, **13**, DOI: 10.3390/antiox13030312.
37. F. Bellanti, A. R. D. Coda and M. I. Trecca, *et al.*, Redox Imbalance in Inflammation: The Interplay of Oxidative and Reductive Stress, *Antioxidants*, 2025, **14**, 656.
38. L. K. Bak, A. Schousboe and H. S. Waagepetersen, The glutamate/GABA-glutamine cycle: aspects of transport, neurotransmitter homeostasis and ammonia transfer, *J. Neurochem.*, 2006, **98**, 641–653.
39. J. V. Andersen, K. H. Markussen and E. Jakobsen, *et al.*, Glutamate metabolism and recycling at the excitatory synapse in health and neurodegeneration, *Neuropharmacology*, 2021, **196**, 108719.
40. J. V. Andersen, The Glutamate/GABA-Glutamine Cycle: Insights, Updates, and Advances, *J. Neurochem.*, 2025, **169**, e70029.
41. D. Zhang, Z. Hua and Z. Li, The role of glutamate and glutamine metabolism and related transporters in nerve cells, *CNS Neurosci. Ther.*, 2024, **30**, e14617.
42. L. K. Bak, A. Schousboe and H. S. Waagepetersen, The glutamate/GABA-glutamine cycle: aspects of transport, neurotransmitter homeostasis and ammonia transfer, *J. Neurochem.*, 2006, **98**, 641–653.
43. R. Wasim, Bioenergetic failure and oxidative stress: mitochondrial contributions to Alzheimer's disease, *Inflammopharmacology*, 2025, 1–17.
44. F. R. Palma, B. N. Gantner and M. J. Sakiyama, *et al.*, ROS production by mitochondria: function or dysfunction?, *Oncogene*, 2024, **43**, 295–303.

45. L. Mosconi, A. Pupi and M. J. De Leon, Brain glucose hypometabolism and oxidative stress in preclinical Alzheimer's disease, *Ann. N. Y. Acad. Sci.*, 2008, **1147**, 180–195.
46. R. A. Quntanilla and C. Tapia-Monsalves, The role of mitochondrial impairment in Alzheimer's disease neurodegeneration: the tau connection, *Curr. Neuropharmacol.*, 2020, **18**, 1076–1091.
47. V. Jakkamsetti, Q. Ma, I. Marin-Valencia, *et al.*, Pyruvate dehydrogenase, pyruvate carboxylase, Krebs cycle, and mitochondrial transport disorders, *Rosenberg's Molecular and Genetic Basis of Neurological and Psychiatric Disease*, Elsevier, 2025, pp. 533–543.
48. J. Song, X. Yang and M. Zhang, *et al.*, Glutamate metabolism in mitochondria is closely related to Alzheimer's disease, *J. Alzheimers Dis.*, 2021, **84**, 557–578.
49. T. Ashleigh, R. H. Swerdlow and M. F. Beal, The role of mitochondrial dysfunction in Alzheimer's disease pathogenesis, *Alzheimers Dement.*, 2023, **19**, 333–342.
50. L. Tilokani, S. Nagashima and V. Paupe, *et al.*, Mitochondrial dynamics: overview of molecular mechanisms, *Essays Biochem.*, 2018, **62**, 341–360.
51. P. K. Chandra, M. K. Panner Selvam and J. A. Castorena-Gonzalez, *et al.*, Fibrinogen in mice cerebral microvessels induces blood–brain barrier dysregulation with aging *via* a dynamin-related protein 1–dependent pathway, *Geroscience*, 2024, **46**, 395–415.
52. Y. Peng, P. Gao and L. Shi, *et al.*, Central and peripheral metabolic defects contribute to the pathogenesis of Alzheimer's disease: targeting mitochondria for diagnosis and prevention, *Antioxid. Redox Signaling*, 2020, **32**, 1188–1236.
53. D. J. Surmeier, J. N. Guzman and J. Sanchez-Padilla, *et al.*, The role of calcium and mitochondrial oxidant stress in the loss of substantia nigra pars compacta dopaminergic neurons in Parkinson's disease, *Neuroscience*, 2011, **198**, 221–231.
54. N. A. Gouda, A. Elkamhawy and J. Cho, Emerging therapeutic strategies for Parkinson's disease and future prospects: A 2021 update, *Biomedicines*, 2022, **10**, 371.
55. R. Perfeito, D. F. Lázaro and T. F. Outeiro, *et al.*, Linking alpha-synuclein phosphorylation to reactive oxygen species formation and mitochondrial dysfunction in SH-SY5Y cells, *Mol. Cell. Neurosci.*, 2014, **62**, 51–59.
56. S. Goyal and R. K. Chaturvedi, Mitochondrial protein import dysfunction in pathogenesis of neurodegenerative diseases, *Mol. Neurobiol.*, 2021, **58**, 1418–1437.
57. H. Zeng, Y. Liu and X. Liu, *et al.*, Interplay of α-synuclein oligomers and endoplasmic reticulum stress in Parkinson's disease: insights into cellular dysfunctions, *Inflammation*, 2025, **48**, 1590–1606.
58. M.-M. Lin, N. Liu and Z.-H. Qin, *et al.*, Mitochondrial-derived damage-associated molecular patterns amplify neuroinflammation in neurodegenerative diseases, *Acta Pharmacol. Sin.*, 2022, **43**, 2439–2447.

59. H. Watanabe, S. Shima and K. Kawabata, *et al.*, Brain network and energy imbalance in Parkinson's disease: linking ATP reduction and α-synuclein pathology, *Front. Mol. Neurosci.*, 2025, **17**, 1507033.
60. E. Obrador, R. Salvador-Palmer and R. López-Blanch, *et al.*, The link between oxidative stress, redox status, bioenergetics and mitochondria in the pathophysiology of ALS, *Int. J. Mol. Sci.*, 2021, **22**, 6352.
61. G. Napolitano, G. Fasciolo and P. Venditti, Mitochondrial management of reactive oxygen species, *Antioxidants*, 2021, **10**, 1824.
62. E. C. Genin, M. Abou-Ali and V. Paquis-Flucklinger, Mitochondria, a key target in amyotrophic lateral sclerosis pathogenesis, *Genes*, 2023, **14**, 1981.
63. S. Kumar, R. Ashraf and A. Ck, Mitochondrial dynamics regulators: implications for therapeutic intervention in cancer, *Cell Biol. Toxicol.*, 2022, **38**, 377–406.
64. R. Weil, E. Laplantine and S. Curic, *et al.*, Role of optineurin in the mitochondrial dysfunction: potential implications in neurodegenerative diseases and cancer, *Front. Immunol.*, 2018, **9**, 1243.
65. S. Madireddy and S. Madireddy, Therapeutic interventions to mitigate mitochondrial dysfunction and oxidative stress–induced damage in patients with bipolar disorder, *Int. J. Mol. Sci.*, 2022, **23**, 1844.
66. D. Nolfi-Donegan, A. Braganza and S. Shiva, Mitochondrial electron transport chain: Oxidative phosphorylation, oxidant production, and methods of measurement, *Redox Biol.*, 2020, **37**, 101674.
67. J. M. N. Duarte, P. F. Schuck and G. L. Wenk, *et al.*, Metabolic disturbances in diseases with neurological involvement, *Aging Dis.*, 2013, **5**, 238.
68. N. B. Pivovarova and S. B. Andrews, Calcium-dependent mitochondrial function and dysfunction in neurons, *FEBS J.*, 2010, **277**, 3622–3636.
69. M. A. Eldeeb, R. A. Thomas and M. A. Ragheb, *et al.*, Mitochondrial quality control in health and in Parkinson's disease, *Physiol. Rev.*, 2022, **102**, 1721–1755.
70. A. D. Cherry and C. A. Piantadosi, Regulation of mitochondrial biogenesis and its intersection with inflammatory responses, *Antioxid. Redox Signaling*, 2015, **22**, 965–976.
71. J. M. Rho and D. Boison, The metabolic basis of epilepsy, *Nat. Rev. Neurol.*, 2022, **18**, 333–347.
72. S. Madireddy and S. Madireddy, Therapeutic strategies to ameliorate neuronal damage in epilepsy by regulating oxidative stress, mitochondrial dysfunction, and neuroinflammation, *Brain Sci.*, 2023, **13**, 784.
73. Y. Zhou, S. Jing and S. Liu, *et al.*, Double-activation of mitochondrial permeability transition pore opening *via* calcium overload and reactive oxygen species for cancer therapy, *J. Nanobiotechnol.*, 2022, **20**, 188.
74. S. M. S. Sears and S. J. Hewett, Influence of glutamate and GABA transport on brain excitatory/inhibitory balance, *Exp. Biol. Med.*, 2021, **246**, 1069–1083.

75. Y. Zong, H. Li and P. Liao, *et al.*, Mitochondrial dysfunction: mechanisms and advances in therapy, *Signal Transduct. Target. Ther.*, 2024, **9**, 124.
76. E.-H. Choi, M.-H. Kim and S.-J. Park, Targeting mitochondrial dysfunction and reactive oxygen species for neurodegenerative disease treatment, *Int. J. Mol. Sci.*, 2024, **25**, 7952.
77. S. Bhardwaj, K. K. Kesari and M. Rachamalla, *et al.*, CRISPR/Cas9 gene editing: New hope for Alzheimer's disease therapeutics, *J. Adv. Res.*, 2022, **40**, 207–221.
78. A. Ore, J. M. Angelastro and C. Giulivi, Integrating mitochondrial biology into innovative cell therapies for neurodegenerative diseases, *Brain Sci.*, 2024, **14**, 899.
79. C. Pal, Targeting mitochondria with small molecules: A promising strategy for combating Parkinson's disease, *Mitochondrion*, 2024, **79**, 101971.
80. J. F. Halling and H. Pilegaard, PGC-1α-mediated regulation of mitochondrial function and physiological implications, *Appl. Physiol. Nutr. Metab.*, 2020, **45**, 927–936.
81. W. Mohamed, J. Kumar and B. S. Alghamdi, *et al.*, Neurodegeneration and inflammation crosstalk: Therapeutic targets and perspectives, *IBRO Neurosci. Rep.*, 2023, **14**, 95–110.
82. H. R. Modi, S. Musyaju and M. Ratcliffe, *et al.*, Mitochondria-targeted antioxidant therapeutics for traumatic brain injury, *Antioxidants*, 2024, **13**, 303.
83. Q. Jiang, J. Yin and J. Chen, *et al.*, Mitochondria-targeted antioxidants: a step towards disease treatment, *Oxid. Med. Cell. Longevity*, 2020, **2020**, 8837893.
84. D. J. Bishop, J. Botella and A. J. Genders, *et al.*, High-intensity exercise and mitochondrial biogenesis: current controversies and future research directions, *Physiology*, 2019, **34**(1), 56–70.
85. L. Chen, Y. Qin and B. Liu, *et al.*, PGC-1 α-mediated mitochondrial quality control: molecular mechanisms and implications for heart failure, *Front. Cell Dev. Biol.*, 2022, **10**, 871357.
86. A.-H. S. Wadan, A. H. Shaaban and M. Z. El-Sadek, *et al.*, Mitochondrial-based therapies for neurodegenerative diseases: a review of the current literature, *Naunyn-Schmiedebergs Arch. Pharmacol.*, 2025, 1–30.
87. D. Pietrzak, K. Kasperek and P. Rękawek, *et al.*, The therapeutic role of ketogenic diet in neurological disorders, *Nutrients*, 2022, **14**, 1952.
88. R. Mayeux and Y. Stern, Epidemiology of Alzheimer disease, *Cold Spring Harbor Perspect. Med.*, 2012, **2**, a006239.
89. E. Forsyth and P. D. Ritzline, An overview of the etiology, diagnosis, and treatment of Alzheimer disease, *Phys. Ther.*, 1998, **78**, 1325–1331.
90. C. L. Masters, R. Bateman and K. Blennow, *et al.*, Alzheimer's disease, *Nat. Rev. Dis. Primers*, 2015, **1**, 1–18.
91. L. G. Apostolova, Alzheimer disease, *Continuum*, 2016, **22**, 419–434.
92. Y. Wang, S. Zhang and J. Zheng, *et al.*, Ketogenic diet and neurological diseases, *Precision Nutr.*, 2025, **4**, e00109.

93. A. Costa and E. Lucarini, Treating chronic stress and chronic pain by manipulating gut microbiota with diet: can we kill two birds with one stone?, *Nutr. Neurosci.*, 2025, **28**, 221–244.
94. I. Shabbir, K. Liu and B. Riaz, *et al.*, Investigating the Therapeutic Potential of the Ketogenic Diet in Modulating Neurodegenerative Pathophysiology: An Interdisciplinary Approach, *Nutrients*, 2025, **17**, 1268.
95. T. Luong, M. Svart and L. C. Gormsen, *et al.*, Ketogenic diet in the management of disease, *Curr. Opin. Clin. Nutr. Metab. Care*, 2025, **28**, 477–482.
96. N. Madrer, N. D. Perera and N. A. Uccelli, *et al.*, Neural Metabolic Networks: Key Elements of Healthy Brain Function, *J. Neurochem.*, 2025, **169**, e70084.
97. I. Bayraktaroglu, N. Ortí-Casañ and D. Van Dam, *et al.*, Systemic inflammation as a central player in the initiation and development of Alzheimer's disease, *Immun. Ageing*, 2025, **22**, 33.
98. V. Topalis, C. Voros and M. Ziaka, Targeting Inflammation in Alzheimer's Disease: Insights Into Pathophysiology and Therapeutic Avenues-A Comprehensive Review, *J. Geriatr. Psychiatry Neurol.*, 2025, 08919887251361578.
99. G. Juszczyk, J. Mikulska and K. Kasperek, *et al.*, Chronic stress and oxidative stress as common factors of the pathogenesis of depression and Alzheimer's disease: The role of antioxidants in prevention and treatment, *Antioxidants*, 2021, **10**, 1439.
100. L. Caberlotto, L. Marchetti and M. Lauria, *et al.*, Integration of transcriptomic and genomic data suggests candidate mechanisms for APOE4-mediated pathogenic action in Alzheimer's disease, *Sci. Rep.*, 2016, **6**, 32583.
101. B. Klimova, M. Novotný and K. Kuca, *et al.*, Effect of an extra-virgin olive oil intake on the delay of cognitive decline: role of secoiridoid oleuropein?, *Neuropsychiatr. Dis. Treat.*, 2019, 3033–3040.
102. E. Lauretti, M. Nenov and O. Dincer, *et al.*, Extra virgin olive oil improves synaptic activity, short-term plasticity, memory, and neuropathology in a tauopathy model, *Aging Cell*, 2020, **19**, e13076.
103. A. Chauhan and V. Chauhan, Beneficial effects of walnuts on cognition and brain health, *Nutrients*, 2020, **12**, 550.
104. L. Kerti, A. V. Witte and A. Winkler, *et al.*, Higher glucose levels associated with lower memory and reduced hippocampal microstructure, *Neurology*, 2013, **81**, 1746–1752.
105. G. Weinstein, P. Maillard and J. J. Himali, *et al.*, Glucose indices are associated with cognitive and structural brain measures in young adults, *Neurology*, 2015, **84**, 2329–2337.
106. M. Ota, J. Matsuo and I. Ishida, *et al.*, Effects of a medium-chain triglyceride-based ketogenic formula on cognitive function in patients with mild-to-moderate Alzheimer's disease, *Neurosci. Lett.*, 2019, **690**, 232–236.
107. M. C. Phillips, L. M. Deprez and G. M. Mortimer, *et al.*, Randomized crossover trial of a modified ketogenic diet in Alzheimer's disease, *Alzheimers Res. Ther.*, 2021, **13**, 51.

108. N. Torosyan, C. Sethanandha and J. D. Grill, *et al.*, Changes in regional cerebral blood flow associated with a 45 day course of the ketogenic agent, caprylidene, in patients with mild to moderate Alzheimer's disease: results of a randomized, double-blinded, pilot study, *Exp. Gerontol.*, 2018, **111**, 118–121.
109. É. Myette-Côté, V. St-Pierre and S. Beaulieu, *et al.*, The effect of a 6-month ketogenic medium-chain triglyceride supplement on plasma cardiometabolic and inflammatory markers in mild cognitive impairment, *Prostaglandins, Leukotrienes Essent. Fatty Acids*, 2021, **169**, 102236.
110. M. Fortier, C. A. Castellano and V. St-Pierre, *et al.*, A ketogenic drink improves cognition in mild cognitive impairment: Results of a 6-month RCT, *Alzheimers Dement.*, 2021, **17**, 543–552.
111. V. Bellou, L. Belbasis and I. Tzoulaki, *et al.*, Environmental risk factors and Parkinson's disease: an umbrella review of meta-analyses, *Parkinsonism Relat. Disord.*, 2016, **23**, 1–9.
112. R. Cacabelos, Parkinson's disease: from pathogenesis to pharmacogenomics, *Int. J. Mol. Sci.*, 2017, **18**, 551.
113. D. Aarsland, J. Kvaløy and K. Andersen, *et al.*, The effect of age of onset of PD on risk of dementia, *J. Neurol.*, 2007, **254**, 38–45.
114. S. Sveinbjornsdottir, The clinical symptoms of Parkinson's disease, *J. Neurochem.*, 2016, **139**, 318–324.
115. H. Ling, L. A. Massey and A. J. Lees, *et al.*, Hypokinesia without decrement distinguishes progressive supranuclear palsy from Parkinson's disease, *Brain*, 2012, **135**, 1141–1153.
116. J. Jankovic, Parkinson's disease: clinical features and diagnosis, *J. Neurol. Neurosurg. Psychiatry*, 2008, **79**, 368–376.
117. C. Gao, J. Liu and Y. Tan, *et al.*, Freezing of gait in Parkinson's disease: pathophysiology, risk factors and treatments, *Transl. Neurodegener.*, 2020, **9**, 12.
118. N. Giladi and A. Nieuwboer, *Understanding and treating freezing of gait in parkinsonism, proposed working definition, and setting the stage*, Wiley Online Library, 2008, pp. S423–S425.
119. S. Lotankar, K. S. Prabhavalkar and L. K. Bhatt, Biomarkers for Parkinson's disease: recent advancement, *Neurosci. Bull.*, 2017, **33**, 585–597.
120. P. Mazzoni, B. Shabbott and J. C. Cortés, Motor control abnormalities in Parkinson's disease, *Cold Spring Harbor Perspect. Med.*, 2012, **2**, a009282.
121. G. Boden, K. Sargrad and C. Homko, *et al.*, Effect of a low-carbohydrate diet on appetite, blood glucose levels, and insulin resistance in obese patients with type 2 diabetes, *Ann. Intern. Med.*, 2005, **142**, 403–411.
122. T. Yasuda, Y. Nakata and H. Mochizuki, α-Synuclein and neuronal cell death, *Mol. Neurobiol.*, 2013, **47**, 466–483.
123. C. Aureli, T. Cassano and A. Masci, *et al.*, 5-S-cysteinyldopamine neurotoxicity: Influence on the expression of α-synuclein and ERp57 in cellular and animal models of Parkinson's disease, *J. Neurosci. Res.*, 2014, **92**, 347–358.

124. S. A. Masino and J. M. Rho, Mechanisms of ketogenic diet action, *Jasper's Basic Mechanisms of the Epilepsies*, 4th edn, 2012.
125. E. Emmanouilidou, G. Minakaki and M. V. Keramioti, *et al.*, GABA transmission *via* ATP-dependent K+ channels regulates α-synuclein secretion in mouse striatum, *Brain*, 2016, **139**, 871–890.
126. S. Zhao, M. Wang and Z. Ma, Therapeutic potential of ATP-sensitive potassium channels in Parkinson's disease, *Brain Res. Bull.*, 2021, **169**, 1–7.
127. M. Alfonsetti, V. Castelli and M. d'Angelo, Are we what we eat? impact of diet on the gut–brain axis in Parkinson's disease, *Nutrients*, 2022, **14**, 380.
128. F. Scheperjans, P. Derkinderen and P. Borghammer, The gut and Parkinson's disease: hype or hope?, *J. Parkinsons Dis.*, 2018, **8**, S31–S39.
129. C. B. Forsyth, K. M. Shannon and J. H. Kordower, *et al.*, Increased intestinal permeability correlates with sigmoid mucosa alpha-synuclein staining and endotoxin exposure markers in early Parkinson's disease, *PLoS One*, 2011, **6**, e28032.
130. A. Paoli, L. Mancin and A. Bianco, *et al.*, Ketogenic diet and microbiota: friends or enemies?, *Genes*, 2019, **10**, 534.
131. M. C. Phillips, D. K. Murtagh and L. J. Gilbertson, *et al.*, Low-fat *versus* ketogenic diet in Parkinson's disease: a pilot randomized controlled trial, *Mov. Disord*, 2018, **33**, 1306–1314.
132. A. Kalani, P. Chaturvedi, D. Brunetti, *et al.*, *Mitochondrial therapy in neurodegeneration*, Frontiers Media SA, 2023, p. 1144093.

CHAPTER 5

Oxidative Stress, Neuroinflammation, and Neuroprotection

VISHAL JAIN

Department of Electrical and Computer Engineering, Neuroscience Institute, Carnegie Mellon University, Pittsburgh, PA, USA
Email: vishalja@andrew.cmu.edu

5.1 Introduction

The human brain, though remarkably efficient, exists in a delicate balance between energy demand and vulnerability. Despite representing only about two percent of total body weight, it consumes roughly one-fifth of the body's oxygen, relying heavily on oxidative metabolism to sustain neuronal activity.[1] This metabolic intensity allows for complex cognitive processes but also renders neural tissue susceptible to oxidative injury. Mitochondria, the powerhouses of neurons, generate reactive oxygen species (ROS) as natural by-products of ATP synthesis. Under normal physiological conditions, ROS and reactive nitrogen species (RNS) play beneficial roles in intracellular signaling, synaptic plasticity, and stress adaptation. Yet, when their generation exceeds the capacity of antioxidant defenses, the resulting imbalance, known as oxidative stress, initiates molecular damage to proteins, lipids, and DNA.[2]

The brain's structural and biochemical characteristics amplify this vulnerability. Neurons are long-lived and irreplaceable, depending on uninterrupted mitochondrial activity and intricate ion regulation. Their membranes, enriched with polyunsaturated fatty acids, are prone to lipid peroxidation, and their antioxidant systems, such as glutathione peroxidase

Chemical Biology No. 26
Chemical Signatures of the Brain: Bridging Biochemistry, Neuroanatomy, and Brain Disorders
Edited by Sachchida Nand Rai and Abhishek Kumar Singh

Published by the Royal Society of Chemistry, www.rsc.org

and catalase, are relatively modest compared to other tissues. Transition metals like iron and copper, essential for normal metabolism, can catalyze free radical reactions, worsening oxidative stress. As these reactive species accumulate, they disrupt signaling pathways, alter gene expression, and impair cellular repair mechanisms. Over time, chronic oxidative stress contributes to the slow attrition of neuronal structure and function, forming the biochemical foundation of aging and neurodegeneration.[3]

Alongside oxidative damage, the central nervous system exhibits a distinct form of inflammation, neuroinflammation, that intertwines tightly with redox imbalance. Once thought to be an immune-privileged organ, the brain is now known to contain a dynamic network of immune-like cells. Microglia, the brain's resident macrophages, constantly survey their environment, responding to injury, infection, and the accumulation of misfolded proteins. When activated, they release cytokines, nitric oxide, and ROS, initiating protective responses. Astrocytes complement this activity by regulating neurotransmitter balance, maintaining the blood–brain barrier, and releasing neurotrophic factors. However, persistent activation of microglia and astrocytes transforms their protective functions into sources of toxicity. Chronic secretion of pro-inflammatory cytokines such as TNF-α and IL-1β maintains a sustained oxidative environment, perpetuating neuronal damage and synaptic dysfunction.[4]

This reciprocal relationship between oxidative stress and neuroinflammation forms a pathological feedback loop. ROS and RNS generated during inflammation activate redox-sensitive transcription factors like NF-κB, leading to the expression of more pro-inflammatory mediators. In parallel, cytokines and activated glia stimulate NADPH oxidase, further increasing ROS production. The outcome is a self-sustaining cycle that underlies a wide array of neurological disorders, including Alzheimer's, Parkinson's, amyotrophic lateral sclerosis, and multiple sclerosis. In Alzheimer's disease, amyloid-β aggregates trigger microglial activation and oxidative burst, which not only injure neurons but also enhance amyloid deposition and tau pathology.[5] In Parkinson's disease, dopaminergic neurons in the substantia nigra face oxidative overload due to dopamine metabolism, mitochondrial dysfunction, and iron accumulation, conditions that converge with neuroinflammatory activation to accelerate neuronal loss.[6]

Aging magnifies these processes. Over the lifespan, mitochondrial efficiency declines, leading to increased ROS generation and decreased ATP output. DNA repair mechanisms become less effective, and the proteasome system responsible for removing damaged proteins weakens.[7] Microglia adopt a "primed" state, responding to minor insults with exaggerated cytokine release, while astrocytes show impaired ability to buffer glutamate or restore redox balance. This chronic, low-grade inflammation, termed "inflammaging", creates a biochemical environment that promotes neurodegenerative progression even in the absence of acute injury. Oxidative stress and neuroinflammation therefore represent both hallmarks of aging and convergent mechanisms of late-life cognitive decline.

The cumulative evidence suggests that preserving brain function requires strategies that maintain redox homeostasis and control inflammation. This has given rise to the concept of *neuroprotection*, an umbrella term encompassing interventions that prevent or slow neuronal injury by targeting oxidative and inflammatory pathways. Neuroprotection extends beyond pharmacological treatments to include lifestyle, dietary, and behavioral interventions that enhance intrinsic resilience. Exercise, for instance, increases cerebral blood flow, boosts mitochondrial biogenesis, and elevates antioxidant enzyme activity while suppressing pro-inflammatory cytokines. Nutritional patterns such as the Mediterranean diet, rich in omega-3 fatty acids, polyphenols, and vitamins, reduce oxidative burden and promote synaptic integrity.[8] Similarly, cognitive and environmental enrichment stimulate neurotrophic factors like BDNF, enhancing neural plasticity and repair.

From a molecular standpoint, several endogenous defense mechanisms underlie neuroprotection. The transcription factor Nrf2 orchestrates the expression of antioxidant and detoxifying enzymes, acting as a master regulator of redox balance.[9] Mitochondrial quality control systems, including mitophagy and fusion–fission dynamics, maintain organelle integrity and prevent the accumulation of ROS-generating defects.[10] Neurotrophic signaling pathways, such as those mediated by BDNF and NGF, foster synaptic resilience, while adaptive stress responses involving sirtuins and heat-shock proteins enhance cellular repair. Strengthening these intrinsic systems forms the cornerstone of long-term neural preservation.

Despite major advances in understanding these mechanisms, translating neuroprotection into clinical success remains difficult. Many antioxidants and anti-inflammatory agents that show promise in animal models fail in human trials, often due to inadequate delivery across the blood–brain barrier or late-stage intervention after irreversible neuronal loss.[11] Moreover, oxidative and inflammatory signaling are integral to normal physiology; excessive suppression can disrupt essential cellular communication. Thus, the therapeutic challenge lies in modulation rather than elimination, fine-tuning redox and immune pathways to restore balance without impairing function.

The growing appreciation of this complexity has shifted modern neuroscience toward an integrative model. Rather than treating oxidative stress and neuroinflammation as isolated phenomena, researchers increasingly view them as interdependent components of a larger systems network that governs brain aging and disease. This network encompasses mitochondria, glial signaling, vascular health, and metabolic regulation, domains that must be targeted together for meaningful neuroprotection. As research continues, the line between prevention and treatment becomes increasingly blurred. Early interventions that reinforce antioxidant defenses, reduce chronic inflammation, and promote neural adaptation hold the greatest potential for maintaining cognitive health across the lifespan.

The purpose of this chapter is to explore the intricate relationship among oxidative stress, neuroinflammation, and neuroprotection in an integrated framework. It examines how redox imbalance initiates cellular injury, how immune mechanisms amplify damage, and how endogenous and external interventions can mitigate or reverse these processes. The goal is not merely to describe pathology but to illuminate the interconnected pathways through which the nervous system defends and repairs itself. By understanding these mechanisms, the field moves closer to developing rational, mechanism-based strategies for preventing and slowing the progression of neurodegenerative diseases, approaches that work with, rather than against, the brain's natural defenses.

5.2 Molecular Mechanism of Oxidative Stress

Oxidative stress represents a fundamental disturbance in cellular redox homeostasis, arising from an imbalance between the production of reactive oxygen species (ROS) and reactive nitrogen species (RNS) and the antioxidant defense systems that normally maintain equilibrium. Within the central nervous system (CNS), this imbalance is particularly detrimental because of the brain's high oxygen consumption, abundant lipid content, and limited antioxidant capacity.[2] Under normal physiological conditions, ROS and RNS act as essential signaling mediators that regulate neuronal plasticity, synaptic transmission, and immune defense. However, when generated in excess, these reactive species induce structural and functional damage to macromolecules such as DNA, proteins, and lipids, triggering cascades that culminate in cell death and neurodegeneration. Understanding the sources of ROS and RNS, the functioning of antioxidant systems, and how their imbalance contributes to neurological disorders is critical for developing targeted neuroprotective interventions.

5.2.1 Sources of Reactive Oxygen Species (ROS)

ROS include superoxide anion ($O_2^{\bullet-}$), hydrogen peroxide (H_2O_2), hydroxyl radical ($^{\bullet}OH$), and singlet oxygen ($^1O_2$). These species originate from both endogenous metabolic reactions and exogenous environmental exposures (Figure 5.1).

5.2.1.1 Mitochondrial ROS Production

Mitochondria are the predominant intracellular sources of ROS under physiological conditions. During oxidative phosphorylation, electrons flow through complexes I–IV of the electron transport chain (ETC). Approximately 0.2–2% of these electrons prematurely reduce molecular oxygen, forming superoxide anions primarily at complexes I (NADH:ubiquinone oxidoreductase) and III (cytochrome bc1 complex).[12] The generated $O_2^{\bullet-}$ is rapidly converted by mitochondrial superoxide dismutase (Mn-SOD or SOD2) into H_2O_2, which can diffuse through mitochondrial membranes. While low concentrations of H_2O_2 participate in redox signaling, excessive accumulation leads to hydroxyl radical formation through Fenton or Haber–Weiss chemistry in the presence of transition metals like Fe^{2+} or Cu^{+}.[13]

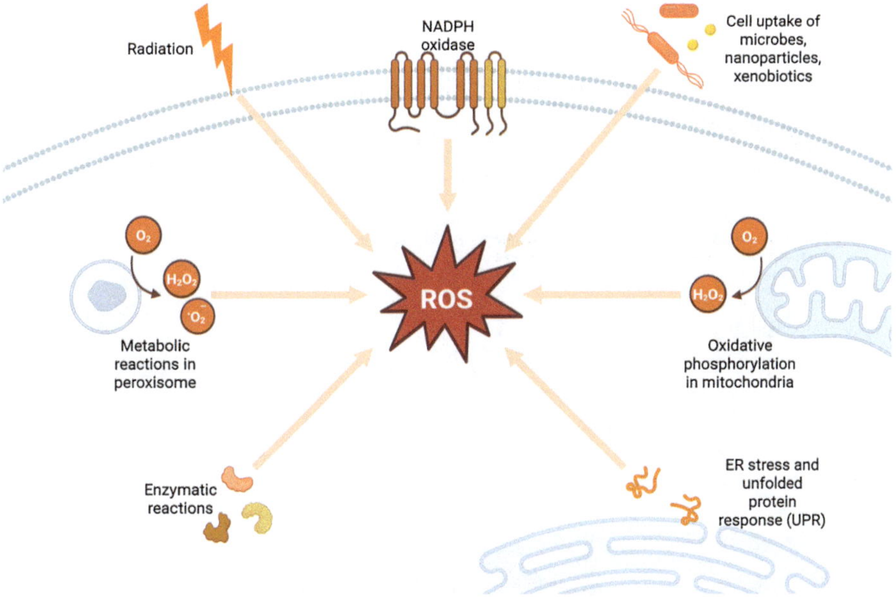

Figure 5.1 Different sources of ROS/RNS generation. Created in BioRender. https://BioRender.com/adpzo9p.

Mitochondrial dysfunction amplifies ROS production. Impairments in ETC complexes, commonly observed in neurodegenerative diseases, cause electron leakage and further ROS generation. For example, inhibition of complex I by rotenone or mutations in mitochondrial DNA (mtDNA) reduce ATP synthesis efficiency, elevating oxidative stress[14] (Figure 5.2). Moreover, defective mitophagy allows accumulation of damaged mitochondria that continuously leak ROS, creating a vicious cycle of oxidative injury.

5.2.1.2 NADPH Oxidases (NOX Family)

NADPH oxidases (NOX1–5, DUOX1/2) are specialized enzymes whose sole function is to produce ROS. In neurons and glial cells, NOX2 and NOX4 are particularly relevant. Microglial NOX2 generates superoxide in response to cytokines, amyloid-β, or pathogen-associated molecular patterns.[15] This superoxide acts both as a signaling molecule and a cytotoxic effector contributing to neuroinflammation. In models of Parkinson's disease and ischemic stroke, NOX activation exacerbates oxidative damage and neuronal death, whereas pharmacological inhibition of NOX reduces infarct size and inflammatory cytokine expression.[16]

5.2.1.3 Peroxisomal and Endoplasmic Reticulum Sources

Peroxisomes play a dual role in ROS metabolism; they generate H_2O_2 as a byproduct of fatty acid β-oxidation while also containing catalase that

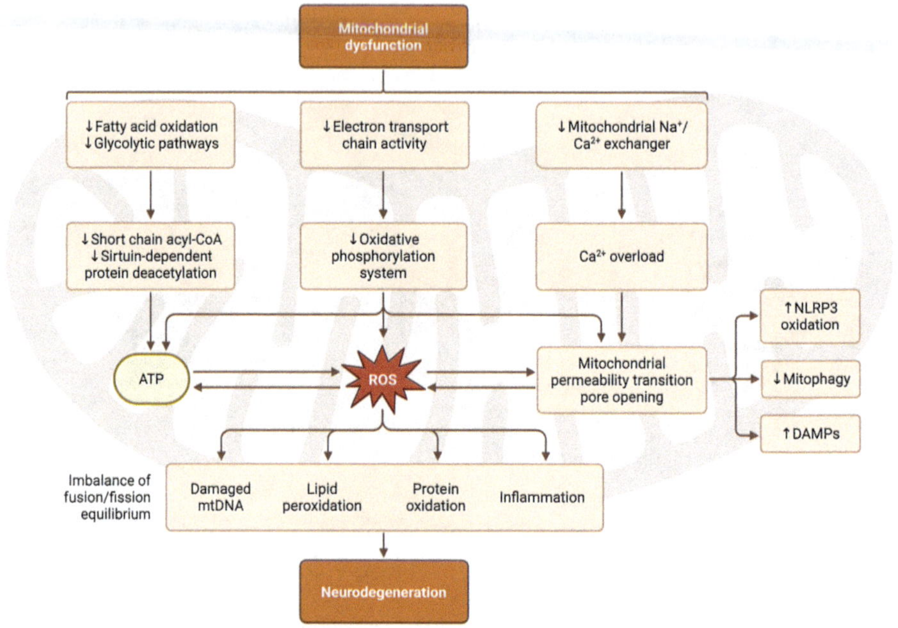

Figure 5.2 Mitochondrial dysfunction during oxidative stress. Created in BioRender. https://BioRender.com/adpzo9p.

decomposes H_2O_2 to water and oxygen. Dysfunctional peroxisomes or reduced catalase activity result in cytosolic accumulation of H_2O_2, enhancing oxidative burden.[17]

The endoplasmic reticulum (ER) also contributes to ROS generation during oxidative protein folding. Protein disulfide isomerase (PDI) and ER oxidoreductin-1 (ERO1) transfer electrons to molecular oxygen, yielding H_2O_2. Prolonged ER stress, which activates the unfolded protein response (UPR), leads to chronic ROS generation and cellular apoptosis.[18]

5.2.1.4 Other Enzymatic and Non-enzymatic Sources

Xanthine oxidase catalyzes the oxidation of hypoxanthine to xanthine and uric acid, generating superoxide and H_2O_2. Under ischemic conditions, this enzyme contributes to reperfusion injury in brain and heart tissues.[19] Cytochrome P450 enzymes in the liver and brain can also produce ROS as metabolic byproducts of xenobiotic detoxification. Non-enzymatic sources include autooxidation of catecholamines such as dopamine, leading to quinone and superoxide formation, a mechanism implicated in dopaminergic neuron loss in Parkinson's disease.[20]

5.2.1.5 Exogenous Sources of ROS

Environmental factors including ionizing radiation, ultraviolet light, pollutants, heavy metals, and cigarette smoke increase ROS generation.[21] These external stressors overwhelm antioxidant defenses and initiate redox-sensitive signaling cascades, thereby amplifying endogenous ROS production.

5.2.1.6 Sources of Reactive Nitrogen Species (RNS)

RNS primarily derive from nitric oxide ($NO^\bullet$), peroxynitrite ($ONOO^-$), and nitrogen dioxide ($NO_2^\bullet$). Nitric oxide synthases (NOS), neuronal (nNOS), endothelial (eNOS), and inducible (iNOS), convert L-arginine into $NO^\bullet$ and citrulline. While physiological $NO^\bullet$ regulates vasodilation and neurotransmission, pathological overproduction, especially by iNOS, triggers oxidative and nitrosative stress.[22]

$NO^\bullet$ reacts rapidly with $O_2^{\bullet-}$ to yield peroxynitrite ($ONOO^-$), a potent oxidant capable of nitrating tyrosine residues, oxidizing DNA, and impairing mitochondrial enzymes. Peroxynitrite-mediated nitration of mitochondrial complex I and manganese SOD (MnSOD) has been observed in Parkinson's and Alzheimer's disease brains, linking nitrosative stress to neurodegeneration.[23] Moreover, activated microglia and astrocytes release both $NO^\bullet$ and ROS, amplifying oxidative–inflammatory feedback loops.[24]

5.2.2 Antioxidant Defense Systems

To maintain redox equilibrium, cells have evolved sophisticated antioxidant defense systems, which are categorized as enzymatic and non-enzymatic (Figure 5.3).

5.2.2.1 Enzymatic Antioxidants

- Superoxide dismutase (SOD): Catalyzes dismutation of $O_2^{\bullet-}$ to H_2O_2 and O_2. Three isoforms exist: Cu/Zn-SOD (cytosolic), Mn-SOD (mitochondrial), and extracellular SOD. SOD mutations cause familial amyotrophic lateral sclerosis (ALS) due to loss of antioxidant function and gain of toxic aggregation properties.[25]
- Catalase (CAT): Located in peroxisomes, decomposes H_2O_2 into water and oxygen. Reduced catalase expression increases vulnerability to oxidative stress in neuronal models.[26]
- Glutathione peroxidase (GPx): Uses reduced glutathione (GSH) to reduce peroxides, forming oxidized glutathione (GSSG). GPx4 specifically prevents lipid peroxidation, thereby inhibiting ferroptosis, an iron-dependent form of cell death linked to neurodegeneration.[27]
- Peroxiredoxins (Prx): Thiol-based peroxidases that detoxify H_2O_2 and peroxynitrite, also modulating redox signaling.

5.2.2.2 Non-enzymatic Antioxidants

- Glutathione (GSH): A tripeptide (γ-glutamyl-cysteinyl-glycine) that serves as a primary redox buffer. It neutralizes free radicals directly and as a cofactor for GPx and glutathione S-transferase.

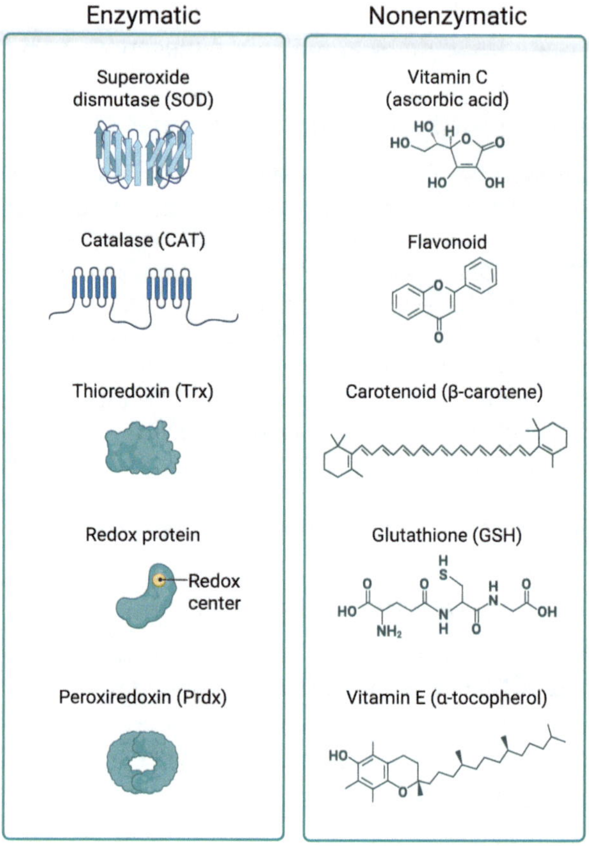

Figure 5.3 Different types of antioxidants. Created in BioRender. https://BioRender.com/adpzo9p.

- Vitamins: Vitamin C (ascorbic acid) reduces ROS in aqueous compartments and regenerates oxidized vitamin E, while vitamin E (α-tocopherol) protects lipid membranes from peroxidation.
- Uric acid and carotenoids: Act as radical scavengers and metal chelators.

5.2.3 Transcriptional Control *via* Nrf2 Pathway

Nuclear factor erythroid-2-related factor 2 (Nrf2) orchestrates antioxidant defense by inducing expression of detoxifying enzymes through the antioxidant response element (ARE). Under basal conditions, Nrf2 is sequestered by Keap1 in the cytoplasm and targeted for degradation. Under oxidative stress, Nrf2 is released, translocates to the nucleus, and activates genes such as heme oxygenase-1 (HO-1), NAD(P)H quinone dehydrogenase 1 (NQO1), and glutamate-cysteine ligase catalytic subunit (GCLC)[9,28] (Figure 5.4).

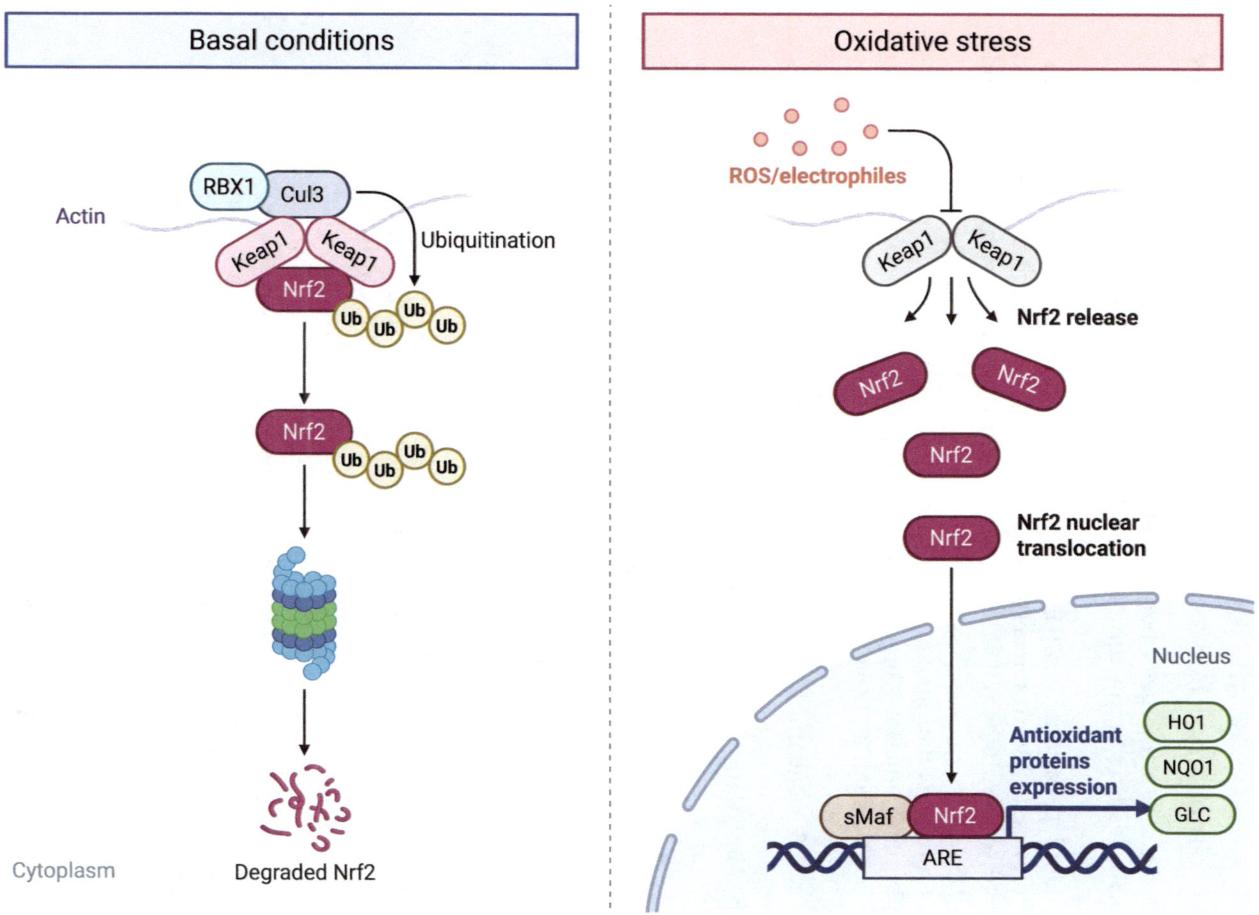

Figure 5.4 Keap1-Nrf2 pathway in basal and oxidative stress condition. Created in BioRender. https://BioRender.com/adpzo9p.

Nrf2 dysfunction has been implicated in Parkinson's and Alzheimer's diseases, where diminished antioxidant gene expression worsens neuronal vulnerability.[29]

5.2.4 Imbalance Between ROS/RNS and Antioxidant Defenses

Oxidative stress emerges when ROS and RNS production exceeds antioxidant capacity, leading to macromolecular damage. The consequences include:

- Lipid peroxidation: ROS attack polyunsaturated fatty acids in membranes, generating reactive aldehydes like malondialdehyde (MDA) and 4-hydroxynonenal (4-HNE), which form adducts with proteins and DNA.[30]
- Protein oxidation: Leads to carbonylation, nitration, and loss of enzymatic function. Aggregated oxidized proteins accumulate in inclusion bodies such as Lewy bodies and neurofibrillary tangles.[31]
- DNA damage: Hydroxyl radicals induce single- and double-strand breaks and 8-oxoguanine formation, triggering mutations and apoptosis.[32]

This imbalance triggers redox-sensitive transcription factors, including NF-κB, AP-1, and HIF-1α, which promote inflammatory cytokine expression and cellular adaptation.[33] However, persistent activation results in chronic inflammation and neuronal degeneration.

5.3 Oxidative Stress and the Nervous System

The nervous system is uniquely sensitive to oxidative stress, perhaps more so than any other organ system. Its dependence on aerobic metabolism is extraordinary, although the brain accounts for only about two percent of total body weight, it consumes nearly one-fifth of the body's oxygen to sustain the continuous activity of billions of neurons.[1] This immense metabolic demand drives rapid ATP turnover through oxidative phosphorylation, an efficient but risky process that inevitably produces reactive oxygen species (ROS) as by-products. Under normal physiological conditions, ROS act as signaling messengers in cellular processes like synaptic plasticity, gene regulation, and metabolic adaptation. However, when their production exceeds the cell's antioxidant defenses, ROS begin to attack proteins, lipids, and nucleic acids, disrupting neuronal homeostasis and triggering neurodegenerative cascades.[2,3]

5.3.1 Intrinsic Vulnerability of Neural Tissue

Several features of the central nervous system (CNS) contribute to its exceptional susceptibility to oxidative damage. Neurons, being post-mitotic cells, are designed for long-term function rather than renewal. Once damaged, they cannot be easily replaced, and the cumulative burden of oxidative insults over a lifetime can have lasting consequences.[34] The neuronal membrane is especially rich in polyunsaturated fatty acids (PUFAs), which enhance fluidity and synaptic function but also make the membrane a prime target for lipid peroxidation. When ROS attack PUFAs, they generate reactive aldehydes such as malondialdehyde (MDA) and 4-hydroxynonenal (4-HNE), which can modify proteins and DNA, forming toxic adducts that propagate further oxidative injury.[30]

The brain's high content of transition metals, such as iron and copper, adds another layer of vulnerability. While these metals are vital for enzymatic reactions and neurotransmitter synthesis, they also catalyze the formation of highly reactive hydroxyl radicals through Fenton chemistry. Iron accumulation, particularly in regions like the substantia nigra and hippocampus, has been implicated in the pathology of several neurodegenerative diseases.[35] Compounding this risk is the brain's relatively modest antioxidant capacity. Compared to other organs such as the liver, the brain expresses lower levels of catalase and glutathione peroxidase, key enzymes responsible for neutralizing hydrogen peroxide.[2] This imbalance tilts the redox state toward oxidative conditions under even modest stress.

5.3.2 Mitochondria: The Core Source of ROS

Mitochondria are central to both the survival and degeneration of neurons. These organelles generate ATP through the electron transport chain (ETC), but about one to two percent of the electrons that traverse the chain leak prematurely, reducing oxygen to form superoxide radicals.[12] Complexes I and III are the main sites of this electron leakage. Normally, these ROS are detoxified by superoxide dismutase (SOD) within the mitochondrial matrix, forming hydrogen peroxide, which is then broken down by glutathione peroxidase or catalase. However, under pathological conditions, such as ischemia, exposure to toxins, or genetic defects in mitochondrial proteins, ROS generation can dramatically increase, overwhelming antioxidant capacity.

Mitochondrial DNA (mtDNA), situated near the inner membrane where ROS are produced, is particularly susceptible to oxidative damage because it lacks protective histones. Mutations in mtDNA compromise ETC function, further increasing electron leakage and ROS formation, creating a destructive feedback loop.[14] Mitochondrial dysfunction not only leads to energy failure but also releases pro-apoptotic factors such as cytochrome c, triggering cell death pathways. Additionally, the dynamic balance between mitochondrial fission and fusion, essential for maintaining mitochondrial quality, is disturbed by oxidative stress (Figure 5.3). Defective mitophagy, the selective removal of damaged mitochondria, has been implicated in Parkinson's disease, particularly in individuals with mutations in the PINK1 or Parkin genes.[10]

5.3.3 Glial Involvement in Oxidative Homeostasis

Although neurons often receive the spotlight in discussions of brain oxidative stress, glial cells, astrocytes, microglia, and oligodendrocytes, play equally critical roles in shaping the redox environment of the CNS. Astrocytes serve as metabolic supporters, supplying neurons with precursors for glutathione (GSH) synthesis, one of the brain's most important antioxidants. Under physiological conditions, they buffer extracellular ROS and maintain ionic balance. However, during chronic stress or inflammation, astrocytes can shift into a reactive phenotype, upregulating NADPH oxidase (NOX) and inducible nitric oxide synthase (iNOS), transforming from protectors into sources of oxidative and nitrosative stress.[36]

Microglia, the brain's resident immune cells, represent another key contributor to redox imbalance. In their resting state, microglia continuously survey the environment for signs of damage. Upon activation, whether due to injury, infection, or accumulation of misfolded proteins, they release superoxide, nitric oxide, and inflammatory cytokines.[37] While acute activation aids in clearing debris, chronic activation results in prolonged ROS and RNS production, amplifying oxidative injury and synaptic dysfunction.[4] This chronic neuroinflammatory response not only damages neurons but also perpetuates glial activation, forming a self-reinforcing cycle of oxidative and inflammatory stress.

5.3.4 Calcium, Excitotoxicity, and ROS

One of the earliest consequences of oxidative imbalance in neurons is the disruption of calcium homeostasis. Calcium ions are essential for neurotransmitter release and neuronal excitability, but excessive accumulation becomes toxic. ROS can impair calcium channels, pumps, and mitochondrial uptake systems, leading to intracellular calcium overload.[38] Overactivation of glutamatergic NMDA receptors, a condition known as excitotoxicity, exacerbates this problem. Excessive calcium entry into neurons stimulates mitochondrial respiration and nitric oxide production, both of which enhance ROS formation.[39]

As calcium accumulates, mitochondria become overloaded, leading to the opening of the mitochondrial permeability transition pore (mPTP). This event results in further ROS release, ATP depletion, and eventual cell death. In stroke and traumatic brain injury, excitotoxicity-driven ROS production is one of the key mechanisms of neuronal loss, linking metabolic disturbance to oxidative degeneration.

5.3.5 Targets of Oxidative Damage

ROS and reactive nitrogen species (RNS) affect virtually every class of biomolecule in the nervous system. Lipid peroxidation compromises membrane integrity and affects synaptic vesicle trafficking, while protein oxidation leads to misfolding, aggregation, and enzymatic inactivation. For example, oxidation of cytoskeletal proteins like actin and tubulin disrupts axonal transport, a process essential for maintaining synaptic function.[40] Peroxynitrite, a potent oxidant formed by the reaction between superoxide and nitric oxide, nitrates tyrosine residues in proteins, altering mitochondrial enzymes and impairing electron transport.[23]

DNA is also a critical target of oxidative stress. The oxidation of guanine produces 8-hydroxy-2′-deoxyguanosine (8-OHdG), a mutagenic lesion that can interfere with transcription and replication fidelity. Persistent DNA damage activates repair pathways that, when overwhelmed, can lead to cell cycle arrest or apoptosis.[32] The cumulative impact of such molecular damage manifests as neuronal dysfunction, loss of synaptic connectivity, and ultimately neurodegeneration.

5.3.6 Redox-sensitive Signaling and Gene Regulation

While excessive ROS are harmful, moderate levels play regulatory roles in cellular signaling. ROS can reversibly modify cysteine residues in transcription factors, kinases, and phosphatases, thus modulating their activity.[33] One of the most well-characterized pathways is the activation of the transcription factor NF-κB, which responds to oxidative and inflammatory stimuli. NF-κB drives the expression of cytokines and adhesion molecules that amplify immune responses, but chronic activation contributes to neuroinflammation and synaptic dysfunction.

Conversely, the Nrf2 pathway serves as the major antioxidant defense mechanism. Under resting conditions, Nrf2 is sequestered in the cytoplasm by its inhibitor Keap1. Upon exposure to oxidative stress, Nrf2 dissociates and translocates to the nucleus, where it binds to antioxidant response elements (AREs) in the promoters of detoxifying genes such as heme oxygenase-1 (HO-1) and NAD(P)H quinone oxidoreductase-1 (NQO1).[9] This transcriptional reprogramming enhances the cell's resilience to oxidative stress (Figure 5.4). However, with aging or chronic disease, the responsiveness of the Nrf2 pathway declines, reducing the brain's ability to adapt to redox challenges.[41]

5.3.7 Disease-linked Manifestations of Oxidative Stress

In many neurological disorders, oxidative stress is not a secondary phenomenon but a driving force of pathology. In Alzheimer's disease (AD), markers of lipid peroxidation and protein oxidation appear early, often before significant amyloid deposition. Amyloid-β (Aβ) peptides interact with mitochondrial membranes, inhibiting complex IV activity and increasing superoxide generation.[5] ROS, in turn, promote Aβ aggregation and tau hyperphosphorylation, reinforcing the pathological loop. Microglial activation around amyloid plaques adds an inflammatory dimension, further heightening oxidative burden.[4]

- In Parkinson's disease (PD), the oxidative degradation of dopamine generates hydrogen peroxide and quinones, which can modify proteins and mitochondrial components. Mitochondrial complex I dysfunction, iron accumulation, and α-synuclein aggregation converge to amplify ROS production.[6] The substantia nigra, with its high iron and dopamine content, thus becomes a hotspot for oxidative injury.
- Amyotrophic lateral sclerosis (ALS) also exemplifies redox-driven neurodegeneration. Mutations in the Cu/Zn superoxide dismutase (SOD1) gene lead to a toxic gain of function, producing excessive ROS and damaging mitochondria.[25] Similarly, glutamate excitotoxicity and impaired astrocytic glutamate uptake increase oxidative stress in motor neurons.
- In multiple sclerosis (MS), inflammatory demyelination is accompanied by elevated ROS and RNS released from activated macrophages and microglia. Peroxynitrite-mediated nitration of myelin proteins and mitochondrial enzymes contributes to demyelination and axonal loss.[42] The same oxidative mechanisms exacerbate secondary progressive forms of MS, where mitochondrial injury and chronic inflammation dominate the pathology.
- Cerebral ischemia and reperfusion injury represent acute oxidative insults. During ischemia, ATP depletion disrupts ion gradients, causing calcium accumulation and ROS generation. When blood flow is restored, oxygen influx triggers a burst of ROS from mitochondria and xanthine oxidase, leading to neuronal apoptosis and necrosis.[11]

5.3.8 Oxidative Stress and Aging

Aging magnifies oxidative burden through cumulative mitochondrial inefficiency and declining antioxidant defenses. Over time, oxidatively damaged proteins and lipids accumulate, mitochondrial DNA mutations increase, and the proteasome's ability to remove damaged components diminishes.[7] The decline in Nrf2 signaling with age further compromises the adaptive antioxidant response, leaving neurons increasingly susceptible to stressors that would be harmless in youth.[41] The progressive imbalance between ROS production and antioxidant defense not only underlies neurodegenerative diseases but also contributes to the gradual cognitive decline characteristic of normal aging.

In sum, oxidative stress in the nervous system represents a convergence of metabolic, molecular, and cellular processes. It arises from the interplay between high energy demand, limited antioxidant capacity, and the delicate balance of glial and neuronal function. The evidence across multiple disorders, including Alzheimer's, Parkinson's, ALS, MS, and stroke, underscores that oxidative imbalance is not an isolated phenomenon but a shared mechanism of neural injury. While transient ROS signaling is vital for neuronal adaptation, chronic oxidative stress erodes the molecular scaffolding of neural function, leading to irreversible damage. Understanding how neurons and glia collectively maintain, lose, and attempt to restore redox equilibrium remains central to unraveling the pathophysiology of the brain across the lifespan and disease.

5.4 Neuroinflammation

Neuroinflammation refers to the inflammatory response that occurs within the central nervous system (CNS) and is primarily mediated by its resident immune cells, microglia and astrocytes. For decades, the brain was considered an "immune-privileged" organ due to the blood–brain barrier (BBB) and a perceived lack of lymphatic drainage. However, this concept has evolved. It is now understood that the CNS is highly immunologically active and maintains a dynamic dialogue with the peripheral immune system.[43] Neuroinflammation, therefore, is not inherently harmful; it serves essential roles in host defense, synaptic pruning, and tissue repair. Yet, when this response becomes chronic or dysregulated, it transitions from a protective mechanism to a destructive process that contributes to neuronal damage, oxidative stress, and neurodegenerative disease.[4]

5.4.1 The Physiological Role of Neuroinflammation

In a healthy brain, neuroinflammation acts as a tightly regulated and transient process. Upon infection, trauma, or accumulation of misfolded proteins, microglia are rapidly activated. They shift from a surveillant, ramified morphology to an amoeboid form capable of phagocytosis and cytokine release. This activation facilitates clearance of cellular debris and pathogens, while also recruiting astrocytes and, in certain cases, peripheral immune cells.[44] Astrocytes, traditionally viewed as metabolic supporters of neurons, also modulate inflammatory tone by releasing neurotrophic factors and scavenging excess glutamate. The resolution of inflammation is just as critical as its initiation, after the acute phase, anti-inflammatory signals and lipid mediators promote a return to homeostasis. Failure in this resolution phase, however, can transform acute neuroinflammation into a chronic, self-perpetuating state.[45]

5.4.2 Microglia: The Primary Effectors of CNS Inflammation

Microglia are the brain's innate immune sentinels, constantly surveying the environment for disturbances. They express an array of pattern recognition receptors (PRRs), including Toll-like receptors (TLRs), NOD-like receptors (NLRs), and scavenger receptors, enabling them to detect damage-associated molecular patterns (DAMPs) and pathogen-associated molecular patterns (PAMPs).[46] Activation of these receptors initiates signaling cascades that

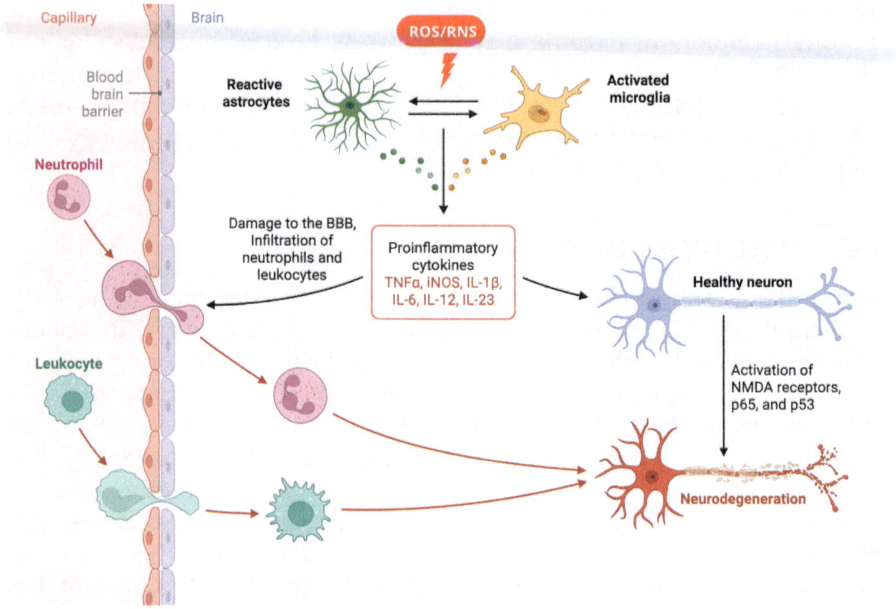

Figure 5.5 Relationship between oxidative stress and neuroinflammation. Created in BioRender. https://BioRender.com/adpzo9p.

lead to the release of proinflammatory cytokines such as interleukin-1β (IL-1β), tumor necrosis factor-alpha (TNF-α), and interleukin-6 (IL-6).

Activated microglia exhibit a spectrum of phenotypes. The classical "M1" state is characterized by proinflammatory and neurotoxic activity, with high production of nitric oxide and reactive oxygen species (ROS). In contrast, the "M2" phenotype is associated with tissue repair, secretion of anti-inflammatory cytokines (IL-10, TGF-β), and promotion of neurogenesis.[47] Although this dichotomy is oversimplified, it highlights the dual nature of microglial function, protective when regulated, but destructive when sustained (Figure 5.5). Chronic activation skews microglia toward a persistent M1-like state, contributing to synaptic stripping, oxidative stress, and neuronal apoptosis.[37]

5.4.3 Astrocytes and Their Regulatory Role

Astrocytes, the most abundant glial cells in the CNS, play an indispensable role in neuroinflammation. They maintain BBB integrity, regulate neurotransmitter levels, and provide metabolic support to neurons. In response to injury or infection, astrocytes become "reactive", a process known as astrogliosis (Figure 5.5). Reactive astrocytes undergo hypertrophy, upregulate glial fibrillary acidic protein (GFAP), and secrete a wide range of cytokines, chemokines, and growth factors.[48]

Astrocytic activation is not merely a passive response to microglial signaling; it is a powerful amplifier of neuroinflammation. The release of cytokines such as IL-6 and monocyte chemoattractant protein-1 (MCP-1) recruits peripheral immune cells into the CNS, while astrocyte-derived ROS and nitric oxide further exacerbate oxidative injury.[49] Astrocytes also communicate bidirectionally with microglia, perpetuating inflammatory loops that prolong glial activation. Over time, this chronic gliosis leads to impaired synaptic function, reduced neurotrophic support, and glial scar formation, hallmarks of long-term CNS inflammation.

5.4.4 Cytokines and Inflammatory Mediators

Cytokines are central to neuroinflammatory signaling. Among them, TNF-α, IL-1β, and IL-6 are the most studied. TNF-α, primarily released by activated microglia, has pleiotropic effects: at physiological levels, it supports synaptic scaling and neuroprotection; in excess, it triggers apoptosis through TNF receptor 1.[50] IL-1β, a master regulator of inflammation, is produced as an inactive precursor that requires cleavage by caspase-1 within the inflammasome complex to become active.[4] Elevated IL-1β impairs long-term potentiation (LTP) and promotes neuronal death, linking inflammation directly to cognitive decline. IL-6, on the other hand, acts as both pro- and anti-inflammatory depending on the context and receptor engagement.

Chemokines, including MCP-1 (CCL2), CXCL8, and fractalkine (CX3CL1), coordinate the migration of immune cells and mediate communication between neurons and glia. The fractalkine–CX3CR1 signaling axis is particularly important for neuron–microglia interaction, serving as a brake on excessive inflammation. Disruption of this pathway has been implicated in Alzheimer's and Parkinson's disease models.[51]

5.4.5 Key Signaling Pathways in Neuroinflammation

Neuroinflammatory signaling involves several converging pathways that regulate cytokine production, oxidative stress, and glial activation. The NF-κB pathway is one of the most prominent. Activation of NF-κB through TLRs or cytokine receptors leads to transcription of inflammatory genes such as TNF-α, IL-1β, iNOS, and COX-2.[33] While transient NF-κB activation promotes defense and repair, chronic stimulation drives sustained inflammation and neuronal injury.

The NLRP3 inflammasome represents another major regulatory hub. It is a multiprotein complex consisting of the NLRP3 receptor, adaptor protein ASC, and caspase-1. Upon activation by signals such as ATP, ROS, or aggregated proteins (*e.g.*, Aβ or α-synuclein), NLRP3 triggers maturation of IL-1β and IL-18, amplifying the inflammatory response.[4] Dysregulation of the inflammasome is now recognized as a key mechanism in neurodegenerative and metabolic brain disorders.

Mitogen-activated protein kinases (MAPKs), including p38, JNK, and ERK, also mediate glial responses to stress. These kinases regulate the production of cytokines and prostaglandins and contribute to neuronal death by modulating apoptotic pathways.[52] Cross-talk between these pathways integrates immune signaling with oxidative and metabolic stress, establishing a complex network that determines whether inflammation resolves or persists.

5.4.6 The Interface Between Neuroinflammation and Oxidative Stress

Oxidative stress and neuroinflammation are intimately connected, forming a self-reinforcing cycle. Activated microglia and astrocytes generate ROS and RNS through the activity of NADPH oxidase and iNOS. These reactive molecules damage nearby neurons and further stimulate glial activation (Figure 5.4). ROS also function as second messengers that activate redox-sensitive transcription factors such as NF-κB and AP-1, perpetuating cytokine release.[37]

Conversely, oxidative damage to neuronal components can trigger inflammatory responses. Lipid peroxidation products, such as 4-HNE act as DAMPs, engage pattern recognition receptors on glia and initiate innate immune responses (Figure 5.3). In diseases like Alzheimer's and Parkinson's, the accumulation of misfolded proteins (Aβ, α-synuclein) induces mitochondrial dysfunction and ROS production, which in turn activate NLRP3 inflammasomes.[4,53] Thus, oxidative stress not only results from inflammation but also acts as a key driver of it.

5.4.7 Breakdown of the Blood–Brain Barrier

A crucial consequence of prolonged neuroinflammation is the compromise of the BBB. Under normal conditions, endothelial cells, astrocyte end-feet, and pericytes form a tightly regulated interface that limits the entry of peripheral immune cells and toxins. Inflammatory cytokines such as TNF-α and IL-1β disrupt tight junction proteins (occludin, claudin-5), increasing permeability.[54] This breakdown allows leukocytes and plasma proteins to infiltrate the parenchyma, intensifying local inflammation.

Microglial-derived matrix metalloproteinases (MMP-2, MMP-9) further degrade the extracellular matrix and tight junctions, exacerbating barrier dysfunction.[55] The resulting influx of peripheral immune cells creates a feed-forward loop where CNS and systemic immune responses amplify one another. In diseases such as multiple sclerosis (MS), this mechanism underlies demyelination and lesion formation.

5.4.8 Neuroinflammation Across Neurological Disorders

- Alzheimer's disease (AD): In AD, microglial activation occurs early in disease progression, even before overt amyloid deposition. Microglia initially attempt to clear Aβ through phagocytosis but eventually become dysfunctional, releasing proinflammatory mediators that exacerbate neuronal damage.[4] Astrocytes surrounding plaques secrete complement components and ROS, contributing to synaptic loss. Elevated cytokines such as IL-1β and TNF-α correlate with cognitive decline and tau pathology.

- Parkinson's disease (PD): In PD, microglia in the substantia nigra display heightened reactivity, releasing TNF-α, IL-6, and ROS in response to α-synuclein aggregates.[56] These inflammatory mediators potentiate dopaminergic neuron death and sustain microglial activation. Postmortem studies show increased expression of iNOS and COX-2 in affected regions, highlighting the inflammatory contribution to PD pathogenesis.
- Multiple sclerosis (MS): MS represents a prototypical inflammatory demyelinating disease. Activated microglia and infiltrating macrophages release ROS, RNS, and cytokines that damage myelin and oligodendrocytes. T cells and B cells crossing the compromised BBB further fuel inflammation through antigen presentation and antibody production.[57] Chronic lesions in progressive MS are characterized by a smoldering form of inflammation, dominated by microglial activation rather than peripheral immune infiltration.
- Traumatic brain injury (TBI): Neuroinflammation also plays a major role in TBI. The primary mechanical insult triggers immediate activation of microglia and astrocytes, followed by secondary waves of inflammation that can persist for months. Elevated levels of IL-1β and TNF-α disrupt synaptic function and contribute to post-traumatic neurodegeneration.[45]

5.4.9 Resolution and Regulation of Neuroinflammation

The resolution of neuroinflammation is an active, orchestrated process involving anti-inflammatory cytokines (IL-10, TGF-β), lipid mediators (resolvins, lipoxins), and regulatory immune cells. Microglia transition from proinflammatory states to reparative phenotypes, clearing debris and promoting tissue repair. Astrocytes contribute by restoring BBB integrity and secreting neurotrophic factors. Failure in these resolution mechanisms results in chronic inflammation, oxidative stress, and progressive neurodegeneration.[58]

Endogenous regulatory mechanisms such as the fractalkine (CX3CL1)/CX3CR1 axis and the CD200/CD200R interaction help maintain microglia in a quiescent state under normal conditions. Disruption of these signaling pathways, common in aging and neurodegenerative diseases, predisposes the brain to exaggerated inflammatory responses.[51] Thus, neuroinflammation is a double-edged sword: indispensable for repair but potentially devastating when unresolved.

Neuroinflammation embodies both the resilience and fragility of the nervous system. As a physiological defense mechanism, it protects neurons from pathogens and injury, orchestrating repair and regeneration. Yet, when sustained or improperly regulated, it transforms into a pathological process that contributes to the very degeneration it seeks to prevent. Microglia and astrocytes, though indispensable for maintaining CNS integrity, become sources of inflammatory and oxidative damage under chronic stimulation.

The interplay between neuroinflammation, oxidative stress, and cellular metabolism underscores the complexity of CNS pathology. Across diverse diseases, ranging from Alzheimer's and Parkinson's to multiple sclerosis and traumatic injury, this shared mechanism of glial activation, cytokine release, and ROS generation forms the backbone of progressive neural injury. Understanding the fine line between protective and pathological inflammation remains central to deciphering brain health and disease.

5.5 Therapeutic Approaches for Oxidative Stress Along with Neuroinflammation

Neuroprotection refers to the ensemble of biological, pharmacological, and behavioral strategies that preserve the structure and function of neurons against injury and degeneration. The concept extends beyond preventing cell death, it encompasses the preservation of synaptic integrity, mitochondrial health, and cellular signaling essential for neuroplasticity and cognitive performance. In the context of oxidative stress and neuroinflammation, neuroprotection aims to restore redox balance, dampen inflammatory cascades, and reinforce endogenous defense mechanisms. The complexity of neural injury mechanisms demands a multidimensional approach, integrating pharmacological interventions, non-pharmacological techniques, dietary and herbal components, and behavioral modifications (Figure 5.6).

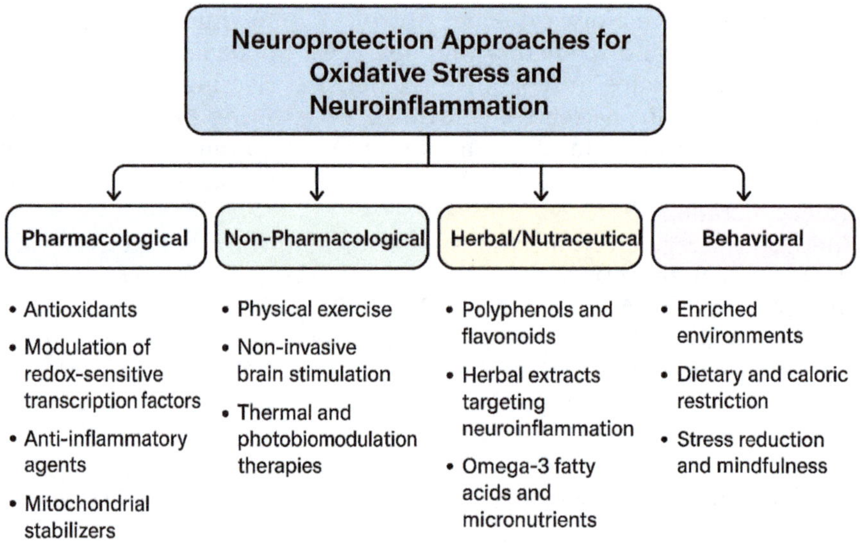

Figure 5.6 Therapeutic approaches against oxidative stress and neuroinflammation.

At its core, neuroprotection relies on maintaining the balance between neuronal excitation and inhibition, oxidative production and detoxification, and immune activation and resolution. Neurons, by design, are metabolically intense and vulnerable to oxidative injury due to their reliance on aerobic metabolism. Any disturbance in mitochondrial efficiency, ion homeostasis, or glial support can initiate degenerative cascades. Endogenous mechanisms such as the Nrf2 antioxidant pathway, neurotrophic signaling (*e.g.*, BDNF, NGF), and heat shock proteins act as frontline defenses. However, with aging and chronic stress, these protective systems weaken, allowing oxidative and inflammatory processes to dominate. Therefore, neuroprotective strategies attempt to either reinforce these endogenous systems or mimic their functions through external interventions.[34]

5.5.1 Pharmacological Approaches

Pharmacological neuroprotection has long been the focus of neuroscience research, with compounds targeting oxidative stress, mitochondrial function, excitotoxicity, and neuroinflammation.

5.5.1.1 Antioxidant-based Drugs

Synthetic and natural antioxidants have been explored to counteract excessive reactive oxygen and nitrogen species. Molecules such as edaravone, a free radical scavenger used in amyotrophic lateral sclerosis (ALS), and *N*-acetylcysteine (NAC), a glutathione precursor, directly enhance intracellular antioxidant capacity.[59] Vitamin E (α-tocopherol) and coenzyme Q10, although not consistently effective in clinical trials, have demonstrated neuroprotective potential in experimental models by stabilizing mitochondrial membranes and preventing lipid peroxidation.[60]

5.5.1.2 Mitochondrial Stabilizers

Because mitochondrial dysfunction is a central source of oxidative stress, agents that preserve mitochondrial integrity are of great interest. Compounds such as creatine, acetyl-L-carnitine, and SS-31 peptide help maintain mitochondrial membrane potential and reduce electron leakage, thus decreasing ROS production.[61] These agents also support ATP synthesis and prevent apoptotic signaling triggered by cytochrome c release.

5.5.1.3 Nrf2 Activators and Redox Modulators

Pharmacological activation of the Nrf2 pathway has emerged as a robust neuroprotective strategy. Molecules like sulforaphane (from cruciferous vegetables), dimethyl fumarate (used in multiple sclerosis), and bardoxolone methyl enhance the transcription of detoxifying enzymes such as heme oxygenase-1 (HO-1) and NAD(P)H quinone oxidoreductase (NQO1).[9] These compounds provide long-lasting protection by boosting the brain's intrinsic antioxidant network rather than simply scavenging free radicals.

5.5.1.4 *Anti-inflammatory Agents*

Given the interdependence of oxidative stress and inflammation, modulating microglial activation and cytokine production offers another avenue for neuroprotection. Nonsteroidal anti-inflammatory drugs (NSAIDs) like ibuprofen and selective COX-2 inhibitors have shown neuroprotective effects in preclinical studies, particularly in reducing amyloid-induced neuroinflammation. However, their chronic use carries systemic risks. More targeted compounds, such as PPAR-γ agonists, modulate inflammatory gene expression without broad immunosuppression.[4]

5.5.1.5 *Ion Channel and Excitotoxicity Modulators*

Excessive glutamate signaling leads to excitotoxicity and calcium overload. NMDA receptor antagonists such as memantine, approved for Alzheimer's disease, work by partially blocking excessive NMDA receptor activation while sparing physiological neurotransmission.[62] Voltage-gated calcium channel blockers like nimodipine also show protective effects by preventing calcium-induced mitochondrial damage and oxidative stress.

These pharmacological approaches demonstrate that successful neuroprotection often requires a multi-targeted design, addressing oxidative stress, inflammation, and excitotoxicity simultaneously rather than focusing on one isolated pathway.

5.5.2 Non-pharmacological Approaches

Non-pharmacological interventions have gained attention for their ability to stimulate endogenous neuroprotective pathways with fewer side effects. These include physical exercise, electrical and sensory stimulation, and light-based therapies that leverage the brain's inherent plasticity and resilience mechanisms.

5.5.2.1 *Exercise and Physical Activity*

Regular aerobic and resistance exercise enhance cerebral blood flow, mitochondrial biogenesis, and antioxidant enzyme expression. Exercise upregulates brain-derived neurotrophic factor (BDNF), promoting synaptic plasticity and neuronal survival.[63] Moreover, it attenuates chronic inflammation by reducing microglial activation and increasing anti-inflammatory cytokines such as IL-10. Even moderate physical activity has been associated with the delayed onset of neurodegenerative symptoms in conditions like Parkinson's and Alzheimer's diseases.

5.5.2.2 Electrical and Sensory Stimulation

Non-invasive brain stimulation, such as transcranial direct current stimulation (tDCS) and repetitive transcranial magnetic stimulation (rTMS), can modulate cortical excitability and neuroplasticity. These techniques induce changes in synaptic strength and activate intracellular pathways associated with antioxidant defense and neurotrophin signaling.[64] Similarly, peripheral sensory stimulation, such as auditory, olfactory, or visual stimulation, has shown potential for modulating deep brain circuits involved in emotional regulation and memory.[65]

5.5.2.3 Photobiomodulation and Hypoxic Conditioning

Photobiomodulation, using red or near-infrared light, stimulates mitochondrial cytochrome c oxidase, enhancing ATP production and reducing oxidative stress. This approach has shown promise in models of traumatic brain injury and neurodegeneration.[66] Controlled intermittent hypoxia, on the other hand, induces adaptive upregulation of antioxidant enzymes and stress-response proteins, promoting neuronal resilience through preconditioning effects.[67]

5.5.2.4 Nutritional Interventions

Dietary approaches such as caloric restriction and ketogenic diets have demonstrated neuroprotective effects through improved mitochondrial function, increased autophagy, and reduced ROS production. Caloric restriction activates sirtuin pathways (particularly SIRT1), enhancing mitochondrial biogenesis and reducing inflammation. The ketogenic diet, by shifting metabolism toward ketone utilization, decreases oxidative stress and stabilizes neuronal energetics.[68]

5.5.3 Herbal and Nutraceutical Approaches

Natural compounds have long been investigated for their ability to counteract oxidative stress and neuroinflammation. Many plant-derived molecules exhibit pleiotropic actions, simultaneously scavenging free radicals, modulating signaling pathways, and regulating gene expression involved in cellular defense.

5.5.3.1 Polyphenols and Flavonoids

Polyphenolic compounds such as curcumin (from turmeric), resveratrol (from grapes), and epigallocatechin gallate (EGCG, from green tea) exert neuroprotective effects *via* antioxidant, anti-inflammatory, and mitochondrial-stabilizing mechanisms.[69–71] Curcumin inhibits NF-κB signaling and upregulates Nrf2-dependent genes, while resveratrol activates sirtuin-1, enhancing mitochondrial function and longevity.[69] Flavonoids also influence synaptic plasticity through modulation of MAPK and CREB pathways, improving cognitive resilience.

5.5.3.2 Omega-3 Fatty Acids

Docosahexaenoic acid (DHA) and eicosapentaenoic acid (EPA), key components of neuronal membranes, possess strong anti-inflammatory and antioxidant properties.[72] Omega-3 supplementation promotes membrane fluidity, reduces lipid peroxidation, and downregulates proinflammatory mediators. Epidemiological studies associate high omega-3 intake with lower incidence of neurodegenerative diseases and improved cognitive outcomes.[72]

5.5.3.3 Adaptogens and Herbal Extracts

Extracts from *Ginkgo biloba*, *Withania somnifera* (ashwagandha), and *Bacopa monnieri* (Brahmi) have been studied for cognitive enhancement and neuroprotection. *Ginkgo biloba* flavonoids improve cerebral circulation and reduce oxidative stress, while withanolides from ashwagandha exhibit anti-inflammatory and neuroregenerative effects.[73,74] *Bacopa* extracts enhance memory and protect against ROS-mediated synaptic damage by increasing antioxidant enzymes like SOD and catalase.[75]

5.5.3.4 Dietary Antioxidants

Vitamins C and E, carotenoids, and selenium function synergistically as dietary antioxidants. Although supplementation outcomes in large clinical trials have been inconsistent, their combined presence in diets rich in fruits, vegetables, and whole grains correlates strongly with reduced cognitive decline and slower progression of neurodegenerative diseases. The "Mediterranean diet," characterized by a high intake of polyphenols, omega-3s, and unsaturated fats, exemplifies this integrated dietary neuroprotection.[8]

5.5.4 Behavioral and Environmental Approaches

The brain's capacity for self-repair and adaptation provides an avenue for behavioral neuroprotection. Interventions such as cognitive stimulation, enriched environments, stress management, and adequate sleep reinforce neural plasticity and metabolic efficiency.

5.5.4.1 Cognitive and Environmental Enrichment

Exposure to stimulating environments, rich in social, sensory, and intellectual engagement, enhances neurogenesis, synaptic density, and expression of neurotrophic factors.[76] Animal studies consistently show that enriched environments mitigate the effects of oxidative stress and inflammation, improving outcomes in models of Alzheimer's and Parkinson's diseases. These effects are mediated partly by increased expression of BDNF and antioxidant enzymes.[77]

5.5.4.2 Sleep and Circadian Regulation

Sleep serves as a critical period for metabolic clearance and redox balance. The glymphatic system, active during slow-wave sleep, facilitates the removal of ROS-induced metabolites and misfolded proteins such as amyloid-β.[78] Chronic sleep deprivation disrupts mitochondrial metabolism, heightens oxidative stress, and impairs memory consolidation.[79] Maintaining circadian rhythm through consistent light–dark exposure and sleep hygiene is therefore essential for long-term neuroprotection.

5.5.4.3 Stress Reduction and Mindfulness Practices

Chronic psychological stress activates the hypothalamic–pituitary–adrenal (HPA) axis, elevating cortisol levels and promoting oxidative and inflammatory damage in brain regions such as the hippocampus and prefrontal cortex. Mindfulness meditation, yoga, and breathing exercises modulate stress hormones and enhance antioxidant enzyme expression, providing measurable neuroprotective effects.[80]

5.5.4.4 Social Engagement and Emotional Health

Loneliness and depression correlate strongly with increased markers of oxidative stress and systemic inflammation. Regular social interaction and positive emotional experiences buffer the effects of stress and contribute to neuroplasticity by sustaining dopaminergic and serotonergic activity.[81] Thus, emotional well-being acts as a biological safeguard against neurodegeneration.

5.5.5 Integrated Perspective on Neuroprotection

Effective neuroprotection requires integration across molecular, cellular, and behavioral levels. Pharmacological agents may correct biochemical imbalances, but their effects are often transient unless supported by lifestyle and environmental interventions that sustain neuronal health. For example, combining antioxidant or Nrf2-activating compounds with regular exercise and cognitive engagement can synergistically enhance mitochondrial resilience and synaptic plasticity. Similarly, diets rich in polyphenols and omega-3 fatty acids complement pharmacological efforts to suppress inflammation and oxidative stress.

The emerging concept of *adaptive neuroprotection* emphasizes that controlled, moderate stress, such as exercise, caloric restriction, or intermittent hypoxia, can induce hormetic responses that strengthen endogenous defense pathways.[82] These adaptive processes, rather than simple suppression of oxidative stress, represent the next frontier in preventing age-related and neurodegenerative decline.

Neuroprotection encompasses a broad and interconnected framework that extends beyond pharmacology. It reflects the brain's inherent capacity to adapt, compensate, and resist damage through coordinated molecular and behavioral mechanisms. The most successful neuroprotective strategies do not act through single targets but rather by restoring balance, between oxidation and reduction, inflammation and resolution, and stress and adaptation.

From pharmacological agents that modulate redox and inflammatory pathways, to lifestyle practices that enhance resilience, neuroprotection embodies the integration of science and behavior. Ultimately, sustaining neural health depends not only on drugs or supplements but on aligning biological defenses with daily life choices that support the brain's natural capacity for renewal and repair.

5.6 Summary

Oxidative stress and neuroinflammation are interdependent and self-perpetuating processes that underlie the pathophysiology of many neurodegenerative and neuropsychiatric disorders. The brain's high oxygen consumption, abundant lipid content, and limited regenerative capacity make it particularly vulnerable to oxidative damage when the balance between reactive oxygen and nitrogen species (ROS/RNS) production and antioxidant defense systems is disrupted. At the molecular level, mitochondrial dysfunction, NADPH oxidase activation, and nitric oxide synthase dysregulation drive the overproduction of ROS and RNS, leading to lipid peroxidation, protein nitration, and DNA oxidation. Endogenous antioxidant systems, *i.e.*, superoxide dismutase, catalase, glutathione peroxidase, and the Nrf2/Keap1 signaling pathway normally counteract oxidative injury, but their impairment results in neuronal damage and impaired synaptic transmission.

Oxidative stress closely interacts with neuroinflammation, where activated microglia and astrocytes release pro-inflammatory cytokines such as TNF-α, IL-1β, and IL-6, leading to blood–brain barrier disruption and neurotoxic signaling cascades. This interaction creates a feed-forward loop: ROS/RNS induce inflammatory gene expression *via* NF-κB and MAPK pathways, while chronic inflammation exacerbates mitochondrial dysfunction and ROS production, amplifying neuronal vulnerability. Neurodegenerative diseases such as Alzheimer's, Parkinson's, and multiple sclerosis exemplify this interplay, where oxidative injury and immune dysregulation converge to drive progressive neuronal death and synaptic failure.

Therapeutic strategies have evolved to target both oxidative and inflammatory cascades simultaneously. Pharmacological interventions such as dimethyl fumarate, minocycline, and MitoQ modulate redox-sensitive transcription factors and glial activation to mitigate neuronal stress. Non-pharmacological interventions, including aerobic exercise, environmental enrichment, and electrical neurostimulation, enhance endogenous antioxidant capacity and promote anti-inflammatory glial phenotypes. Herbal and nutraceutical compounds such as curcumin, resveratrol, and ashwagandha provide multi-targeted antioxidant and anti-inflammatory effects, while lifestyle modifications like caloric restriction, intermittent fasting, and mindfulness reduce systemic oxidative burden. Nanocarrier-based drug delivery, exosome vectors, and adaptive neurostimulation have emerged as promising translational tools to enhance therapeutic efficacy and brain bioavailability.

5.7 Limitations and Future Directions

While the conceptual framework and therapeutic options are promising, multiple limitations hinder effective translation to clinical practice. First, the extrinsic complexity of oxidative and inflammatory networks in the CNS with myriad ROS/RNS species, glial phenotypes, immune signaling pathways, and feedback loops makes single-target interventions often insufficient. Second, preclinical to clinical translation remains problematic: most animal models lack the age-related, comorbid, and heterogeneous environment of human neurodegenerative disease, leading to failures of antioxidant and anti-inflammatory agents in human trials. Third, pharmacokinetic and bioavailability challenges are frequently encountered: many compounds with neuroprotective promise cannot penetrate the BBB, are rapidly metabolized, or do not reach effective concentrations in neural tissue. Fourth, there is a lack of reliable biomarkers for *in vivo* monitoring of oxidative stress and neuroinflammation, current assays (*e.g.*, 8-oxo-dG, malondialdehyde, cytokines) lack spatial/temporal resolution and CNS specificity. Fifth, issues around safety and chronic dosing emerge: large-scale antioxidant supplementation may disrupt physiologic redox signaling or immune repair functions (redox paradox), and prolonged inflammation suppression may interfere with host defense. Finally, inter-individual variability in genetics, metabolic status, lifestyle, and baseline antioxidant/immune reserve creates heterogeneity in susceptibility and treatment response, suggesting that one-size-fits-all strategies are unlikely to succeed.

To advance meaningful neuroprotection, future research must adopt integrative and personalized approaches. First, systems-level analyses using multi-omics (transcriptomics, metabolomics, redox-proteomics), network modelling and machine learning can identify central nodes linking oxidative stress and immune signaling, offering better therapeutic targets. Second, targeted delivery technologies including nanocarriers, exosome-based vectors, intranasal or focused-ultrasound mediated BBB transport can improve brain bioavailability and reduce off-target toxicity. Third, adaptive neurostimulation paradigms, integrating electrophysiological and metabolic feedback, may allow dynamic modulation of redox–inflammatory balance and glial phenotypes. Fourth, personalized redox–immune medicine guided by genomic, metabolomic and imaging biomarkers could optimize patient selection, dose-timing and combinatorial therapy strategies. Fifth, combined multimodal therapies pairing pharmacological antioxidants/anti-inflammatories with lifestyle-based hermetic interventions (exercise, enriched environments, caloric restriction) offer synergistic potential to restore endogenous defense mechanisms. Lastly, large-scale, longitudinal human trials deploying multimodal biomarker panels (neuroimaging, blood/CSF redox/inflammatory markers, cognition metrics) are essential to validate preclinical findings and accelerate translation. In sum, the dual regulation of oxidative stress and neuroinflammation remains a central axis for preserving neural integrity, and future progress will depend on integrated molecular, technological, and behavioral strategies to sustain brain redox equilibrium and immune homeostasis across the lifespan. Large-scale human trials with rigors, using neuroimaging, redox signatures, and cognitive metrics as multimodal biomarkers, will be necessary to confirm the clinical relevance of preclinical findings.

5.8 Concluding Remarks

Such duality in the regulation of oxidative stress and neuroinflammation stands out as a very important approach to maintaining neural integrity. Advances will come through the development of integrated therapeutic systems that incorporate molecular, technological, and behavioral approaches. Current interventions demonstrate partial efficacy but are likely to be modified by precision targeting, biomarker-guided dosing, and multi-domain synergy to redefine the management of neurodegenerative diseases. Neuroprotection, previously thought of as a reactive field, is becoming a proactive one focused on sustaining brain redox equilibrium and immunological homeostasis across the human lifespan.

Acknowledgements

The author gratefully acknowledges his current lab supervisor, Prof. Maysam Chamanzar, for providing access to the BioRender lab account, to create the figures for this chapter.

References

1. D. Attwell and S. B. Laughlin, An energy budget for signaling in the grey matter of the brain, *J. Cereb. Blood Flow Metab.*, 2001, **21**, 1133–1145.
2. B. Halliwell and M. C. John, *Gutteridge Free Radicals in Biology and Medicine*, 5th edn, 2015.
3. J. N. Cobley, M. L. Fiorello and D. M. Bailey, 13 reasons why the brain is susceptible to oxidative stress, *Redox Biol.*, 2018, **15**, 490–503.

4. M. T. Heneka, M. P. Kummer and E. Latz, Innate immune activation in neurodegenerative disease, *Nat. Rev. Immunol.*, 2014, **14**, 463–477.
5. D. A. Butterfield and D. Boyd-Kimball, Oxidative Stress, Amyloid-beta Peptide, and Altered Key Molecular Pathways in the Pathogenesis and Progression of Alzheimer's Disease, *J. Alzheimers Dis.*, 2018, **62**, 1345–1367.
6. V. Dias, E. Junn and M. M. Mouradian, The role of oxidative stress in Parkinson's disease, *J. Parkinsons Dis.*, 2013, **3**, 461–491.
7. C. Lopez-Otin, M. A. Blasco, L. Partridge, M. Serrano and G. Kroemer, The hallmarks of aging, *Cell*, 2013, **153**, 1194–1217.
8. N. Scarmeas, C. A. Anastasiou and M. Yannakoulia, Nutrition and prevention of cognitive impairment, *Lancet Neurol.*, 2018, **17**, 1006–1015.
9. T. W. Kensler, N. Wakabayashi and S. Biswal, Cell survival responses to environmental stresses *via* the Keap1-Nrf2-ARE pathway, *Annu. Rev. Pharmacol. Toxicol.*, 2007, **47**, 89–116.
10. A. M. Pickrell and R. J. Youle, The roles of PINK1, parkin, and mitochondrial fidelity in Parkinson's disease, *Neuron*, 2015, **85**, 257–273.
11. C. L. Allen and U. Bayraktutan, Oxidative stress and its role in the pathogenesis of ischaemic stroke, *Int. J. Stroke*, 2009, **4**, 461–470.
12. M. P. Murphy, How mitochondria produce reactive oxygen species, *Biochem. J.*, 2009, **417**, 1–13.
13. C. C. Winterbourn, Reconciling the chemistry and biology of reactive oxygen species, *Nat. Chem. Biol.*, 2008, **4**, 278–286.
14. M. T. Lin and M. F. Beal, Mitochondrial dysfunction and oxidative stress in neurodegenerative diseases, *Nature*, 2006, **443**, 787–795.
15. S. Sorce and K. H. Krause, NOX enzymes in the central nervous system: from signaling to disease, *Antioxid. Redox Signal.*, 2009, **11**, 2481–2504.
16. J. D. Lambeth and A. S. Neish, Nox enzymes and new thinking on reactive oxygen: a double-edged sword revisited, *Annu. Rev. Pathol.*, 2014, **9**, 119–145.
17. M. Schrader and H. D. Fahimi, Peroxisomes and oxidative stress, *Biochim. Biophys. Acta*, 2006, **1763**, 1755–1766.
18. B. P. Tu and J. S. Weissman, Oxidative protein folding in eukaryotes: mechanisms and consequences, *J. Cell Biol.*, 2004, **164**, 341–346.
19. P. Pacher, S. Batkai and G. Kunos, The endocannabinoid system as an emerging target of pharmacotherapy, *Pharmacol. Rev.*, 2006, **58**, 389–462.
20. D. G. Graham, Oxidative pathways for catecholamines in the genesis of neuromelanin and cytotoxic quinones, *Mol. Pharmacol.*, 1978, **14**, 633–643.
21. D. J. Betteridge, What is oxidative stress?, *Metabolism*, 2000, **49**, 3–8.
22. R. Radi, Oxygen radicals, nitric oxide, and peroxynitrite: Redox pathways in molecular medicine, *Proc. Natl. Acad. Sci. U. S. A.*, 2018, **115**, 5839–5848.
23. J. S. Beckman and W. H. Koppenol, Nitric oxide, superoxide, and peroxynitrite: the good, the bad, and ugly, *Am. J. Physiol.*, 1996, **271**, C1424–C1437.

24. C. A. Colton and D. L. Gilbert, Microglia, an *in vivo* source of reactive oxygen species in the brain, *Adv. Neurol.*, 1993, **59**, 321–326.
25. D. R. Rosen, *et al.*, Mutations in Cu/Zn superoxide dismutase gene are associated with familial amyotrophic lateral sclerosis, *Nature*, 1993, **362**, 59–62.
26. P. Chelikani, I. Fita and P. C. Loewen, Diversity of structures and properties among catalases, *Cell. Mol. Life Sci.*, 2004, **61**, 192–208.
27. A. Seiler, *et al.*, Glutathione peroxidase 4 senses and translates oxidative stress into 12/15-lipoxygenase dependent- and AIF-mediated cell death, *Cell Metab.*, 2008, **8**, 237–248.
28. Q. Ma, Role of nrf2 in oxidative stress and toxicity, *Annu. Rev. Pharmacol. Toxicol.*, 2013, **53**, 401–426.
29. D. A. Johnson and J. A. Johnson, Nrf2–a therapeutic target for the treatment of neurodegenerative diseases, *Free Radic. Biol. Med.*, 2015, **88**, 253–267.
30. A. Ayala, M. F. Munoz and S. Arguelles, Lipid peroxidation: production, metabolism, and signaling mechanisms of malondialdehyde and 4-hydroxy-2-nonenal, *Oxid. Med. Cell. Longev.*, 2014, **2014**, 360438.
31. D. A. Butterfield and B. Halliwell, Oxidative stress, dysfunctional glucose metabolism and Alzheimer disease, *Nat. Rev. Neurosci.*, 2019, **20**, 148–160.
32. V. A. Bohr, DNA damage and its processing. relation to human disease, *J. Inherit. Metab. Dis.*, 2002, **25**, 215–222.
33. M. J. Morgan and Z. G. Liu, Crosstalk of reactive oxygen species and NF-kappaB signaling, *Cell Res.*, 2011, **21**, 103–115.
34. P. R. Angelova and A. Y. Abramov, Role of mitochondrial ROS in the brain: from physiology to neurodegeneration, *FEBS Lett.*, 2018, **592**, 692–702.
35. A. Singh, R. Kukreti, L. Saso and S. Kukreti, Oxidative Stress: A Key Modulator in Neurodegenerative Diseases, *Molecules*, 2019, **24**(8), 1583.
36. R. Dringen, J. M. Gutterer and J. Hirrlinger, Glutathione metabolism in brain metabolic interaction between astrocytes and neurons in the defense against reactive oxygen species, *Eur. J. Biochem.*, 2000, **267**, 4912–4916.
37. M. L. Block and J. S. Hong, Chronic microglial activation and progressive dopaminergic neurotoxicity, *Biochem. Soc. Trans.*, 2007, **35**, 1127–1132.
38. D. G. Nicholls, Mitochondrial calcium function and dysfunction in the central nervous system, *Biochim. Biophys. Acta*, 2009, **1787**, 1416–1424.
39. X. X. Dong, Y. Wang and Z. H. Qin, Molecular mechanisms of excitotoxicity and their relevance to pathogenesis of neurodegenerative diseases, *Acta Pharmacol. Sin.*, 2009, **30**, 379–387.
40. G. Plascencia-Villa and G. Perry, Roles of Oxidative Stress in Synaptic Dysfunction and Neuronal Cell Death in Alzheimer's Disease, *Antioxidants*, 2023, **12**(8), 1628.
41. J. Zhu, *et al.*, An overview of chemical inhibitors of the Nrf2-ARE signaling pathway and their potential applications in cancer therapy, *Free Radic. Biol. Med.*, 2016, **99**, 544–556.

42. J. van Horssen, M. E. Witte, G. Schreibelt and H. E. de Vries, Radical changes in multiple sclerosis pathogenesis, *Biochim. Biophys. Acta*, 2011, **1812**, 141–150.
43. R. M. Ransohoff, How neuroinflammation contributes to neurodegeneration, *Science*, 2016, **353**, 777–783.
44. M. Colonna and O. Butovsky, Microglia Function in the Central Nervous System During Health and Neurodegeneration, *Annu. Rev. Immunol.*, 2017, **35**, 441–468.
45. A. Kumar and D. J. Loane, Neuroinflammation after traumatic brain injury: opportunities for therapeutic intervention, *Brain, Behav., Immun.*, 2012, **26**, 1191–1201.
46. H. Kettenmann, U. K. Hanisch, M. Noda and A. Verkhratsky, Physiology of microglia, *Physiol. Rev.*, 2011, **91**, 461–553.
47. J. D. Cherry, J. A. Olschowka and M. K. O'Banion, Neuroinflammation and M2 microglia: the good, the bad, and the inflamed, *J. Neuroinflammation*, 2014, **11**, 98.
48. M. V. Sofroniew, Multiple roles for astrocytes as effectors of cytokines and inflammatory mediators, *Neuroscientist*, 2014, **20**, 160–172.
49. S. A. Liddelow, *et al.*, Neurotoxic reactive astrocytes are induced by activated microglia, *Nature*, 2017, **541**, 481–487.
50. M. K. McCoy and M. G. Tansey, TNF signaling inhibition in the CNS: implications for normal brain function and neurodegenerative disease, *J. Neuroinflammation*, 2008, **5**, 45.
51. R. C. Paolicelli, K. Bisht and M. E. Tremblay, Fractalkine regulation of microglial physiology and consequences on the brain and behavior, *Front. Cell. Neurosci.*, 2014, **8**, 129.
52. B. Kaminska, A. Gozdz, M. Zawadzka, A. Ellert-Miklaszewska and M. Lipko, MAPK signal transduction underlying brain inflammation and gliosis as therapeutic target, *Anat. Rec.*, 2009, **292**, 1902–1913.
53. S. R. Subramaniam and M. F. Chesselet, Mitochondrial dysfunction and oxidative stress in Parkinson's disease, *Prog. Neurobiol.*, 2013, **106–107**, 17–32.
54. A. Varatharaj and I. Galea, The blood-brain barrier in systemic inflammation, *Brain, Behav., Immun.*, 2017, **60**, 1–12.
55. G. A. Rosenberg, Matrix metalloproteinases and their multiple roles in neurodegenerative diseases, *Lancet Neurol.*, 2009, **8**, 205–216.
56. E. C. Hirsch and S. Hunot, Neuroinflammation in Parkinson's disease: a target for neuroprotection?, *Lancet Neurol.*, 2009, **8**, 382–397.
57. H. Lassmann, Multiple Sclerosis Pathology, *Cold Spring Harb. Perspect. Med.*, 2018, **8**(3), a028936.
58. M. Fricker, A. M. Tolkovsky, V. Borutaite, M. Coleman and G. C. Brown, Neuronal Cell Death, *Physiol. Rev.*, 2018, **98**(2), 813–880.
59. M. C. D. S. Tenório, N. G. Graciliano, F. A. Moura, A. C. M. Oliveira and M. O. F. Goulart, N-Acetylcysteine (NAC): Impacts on Human Health, *Antioxidants*, 2021, **10**(6), 967.

60. S. Kasparova, *et al.*, Effect of coenzyme Q10 and vitamin E on brain energy metabolism in the animal model of Huntington's disease, *Neurochem. Int.*, 2006, **48**, 93–99.
61. K. Barhwal, *et al.*, Acetyl-l-carnitine (ALCAR) prevents hypobaric hypoxia-induced spatial memory impairment through extracellular related kinase-mediated nuclear factor erythroid 2-related factor 2 phosphorylation, *Neuroscience*, 2009, **161**, 501–514.
62. C. G. Parsons, A. Stoffler and W. Danysz, Memantine: a NMDA receptor antagonist that improves memory by restoration of homeostasis in the glutamatergic system–too little activation is bad, too much is even worse, *Neuropharmacology*, 2007, **53**, 699–723.
63. F. Gomez-Pinilla and C. Hillman, The influence of exercise on cognitive abilities, *Compr. Physiol.*, 2013, **3**, 403–428.
64. M. A. Nitsche, *et al.*, Transcranial direct current stimulation: State of the art 2008, *Brain Stimul.*, 2008, **1**, 206–223.
65. A. M. Luckey, K. Adcock and S. Vanneste, Peripheral nerve stimulation: A neuromodulation-based approach, *Neurosci. Biobehav. Rev.*, 2023, **149**, 105180.
66. F. Salehpour, *et al.*, Near-infrared photobiomodulation combined with coenzyme Q(10) for depression in a mouse model of restraint stress: reduction in oxidative stress, neuroinflammation, and apoptosis, *Brain Res. Bull.*, 2019, **144**, 213–222.
67. A. Navarrete-Opazo and G. S. Mitchell, Therapeutic potential of intermittent hypoxia: a matter of dose, *Am. J. Physiol.: Regul., Integr. Comp. Physiol.*, 2014, **307**, R1181–R1197.
68. M. P. Mattson, K. Moehl, N. Ghena, M. Schmaedick and A. Cheng, Intermittent metabolic switching, neuroplasticity and brain health, *Nat. Rev. Neurosci.*, 2018, **19**, 63–80.
69. P. Garodia, M. Hegde, A. B. Kunnumakkara and B. B. Aggarwal, Curcumin, inflammation, and neurological disorders: How are they linked?, *Integr. Med. Res.*, 2023, **12**, 100968.
70. C. Moussa, *et al.*, Resveratrol regulates neuro-inflammation and induces adaptive immunity in Alzheimer's disease, *J. Neuroinflammation*, 2017, **14**, 1.
71. A. Payne, S. Nahashon, E. Taka, G. M. Adinew and K. F. A. Soliman, Epigallocatechin-3-Gallate (EGCG): New Therapeutic Perspectives for Neuroprotection, Aging, and Neuroinflammation for the Modern Age, *Biomolecules*, 2022, **12**(3), 371.
72. M. Banaszak, *et al.*, Role of Omega-3 fatty acids eicosapentaenoic (EPA) and docosahexaenoic (DHA) as modulatory and anti-inflammatory agents in noncommunicable diet-related diseases - Reports from the last 10 years, *Clin. Nutr. ESPEN*, 2024, **63**, 240–258.
73. I. Baitharu, *et al.*, Withania somnifera root extract ameliorates hypobaric hypoxia induced memory impairment in rats, *J. Ethnopharmacol.*, 2013, **145**, 431–441.

74. N. Kushwah, *et al.*, Ginkgo biloba L. Prevents Hypobaric Hypoxia-Induced Spatial Memory Deficit Through Small Conductance Calcium-Activated Potassium Channel Inhibition: The Role of ERK/CaMKII/CREB Signaling, *Front. Pharmacol.*, 2021, **12**, 669701.
75. U. K. Meena and A. K. Maurya, Bacopa monnieri Extract Diminish Hypoxia-Induced Anxiety by Regulating HIF-1alpha Signaling and Enhancing the Antioxidant Defense System in Hippocampus, *Neuromol. Med.*, 2025, **27**, 11.
76. J. Nithianantharajah and A. J. Hannan, Enriched environments, experience-dependent plasticity and disorders of the nervous system, *Nat. Rev. Neurosci.*, 2006, **7**, 697–709.
77. V. Jain, *et al.*, Enriched environment prevents hypobaric hypoxia induced neurodegeneration and is independent of antioxidant signaling, *Cell. Mol. Neurobiol.*, 2012, **32**, 599–611.
78. L. Xie, *et al.*, Sleep drives metabolite clearance from the adult brain, *Science*, 2013, **342**, 373–377.
79. M. Wadhwa, *et al.*, Caffeine and Modafinil Ameliorate the Neuroinflammation and Anxious Behavior in Rats during Sleep Deprivation by Inhibiting the Microglia Activation, *Front. Cell. Neurosci.*, 2018, **12**, 49.
80. Y. Y. Tang, B. K. Holzel and M. I. Posner, The neuroscience of mindfulness meditation, *Nat. Rev. Neurosci.*, 2015, **16**, 213–225.
81. J. T. Cacioppo and S. Cacioppo, Social Relationships and Health: The Toxic Effects of Perceived Social Isolation, *Soc. Personal. Psychol. Compass*, 2014, **8**, 58–72.
82. M. P. Mattson, Lifelong brain health is a lifelong challenge: from evolutionary principles to empirical evidence, *Ageing Res. Rev.*, 2015, **20**, 37–45.

CHAPTER 6

Lipid Signaling and Brain Disorders

MUKESH KUMAR,*[a] NITISH KUMAR SINGH,[b] MAYANK GANGWAR,[c] VIVEK CHANDRA VERMA,[d] SYED MOHAMMAD NASAR ATA[e] AND MUDASSIR JAN MAKHDOOMI[f]

[a] Postdoc Research Fellow, Integrative Neurochemistry Laboratory, Harvard Medical School, McLean Hospital, USA; [b] Department of Biosciences, Manipal University Jaipur, Rajasthan, India; [c] Clinical Studies and Trials Unit, Division of Development/Indian Council of Medical Research, Department of Health Research (Min. of Health & FW), India; [d] Department of Biochemistry, Panjab University, India; [e] Department of Neurology, Henry Ford Health System, USA; [f] Department of Pediatrics, All India Institute of Medical Sciences (AIIMS), India
*Email: mkumar12@mclean.harvard.edu

6.1 Introduction to Lipid Signaling

Lipids are no longer regarded merely as structural components of cellular membranes or energy reservoirs; they have emerged as dynamic signaling molecules that are essential for maintaining neuronal function, synaptic integrity, and brain homeostasis. The human brain, though representing only about 2% of total body mass, is composed of roughly 50% to 60% lipids by dry weight, emphasizing the centrality of lipid metabolism in neural physiology. Lipid-derived messengers such as eicosanoids, sphingolipids, phosphoinositides, and endocannabinoids regulate neurogenesis, synaptic transmission, inflammation, oxidative stress response, and neuronal

Chemical Biology No. 26
Chemical Signatures of the Brain: Bridging Biochemistry, Neuroanatomy, and Brain Disorders
Edited by Sachchida Nand Rai and Abhishek Kumar Singh

Published by the Royal Society of Chemistry, www.rsc.org

survival.[1] Aberrations in these lipid signaling pathways are implicated in the pathogenesis of numerous neurological and psychiatric disorders, including Alzheimer's disease, Parkinson's disease, multiple sclerosis, depression, and schizophrenia. Lipid signaling operates through biochemical communications mediated by lipid molecules or their metabolites, which interact with receptors, ion channels, and enzymes to modulate cellular responses. These messengers are synthesized on demand from membrane phospholipid precursors and act locally due to their hydrophobic nature. Phosphoinositides such as PIP2 and PIP3, derived from phosphatidylinositol, play key roles in intracellular signaling cascades, membrane trafficking, and cytoskeletal dynamics, particularly in the PI3K-Akt and PLC-IP3-DAG pathways that govern synaptic plasticity and neuroprotection.[2] Sphingolipids, including ceramides and sphingosine-1-phosphate (S1P), regulate apoptosis, proliferation, and inflammation, with S1P signaling critical for myelination and oligodendrocyte survival. Eicosanoids—prostaglandins, leukotrienes, and thromboxanes—generated from arachidonic acid, modulate neuroinflammation and synaptic signaling, while endocannabinoids such as anandamide (AEA) and 2-arachidonoylglycerol (2-AG) influence neurotransmitter release, mood, and neuroimmune balance through CB1 and CB2 receptors.[3] These lipid molecules orchestrate communication across neurons, astrocytes, and microglia, forming a complex signaling network vital for brain function. Lipid rafts, composed of cholesterol and sphingolipids, serve as microdomains that cluster G-protein-coupled receptors, ion channels, and kinases, influencing receptor localization and signal transduction efficiency. Alterations in lipid composition can affect neurotransmission and cognitive processes. Phospholipase enzymes such as PLA2, PLC, and PLD hydrolyze phospholipids to generate second messengers like DAG and IP3, which modulate calcium signaling and activate protein kinase C, a regulator of neuronal excitability. Lipid signaling also interfaces with major intracellular pathways such as mTOR, MAPK, and NF-κB, impacting inflammation, oxidative stress, and gene expression.[4] Beyond signaling, lipids are structural and energetic foundations of cellular life, forming membrane bilayers that control permeability and compartmentalization, while triglycerides stored in lipid droplets serve as dense energy reserves mobilized through β-oxidation.[5] Lipids also mediate stress responses *via* peroxidation of polyunsaturated fatty acids, activating antioxidant defenses through PPAR and Nrf2 pathways, and participate in autophagy through conjugation of phosphatidylethanolamine with LC3. Additionally, they modulate gene expression by binding nuclear receptors like PPARs and LXRs.[6] In the brain, lipids enable intricate signaling processes underlying cognition and behavior. The nervous system functions through electrochemical communication, where action potentials trigger neurotransmitter release—glutamate, GABA, dopamine, serotonin, and acetylcholine—at synapses to regulate excitatory or inhibitory responses. Intracellular signaling pathways such as PI3K/Akt, MAPK/ERK, cAMP/PKA, and calcium/calmodulin translate these external stimuli into long-term cellular adaptations governing neuroplasticity, learning, and memory.[7] The CREB pathway, calcium signaling, and lipid rafts coordinate receptor clustering and synaptic efficiency. Glial cells, including astrocytes and microglia, further modulate neurotransmission, energy metabolism, and immune responses.[8] The brain's integration of neuroendocrine and autonomic pathways enables homeostatic control of energy balance through insulin, leptin, and lipid-based signals. Disruptions in these pathways contribute to neurodegenerative, psychiatric, and metabolic disorders. Thus, lipids act as the biochemical nexus linking cellular structure, energy metabolism, and signaling, orchestrating the physiological and molecular foundations of neural function. Understanding lipid signaling is therefore critical not only for elucidating brain physiology but also for developing novel diagnostic and therapeutic strategies targeting neurological and neuropsychiatric diseases.[9]

6.2 Types of Lipids Involved in Signaling

Lipids are an exceptionally diverse group of biomolecules that play vital roles not only in maintaining the structural integrity of cellular membranes but also as powerful signaling mediators in both physiological and pathological processes. In the nervous system and other tissues, specific classes of lipids function as second messengers, ligands for receptors, or modulators of enzymatic pathways that regulate growth, differentiation, apoptosis, and inflammatory responses.[10] Unlike traditional neurotransmitters or peptide hormones, many lipid signaling molecules are synthesized on demand from membrane precursors in response to cellular stimuli, allowing precise spatiotemporal control of signaling events. The main types of signaling lipids include phosphoinositides, sphingolipids, eicosanoids, endocannabinoids, and lysophospholipids, each with distinct biochemical properties and physiological roles (Table 6.1).

6.3 Lipid Signaling Pathways in the Brain

Lipid signaling pathways in the brain form an intricate network that integrates structural, metabolic, and functional aspects of neuronal physiology. Understanding these pathways not only illuminates the biochemical basis of brain function but also unveils novel therapeutic targets for neurological and psychiatric disorders driven by lipid dysregulation.[15] In neurology, GPCRs represent the pivotal molecular interface between lipid signaling and neural function. Lipid mediators such as eicosanoids, endocannabinoids, sphingolipids, and lysophospholipids fine-tune neural activity, inflammation, and repair *via* GPCR-dependent mechanisms.[16] These interactions sustain brain homeostasis, yet when dysregulated, they contribute to pathologies ranging from neuroinflammation and epilepsy to Alzheimer's and multiple sclerosis. By deciphering the GPCR–lipid communication network, neuroscience is entering a new era of therapeutic innovation—one where

Table 6.1 Neurological roles and metabolic significance of major lipid classes.[11–14]

Lipid type/class	Structure/neurological function	Metabolic pathways	Neurological implications
Phospholipids	Glycerol backbone + two fatty acids + phosphate-containing head (*e.g.*, phosphatidylcholine, phosphatidylserine, phosphatidylethanolamine) Major structural components of neuronal membranes; influence membrane fluidity and electrical conductivity; involved in signal transduction and vesicle fusion	Hydrolyzed/remodeled by phospholipases A_2, C, D; resynthesized *via* Kennedy pathway	Altered phospholipid metabolism linked with Alzheimer's, Parkinson's, and mood disorders; phosphatidylserine supplementation improves memory and cognition
Sphingolipids	Sphingosine backbone + fatty acid (forming ceramide) ± head groups (sphingomyelin, glycosphingolipids, gangliosides) Crucial for myelin sheath integrity, neuronal recognition, and axonal stability; mediate cell–cell signaling	Synthesized from serine + palmitoyl-CoA (*via* serine palmitoyltransferase); converted to ceramide→sphingomyelin or glycosphingolipids	Defects cause lysosomal storage diseases (Tay–Sachs, Gaucher's, Niemann–Pick); dysregulation linked to demyelinating diseases and neurodegeneration
Eicosanoids and their metabolic pathways	20-Carbon polyunsaturated fatty acids derived from arachidonic acid (*e.g.*, prostaglandins, leukotrienes, thromboxanes). Regulate neuroinflammation, synaptic activity, and cerebral blood flow	Cyclooxygenase (COX-1/2), lipoxygenase (LOX), and cytochrome P450 pathways convert arachidonic acid to eicosanoids	Overactivation of COX/LOX leads to oxidative stress and neuroinflammation in Alzheimer's, multiple sclerosis, and ischemic injury

Table 6.1 (*Continued*)

Lipid type/class	Structure/neurological function	Metabolic pathways	Neurological implications
Endocannabinoids	Lipid neurotransmitters (anandamide, 2-arachidonoylglycerol) derived from arachidonic acid-containing phospholipids. Modulate neurotransmitter release, synaptic plasticity, appetite, pain, and stress responses; maintain neural homeostasis	Synthesized "on-demand" by NAPE-PLD and DAGL; degraded by FAAH (anandamide) and MAGL (2-AG)	Dysregulation implicated in anxiety, depression, epilepsy, and neurodegenerative disorders; target for therapeutic cannabinoids
Cholesterol and lipid rafts	Sterol with rigid ring structure; forms lipid-rich microdomains ("rafts") in neuronal membranes and myelin. Maintains membrane integrity, supports synapse formation, and organizes receptors within lipid rafts	Synthesized *de novo via* HMG-CoA reductase pathway; distributed in CNS by ApoE-containing lipoproteins	Cholesterol imbalance alters amyloid precursor processing in Alzheimer's; ApoE4 allele increases AD risk; essential for myelination and synaptic signaling

modulation of lipid signaling through GPCRs can restore balance to disrupted neural circuits and promote long-term neuroprotection.[17]

6.3.1 G-protein-coupled Receptors (GPCRs) and Lipid Mediators in Neurology

G-protein-coupled receptors (GPCRs) form one of the largest and most versatile families of membrane receptors in the human body. They play a central role in translating extracellular signals into intracellular responses, orchestrating a wide array of physiological processes including perception, behavior, mood, and cognition. In the brain, GPCRs act as molecular sentinels that respond to neurotransmitters, neuropeptides, hormones, and importantly, lipid-derived signaling molecules such as prostaglandins, endocannabinoids, sphingolipids, and lysophospholipids.[18] The interaction between GPCRs and lipid mediators represents a key dimension of neuromodulation, influencing synaptic transmission, neuroinflammation, neuroprotection, and even neurodegeneration. GPCRs share a characteristic seven-transmembrane α-helical structure that spans the plasma membrane. The extracellular domain recognizes and binds to specific ligands, while the intracellular region interacts with heterotrimeric G-proteins composed of α, β, and γ subunits.[19] Upon ligand binding, the receptor undergoes a conformational change that facilitates the exchange of GDP for GTP on the Gα subunit. This activation leads to the dissociation of Gα and Gβγ, which then interact with various effector proteins such as adenylyl cyclases, phospholipases, and ion channels.[20]

The signaling cascades triggered by GPCR activation primarily involve:

1. **Gαs** → stimulates adenylyl cyclase → increases cAMP → activates protein kinase A (PKA).
2. **Gαi/o** → inhibits adenylyl cyclase → decreases cAMP levels → modulates ion channels and neurotransmitter release.
3. **Gαq/11** → activates phospholipase C-β (PLC-β) → hydrolyzes phosphatidylinositol 4,5-bisphosphate (PIP_2) → generates diacylglycerol (DAG) and inositol trisphosphate (IP_3) → triggers intracellular calcium release and PKC activation.

These downstream events culminate in changes in neuronal excitability, gene transcription, synaptic plasticity, and metabolic regulation. GPCR signaling is further fine-tuned by receptor phosphorylation and β-arrestin recruitment, which modulates desensitization, internalization, and alternative signaling routes independent of G-proteins.[21] Lipids are not merely structural elements but dynamic signaling molecules that can activate specific GPCRs. These lipid mediators include eicosanoids, endocannabinoids, lysophospholipids, sphingolipids, and other oxidized derivatives of fatty acids.[22] Each class acts through distinct GPCR subtypes to modulate neural communication and neuroimmune responses.

6.3.1.1 Prostaglandin GPCRs

Prostaglandins exert their effects through specific receptors—Prostaglandin E Receptor (EP1–EP4), IP for prostacyclin, and TP for thromboxane A_2. These receptors are widely distributed in the brain, especially in the cortex, hippocampus, and hypothalamus. Activation of EP2 and EP4 receptors (coupled to Gs) elevates cAMP and can protect neurons from excitotoxic damage.[23] Conversely, EP1 (coupled to Gq) increases intracellular calcium and has been implicated in neurotoxicity during ischemic stroke.

6.3.1.2 Leukotriene GPCRs

Leukotriene receptors (BLT1/2 and CysLT1/2) modulate microglial activation and blood–brain barrier (BBB) permeability. Elevated leukotriene signaling contributes to neuroinflammatory processes seen in multiple sclerosis and traumatic brain injury. Pharmacological blockade of CysLT1 has been shown to ameliorate cognitive decline in Alzheimer's disease models, highlighting its therapeutic potential.[24]

Through these eicosanoid-activated GPCRs, lipid mediators tightly regulate inflammation, pain perception, sleep, and synaptic activity, balancing neuroprotection and neurotoxicity.

6.3.2 Endocannabinoid Signaling *via* GPCRs

The endocannabinoid system (ECS) is one of the best-characterized lipid–GPCR networks in the brain. The key endogenous cannabinoids—anandamide (AEA) and 2-arachidonoylglycerol (2-AG)—act as retrograde neurotransmitters that diffuse back across the synapse to modulate presynaptic neurotransmitter release.[24]

6.3.2.1 CB1 Receptors

Predominantly expressed in neurons, CB1 receptors couple to Gi/o proteins, leading to inhibition of adenylyl cyclase, reduced calcium influx, and decreased neurotransmitter release (*e.g.*, glutamate, GABA, dopamine).[25] This mechanism regulates synaptic plasticity, learning, pain modulation, and emotional balance. CB1-mediated suppression of neurotransmission contributes to neuroprotection during excitotoxic stress.

6.3.2.2 CB2 Receptors

These are primarily located on microglia and immune cells in the CNS. Their activation dampens neuroinflammatory responses by reducing cytokine release and oxidative stress.[25] CB2 signaling has been implicated in neurodegenerative conditions such as Parkinson's and Alzheimer's diseases, where microglial overactivation contributes to neuronal loss.

6.3.2.3 GPR55 and GPR18

Often termed "non-classical" cannabinoid receptors, these orphan GPCRs also respond to lipid mediators related to endocannabinoids. GPR55 influences neuronal excitability and has been linked to neuropathic pain, while GPR18 may participate in microglial migration and neuroimmune signaling.[26]

Thus, the ECS–GPCR interplay provides a molecular framework for adaptive control of neural circuits, representing a promising target for treating epilepsy, anxiety, depression, and neurodegenerative diseases.

6.3.3 Lysophospholipid GPCRs: LPA and S1P Receptors

Lysophosphatidic acid (LPA) and sphingosine-1-phosphate (S1P) are potent lipid mediators that signal through specific GPCRs—LPA_{1-6} and $S1P_{1-5}$, respectively. These receptors regulate critical neurodevelopmental and neuroprotective functions. LPA_1 is highly expressed in the developing brain, controlling neural progenitor proliferation, migration, and differentiation.[27] Dysregulated LPA signaling is linked to aberrant cortical development and hydrocephalus. In adults, LPA contributes to neuropathic pain and demyelination. $S1P_1$ and $S1P_5$ receptors are abundant in oligodendrocytes, where they influence myelination and remyelination. Pharmacological modulation of S1P signaling, as seen with the drug fingolimod (FTY720)—a functional S1P receptor modulator—is a successful therapy for multiple sclerosis. By preventing lymphocyte egress from lymph nodes and directly protecting neural cells, fingolimod exemplifies how GPCR–lipid interactions can be clinically exploited. Recent studies have identified GPCRs that respond to free fatty acids (FFAs) and their oxidized derivatives. These include GPR40 (FFAR1), GPR41 (FFAR3), and GPR120, all expressed in brain regions such as the hypothalamus and hippocampus. They participate in metabolic regulation, appetite control, and cognitive functions.[28] GPR120 activation by omega-3 fatty acids exerts anti-inflammatory and neuroprotective effects by inhibiting microglial activation. Furthermore, oxidized linoleic acid metabolites (OXLAMs) activate GPR132 and GPR31, linking oxidative stress and lipid peroxidation to neuronal injury and neuroinflammatory signaling in neurodegenerative diseases.

6.3.4 GPCR–Lipid Mediator Crosstalk and Neuroinflammation

Neuroinflammation represents a double-edged sword in the brain—necessary for defense and repair, yet potentially destructive when uncontrolled. Lipid mediators acting through GPCRs serve as both initiators and resolvers of inflammation. For instance, eicosanoid and lysophospholipid signaling activate proinflammatory cascades in microglia, whereas endocannabinoid and specialized pro-resolving mediators (SPMs) engage GPCRs that terminate inflammation and promote tissue recovery.[29] Examples include (i) activation of CB2 and GPR37 by lipid mediators leading to reduced microglial activation. (ii) Engagement of ALX/FPR2 by lipoxins and resolvins, stimulating anti-inflammatory signaling. This balance between pro- and anti-inflammatory GPCR pathways determines neuronal survival and functional restoration following injury or disease. The intricate interplay between GPCRs and lipid mediators provides fertile ground for developing new neurotherapeutics. Current strategies target GPCR–lipid interactions to modulate neuroinflammation, neurodegeneration, and synaptic dysfunction. Emerging research also suggests that GPCRs may form heterodimers with other receptors (*e.g.*, CB1–D2 or S1P–LPA), resulting in unique signaling properties.[30] Understanding these receptor complexes could unveil new therapeutic avenues. Additionally, advances in lipidomics and structural biology illuminate how specific lipid ligands fit into GPCR binding pockets, enabling rational drug design with high selectivity and reduced side effects.[31]

6.4 Lipid–Protein Interactions in Neural Function

Lipid–protein interactions are fundamental to the organization, dynamics, and function of neural membranes. In the brain, where nearly half of the dry weight is lipid, the interplay between membrane lipids and proteins dictates neuronal excitability, signal transduction, and synaptic communication. Neural membranes are not static bilayers, but fluid mosaics composed of phospholipids, sphingolipids, cholesterol, and embedded proteins whose behavior is tightly coupled to their lipid microenvironment.[32] These interactions determine the spatial localization and conformational state of ion channels, neurotransmitter receptors, and signaling complexes essential for neuronal communication. Cholesterol, for instance, modulates the activity of major neurotransmitter receptors such as NMDA, AMPA, and GABA receptors by influencing their membrane microenvironment. It stabilizes receptor conformations and facilitates the assembly of receptor–scaffold complexes within lipid rafts—cholesterol- and sphingolipid-rich membrane domains that act as signaling platforms. Within these rafts, proteins such as synaptophysin, flotillins, and kinases cluster to coordinate neurotransmission and endocytosis.[33] Disruption of lipid–protein balance, such as cholesterol depletion, leads to altered receptor function, impaired neurotransmitter release, and cognitive decline, as seen in neurodegenerative conditions like Alzheimer's disease. Phospholipids also contribute to protein anchoring and signaling. Phosphatidylinositol 4,5-bisphosphate (PIP_2) interacts with a wide array of ion channels and cytoskeletal proteins, regulating their gating and membrane trafficking.[34] Similarly, phosphatidylserine (PS) provides a negatively charged surface for protein docking, including protein kinase C (PKC) and small GTPases, facilitating synaptic vesicle fusion and membrane repair. Sphingolipids, particularly sphingomyelin and gangliosides, interact with myelin-associated proteins to maintain axonal insulation and promote neuronal survival. On a dynamic level, lipid–protein interactions regulate the curvature and fusion of synaptic membranes during vesicle exocytosis and endocytosis. Proteins such as synaptotagmin and syntaxin require specific lipid environments to mediate calcium-triggered vesicle fusion. Moreover, lipid-binding domains like pleckstrin homology (PH), C2, and FYVE motifs allow proteins to sense and respond to lipid signals, thus integrating lipid metabolism with protein-mediated signaling pathways.[35] Disruption of these interactions has profound neurological implications. Mutations in lipid-binding proteins or alterations in lipid composition can destabilize membrane domains, impair synaptic plasticity, and contribute to disorders such as Parkinson's disease, Huntington's disease, and multiple sclerosis.[36] Overall, lipid–protein interactions represent a critical nexus through which the biochemical flexibility of lipids is translated into precise, coordinated neural functions.

6.4.1 Intracellular Signaling Cascades Involving Lipids

Lipids are central mediators of intracellular signaling cascades in neurons, translating external stimuli into complex biochemical responses that regulate growth, metabolism, and communication. Beyond their structural roles, membrane lipids serve as dynamic precursors of second messengers and allosteric modulators of signaling proteins. The generation and turnover of lipid-derived messengers such as diacylglycerol (DAG), inositol trisphosphate (IP_3), ceramide, and phosphatidic acid are essential for maintaining synaptic function and neuronal survival.[37] A primary lipid-dependent pathway is the phosphoinositide cascade, initiated by activation of G-protein-coupled receptors (GPCRs) or receptor tyrosine kinases (RTKs). Phospholipase C (PLC) hydrolyzes phosphatidylinositol 4,5-bisphosphate (PIP_2) to yield DAG and IP_3. DAG remains within the membrane, activating protein kinase C (PKC), which phosphorylates numerous neuronal substrates involved in neurotransmission, cytoskeletal remodeling, and gene expression. Simultaneously, IP_3 diffuses through the cytosol to release calcium from the endoplasmic reticulum, triggering calcium-dependent signaling that modulates synaptic activity and long-term potentiation (LTP). This dual action exemplifies how lipid cleavage generates both membrane-associated and cytosolic messengers that coordinate neuronal responses. Another major lipid signaling route involves phosphatidylinositol 3-kinase (PI3K)/Akt pathway.[38] PI3K phosphorylates PIP_2 to form PIP_3, a docking site for proteins containing pleckstrin homology domains such as Akt. Once activated, Akt promotes neuronal survival by inhibiting pro-apoptotic factors and supporting metabolic activity. Dysregulation of PI3K/Akt signaling contributes to neurodegenerative diseases and insulin-resistance-related cognitive impairment. Sphingolipid metabolites, particularly ceramide and sphingosine-1-phosphate (S1P), constitute another lipid signaling axis (Figure 6.1). Ceramide acts as a pro-apoptotic messenger, activating protein phosphatases and caspases, while S1P promotes cell proliferation and survival

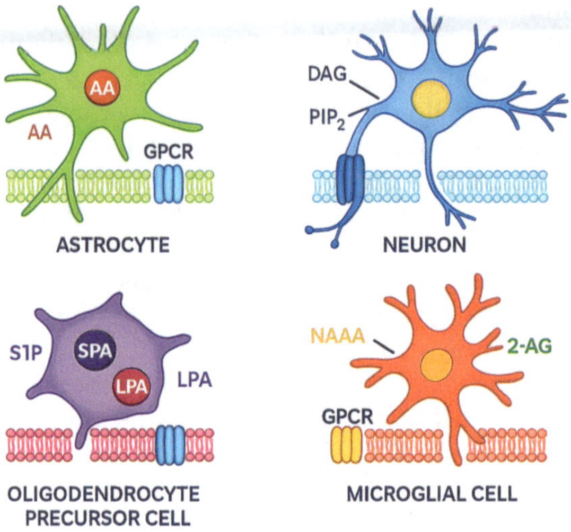

Figure 6.1 A diagram showing the major lipid signaling pathways in a neuron. The cell membrane contains phospholipids and signaling molecules. PIP_2 is converted by PLC into DAG and IP_3. IP_3 is shown triggering calcium release from the endoplasmic reticulum, while DAG activates PKC. Additional pathways show arachidonic acid leading to eicosanoids, sphingosine 1 phosphate and lysophosphatidic acid acting through GPCRs, and endocannabinoids 2AG and AEA regulated by NAAA.

through specific GPCRs ($S1P_{1-5}$). The balance between these two metabolites, often termed the "sphingolipid rheostat" determines neuronal fate under stress conditions.[39] Fingolimod, an S1P receptor modulator, exploits this pathway therapeutically in multiple sclerosis to protect neurons and modulate immune responses. Additionally, endocannabinoid signaling represents a unique lipid-mediated cascade. Endocannabinoids like anandamide and 2-arachidonoylglycerol, derived from membrane phospholipids, activate CB1 receptors on presynaptic terminals, inhibiting neurotransmitter release. This retrograde signaling modulates synaptic plasticity, anxiety, and pain perception. Overall, intracellular lipid signaling integrates metabolic, electrical, and molecular cues to sustain brain homeostasis. The spatial and temporal precision of lipid messenger production enables rapid yet regulated neuronal communication. Aberrations in these lipid cascades disrupt signaling fidelity and contribute to disorders such as epilepsy, Alzheimer's disease, and ischemic brain injury, underscoring their vital role in neurological health.[40]

6.4.2 Impact of Lipid Modifications on Synaptic Plasticity

Synaptic plasticity—the capacity of synapses to strengthen or weaken over time—is the molecular foundation of learning and memory. Lipid modifications, encompassing both changes in membrane composition and post-translational lipidation of proteins, play a decisive role in shaping synaptic structure and signaling efficiency. The lipid environment influences receptor distribution, vesicle dynamics, and intracellular signaling, thereby determining the strength and adaptability of synaptic connections. Membrane lipids such as phosphatidylserine, cholesterol, and sphingolipids directly affect receptor mobility and clustering within the synapse. Cholesterol-rich lipid rafts serve as platforms for neurotransmitter receptors, scaffolding proteins, and kinases that govern long-term potentiation (LTP) and long-term depression (LTD).[41] Alterations in cholesterol levels disrupt NMDA and AMPA receptor trafficking, impairing LTP induction. Similarly, gangliosides modulate receptor–ligand interactions and neurite outgrowth, contributing to synaptic stability and repair. Beyond membrane composition, lipid modifications of proteins—including palmitoylation, prenylation, and myristoylation—are crucial for targeting synaptic proteins to appropriate membrane domains. Palmitoylation, the reversible attachment of palmitic acid to cysteine residues, enhances the hydrophobicity of proteins such as PSD-95, SNAP-25, and G-proteins, ensuring their localization to synaptic membranes.[42] Cycles of palmitoylation and depalmitoylation dynamically regulate receptor anchoring and signal propagation during synaptic remodeling. Dysregulated palmitoylation is implicated in neurodevelopmental and psychiatric disorders like schizophrenia and autism. Endocannabinoid signaling, another lipid-driven process, exerts a profound influence on synaptic plasticity. By retrogradely inhibiting neurotransmitter release through CB1 receptor activation, endocannabinoids mediate a form of long-term depression (eCB-LTD) that balances excitatory and inhibitory inputs.[43] This process fine-tunes synaptic responsiveness during memory formation and stress adaptation. Furthermore, the metabolism of phosphoinositides within the postsynaptic membrane regulates actin cytoskeleton remodeling—a critical determinant of dendritic spine morphology. Phosphatidylinositol 4,5-bisphosphate (PIP_2) and its phosphorylated derivatives recruit actin-binding proteins that stabilize synaptic structures. Alterations in PIP_2 turnover disrupt spine dynamics, contributing to cognitive decline in neurodegenerative diseases. Mitochondrial lipid signaling also contributes indirectly to synaptic plasticity by regulating energy supply and reactive oxygen species balance during synaptic activation. Cardiolipin, a mitochondria-specific phospholipid, stabilizes respiratory chain complexes and modulates calcium buffering, thereby influencing synaptic efficacy and neuronal survival.[44] In sum, lipid modifications operate as dynamic regulators of synaptic architecture and signaling. By modulating protein localization, receptor trafficking, and cytoskeletal organization, they ensure the adaptability and resilience of neural networks. Dysfunctions in lipid metabolism or lipid–protein coupling compromise synaptic plasticity, leading to cognitive impairments and neurodegenerative pathologies. Understanding these lipid-dependent mechanisms opens new avenues for therapeutic modulation of learning, memory, and neural repair.

6.5 Lipid Dysregulation in Brain Disorders

Beyond their structural roles, lipids act as dynamic signaling molecules that influence neuronal survival, synaptic plasticity, and neuroinflammatory processes. The brain's unique lipid composition—rich in phospholipids, sphingolipids, cholesterol, and polyunsaturated fatty acids—ensures precise regulation of communication among neurons and glial cells. However, disruptions in lipid metabolism or homeostasis, collectively referred to as lipid dysregulation, can profoundly affect neural function. Mounting evidence implicates lipid imbalance as a central factor in the pathogenesis of various neurological and psychiatric disorders, including Alzheimer's disease (AD), Parkinson's disease (PD), multiple sclerosis (MS), Huntington's disease (HD), and major depressive disorder.[45] At the cellular level, lipid dysregulation disturbs membrane architecture and signaling fidelity. Phospholipids and sphingolipids not only form the structural framework of neuronal membranes but also serve as reservoirs for second messengers involved in neurotransmission and intracellular signaling. Alterations in their composition can impair receptor function, synaptic vesicle recycling, and axonal transport. For instance, reduced levels of phosphatidylserine and phosphatidylethanolamine affect membrane fluidity and neurotransmitter release, while excess ceramide accumulation triggers neuronal apoptosis through activation of proinflammatory and proapoptotic pathways. Similarly, defective sphingolipid metabolism contributes to demyelination and axonal degeneration in disorders like MS and Krabbe disease. Cholesterol homeostasis in the brain is another critical determinant of neuronal health. Unlike peripheral tissues, the brain synthesizes its own cholesterol, which is essential for synapse formation and myelin stability. Disruption in cholesterol turnover—often due to mutations in apolipoprotein E (ApoE), ATP-binding cassette transporters, or enzymes of the mevalonate pathway—has been strongly linked to Alzheimer's pathology.[46] The ApoE4 isoform impairs cholesterol transport and amyloid-β clearance, exacerbating plaque formation and neurodegeneration. Moreover, oxidized cholesterol derivatives, or oxysterols, can cross the blood–brain barrier and act as neurotoxic signaling molecules that promote inflammation and oxidative stress. Another key dimension of lipid dysregulation involves neuroinflammation and oxidative damage. Dysregulated eicosanoid and endocannabinoid pathways result in chronic activation of microglia and astrocytes, leading to the release of proinflammatory cytokines and reactive oxygen species.[47] This sustained inflammatory environment damages neuronal membranes, disrupts lipid rafts, and impairs neurotransmission. For example, excessive activation of cyclooxygenase (COX) pathways elevates prostaglandin production, contributing to neuronal injury in ischemic stroke and Alzheimer's disease. Furthermore, lipid peroxidation—a process in which polyunsaturated fatty acids are oxidized—produces reactive aldehydes like malondialdehyde and 4-hydroxynonenal that modify proteins and nucleic acids, aggravating neurodegenerative cascades.[48] These oxidative modifications disrupt

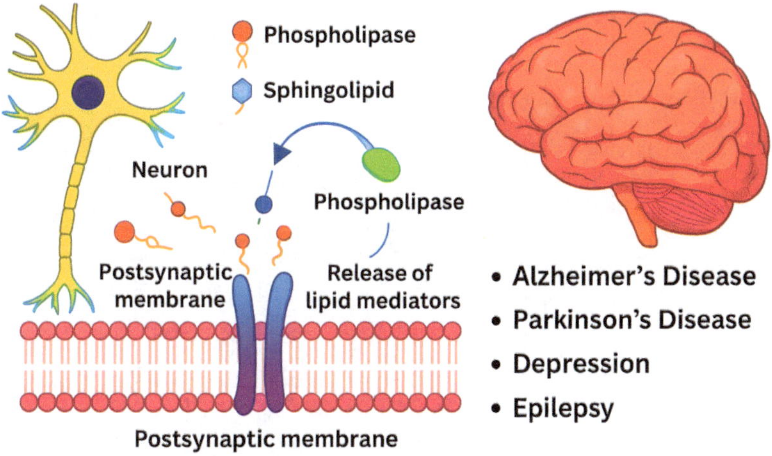

Figure 6.2 A diagram showing lipid signaling in neurons and its association with neurological disorders. A presynaptic neuron communicates with a postsynaptic membrane. Phospholipase enzymes and sphingolipids are shown acting on membrane lipids, with arrows indicating the production and release of lipid mediators. On the right, a brain illustration is accompanied by the labels Alzheimer's disease, Parkinson's disease, Depression, and Epilepsy, indicating disorders associated with altered lipid signaling.

mitochondrial lipid membranes, impairing ATP production and neuronal viability. In summary, lipid dysregulation in the brain represents a multifaceted pathological mechanism encompassing altered synthesis, degradation, and signaling of diverse lipid species (Figure 6.2). Its consequences extend beyond structural damage to affect neurotransmission, inflammation, and cellular metabolism.[49] Understanding these lipid abnormalities not only enhances our comprehension of neurological disease mechanisms but also offers novel therapeutic opportunities aimed at restoring lipid balance and preserving neural function (Table 6.2).

6.6 Lipid-based Biomarkers and Therapeutics

Lipid profiling, or lipidomics, has emerged as a powerful tool for elucidating the complex lipid alterations that underlie brain disorders. Given that the brain contains the highest lipid concentration of any organ, disruptions in lipid metabolism are closely associated with neuronal dysfunction, neuroinflammation, and neurodegeneration.[50] Lipidomics enables comprehensive identification and quantification of lipid species—such as phospholipids, sphingolipids, sterols, and fatty acids—providing valuable insights into disease mechanisms, progression, and therapeutic response. Advances in analytical technologies, including mass spectrometry (MS), liquid chromatography (LC-MS), and nuclear magnetic resonance (NMR), have revolutionized lipid profiling by allowing high-resolution detection of

Table 6.2 Lipid dysregulation and its role in major brain disorders.[44–48]

Neurological disorder	Key lipid alterations/mechanisms	Affected pathways/molecules	Neurological impact	Clinical/therapeutic implications
Alzheimer's disease and lipid signaling	Altered cholesterol metabolism, disrupted phospholipid balance, increased ceramide and sphingomyelin levels	ApoE–cholesterol transport pathway, phosphatidylinositol signaling, sphingolipid–ceramide axis	Cholesterol and sphingolipid accumulation promote amyloid-β aggregation and tau hyperphosphorylation; lipid raft disruption impairs synaptic signaling	Targeting cholesterol metabolism (statins, ApoE modulators), omega-3 fatty acid supplementation, and inhibition of ceramide synthesis show neuroprotective effects
Parkinson's disease: lipid imbalance and neurodegeneration	Decreased polyunsaturated fatty acids (PUFAs), increased oxidized lipids, disrupted sphingolipid and glycosphingolipid metabolism	α-Synuclein–lipid interactions, mitochondrial lipid oxidation, ceramide and glucosylceramide accumulation	Lipid oxidation enhances α-synuclein aggregation; ceramide imbalance induces mitochondrial dysfunction and dopaminergic neuron death	Glucocerebrosidase (GBA) enzyme modulators, antioxidants, and PUFA-based dietary therapies may slow disease progression
Schizophrenia and altered lipid metabolism	Reduced membrane phospholipids (phosphatidylserine, phosphatidylethanolamine), altered arachidonic acid turnover, dysregulated cholesterol synthesis	Phospholipase A_2 signaling, arachidonic acid–COX pathway, myelin lipid composition changes	Impaired membrane fluidity and receptor signaling affect neurotransmission (dopamine, glutamate); altered myelination contributes to cognitive and sensory deficits	Phospholipid replacement therapies, omega-3 fatty acid supplementation, and modulation of PLA_2 activity explored as adjunct treatments

Mood disorders and lipid signaling dysfunction	Imbalance of omega-3/omega-6 fatty acids, altered endocannabinoid tone, dysregulated phosphoinositide and eicosanoid signaling	Endocannabinoid receptors (CB_1/CB_2), COX-derived prostaglandins, phosphatidylinositol pathway	Impaired lipid signaling alters monoamine transmission, HPA-axis regulation, and synaptic plasticity, leading to depression and anxiety	Nutritional interventions with omega-3 fatty acids, FAAH inhibitors, and mood stabilizers targeting phosphoinositide metabolism (*e.g.*, lithium) improve symptoms
Epilepsy and lipid-mediated pathways	Altered membrane phospholipid composition, increased lysophosphatidic acid (LPA), and oxidative lipid stress	Phospholipase C and A_2 pathways, endocannabinoid signaling, mitochondrial lipid oxidation	Lipid imbalance disrupts ion channel function and neurotransmitter release, lowering seizure threshold and promoting excitotoxicity	Ketogenic diet restores lipid balance; modulation of endocannabinoid or phospholipase activity offers potential anticonvulsant strategies

lipids even in minute brain regions or biofluids like cerebrospinal fluid (CSF) and plasma. In Alzheimer's disease (AD), lipid profiling has revealed consistent alterations in phosphatidylcholine, sphingomyelin, and ceramide species, correlating with amyloid-β deposition and tau pathology.[51] Ceramide levels in plasma and CSF have emerged as potential early biomarkers for predicting cognitive decline, reflecting increased neuronal stress and apoptosis. Similarly, cholesterol and its oxidized derivatives (oxysterols) are elevated in AD patients, signifying dysregulated lipid transport across the blood–brain barrier. In Parkinson's disease (PD), reductions in polyunsaturated fatty acids (PUFAs) and increased oxidized phospholipids have been linked to α-synuclein aggregation and mitochondrial dysfunction. Lipidomic patterns also distinguish patients from schizophrenia, depression, and multiple sclerosis (MS), suggesting disease-specific lipid signatures. Moreover, lipid profiling provides dynamic insights into treatment efficacy and disease monitoring. For instance, alterations in sphingolipid ratios can indicate a therapeutic response in MS patients treated with sphingosine-1-phosphate receptor modulators like fingolimod. Plasma phosphatidylserine levels also correlate with antidepressant response, underscoring lipid biomarkers' value in precision psychiatry.[52] Importantly, lipid profiling integrates with other "omics" approaches—such as proteomics and metabolomics—to reveal lipid–protein and lipid–gene interactions. Systems-level analysis helps map lipid pathways perturbed during neuroinflammation or oxidative stress, thereby identifying potential drug targets. Future directions involve using single-cell lipidomics and imaging mass spectrometry to visualize lipid changes in specific brain cell types or regions.[53] Overall, lipid profiling not only enhances understanding of the biochemical underpinnings of brain disorders but also serves as a noninvasive diagnostic and prognostic tool. Its clinical translation may enable early detection, patient stratification, and therapeutic monitoring—paving the way for lipid-based personalized medicine in neurology.[54]

6.6.1 Targeting Lipid Pathways for Treatment

Therapeutic targeting of lipid pathways represents an innovative and promising strategy for managing neurological disorders. Since lipids govern membrane dynamics, signaling cascades, and inflammatory processes, modulating their synthesis, degradation, or receptor-mediated signaling can restore homeostasis in diseased neural tissues. Several lipid-regulating enzymes and transporters are now recognized as viable pharmacological targets in neurodegeneration, psychiatric disorders, and neuroinflammatory conditions. One major therapeutic approach involves the manipulation of sphingolipid metabolism. The balance between ceramide and sphingosine-1-phosphate (S1P)—often termed the "sphingolipid rheostat"—determines neuronal survival or death.[55] Excess ceramide induces apoptosis and oxidative stress, while S1P promotes survival and anti-inflammatory signaling. Drugs like fingolimod (FTY720) act as S1P receptor modulators, sequestering autoreactive lymphocytes and exerting direct neuroprotective effects in multiple sclerosis (MS). Beyond MS, S1P receptor targeting is being investigated for neuroprotection in Alzheimer's and ischemic stroke. Phosphoinositide signaling pathways also offer therapeutic potential. Modulating enzymes such as phospholipase A_2 (PLA_2) and phospholipase C (PLC) can control the generation of lipid second messengers like diacylglycerol (DAG) and inositol trisphosphate (IP_3), which regulate calcium flux and protein kinase activation.[56] Overactive PLA_2 has been implicated in neuroinflammation and oxidative stress; thus, selective PLA_2 inhibitors are being explored to mitigate neurodegeneration and epilepsy. Another strategy targets cholesterol metabolism, crucial for synaptogenesis and membrane repair. Agents that regulate cholesterol synthesis (statins), efflux (ApoE modulators), and esterification (ACAT inhibitors) have shown neuroprotective benefits by reducing amyloid-β accumulation and inflammation.[57] Similarly, omega-3 fatty acids and lipid nanoparticles are being employed to correct membrane lipid imbalance and deliver neuroactive compounds effectively across the blood–brain barrier. Furthermore, modulating endocannabinoid signaling *via* inhibition of fatty acid amide hydrolase (FAAH) or monoacylglycerol lipase (MAGL) enhances endogenous cannabinoid tone, promoting synaptic plasticity, stress resilience, and neuroprotection.[58] These mechanisms highlight how targeting lipid turnover enzymes can modulate neurotransmission and emotional regulation. Emerging research also explores lipid-based nanotherapeutics, where liposomes, micelles, and solid lipid nanoparticles serve as vehicles for drug delivery or as active modulators of lipid metabolism. Such systems improve bioavailability and target specificity of neuroprotective agents. Collectively, therapeutic manipulation of lipid pathways offers multifaceted control over neurodegeneration, inflammation, and neurotransmission. Future drug discovery in neurology is increasingly focusing on enzyme inhibitors, receptor modulators, and lipid mimetics that restore lipid homeostasis—shifting treatment paradigms from symptom management to molecular correction of lipid dysfunction. Lipid-lowering drugs, primarily developed for cardiovascular disorders, have shown significant neuroprotective potential due to their influence on lipid metabolism, inflammation, and oxidative stress within the brain.[59] Among these, statins, fibrates, and omega-3 fatty acids are the most studied for their ability to modulate cholesterol and fatty acid dynamics that impact neural health. Statins, which inhibit 3-hydroxy-3-methylglutaryl-CoA (HMG-CoA) reductase, lower peripheral cholesterol synthesis and exhibit pleiotropic effects in the brain. They reduce oxidative stress, enhance endothelial function, and suppress neuroinflammation by inhibiting isoprenoid-dependent protein prenylation involved in inflammatory signaling.[60] In Alzheimer's disease, statins have been shown to lower amyloid-β production by reducing cholesterol in neuronal lipid rafts, thereby altering amyloid precursor protein (APP) processing. Additionally, statins upregulate nitric oxide synthesis, improving cerebral blood flow and protecting neurons from ischemic injury.

Fibrates, which activate peroxisome proliferator-activated receptor-alpha (PPAR-α), influence lipid oxidation and energy metabolism. Their activation modulates neuroinflammatory gene expression, reduces microglial activation, and improves mitochondrial function. PPAR-α agonists like fenofibrate have demonstrated cognitive and anti-inflammatory benefits in experimental models of neurodegeneration and epilepsy.[61] Omega-3 polyunsaturated fatty acids (PUFAs), particularly eicosapentaenoic acid (EPA) and docosahexaenoic acid (DHA), are essential for neuronal membrane integrity, synaptic signaling, and anti-inflammatory lipid mediator production. DHA is particularly abundant in the brain and supports neurogenesis, synaptic plasticity, and neurotransmission. Supplementation with omega-3 PUFAs mitigates depressive symptoms, improves cognitive performance, and reduces neuroinflammation by generating specialized pro-resolving mediators (SPMs) such as resolvins and protectins. Lipid-lowering drugs also exhibit indirect neuroprotective effects by mitigating vascular risk factors that contribute to stroke and vascular dementia. Statins and fibrates improve endothelial integrity, reduce atherosclerotic plaque formation, and limit oxidative lipid damage within cerebral vessels. Their anti-inflammatory effects extend to microglia, attenuating chronic neuroinflammation that exacerbates neuronal death. Despite promising evidence, lipid-lowering therapy must balance central and peripheral lipid needs, as excessive depletion of cholesterol can impair synaptic function and myelination.[62] Current research emphasizes the development of brain-penetrant statins and selective PPAR modulators that optimize neuroprotection without systemic adverse effects. In conclusion, lipid-lowering agents transcend their cardiovascular roles by modulating critical lipid-dependent pathways in the brain. Through anti-inflammatory, antioxidative, and membrane-stabilizing actions, they hold promise as adjunct therapies in Alzheimer's disease, Parkinson's disease, and other neurodegenerative conditions characterized by lipid dysregulation.[63]

6.6.2 The Endocannabinoid System as a Potential Therapeutic Target

The endocannabinoid system (ECS) represents a pivotal lipid signaling network that regulates synaptic transmission, neuroprotection, mood, and pain processing. It consists of endogenous lipid mediators—mainly anandamide (AEA) and 2-arachidonoylglycerol (2-AG)—their synthetic and degradative enzymes, and cannabinoid receptors CB_1 and CB_2. Dysregulation of the ECS contributes to numerous neurological disorders, making it an attractive therapeutic target. CB_1 receptors are highly expressed in neurons, particularly in presynaptic terminals, where they inhibit neurotransmitter release by reducing calcium influx.[64] This retrograde signaling mechanism modulates synaptic plasticity and prevents excitotoxicity. CB_2 receptors, though primarily expressed in immune cells, are also present in microglia and astrocytes, where their activation attenuates neuroinflammatory responses. Together, CB_1 and CB_2 maintain neural homeostasis through a balance of excitatory and inhibitory modulation. Therapeutically, enhancing endocannabinoid signaling can protect neurons from degeneration and oxidative stress. In Alzheimer's disease, CB_1 activation reduces amyloid-β-induced neurotoxicity and improves cognitive function, while CB_2 activation suppresses microglial activation and cytokine release. In Parkinson's disease, CB_1 receptors regulate dopaminergic signaling, and CB_2-mediated anti-inflammatory effects alleviate neuroinflammation. ECS modulation also holds promise in epilepsy, depression, and anxiety disorders, where altered endocannabinoid tone affects neuronal excitability and mood regulation.[65] Pharmacological approaches to target the ECS include enzyme inhibitors, receptor agonists, and reuptake modulators. Inhibiting fatty acid amide hydrolase (FAAH) and monoacylglycerol lipase (MAGL) enzymes responsible for endocannabinoid degradation increases endogenous cannabinoid levels, enhancing neuroprotective signaling without the psychotropic effects associated with exogenous cannabinoids. Selective CB_2 agonists are being developed to harness anti-inflammatory benefits while minimizing CB_1-mediated psychoactivity. Moreover, ECS modulation interacts with other lipid pathways, including sphingolipid and eicosanoid signaling, amplifying its influence on neuroinflammation and metabolic balance. Endocannabinoids can also activate non-classical receptors such as GPR55 and TRPV1, broadening their therapeutic scope in neuropathic pain and metabolic syndromes affecting the brain. Clinical studies using cannabinoid-based drugs like nabiximols (Sativex) and cannabidiol (CBD) have demonstrated benefits in spasticity, epilepsy, and neurodegenerative symptoms.[66] Importantly, CBD modulates ECS activity indirectly, offering neuroprotection and anxiolytic effects without psychotropic consequences. Overall, the ECS functions as a master regulator of lipid-mediated neural communication. Targeting this system through selective agonists, enzyme inhibitors, and natural cannabinoids offers a multifaceted approach to restoring synaptic balance, reducing inflammation, and promoting neuronal survival. Future therapies will likely combine ECS modulation with other lipid-targeted interventions to achieve synergistic neuroprotective outcomes.

6.7 Advances in Lipidomics and Brain Research

The word lipidomics was introduced by Han and Gross. It refers to the high-throughput investigation and comprehensive characterization of a network of cellular lipids and their interaction with proteins as well as the regulation of genes in biological systems. The term "lipidome" is applied to the entire lipid profile of a cell, tissue, organism or ecosystem and is a sub-class of the "metabolome" which comprises the three other main classes of biological molecules: proteins/amino-acids, sugars and nucleic acids.[67] Lipidomics is a comparatively novel area of research that has been spurred on by the availability of advanced technologies like mass spectrometry (MS), nuclear magnetic resonance (NMR) spectroscopy, fluorescence spectroscopy, dual polarization interferometry and computational techniques, along with the awareness of the significance of

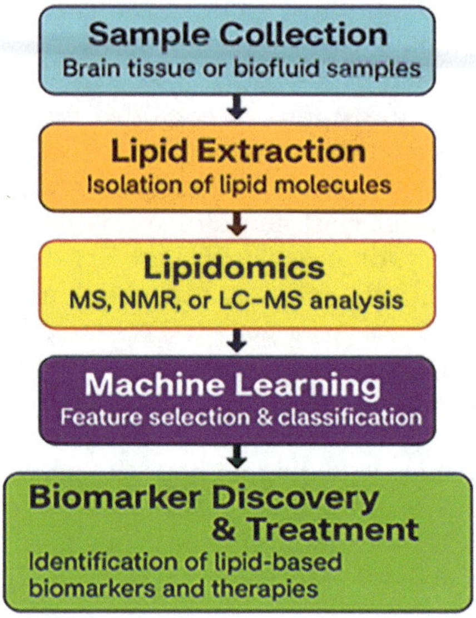

Figure 6.3 A flowchart showing the lipidomics workflow used in brain research. The process begins with sample collection, followed by lipid extraction to isolate lipid molecules. The extracted lipids undergo lipidomics analysis using methods labeled MS, NMR, and LC–MS. The resulting data are processed using machine learning for feature selection and classification. The workflow ends with biomarker discovery and treatment, indicating the identification of lipid-based biomarkers and therapeutic targets.

lipids in various neurodegenerative disorders and metabolic disorders like obesity, atherosclerosis, stroke, hypertension and diabetes (Figure 6.3).[68]

In neuroscience, lipidomics can be employed to identify biomarkers of early diagnosis and prognosis of brain illness and clarify the biological roles of the brain as well as that of its individual parts. Lipids and their dysregulation of different types have been shown to play a role in neurodegenerative disorders as well as mental disorders.

6.7.1 Alzheimer's Disease

AD is the most prevalent neurodegenerative disease and the most prevalent cause of dementia. It is widely recognized that AD histopathological alterations are linked with amyloid plaques and neurofibrillary tangles. Nevertheless, the relationship between the symptoms of AD (*e.g.*, memory decline and cognitive impairment) and their etiology remains unclear.[69] Lipid changes research is focusing on the molecular pathways involved in amyloid precursor protein (APP) processing, amyloid beta (Aβ) generation, and Aβ aggregation.[62] AD etiology revealed a dysregulation of cholesterol transport from astrocytes to neurons. The primary brain cholesterol carrier is apolipoprotein E (APOE), which has the ability to bind and clear Aβ peptides.

Besides, the protein functions of β-site APP cleaving enzyme 1 (BACE1) and γ-secretase are affected by cholesterol content, implying the role of cholesterol metabolism in the development of AD. Ceramides stabilize BACE1 and extend its half-life, thus enhancing the rate of Aβ production. Sphingolipid metabolism in the brain is strongly linked to the production of an Aβ oligomer, Aβ42.[70] Glycosphingolipids also play a role in amyloid fibril formation. All in all, sphingolipids, vital constituents of cell membranes, determine APP processing, Aβ generation, and the ensuing process, amyloid aggregation. Phospholipids are implicated in the function of γ-secretase and APP-processing-related proteins. All the above studies of AD patients' brains have also described alterations in different kinds of phospholipids such as PC, PE, and phosphatidylinositol, as well as phospholipid-metabolizing enzymes like phospholipase C (PLC) and phospholipase D (PLD). The GM1-Aβ complexes were identified within the brains of early AD patients, and the density of the complexes correlated with the Aβ oligomers found in the CSF.[71] Phosphatidylinositol-4,5-bisphosphate, a thoroughly described derivative of phosphatidylinositol, is a substrate for PLC hydrolysis. A few studies have indicated that the inhibition of PLC results in decreased phosphatidylinositol-4,5-bisphosphate turnover, hence resulting in decreased Aβ42 secretion. However, inhibition of the PLC activity has been reported to inhibit the activity of non-amyloidogenic α-secretase. Hence, whether PLC signaling is advantageous in inhibiting amyloidogenesis in AD remains undetermined. Recent research has made it clear that there are changes in the concentration of plasmalogen PE (PE-P) present in the white and gray matter within the brains of AD patients. The brain white matter concentration of PE-P reduces by around 40% during early AD, while that in the gray matter improves from 10 to 30% with the development of the disease.[72] This result implies that PE-P may be a viable lipid biomarker for AD.

6.7.2 Parkinson's Disease

PD is the second most common neurodegenerative disorder that afflicts individuals older than 60 years of age. The characteristic aspect of PD is the deposition of α-synuclein (α-syn), a neuronal protein, that influences the modulation of synaptic vesicles and neurotransmitter release, the resultant formation of Lewy bodies (*i.e.*, fibrillized α-syn), and the degeneration of dopaminergic neurons. The N-terminal portion of α-syn contains basic residues, such as KTKEGV peptide repeats, which signify high lipid membrane affinity.[72] Past research has indicated that α-syn prefers to bind with negatively charged phospholipids, and thus it is accumulated on the phospholipid surface of lipid droplets with a high concentration of triacylglycerols. Conversely, mutant α-syn, such as A30P, has little lipid-binding preference. While the A53T mutant of α-syn binds to lipid droplets, it causes the hydrolysis of the accumulated triacylglycerols.[73] The primary visual cortex of HD patients has decreased TAG levels.

6.7.3 Huntington's Disease

Less information is available regarding lipid change in HD, compared to the lipidomic research in AD and PD. There are variations between HD patients and controls in the amounts of enzymes that metabolize sphingosine-1-phosphate (S1P). Individuals afflicted with HD showed up-regulation of S1P lyase 1 and down-regulation of sphingosine kinase 1 expression, suggesting an imbalance in sphingolipid metabolism in HD patients. Furthermore, human research on HD has explored disruption of the levels of neural cholesterol and cholesteryl ester (CE).[74]

6.7.4 Schizophrenia

A couple of investigations have found that the cerebral cortical tissues of SCZ patients have abnormal phospholipid membrane compositions. These alterations in the neuronal cell membrane lipid composition could influence neurotransmitter storage and release. In addition, brain tissue and blood samples of SCZ patients have been found to exhibit alterations in lipidomes. A total of 10.4% of SCZ patients' prefrontal cortex lipids were found to have significantly changed concentrations compared to healthy controls.[75] Analysis of serum samples for SCZ biomarkers has shown that the concentration of all PEs, most PCs, and lysophosphatidylcholines is reduced, and that of SMs and the majority of lysophosphatidylethanolamines is increased in SCZ patients *versus* controls. In particular, six lipids, LPC (18:0), LPC (20:2), PC (18:2/18:2), PC (O-16:0/18:2), LPE (20:4), and PE (P-18:0/18:2), were demonstrated to be differentially expressed in SCZ patients, suggesting that they could be utilized as disease biomarkers.[76]

6.8 Challenges in Lipid-based Drug Development

Lipids do not only play a role in the prevention of neurological diseases but also play a valuable role in contemporary medicine as a means of delivering drugs to specific sites of disease. Lipids can pass through the blood–brain barrier, and their brain concentrations have been associated with dietary consumption. Solid lipid nanoparticles (SLNs) and nanostructured lipid carriers (NLCs) constitute one of the new lipid-based drug delivery systems designed particularly to combat and resolve the problems inherent in conventional drug delivery systems. Liposomes, being lipid-based nanocarriers, are increasingly used as drug delivery systems for neurological diseases.[77] Their capacity to encapsulate hydrophobic drugs, shield them against degradation, and enhance targeted delivery across the blood–brain barrier (BBB) qualifies them as potential candidates for the treatment of brain diseases. Liposomes can be engineered to entrap anti-inflammatory agents, neuroprotective agents, or gene therapies, with sustained and localized therapy in the brain. Importantly, liposomal preparations of curcumin and cannabinoids have shown anti-inflammatory and neuroprotective properties in AD and PD models. Furthermore, liposomes can be designed to specifically target certain cell types in the brain, *e.g.*, neurons, astrocytes, and microglia, allowing for personalized therapeutic interventions. Liposomes may also be used to deliver lipid-lowering drugs or lipid-modulating molecules to re-establish normal lipid homeostasis in the brain, thus reducing neuroinflammation and cognitive dysfunction.[78] Lipid metabolism and lipid droplets (LDs) are pivotal in controlling neuroinflammation and BBB integrity in neurological diseases. Dysregulation of lipid homeostasis is responsible for chronic inflammation, neuronal injury, and BBB dysfunction to worsen stroke, AD, PD, and MS. Modulating the pathways of lipid metabolism, LD dynamics, and employing lipid-based nanocarriers such as liposomes holds valuable therapeutic potential for reducing inflammation, restoring lipid balance and enhancing drug delivery to the brain. Subsequent studies should aim to create definitive interventions that modulate lipid metabolism and LD formation and provide novel prospects for the management of neurodegenerative diseases.

Drawbacks related to the application of cationic lipids in liposomes and lipoplexes, such as fast clearance from the reticuloendothelial system (RES) through opsonization and complement system activation and toxicity problems related to cell apoptotic pathways, and increased reactive oxygen species (ROS) levels, have led to the creation of ionizable cationic lipids and lipid nanoparticles (LNPs). The BBB remains one of the key barriers to lipidic delivery systems.[79] Most of the recent and current research on the delivery of small molecules or biomolecules in the CNS with the help of nanotechnology has been directed toward liposomal nanovesicles. Nevertheless, other vehicles, *e.g.*, LNPs are potentially more suitable for the delivery of genetic material in the brain based on improved encapsulation of a fragile RNA load.[80]

6.8.1 The Potential of Personalized Lipid Therapies in Neurology

Personalized neurology is the translation of concepts of personalized medicine into more effective therapies with less adverse effects. In addition to genomic tools like neurogenomics, neuroproteomics and neurometabolomics, nongenomic strategies like cell/gene therapy and nanobiotechnology are utilized also.[81] Neurogenomics is enhancing knowledge of neurologic diseases, which is a fundamental foundation of the establishment of intelligent therapies.[82] There is a requirement for biomarkers widely available for the identification and monitoring of the neurobiological and pathophysiological progression of mental health disorders. High-throughput omics technologies based on the usage of liquid chromatography–mass spectrometry (LC–MS) are motivating an explosion in the production of biological data, offering a systems-level understanding of physiological and pathogenic states.[83] Biomarkers and diagnostic integration with therapeutics are crucial for the choice and monitoring of treatments of neurological disease.[84] Precision medicine is an emerging method to address the underlying cause of any disease/condition by considering a person's biological susceptibility and their reaction to therapy.[85] Precision medicine offers profound significance to the domain of psychiatry as it disrupts the conventional use of clinical principles in mental illnesses with novel diagnostic and therapeutic paradigms that investigate the pathophysiology.[86] The work of tailored lipid therapies in neurology is colossal and needs several meta-data sets across biological systems to be correlated with multimodal information (*e.g.*, gender, age, clinical features, neuropsychological features, or imaging from the brain, and genome-wide association studies).

6.9 Conclusion

Lipids are no longer viewed merely as structural components of neuronal membranes but as dynamic regulators of brain physiology and pathology. Their complex roles in membrane architecture, intracellular signaling, neuroinflammation, and synaptic function underscore their importance in maintaining neural homeostasis. The exploration of lipid metabolism and signaling pathways has illuminated a vast biochemical landscape linking lipid dysregulation to major neurological disorders, including Alzheimer's disease, Parkinson's disease, schizophrenia, epilepsy, and mood disorders. Emerging evidence indicates that even subtle disturbances in lipid composition or turnover can initiate cascades leading to oxidative stress, mitochondrial dysfunction, impaired neurotransmission, and neuronal death. Consequently, lipids serve as both pathogenic mediators and therapeutic opportunities in the realm of neurobiology. The integration of lipidomics and molecular neuroscience has advanced our understanding of lipid dynamics in brain health and disease. High-resolution profiling of lipids across different neural cell types and regions has revealed disease-specific lipid signatures that can serve as biomarkers for diagnosis, prognosis, and therapeutic monitoring. These lipid-based biomarkers hold tremendous potential for early detection of neurodegenerative and psychiatric disorders, allowing clinicians to intervene before irreversible neuronal loss occurs. Moreover, lipidomics-driven research has uncovered the mechanistic links between lipid signaling and protein misfolding, neuroinflammation, and synaptic dysfunction—hallmarks of many brain pathologies. Therapeutically, targeting lipid pathways represents a frontier in precision neurology. Strategies that modulate sphingolipid metabolism, phosphoinositide signaling, and cholesterol transport are demonstrating efficacy in experimental and clinical contexts. The development of S1P receptor modulators like fingolimod for multiple sclerosis exemplifies the translation of lipid biology into practical neurotherapy. Similarly, endocannabinoid system modulation, through enzyme inhibitors or receptor agonists, is being leveraged to restore synaptic balance, reduce inflammation, and protect against excitotoxicity. The potential of lipid-lowering drugs such as statins and PPAR agonists extends beyond vascular protection, offering anti-inflammatory and neuroprotective effects that complement traditional pharmacological approaches. Furthermore, nutritional lipids, particularly omega-3 polyunsaturated fatty acids, have garnered significant attention for their roles in neurodevelopment, plasticity, and cognitive resilience. Their incorporation into neuronal membranes enhances receptor function, synaptic fluidity, and anti-inflammatory signaling, highlighting the interplay between diet, metabolism, and mental health. The rise of lipid-based nanotherapeutics also opens avenues for targeted drug delivery across the blood–brain barrier, a persistent challenge in neuropharmacology. Despite these advances, challenges remain. The brain's lipidome is extraordinarily diverse and dynamic, influenced by genetics, diet, environment, and aging. Deciphering the precise causal relationships between lipid alterations and disease requires longitudinal studies integrating lipidomics with genomics and proteomics. Moreover, therapeutic modulation of lipid pathways must be approached cautiously to avoid systemic metabolic disturbances. In conclusion, lipid biology offers a unifying framework for understanding and treating brain disorders. From serving as biomarkers that reflect disease progression to functioning as therapeutic targets that restore cellular balance, lipids occupy a central position in modern neuroscience. Continued interdisciplinary research bridging lipidomics, neurobiology, and clinical practice will not only deepen our insight into the lipid–brain axis but also pave the way for innovative, lipid-centered strategies to preserve cognitive health and combat neurodegeneration.

Abbreviations

AA	Arachidonic acid
AEA	Arachidonoylethanolamide (anandamide)
Akt	Protein kinase B (PKB)
AMPA	α-Amino-3-hydroxy-5-methyl-4-isoxazolepropionic acid (as in AMPA receptor)
APP	Amyloid precursor protein
BBB	Blood–brain barrier
cAMP	Cyclic adenosine monophosphate
CE	Cholesteryl ester
COX	Cyclooxygenase
DAG	Diacylglycerol
DHA	Docosahexaenoic acid
ECS	Endocannabinoid system
FAAH	Fatty acid amide hydrolase
FPR2	Formyl peptide receptor 2
GPCR	G-protein-coupled receptor
HMG-CoA	3-Hydroxy-3-methylglutaryl coenzyme A
IP_3	Inositol 1,4,5-trisphosphate
LOX	Lipoxygenase
LPA	Lysophosphatidic acid
LXRs	Liver X receptors
MAGL	Monoacylglycerol lipase

MAPK	Mitogen-activated protein kinase
mTOR	Mechanistic (mammalian) target of rapamycin
NAAA	*N*-Acylethanolamine acid amidase
NF-κB	Nuclear factor kappa-light-chain-enhancer of activated B cells
NLC	Nanostructured lipid carrier(s)
NMDA	*N*-Methyl-D-aspartate (as in NMDA-type glutamate receptor)
OXLAMs	Oxidized linoleic acid metabolites
PI3K	Phosphoinositide 3-kinase
PKC	Protein kinase C
PLA_2/PLA2	Phospholipase A_2
PLC	Phospholipase C
PLD	Phospholipase D
PPAR	Peroxisome proliferator-activated receptor
PUFAs	Polyunsaturated fatty acids
RES	Reticuloendothelial system
S1P receptor modulators	Sphingosine-1-phosphate receptor modulators
SLN	Solid lipid nanoparticle(s)
SPMs	Specialized pro-resolving mediators

References

1. C. N. Barber and D. M. Raben, Lipid metabolism crosstalk in the brain: glia and neurons, *Front. Cell. Neurosci.*, 2019, **13**, 212.
2. M. Westra, Y. Gutierrez and H. D. MacGillavry, Contribution of membrane lipids to postsynaptic protein organization, *Front. Synaptic Neurosci.*, 2021, **13**, 790773.
3. R. Ahrends, S. R. Ellis, S. H. Verhelst and M. R. Kreutz, Synaptoneurolipidomics: lipidomics in the study of synaptic function, *Trends Biochem. Sci.*, 2025, **50**(2), 156–170.
4. O. V. Galkina, O. V. Vetrovoy, I. E. Krasovskaya and N. D. Eschenko, Role of lipids in regulation of neuroglial interactions, *Biochemistry*, 2023, **88**(3), 337–352.
5. F. Mochel, Lipids and synaptic functions, *J. Inherited Metab. Dis.*, 2018, **41**(6), 1117–1122.
6. T. J. Tracey, F. J. Steyn, E. J. Wolvetang and S. T. Ngo, Neuronal lipid metabolism: multiple pathways driving functional outcomes in health and disease, *Front. Mol. Neurosci.*, 2018, **11**, 10.
7. S. A. Bennett, N. Valenzuela, H. Xu, B. Franko, S. Fai and D. Figeys, Using neurolipidomics to identify phospholipid mediators of synaptic (dys) function in Alzheimer's Disease, *Front. Physiol.*, 2013, **4**, 168.
8. D. Fitzner, J. M. Bader, H. Penkert, C. G. Bergner, M. Su, M. T. Weil, M. A. Surma, M. Mann, C. Klose and M. Simons, Cell-type-and brain-region-resolved mouse brain lipidome, *Cell Rep.*, 2020, **32**(11), 108132.

9. N. G. Bazan, D. T. Stark and N. A. Petasis, Lipid mediators: Eicosanoids, docosanoids and platelet-activating factor, *Basic Neurochemistry*, 2012, pp. 643–662.
10. Y. C. Yung, N. C. Stoddard, H. Mirendil and J. Chun, Lysophosphatidic acid signaling in the nervous system, *Neuron*, 2015, **85**(4), 669–682.
11. I. Ishii, N. Fukushima, X. Ye and J. Chun, Lysophospholipids and Their G-Coupled Protein Signaling in Alzheimer's Disease: From Physiological Performance to Pathological Impairment, *Front. Mol. Neurosci.*, 2020, **13**, 58.
12. R. G. Biringer, A review of non-prostanoid, eicosanoid receptors: expression, characterization, regulation, and mechanism of action, *J. Cell Commun. Signal.*, 2022, **16**(1), 5–46.
13. İ Ekin, Eicosanoids in human physiology: polyunsaturated fatty acid sources, biosynthesis, functions, and therapeutic implications, *Int. Anatolia Acad. Online J. Health Sci.*, 2023, **9**(2), 117–133.
14. R. G. Biringer, The enzymology of human eicosanoid pathways: The lipoxygenase branches, *Mol. Biol. Rep.*, 2020, **47**(9), 7189–7207.
15. L. H. Geraldo, T. C. Spohr, R. F. Amaral, A. C. Fonseca, C. Garcia, F. D. Mendes, C. Freitas, M. F. dosSantos and F. R. Lima, Role of lysophosphatidic acid and its receptors in health and disease: novel therapeutic strategies, *Signal Transduct. Target. Ther.*, 2021, **6**(1), 45.
16. T. J. Tracey, F. J. Steyn, E. J. Wolvetang and S. T. Ngo, Neuronal lipid metabolism: multiple pathways driving functional outcomes in health and disease, *Front. Mol. Neurosci.*, 2018, **11**, 10.
17. F. Yin, Lipid metabolism and Alzheimer's disease: clinical evidence, mechanistic link and therapeutic promise, *FEBS J.*, 2023, **290**(6), 1420–1453.
18. C. Gao, J. Jiang, Y. Tan and S. Chen, Microglia in neurodegenerative diseases: mechanism and potential therapeutic targets, *Signal Transduct. Target. Ther.*, 2023, **8**(1), 359.
19. J. Cornell, S. Salinas, H. Y. Huang and M. Zhou, Microglia regulation of synaptic plasticity and learning and memory, *Neural Regen. Res.*, 2022, **17**(4), 705–716.
20. Y. C. Yung, N. C. Stoddard, H. Mirendil and J. Chun, Lysophosphatidic acid signaling in the nervous system, *Neuron*, 2015, **85**(4), 669–682.
21. L. Zhang, Y. Zhou, Z. Yang, L. Jiang, X. Yan, W. Zhu, Y. Shen, B. Wang, J. Li and J. Song, Lipid droplets in central nervous system and functional profiles of brain cells containing lipid droplets in various diseases, *J. Neuroinflammation*, 2025, **22**(1), 7.
22. F. Yin, Lipid metabolism and Alzheimer's disease: clinical evidence, mechanistic link and therapeutic promise, *FEBS J.*, 2023, **290**(6), 1420–1453.
23. A. Shamim, T. Mahmood, F. Ahsan, A. Kumar and P. Bagga, Lipids: An insight into the neurodegenerative disorders, *Clin. Nutr. Exp.*, 2018, **20**, 1–9.
24. G. Hussain, H. Anwar, A. Rasul, A. Imran, M. Qasim, S. Zafar, M. Imran, S. K. Kamran, N. Aziz, A. Razzaq and W. Ahmad, Lipids as biomarkers of brain disorders, *Crit. Rev. Food Sci. Nutr.*, 2020, **60**(3), 351–374.

25. R. Rajmohan and P. H. Reddy, Amyloid-beta and phosphorylated tau accumulations cause abnormalities at synapses of Alzheimer's disease neurons, *J. Alzheimers Dis.*, 2017, **57**(4), 975–999.
26. D. M. Oliver and P. H. Reddy, Molecular basis of Alzheimer's disease: focus on mitochondria, *J. Alzheimers Dis.*, 2019, **72**(s1), S95–S116.
27. A. Di Pardo, E. Amico, A. Basit, A. Armirotti, P. Joshi, M. D. Neely, R. Vuono, S. Castaldo, A. F. Digilio, F. Scalabrì and G. Pepe, Defective sphingosine-1-phosphate metabolism is a druggable target in Huntington's disease, *Sci. Rep.*, 2017, **7**(1), 5280.
28. A. Di Pardo, G. Pepe, S. Castaldo, F. Marracino, L. Capocci, E. Amico, M. Madonna, S. Giova, S. K. Jeong, B. M. Park and B. D. Park, Stimulation of sphingosine kinase 1 (SPHK1) is beneficial in a Huntington's disease pre-clinical model, *Front. Mol. Neurosci.*, 2019, **12**, 100.
29. J. H. Yoon, Y. Seo, Y. S. Jo, S. Lee, E. Cho, A. Cazenave-Gassiot, Y. S. Shin, M. H. Moon, H. J. An, M. R. Wenk and P. G. Suh, Brain lipidomics: From functional landscape to clinical significance, *Sci. Adv.*, 2022, **8**(37), eadc9317.
30. M. Girych, W. Kulig, G. Enkavi and I. Vattulainen, How neuromembrane lipids modulate membrane proteins: insights from G-protein-coupled receptors (GPCRs) and receptor tyrosine kinases (RTKs), *Cold Spring Harbor Perspect. Biol.*, 2023, **15**(10), a041419.
31. R. Baccouch, E. Rascol, K. Stoklosa and I. D. Alves, The role of the lipid environment in the activity of G protein coupled receptors, *Biophys. Chem.*, 2022, **285**, 106794.
32. A. J. Gonzalez-Hernandez, H. Munguba and J. Levitz, Emerging modes of regulation of neuromodulatory G protein-coupled receptors, *Trends Neurosci.*, 2024, **47**(8), 635–650.
33. B. R. Pluimer, M. Colt and Z. Zhao, G protein-coupled receptors in the mammalian blood-brain barrier, *Front. Cell. Neurosci.*, 2020, **14**, 139.
34. P. M. Dijkman and A. Watts, Lipid modulation of early G protein-coupled receptor signalling events, *Biochim. Biophys. Acta, Biomembr.*, 2015, **1848**(11), 2889–2897.
35. A. R. Carotenuto, L. Lunghi, V. Piccolo, M. Babaei, K. Dayal, N. Pugno, M. Zingales, L. Deseri and M. Fraldi, Mechanobiology predicts raft formations triggered by ligand-receptor activity across the cell membrane, *J. Mech. Phys. Solids*, 2020, **141**, 103974.
36. Y. D. Paila and A. Chattopadhyay, Membrane cholesterol in the function and organization of G-protein coupled receptors, *Cholesterol Binding and Cholesterol Transport Proteins: Structure and Function in Health and Disease*, 2010, pp. 439–466.
37. J. C. Park, A. Luebbers, M. Dao, A. Semeano, A. M. Nguyen, M. P. Papakonstantinou, S. Broselid, H. Yano, K. A. Martemyanov and M. Garcia-Marcos, Fine-tuning GPCR-mediated neuromodulation by biasing signaling through different G protein subunits, *Mol. Cell*, 2023, **83**(14), 2540–2558.

38. A. S. Husted, M. Trauelsen, O. Rudenko, S. A. Hjorth and T. W. Schwartz, GPCR-mediated signaling of metabolites, *Cell Metab.*, 2017, **25**(4), 777–796.
39. B. I. Sejdiu and D. P. Tieleman, Lipid-protein interactions are a unique property and defining feature of G protein-coupled receptors, *Biophys. J.*, 2020, **118**(8), 1887–1900.
40. J. Peng, D. Liang and Z. Zhang, Palmitoylation of synaptic proteins: roles in functional regulation and pathogenesis of neurodegenerative diseases, *Cell. Mol. Biol. Lett.*, 2024, **29**(1), 108.
41. M. Westra, Y. Gutierrez and H. D. MacGillavry, Contribution of membrane lipids to postsynaptic protein organization, *Front. Synaptic Neurosci.*, 2021, **13**, 790773.
42. G. Hussain, J. Wang, A. Rasul, H. Anwar, A. Imran, M. Qasim, S. Zafar, S. K. Kamran, A. Razzaq, N. Aziz and W. Ahmad, Role of cholesterol and sphingolipids in brain development and neurological diseases, *Lipids Health Dis.*, 2019, **18**(1), 26.
43. S. Li and Z. H. Sheng, Fuelling synapses *via* lipid metabolism, *Nat. Metab.*, 2025, **7**(7), 1310–1311.
44. T. Zhang, *et al.*, Regulation of synaptic function and lipid metabolism, *Neural Regen. Res.*, 2026, **21**(3), 1037–1057.
45. B. Uranbileg, H. Isago, E. Sakai, M. Kubota, Y. Saito and M. Kurano, Alzheimer's disease manifests abnormal sphingolipid metabolism, *Front. Aging Neurosci.*, 2024, **16**, 1368839.
46. S. He, Z. Xu and X. Han, Lipidome disruption in Alzheimer's disease brain: detection, pathological mechanisms, and therapeutic implications, *Mol. Neurodegener.*, 2025, **20**(1), 11.
47. D. Yang, X. Wang, L. Zhang, Y. Fang, Q. Zheng, X. Liu, W. Yu, S. Chen, J. Ying and F. Hua, Lipid metabolism and storage in neuroglia: role in brain development and neurodegenerative diseases, *Cell Biosci.*, 2022, **12**(1), 106.
48. T. C. Dakal, F. Xiao, C. K. Bhusal, P. C. Sabapathy, R. Segal, J. Chen and X. Bai, Lipids dysregulation in diseases: core concepts, targets and treatment strategies, *Lipids Health Dis.*, 2025, **24**(1), 61.
49. Y. Cao, L. W. Zhao, Z. X. Chen and S. H. Li, New insights in lipid metabolism: Potential therapeutic targets for the treatment of Alzheimer's disease, *Front. Neurosci.*, 2024, **18**, 1430465.
50. P. Katsel, C. Li and V. Haroutunian, Gene expression alterations in the sphingolipid metabolism pathways during progression of dementia and Alzheimer's disease: a shift toward ceramide accumulation at the earliest recognizable stages of Alzheimer's disease?, *Neurochem. Res.*, 2007, **32**(4), 845–856.
51. J. M. Dietschy and S. D. Turley, Cholesterol metabolism in the brain, *Curr. Opin. Lipidol.*, 2001, **12**(2), 105–112.
52. X. Y. Chua, F. Torta, J. R. Chong, N. Venketasubramanian, S. Hilal, M. R. Wenk, C. P. Chen, T. V. Arumugam, D. R. Herr and M. K. Lai, Lipidomics

profiling reveals distinct patterns of plasma sphingolipid alterations in Alzheimer's disease and vascular dementia, *Alzheimers Res. Ther.*, 2023, **15**(1), 214.
53. B. Das and R. Yan, A close look at BACE1 inhibitors for Alzheimer's disease treatment, *CNS Drugs*, 2019, **33**(3), 251–263.
54. N. Bernoud-Hubac, A. Lo Van, A. N. Lazar and M. Lagarde, Ischemic brain injury: involvement of lipids in the pathophysiology of stroke and therapeutic strategies, *Antioxidants*, 2024, **13**(6), 634.
55. V. Campos-Peña, P. Pichardo-Rojas, T. Sánchez-Barbosa, E. Ortíz-Islas, C. E. Rodríguez-Pérez and P. Montes, *et al.*, Amyloid β, lipid metabolism, basal cholinergic system, and therapeutics in Alzheimer's disease, *Int. J. Mol. Sci.*, 2022, **23**(20), 12092.
56. Y. Cao, L. W. Zhao, Z. X. Chen and S. H. Li, New insights in lipid metabolism: Potential therapeutic targets for the treatment of Alzheimer's disease, *Front. Neurosci.*, 2024, **18**, 1430465.
57. C. Cinti, M. G. Trivella, M. Joulie, H. Ayoub and M. Frenzel, The roadmap toward personalized medicine: challenges and opportunities, *J. Pers. Med.*, 2024, **14**(6), 546.
58. B. S. B. Correia, J. V. Nani, R. Waladares Ricardo, D. Stanisic, T. B. B. C. Costa, M. A. Hayashi and L. Tasic, Effects of psychostimulants and antipsychotics on serum lipids in an animal model for schizophrenia, *Biomedicines*, 2021, **9**(3), 235.
59. A. Di Pardo, E. Amico, A. Basit, A. Armirotti, P. Joshi and M. D. Neely, *et al.*, Defective sphingosine-1-phosphate metabolism is a druggable target in Huntington's disease, *Sci. Rep.*, 2017, **7**(1), 5280.
60. B. Dudok and I. Soltesz, Imaging the endocannabinoid signaling system, *J. Neurosci. Methods*, 2022, **367**, 109451.
61. G. R. Ediriweera, A. J. Sivaram, G. Cowin, M. L. Brown, L. Mcalary and J. S. Lum, *et al.*, Lipid nanoparticles and transcranial focused ultrasound enhance the delivery of SOD1 antisense oligonucleotides to the murine brain for ALS therapy, *J. Controlled Release*, 2025, **378**, 221–235.
62. N. Fabelo, V. Martín, R. Marín, D. Moreno, I. Ferrer and M. Díaz, Altered lipid composition in cortical lipid rafts occurs at early stages of sporadic Alzheimer's disease and facilitates APP/BACE1 interactions, *Neurobiol. Aging*, 2014, **35**(8), 1801–1812.
63. B. S. Fernandes, L. M. Williams, J. Steiner, M. Leboyer, A. F. Carvalho and M. Berk, The new field of 'precision psychiatry', *BMC Med.*, 2017, **15**(1), 80.
64. T. Guo, D. Zhang, Y. Zeng, T. Y. Huang, H. Xu and Y. Zhao, Molecular and cellular mechanisms underlying the pathogenesis of Alzheimer's disease, *Mol. Neurodegener.*, 2020, **15**(1), 40.
65. D. Haspula and M. A. Clark, Cannabinoid receptors: an update on cell signaling, pathophysiological roles and therapeutic opportunities in neurological, cardiovascular, and inflammatory diseases, *Int. J. Mol. Sci.*, 2020, **21**(20), 7693.

66. S. Hong, B. L. Ostaszewski, T. Yang, T. T. O'Malley, M. Jin and K. Yanagisawa, *et al.*, Soluble Aβ oligomers are rapidly sequestered from brain ISF *in vivo* and bind GM1 ganglioside on cellular membranes, *Neuron*, 2014, **82**(2), 308–319.
67. K. K. Jain, Personalized therapy of neurological disorders, in *Textbook of Personalized Medicine*, Springer International Publishing, Cham, 2020, pp. 213–262.
68. S. Kamano, D. Ozawa, K. Ikenaka and Y. Nagai, Role of Lipids in the Pathogenesis of Parkinson's Disease, *Int. J. Mol. Sci.*, 2024, **25**(16), 8935.
69. H. Kittaka, J. DeBrecht and S. K. Mishra, Differential contribution of sensory transient receptor potential channels in response to the bioactive lipid sphingosine-1-phosphate, *Mol. Pain*, 2020, **16**, 1744806920903515.
70. A. Kline, Apolipoprotein E, amyloid-β clearance and therapeutic opportunities in Alzheimer's disease, *Alzheimers Res. Ther.*, 2012, **4**(4), 32.
71. J. A. Kulkarni, D. Witzigmann, S. B. Thomson, S. Chen, B. R. Leavitt, P. R. Cullis and R. Van Der Meel, The current landscape of nucleic acid therapeutics, *Nat. Nanotechnol.*, 2021, **16**(6), 630–643.
72. P. Kumari, D. Ghosh, A. Vanas, Y. Fleischmann, T. Wiegand and G. Jeschke, *et al.*, Structural insights into α-synuclein monomer–fibril interactions, *Proc. Natl. Acad. Sci. U. S. A.*, 2021, **118**(10), e2012171118.
73. N. Landman, S. Y. Jeong, S. Y. Shin, S. V. Voronov, G. Serban and M. S. Kang, *et al.*, Presenilin mutations linked to familial Alzheimer's disease cause an imbalance in phosphatidylinositol 4, 5-bisphosphate metabolism, *Proc. Natl. Acad. Sci. U. S. A.*, 2006, **103**(51), 19524–19529.
74. I. Manuel, L. Lombardero, A. Llorente-Ovejero and R. Rodríguez-Puertas, Neuropeptides and neurolipids: what they are and how they relate to Alzheimer's disease, in *Genetics, Neurology, Behavior, and Diet in Dementia*, Academic Press, 2020, pp. 423–439.
75. O. C. Marian, C. Tran and A. S. Don, Altered lipid metabolic homeostasis in the pathogenesis of Alzheimer's disease, in *Lipid signaling and metabolism*, Academic Press, 2020, pp. 469–504.
76. C. Martínez-Aguirre, M. de los Ángeles Nuñez-Lumbreras and L. Rocha, The endocannabinoid system in health and disease: Features in epilepsy, in *Neurobiology and Physiology of the Endocannabinoid System*, Academic Press, 2023, pp. 215–227.
77. R. Moreira, C. Nóbrega, L. P. de Almeida and L. Mendonça, Brain-targeted drug delivery-nanovesicles directed to specific brain cells by brain-targeting ligands, *J. Nanobiotechnol.*, 2024, **22**(1), 260.
78. G. R. Phillips, S. E. Hancock, S. H. Brown, A. M. Jenner, F. Kreilaus, K. A. Newell and T. W. Mitchell, Cholesteryl ester levels are elevated in the caudate and putamen of Huntington's disease patients, *Sci. Rep.*, 2020, **10**(1), 20314.
79. S. Pietropaolo and G. Marsicano, The role of the endocannabinoid system as a therapeutic target for autism spectrum disorder: Lessons from

behavioral studies on mouse models, *Neurosci. Biobehav. Rev.*, 2022, **132**, 664–678.
80. A. Shafie, A. A. Ashour, S. Anwar, F. Anjum and M. I. Hassan, Exploring molecular mechanisms, therapeutic strategies, and clinical manifestations of Huntington's disease, *Arch. Pharmacal Res.*, 2024, **47**(6), 571–595.
81. A. Shamim, T. Mahmood, F. Ahsan, A. Kumar and P. Bagga, Lipids: An insight into the neurodegenerative disorders, *Clin. Nutr. Exp.*, 2018, **20**, 1–19.
82. T. Smith, K. J. Knudsen and S. A. Ritchie, A novel inducible animal model for studying chronic plasmalogen deficiency associated with Alzheimer's disease, *Brain Res.*, 2024, **1843**, 149132.
83. T. J. Tracey, S. E. Kirk, F. J. Steyn and S. T. Ngo, The role of lipids in the central nervous system and their pathological implications in amyotrophic lateral sclerosis, in *Seminars in Cell & Developmental Biology*, Academic Press, 2021, vol. 112, pp. 69–81.
84. M. Tsakiri, C. Zivko, C. Demetzos and V. Mahairaki, Lipid-based nanoparticles and RNA as innovative neuro-therapeutics, *Front. Pharmacol.*, 2022, **13**, 900610.
85. G. van Echten-Deckert, The role of sphingosine 1-phosphate metabolism in brain health and disease, *Pharmacol. Ther.*, 2023, **244**, 108381.
86. D. B. Vieira and L. F. Gamarra, Getting into the brain: liposome-based strategies for effective drug delivery across the blood–brain barrier, *Int. J. Nanomed.*, 2016, 5381–5414.

CHAPTER 7

Epigenetic and Transcriptomic Regulations in Brain Disorders

AJAY PAL*[a] AND JAISHRIRAM RATHORED[b]

[a] Center for Neural Dev. & Repair, Temple University School of Medicine, Philadelphia, PA, USA; [b] Central Research Laboratory and Molecular Diagnostics, Datta Meghe Institute of Higher Education & Research, Maharashtra, India
*Email: ajay.pal@temple.edu

7.1 Introduction

The human brain is a complex and dynamic organ, requiring precise regulation of gene expression across spatial and temporal scales. Unlike static genomic information, epigenetic and transcriptomic mechanisms provide layers of regulatory control that adapt neuronal function to internal and external cues. These mechanisms govern how genes are activated, silenced, or modulated without altering the DNA sequence, and they are particularly critical in the brain, where cellular diversity, plasticity, and long-term memory rely heavily on fine-tuned gene expression.[1] Epigenetic and transcriptomic mechanisms dynamically regulate gene activity in response to environmental, physiological, and developmental signals.[2,3] Neuronal activity, learning, and stress directly reshape gene expression through changes in chromatin structure and transcriptional programs.[4–6] These mechanisms do not alter DNA sequences but control how genes are turned on or off, often through DNA methylation, histone modifications, and non-coding RNAs.[7–9] Early-life factors, such as maternal stress or diet, can leave lasting epigenetic marks on the developing brain.[9–12] These changes influence neuronal

Chemical Biology No. 26
Chemical Signatures of the Brain: Bridging Biochemistry, Neuroanatomy, and Brain Disorders
Edited by Sachchida Nand Rai and Abhishek Kumar Singh

Published by the Royal Society of Chemistry, www.rsc.org

function, behaviour, and disease vulnerability as widespread histone acetylation differences have been identified in individuals with autism spectrum disorder, linking specific epigenetic signatures to neurodevelopmental pathology.[12,13] These findings confirm that epigenetic and transcriptomic regulation acts as a flexible, responsive interface between the genome and the environment.[13,14] The complexity of the human brain depends not only on its genetic code but also on dynamic epigenetic and transcriptomic processes.[2,3] It has been reported that major developmental shifts in gene expression and epigenetic patterns occur across brain regions and cell types from prenatal stages through adolescence, with risk genes for brain disorders clustering in specific cell types and gene networks.[2,13,15] These dynamic activities exceptionally regulate spatiotemporal gene expression across the central nervous system (CNS).[16] These molecular regulators provide the flexibility required for neural development, neuronal differentiation, maturation, synaptic plasticity, and behavioural adaptation.[17] Unlike the genetic mutation, which is mainly static, epigenetic and transcriptomic regulations allow for the modulation of gene activity in response to environmental, physiological, and conditional stimuli.[15] Gene expression and regulation play a crucial role in shaping an organism's behaviour.[18] Epigenetic mechanisms control gene activity without changing the genetic code.[12] Instead, they add chemical modifications to DNA, or the proteins associated with it.[12] These dynamic modifications act as a bridge, linking environmental signals to the underlying genotype.[15]

Epigenetics refers to heritable yet reversible modifications, such as DNA methylation, histone modifications, and chromatin remodelling, that alter chromatin structure and gene accessibility without changing DNA sequence.[13,15] Transcriptomics encompasses all mechanisms regulating RNA production and processing, including transcriptional activation, splicing, editing, transport, and degradation.[19] Together, these processes compose the expression of neuronal and glial phenotypes in health and disease.[11,20] Disruption of these finely tuned systems has been increasingly associated with a wide range of brain disorders.[20–23] Neurodevelopmental conditions like autism spectrum disorder (ASD) and Rett syndrome exhibit aberrant DNA methylation and histone changes that disrupt developmental trajectories.[22,24–28] Psychiatric conditions,[29] including schizophrenia and major depression, show transcriptomic alterations linked to disrupted synaptic signalling and neuroinflammation.[24,30–34] Neurodegenerative diseases such as Alzheimer's and Parkinson's reveal extensive epigenomic reprogramming associated with disease progression and cellular dysfunction.[30,35–38] This chapter offers a comprehensive review of current knowledge in epigenetic and transcriptomic regulation, their interplay, disease-specific signatures, and the technological advancements that have enabled their study. Emphasis is placed on integrative multi-omics and therapeutic potential, aiming to bridge mechanistic insight with clinical translation.

7.2 Overview of Epigenetic Mechanisms

Epigenetic regulation involves molecular processes that alter gene activity without changing the DNA sequence.[20,39,40] In the brain, these processes preserve neuronal identity, shape neural circuits, and enable synaptic plasticity.[20,41] They depend on precise, activity-dependent gene expressions and include DNA methylation, histone modifications, non-coding RNAs (ncRNAs), and ATP-dependent chromatin remodelling.[20,41] Waddington introduced the concept of *epigenesis* to describe mechanisms that operate above the gene level and direct cell fate.[39–41] Biochemically, the most widely studied mechanisms are DNA methylation and chromatin regulation through histone modifications.[20,41,42]

Although initially studied in development, epigenetic mechanisms also respond dynamically to experience.[40] Environmental toxins, maternal behaviour, psychological stress, learning, drug exposure, and trauma can all modify the chemical and three-dimensional structure of DNA in the nervous system.[31,43–50] These epigenomic changes alter gene readout in neurons and can produce long-lasting and sometimes permanent changes in brain function.[20,30,51] There is an ongoing debate regarding whether ncRNAs should be classified as true epigenetic regulators.[20] Nevertheless, researchers such as Shelley Berger and colleagues argue that these molecules play a pivotal role in initiating epigenetic modifications.[52] Their established functions in the brain place ncRNAs alongside DNA methylation and histone modifications as central components of epigenetic regulation.[20] Each mechanism operates through distinct yet interconnected chemical and structural pathways.[20,39,41] More recently, the field has expanded to include RNA modifications, particularly those generating regulatory ncRNAs, thereby adding further complexity to gene regulation within the nervous system.[20,30]

All major epigenetic modifications share two key biochemical features. They regulate transcription without altering the DNA sequence.[39] They can also, at least in principle, self-regenerate and self-perpetuate.[41] This capacity allows them to store stable changes in gene function that persist through cell division and, in some cases, across generations. As Crick[53] and Lisman[54] noted, such self-perpetuating changes act as a form of cellular information storage, a role for which epigenetic mechanisms are uniquely suited.[39,55]

7.2.1 DNA Methylation

DNA methylation is the addition of a methyl group to the 5′ carbon of cytosine, most often at CpG sites.[39,41,56,57] DNA methyltransferases (DNMT1, DNMT3A, DNMT3B)[23,58] catalyse this reaction.[39] In promoter regions, methylation typically represses transcription by blocking the access of transcription factors and RNA polymerase.[56,57,59] In gene bodies, methylation can support active transcription.[57,59] CpG sites occur where cytosine is followed by guanine in the DNA sequence.[57,59] They often cluster in CpG island regions with high GC content, an observed/expected CpG ratio greater than 0.6, and a length of over 200 base pairs.[59,60] CpG islands are typically protected from methylation, but intragenic CpG islands within active genes can remain unmethylated.[41,57,59,60] These regions can influence mRNA isoform length and expand transcriptome diversity.[57,59] Methylation patterns in CpG islands differ between neural and non-neural tissues, and brain-specific methylation databases now exist.[41,59–61]

DNA methylation is an ancient regulatory mechanism conserved across life.[20,39,57] It protects against transposable elements, regulates gene expression, and contributes to tissue identity.[20,41,58] In neurons, methylation patterns shift in response to activity, shaping memory formation and synaptic remodelling.[56,58,62–64] Abnormal methylation of *OXTR* and *MeCP2* is linked to ASD and Rett syndrome, while broader methylation defects occur in neurodegenerative diseases.[20,30,63] Specialized proteins maintain and interpret methylation marks.[62] DNMTs add methyl groups, methyl-CpG binding proteins (MBPs) read them, and ten-eleven translocation (TET) enzymes remove them. Common methylation forms include 5-methylcytosine (5mC) and its oxidized derivatives, such as 5-hydroxymethylcytosine.[20,58,62,65] DNA methylation can also occur on adenine, most notably *N*6-methyladenine (6mA), which recent studies link to neurodegeneration.[20,30,58,65]

7.2.2 Histone Modifications

Histone proteins are post-translationally modified at their N-terminal tails, influencing chromatin accessibility.[20,66,67] Acetylation (by histone acetyltransferases; HATs) typically promotes transcriptional activation, while deacetylation by histone deacetylases (by HDACs) condenses chromatin.[67] Methylation at specific lysine residues can either activate (*e.g.*, H3K4me3) or repress (*e.g.*, H3K27me3) transcription.[20,67] Neuronal histone modifications are tightly linked to development, learning, and plasticity.[66,67] Dysregulation of these marks, such as decreased H3K4me3 in schizophrenia, highlights their role in pathophysiology.[20]

Histone modifications are key regulators of gene expression.[41,66,67] They determine which genes are active or silent in each cell, shaping cellular function and, in broader contexts, influencing processes such as aging.[41,66] Histones are protein spools that package DNA into chromatin.[66,68] Their structure and interaction with DNA change when specific chemical groups are added or removed.[20] Post-translational modifications of histones include acetylation, methylation, phosphorylation, ubiquitination, sumoylation, lactylation, serotonylation, and crotonylation.[20,41,66,68–74] Acetylation by HATs opens chromatin and promotes transcription, while deacetylation by HDACs compacts chromatin and represses gene activity.[66] Methylation effects depend on the residue and degree of modification: lysine trimethylation at H3K4 activates transcription, whereas H3K27me3 represses it.[66] Other marks, such as phosphorylation or ubiquitination, regulate chromatin accessibility and recruit specific transcriptional machinery.[66,70,75]

Histone proteins form octameric cores around which DNA is wrapped, creating nucleosomes, the fundamental units of chromatin.[66] Post-translational modifications alter the chromatin structure to either enhance or inhibit transcriptional readout.[20,76] These modifications act in combinations at specific residues, forming a "histone code" that the cell's transcriptional machinery can interpret to fine-tune gene activity.[66,76] Regulation also occurs through histone subunit exchange.[76] Variants such as H2A.Z and H3.3 replace canonical histones within nucleosomes, modifying transcriptional output in ways most similar to the post-translational changes.[66,76] Both modification patterns and histone variant incorporation work together to control gene expression in a context-dependent manner.[20,41,66] The combination of chemical modifications and histone variant dynamics creates a flexible and responsive chromatin landscape.[77] In neurons, this system supports developmental programs, learning, memory, and plasticity.[41,66] The disruption of histone modification patterns, such as reduced H3K4me3 in schizophrenia, links directly to brain disorders.[20,66] The neuronal "histone code" represents a form of molecular information processing, one that is only beginning to be fully understood in the context of brain function.[66,77]

7.2.3 Non-coding RNAs

NcRNAs are functional RNA molecules transcribed from DNA but not translated into proteins.[78–81] NcRNAs, especially miRNAs, lncRNAs, and circRNAs, exert potent regulatory effects.[78,80–82] They regulate gene expression at the transcriptional, post-transcriptional, translational, and even protein localization levels.[79,82–84] These molecules modulate gene expression at transcriptional and post-transcriptional levels by interacting with chromatin remodels or targeting specific mRNAs.[20,79,85] For instance, lncRNA MEG3 recruits PRC2 to repress gene transcription, while miR-132 regulates dendritic spine formation.[80] Many ncRNAs also act in epigenetic control by modifying chromatin states and interacting with DNA methylation and histone modification pathways.[80,82] NcRNAs fall into two main categories: housekeeping ncRNAs, which perform essential cellular functions, and regulatory ncRNAs, which modulate gene expression.[79,81] Regulatory ncRNAs are further divided by size and can guide chromatin remodels to specific genomic loci.[85] Short non-coding RNAs include small interfering RNAs (siRNAs), microRNAs (miRNAs),[86] and PIWI-interacting RNAs (piRNAs).[20,85] miRNAs, siRNAs, piRNAs, and snRNAs act with sequence specificity, enabling precise regulation of transcription.[87] For example, piRNAs silence transposable elements, while specific lncRNAs recruit remodelling complexes to modify chromatin states in neurons.[88] Long ncRNAs (lncRNAs) exceed 200 nucleotides in length.[78,81]

SiRNAs are 19–24 nucleotides long, originate from double-stranded RNA, and silence gene transcription.[20,89] MiRNAs, also 19–24 nucleotides long, derive from hairpin-shaped primary transcripts and inhibit mRNA translation or promote degradation.[84,89] In the nervous system, miRNAs such as miR-132 regulate dendritic spine formation, while miR-124 controls neuronal differentiation.[79,86,89,90] Many miRNAs are linked to neural stem cell fate, apoptosis, and neurodegenerative disorders.[84,90] PiRNAs, 26–31 nucleotides in length, originate from long precursor transcripts and repress transposons through transcriptional or post-transcriptional silencing.[79,83,91]

LncRNAs have diverse origins and functions.[20,92] They regulate chromatin remodelling, recruit transcriptional complexes, and modulate mRNA splicing and stability.[84,92] For example, the lncRNA MEG3 recruits PRC2 to silence target genes, and GOMAFU influences neuronal splicing programs.[84] LncRNAs also respond rapidly to neuronal stimuli such as BDNF, indicating a role in activity-dependent transcription.[79,92]

Circular RNAs (circRNAs) arise through back-splicing of pre-mRNA.[93] They often act as sponges for miRNAs, reducing their availability to target mRNAs, and can modulate neuronal signalling pathways.[82,94,95] In the brain, circRNAs accumulate with age and are being studied as biomarkers for neurodegenerative diseases.[79,83,94]

Together, these ncRNA classes form an interconnected network that fine-tunes neuronal gene expression.[78,79] By working alongside DNA methylation and histone modifications, they help maintain neural identity, support synaptic plasticity, and adapt gene activity to experience.[79,81]

7.2.4 Chromatin Remodelling

ATP-dependent chromatin remodelling complexes (*e.g.*, SWI/SNF, chromodomain helicase DNA-binding protein 8; CHD8) restructure nucleosomes to modulate DNA accessibility.[39] These complexes are essential for neuronal lineage specification and activity-driven transcription.[20,41] CHD8 mutations, frequently observed in ASD, lead to widespread dysregulation of neurodevelopmental gene networks.[20,42] Together, these mechanisms orchestrate a flexible yet stable epigenetic landscape essential for brain function and plasticity.[96] Disruption of any one component can trigger cascading effects that underlie many brain disorders.[30,42] ATP-dependent chromatin remodelling alters nucleosome structure to regulate DNA accessibility. Remodelling complexes use ATP hydrolysis to reposition, eject, or restructure nucleosomes, thereby controlling transcriptional access to specific genes. In neurons, these mechanisms are essential for lineage specification, synaptic plasticity, and activity-dependent transcription.[30,96,97]

The SWI/SNF family of remodelling complexes changes the affinity between DNA and the histone core, loosening chromatin structure to promote transcription.[96] Mutations in SWI/SNF subunits are linked to neurodevelopmental disorders and intellectual disabilities.[98] The CHD8 and other remodelling factors regulate large sets of neurodevelopmental genes, with CHD8 mutations strongly associated with ASD.[39,41]

The REST/CoREST/Sin3A complex plays a pivotal role in determining neuronal *versus* non-neuronal identity.[98] In non-neuronal cells, REST silences neuronal genes, while in neurons it is tightly regulated to allow expression of genes required for synaptic function.[98]

Other genome-altering processes intersect with chromatin remodelling.[39] LINE-1 (L1) retrotransposons can copy and insert themselves into new genomic locations, changing transcriptional landscapes and contributing to neuronal genomic diversity.[88] Although L1 insertion changes DNA sequence and is therefore not a classical epigenetic mechanism, its functional similarity to epigenetic regulation has led to its inclusion in broader neuroepigenetic discussions.[88] This process may drive cellular heterogeneity in the brain through insertional mutagenesis.[41]

An additional proposed mechanism involves prion-like proteins as persistent regulators of synaptic function.[41] In *Aplysia*, a prion-like form of cytoplasmic polyadenylation element-binding protein (ApCPEB) maintains long-term changes in synaptic efficacy during memory storage.[30] Such proteins are self-perpetuating, heritable, and capable of sustaining altered gene function, placing them conceptually within the epigenetic framework.[30,97]

Together, these chromatin remodelling pathways form a dynamic regulatory layer in neurons.[51] They integrate ATP-dependent structural changes, ncRNA targeting, mobile genetic elements, and potentially prion-like protein activity to maintain both stability and flexibility in gene expression, supporting brain function and plasticity.[51,97] As depicted schematically in Figure 7.1, the principal epigenetic mechanisms governing neural gene regulation include DNA methylation at CpG sites, histone modifications such as H3K9 acetylation and H3K27 trimethylation, chromatin remodelling between open and closed states, and non-coding RNA interactions that modulate transcriptional activity.

7.3 Transcriptomic Regulation in the Brain

Transcriptomic regulation refers to the control of gene expression at the RNA level, encompassing transcription initiation, RNA splicing, editing, transport, stability, and translation.[99] In the brain, where cell type diversity and plasticity are extreme, transcriptomic precision is essential.[100] Neurons, astrocytes, oligodendrocytes, and microglia each maintain unique transcriptomic signatures, which change dynamically in response to activity, development, and disease.[101] This regulatory landscape is shaped by transcription factors, RNA-binding proteins, splicing machinery, and a wide array of non-coding RNAs.[102]

Epigenetic changes, neuronal signalling pathways, and extracellular environmental stimuli all have an impact on transcriptome regulation in the brain, adding to this complex framework.[103] Epigenetic processes, including DNA methylation and histone modifications, work together with transcriptional regulators to produce cell type-specific gene expression patterns, resulting in functional specialization.[104] Furthermore, neural activity induces fast transcriptome reprogramming *via* activity-dependent transcription factors and RNA alterations, allowing adaptive responses to synaptic stimuli.[105]

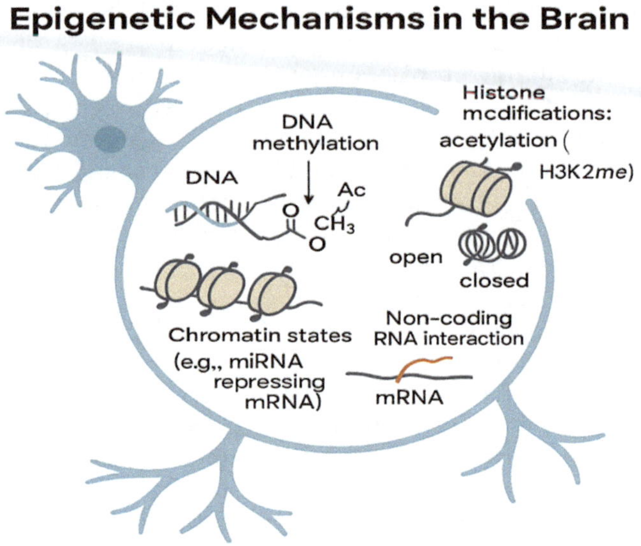

Figure 7.1 Epigenetic mechanisms in the brain. A schematic overview illustrating major epigenetic processes that regulate gene expression in neurons and glial cells. Within the cutaway of a brain cell nucleus, genomic regions are shown undergoing distinct epigenetic modifications: DNA methylation at CpG sites, which typically represses transcription; histone modifications, including acetylation (H3K9ac) associated with active chromatin and methylation (H3K27me3) linked to repressed chromatin; and chromatin remodelling, representing transitions between open (euchromatin) and closed (heterochromatin) states. Additionally, non-coding RNA interactions are depicted, where microRNAs (miRNAs) bind to target mRNAs to inhibit translation or promote degradation. Together, these mechanisms orchestrate dynamic and cell-type-specific transcriptional programs essential for neural development, plasticity, and disease processes.

Alternative splicing is crucial for producing the different protein isoforms required for complex brain activities, while RNA stability and degradation mechanisms regulate transcript availability.[106] NcRNAs, which include miRNAs, lncRNAs, and circRNAs, act as post-transcriptional regulators, influencing translation efficiency and transcript half-life in a geographically and temporally constrained manner.[12,107]

Furthermore, glial cells actively modulate transcriptomes, promoting neuronal homeostasis and affecting neuroinflammatory responses.[108] Disruptions in these finely regulated processes have been linked to a variety of neurological ailments, including neurodegenerative diseases, psychiatric conditions, and developmental anomalies.[109,110] Thus, the brain's transcriptome landscape is a highly dynamic and multilayered regulatory system that combines innate genetic instructions and extrinsic stimuli to maintain neuronal diversity, flexibility, and overall central nervous system function.[111]

7.3.1 Transcription Factors and Co-regulators

Transcription factors (TFs) are DNA-binding proteins that control the rate of transcription of target genes.[112] They work in combination with co-activators and co-repressors to recruit or block the transcriptional machinery. In the brain, lineage-specific TFs such as NEUROD1, SOX2, REST, and FOXP2 orchestrate neuronal differentiation and circuit formation.[113]

REST (RE1-Silencing Transcription Factor), also known as NRSF, represses neuronal gene expression in non-neuronal cells by binding to RE1 (Repressor Element 1) sites and recruiting co-repressors and histone modifiers like HDACs, CoREST, and chromatin-remodelling complexes.[114] This process leads to histone deacetylation and chromatin compaction, creating a repressive environment that prevents the transcription of genes characteristic of a neuronal phenotype.[115] In neurons, its activity must be finely tuned; dysregulation of REST has been implicated in epilepsy, Huntington's disease, and Alzheimer's disease.[116,117]

7.3.2 RNA Splicing and Isoform Diversity

Alternative splicing allows a single gene to produce multiple transcript isoforms, greatly expanding proteomic diversity.[118] Alternative splicing allows a single gene to create diverse messenger RNA (mRNA) transcripts and protein isoforms, significantly increasing the complexity of the proteome beyond what the genome alone could achieve.[118,119] This process involves the selective inclusion or exclusion of specific gene segments, known as exons, during mRNA processing, resulting in different functional proteins from the same genetic information.[120] The regulation of this process is crucial for cellular function and is disrupted in various diseases, including cancer and neurological disorders.[119] In the human brain, it is estimated that over 90% of multi-exon genes are alternatively spliced.[121,122] Splicing decisions are controlled by spliceosomal components and RNA-binding proteins such as RBFOX, PTBP1/2, and NOVA1/2.[123]

Dysregulation of splicing is a recurring theme in brain disorders. In frontotemporal dementia (FTD) and amyotrophic lateral sclerosis (ALS), for example, altered splicing of the TAU or TDP-43 genes leads to toxic isoforms that disrupt cellular homeostasis.[124–126] Splicing misregulation also plays a central role in autism and schizophrenia, particularly in genes involved in synaptic architecture and neuronal signalling.[111,127]

7.3.3 Non-coding RNAs in Transcriptomic Control

Non-coding RNAs (ncRNAs) do not code for proteins but play crucial roles in post-transcriptional regulation:[128] ncRNAs in the brain and their roles include microRNAs (miRNAs) like miR-132, which is critical for synaptic plasticity and memory, and long non-coding RNAs (lncRNAs) such as BACE1-AS, which influences the production of beta-amyloid in Alzheimer's disease

(AD). Dysregulation of these ncRNAs has been linked to several neurological disorders, including AD, Parkinson's disease (PD), Huntington's disease (HD), and epilepsy.

7.3.3.1 MicroRNAs (miRNAs, ~22 nucleotides)

MiRNAs are short ncRNAs that regulate gene expression by binding to complementary mRNA sequences, leading to its degradation or translational inhibition.[129]

7.3.3.1.1 Functions in Brain Health. Synaptic plasticity and memory: MiR-132 is a key regulator of synaptic plasticity and is essential for memory formation. It modulates the function of CREB and MEF2 transcription factors, which are important for dendritic spine development.[130]

Neurogenesis and neuronal development: The most abundant miRNA in the brain, miR-124, regulates neurogenesis, the process by which new neurons are formed.[131,132] MiR-9 is one of the most highly expressed microRNAs in the developing and adult vertebrate brain, and promotes neuronal differentiation and regulates neuronal migration.[133] Functional analyses in different vertebrate species have revealed a prominent role of this microRNA in balancing proliferation in embryonic neural progenitor populations.[133]

7.3.3.2 Long Non-coding RNAs (lncRNAs, >200 Nucleotides)

LncRNAs are over 200 nucleotides in length and are abundant in the central nervous system, where they help regulate transcription, chromatin remodelling, and mRNA stability.[134]

7.3.3.2.1 Functions in Brain Health. Neurogenesis and neuronal differentiation: LncRNAs like EVF-2 and Gomafu are involved in neural stem cell differentiation and neurogenesis.[135] The lncRNA BDNF-AS regulates the brain-derived neurotrophic factor (BDNF) gene, which is critical for synaptic function.[136]

Synaptic plasticity: MALAT-1, found in hippocampal neurons, plays a role in synaptogenesis.[137,138] Other lncRNAs are enriched near synapses and are linked to learning and memory.[139,140]

7.3.3.3 Circular RNAs (circRNAs)

CircRNAs are a class of ncRNAs that form a closed circular loop, making them highly stable and resistant to degradation. In the brain, they often act as miRNA sponges and are implicated in disease, modulating gene expression indirectly.[141,142]

7.3.3.3.1 Functions in Brain Health. Synaptic function: CircRNAs, like circHOMER1, have been shown to regulate genes involved in synaptic plasticity, influencing neuronal function.[143,144]

Neurodevelopment: Brain-expressed circRNAs regulate neuronal development, with some showing age-dependent accumulation.[145,146]

These are the unique repertoire of ncRNAs that regulate neuronal development, synaptic plasticity, and memory formation. Non-coding RNAs (ncRNAs), including miRNAs (*e.g.*, miR-132 for synaptic plasticity and memory), lncRNAs (*e.g.*, BACE1-AS for synaptic function and chromatin remodelling), and circRNAs (*e.g.*, circHOMER1 as miRNA sponges for neuronal development), are vital for brain health processes like neurogenesis and gene regulation. Their dysregulation contributes to neurological disorders such as Alzheimer's (reduced miR-132, elevated BACE1-AS boosting beta-amyloid), Parkinson's, Huntington's, epilepsy, and ALS, leading to neuronal dysfunction, inflammation, and protein aggregation.[128] As illustrated in Figure 7.2, transcriptomic regulation follows a tightly orchestrated RNA

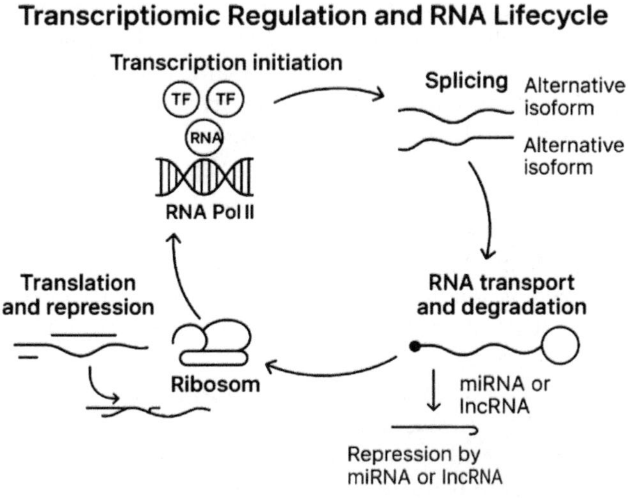

Figure 7.2 Transcriptomic regulation and RNA lifecycle. Illustration of the key stages governing RNA biogenesis and regulation in neural cells. The figure traces the transcriptomic lifecycle from transcription initiation, where transcription factors and RNA polymerase II assemble at promoter regions to drive pre-mRNA synthesis, through co-transcriptional and post-transcriptional processing, including splicing and generation of alternative isoforms that expand transcript diversity. Processed mRNAs are then exported to the cytoplasm, where their stability and localization are tightly controlled. The translation stage is regulated by ribosome engagement and can be repressed by non-coding RNAs, such as microRNAs (miRNAs) and long non-coding RNAs (lncRNAs), which bind complementary mRNA sequences to inhibit translation or promote decay. Arrows and icons depict this dynamic, cyclic RNA lifecycle, emphasizing the interconnected regulatory mechanisms that fine-tune gene expression in the brain.

lifecycle encompassing transcription initiation by RNA polymerase II, alternative splicing, mRNA export and stability control, and translational modulation by microRNAs and long non-coding RNAs.

7.4 Epigenetic Dysregulation in Brain Disorders

Epigenetic dysregulation has emerged as a critical contributor to the onset and progression of diverse brain disorders, including neurodegenerative diseases, psychiatric illnesses, and neurodevelopmental conditions.[147] These changes often precede overt symptoms, suggesting a foundational role in disease etiology.[147] Unlike genetic mutations, epigenetic modifications are potentially reversible, making them attractive therapeutic targets.[148] This section discusses how aberrant DNA methylation, histone modifications, and non-coding RNA expression contribute to neurodevelopmental, neuropsychiatric, and neurodegenerative disorders.[147,148]

Aberrant DNA methylation patterns in the brain can impair promoter and enhancer function, resulting in the improper silencing or activation of genes required for synaptic maintenance, neuronal development, and glial function.[149] In addition, abnormalities in histone acetylation, methylation, or phosphorylation can change chromatin structure, affecting transcriptional programs necessary for brain circuit stability and plasticity.[150] Mutations in epigenetic regulators, environmental exposures, metabolic stress, or inflammatory signals can all cause these changes.[99]

Dysregulated microRNAs can destabilize transcripts required for neurotransmitter release, cytoskeletal organization, or neuroprotective pathways, whereas aberrant long non-coding RNAs can attract chromatin modifiers to pathogenic gene loci.[151,152] These molecular abnormalities frequently interact synergistically, resulting in long-lasting changes in cellular identity and function that lead to cognitive impairments, emotional dysregulation, and motor dysfunction.[102]

Importantly, many of these epigenetic alterations can be detected before permanent neuronal death occurs, making them useful as early biomarkers for disease diagnosis and monitoring.[151] Therapeutic methods to these systems, such as enzyme inhibitors, RNA-based modulators, and locus-specific epigenome editing, are now being investigated.[148] Such techniques, which restore proper gene expression patterns, show promise for delaying or reversing disease processes in a variety of neurological and psychiatric diseases.[113]

7.4.1 Neurodevelopmental Disorders

Autism spectrum disorders (ASD) are associated with aberrant DNA methylation and altered histone modifications, especially in genes regulating neuronal plasticity.[153] Rett syndrome, a severe neurodevelopmental disorder, arises from mutations in the MECP2 gene, which is essential for binding to methylated DNA and maintaining proper epigenetic function, thereby disrupting gene silencing and impairing neuronal development.[154] Angelman and Prader–Willi syndromes result from perturbations in the epigenetic process of genomic imprinting on chromosome 15q11-13.[155] Certain forms of epilepsy are linked to genetic variations in chromatin-binding proteins and other epigenetic factors.[153,156]

7.4.1.1 Autism Spectrum Disorder (ASD)

ASD is characterized by impaired social communication and repetitive behaviours, with high genetic and epigenetic heterogeneity.[157] Aberrant DNA methylation patterns have been found in genes regulating synaptic function, neurogenesis, and immune signalling.[158] For instance, hypermethylation of the OXTR (oxytocin receptor) gene correlates with reduced expression and impaired social behaviour.[159]

Mutations in epigenetic regulators such as MeCP2, CHD8, and EHMT1 are recurrent in ASD cases. CHD8, a chromatin remodeller, controls neurodevelopmental gene networks; loss-of-function mutations lead to widespread transcriptional dysregulation during corticogenesis.[160–163] MeCP2, critical for maintaining DNA methylation-dependent repression, is central to Rett syndrome, a syndromic autism disorder.[164,165]

7.4.1.2 Rett Syndrome

Rett syndrome is a severe neurological disorder, predominantly affecting females, caused by mutations in the MeCP2 gene, which encodes a methyl-CpG-binding protein that silences genes by recruiting HDACs and other repressors.[166] This gene produces the MeCP2 protein, crucial for proper gene regulation and brain development.[167] A non-functional MeCP2 protein leads to widespread epigenetic dysregulation, resulting in neuronal problems, including impaired synaptic function, neuronal atrophy, and severe cognitive decline.[167,168]

7.4.1.3 Angelman and Prader–Willi Syndromes

Angelman and Prader–Willi syndromes arise from disruptions in the genomic imprinting process on chromosome 15q11-13.[155] This specific region contains genes that are expressed in a parent-of-origin-dependent manner.[155] A paternal deletion or malfunction leads to Prader–Willi syndrome, while a maternal deletion or malfunction results in Angelman syndrome.[155] These disorders are caused by a perturbation in the normal epigenetic process of genomic imprinting on chromosome 15q11-13.[155,169]

7.4.1.4 Epilepsy

Certain types of epilepsy are linked to genetic variations in chromatin-binding proteins and other epigenetic factors.[156] Certain epilepsies are linked to genetic variations in genes that encode chromatin-binding proteins and other epigenetic factors, with mutations in genes like CHD2 and ARX leading to disorders such as developmental epileptic encephalopathy.[170] These variations can disrupt normal neurodevelopment and cellular functions, contributing to the pathological mechanisms of epilepsy.[156]

7.4.2 Neuropsychiatric Disorders

Schizophrenia and bipolar disorder are tied to abnormal DNA methylation patterns and histone modifications in genes involved in neural development and neurotransmission; for instance, research on twins with these conditions has identified distinct epigenetic changes that suggest a role in disease etiology.[171,172] Depression and post-traumatic stress disorder (PTSD) can stem from chronic stress and trauma, which induce epigenetic alterations in genes regulating the stress response, thereby heightening vulnerability to these conditions.

7.4.2.1 Schizophrenia

Schizophrenia involves disruptions in dopaminergic, glutamatergic, and GABAergic signalling pathways.[171] Epigenomic studies have identified differential DNA methylation in genes such as RELN, GAD1, and COMT, which are essential for synaptic signalling and neurotransmitter metabolism.[172,173] Research, including studies on twins, shows distinct epigenetic changes in affected individuals compared to controls, particularly in brain regions like the frontal cortex, affecting genes involved in neurodevelopment, neurotransmission, and stress response.[173] These epigenetic alterations, though complex and requiring further research, offer a plausible biological mechanism for how genetic predispositions and environmental factors interact to contribute to these conditions.[174]

Histone modifications are also perturbed; for example, decreased H3K4me3 (an active mark) at promoters of neuroplasticity genes has been observed in postmortem brains of patients.[175] These changes may originate early in development and persist into adulthood, possibly contributing to long-term cognitive and behavioural abnormalities.[173]

7.4.2.2 Major Depressive Disorder (MDD)

MDD is associated with stress-induced epigenetic reprogramming, especially in the hippocampus and prefrontal cortex. Chronic stress elevates levels of DNMTs and repressive histone marks (*e.g.*, H3K27me3), leading to silencing of brain-derived neurotrophic factors (BDNF) and other neurotrophic genes.[176,177]

Epigenetic changes in glucocorticoid receptor genes (*e.g.*, NR3C1) have been linked to early-life adversity, potentially predisposing individuals to lifelong vulnerability to depression.[178] Studies show that early-life adversity, such as childhood maltreatment, can lead to epigenetic changes, particularly DNA methylation, in the NR3C1 gene (which codes for the glucocorticoid receptor).[178,179] These epigenetic modifications are associated with altered activity in the hypothalamic–pituitary–adrenal (HPA) axis and can disrupt glucocorticoid signalling, increasing vulnerability to depression and other stress-related disorders later in life.[178] Antidepressants such as selective serotonin reuptake inhibitors (SSRIs) and HDAC inhibitors may exert therapeutic effects partly *via* epigenetic remodelling.[180,181]

7.4.3 Neurodegenerative Diseases

7.4.3.1 Alzheimer's Disease (AD)

Alzheimer's disease features progressive cognitive decline, amyloid-β accumulation, tau pathology, and neuroinflammation. Epigenetic alterations, particularly global DNA hypomethylation and local promoter hypermethylation, are frequently observed.[182,183]

Genes related to amyloid processing (APP, BACE1), tau phosphorylation (MAPT), and neuroinflammation (IL1B, TNFα) exhibit methylation or histone modification changes in AD brains.[184] Age-associated epigenetic drift can contribute to AD vulnerability by progressively disrupting gene expression, leading to a breakdown of neuronal function through the gradual loss of transcriptional control.[185,186]

7.4.3.2 Parkinson's Disease (PD)

PD is characterized by dopaminergic neuron loss in the substantia nigra and α-synuclein accumulation.[187] Studies have identified hypomethylation in the SNCA gene promoter, correlating with overexpression of α-synuclein, a key pathogenic driver.[187,188] Histone acetylation imbalances and altered miRNA profiles (*e.g.*, miR-34b/c, miR-133b) further contribute to dopaminergic dysfunction and neuroinflammation.[189]

7.4.3.3 Huntington's Disease (HD)

HD is an autosomal dominant disorder caused by CAG repeat expansion in the HTT gene.[190] Beyond genetic mutation, epigenetic deregulation plays a substantial role in disease progression.[191,192] Mutant huntingtin protein interacts abnormally with histone acetyltransferases (*e.g.*, CBP), leading to widespread histone hypoacetylation and transcriptional repression.[192–194] Histone deacetylase (HDAC) inhibitors have shown promise in preclinical models of HD by restoring normal gene expression, reducing neuroinflammation, and improving neuronal survival, leading to better motor and cognitive function in HD animal models.[195] While HDAC inhibitors are still in development, particularly due to potential side effects, research is ongoing to find more targeted and less toxic approaches to leverage their

Table 7.1 Key epigenetic regulators implicated in neural gene expression and brain disorders: this table summarizes major epigenetic modifiers that govern chromatin structure and transcriptional control in the brain. Listed are representative enzymes and regulatory proteins, their molecular types, and functional roles in maintaining neural gene expression. DNA methyltransferases (DNMTs) preserve or establish methylation patterns; histone-modifying enzymes (such as EZH2, HDACs, CBP/p300, and G9a) dynamically regulate chromatin accessibility and transcription; and chromatin remodellers (*e.g.*, CHD8) reposition nucleosomes to fine-tune gene expression. Dysregulation of these factors is associated with diverse neurodevelopmental and neuropsychiatric disorders, including autism spectrum disorder, schizophrenia, depression, and Alzheimer's disease.

Epigenetic modifier	Type	Functional role in the brain	Associated disorders
DNMT1, DNMT3A/B	DNA methyltransferases	Maintenance and *de novo* methylation	ASD, schizophrenia, Alzheimer's
MECP2	Methyl-CpG binding protein	Transcriptional repression, chromatin compaction	Rett syndrome, depression
CHD8	Chromatin remodeller	Nucleosome positioning, gene regulation	Autism spectrum disorder
EZH2	Histone methyltransferase	Catalyses H3K27me3, gene repression	Glioma, neuroblastoma
HDAC1/2/3	Histone deacetylases	Remove acetyl groups, repress transcription	Alzheimer's, Huntington's, depression
CBP/p300	Histone acetyltransferases	Enhance transcription *via* H3/H4 acetylation	Learning/memory deficits, Rubinstein–Taybi syndrome
REST	Transcriptional repressor	Silences neuronal genes in non-neuronal cells	Epilepsy, neurodevelopmental delay
G9a (EHMT2)	Histone methyltransferase	Catalyses H3K9me2, gene repression	Depression, intellectual disability

neuroprotective benefits for HD treatment.[196] HDAC inhibitors have shown promise in preclinical models to restore gene expression and improve neuronal survival.[195–198] As summarized in Table 7.1, key epigenetic regulators implicated in neural gene expression include DNA methyltransferases, histone-modifying enzymes, and chromatin remodellers, whose coordinated activity sustains neuronal identity and plasticity; perturbations in these pathways are linked to multiple neurodevelopmental and neuropsychiatric disorders.

7.5 Transcriptomic Alterations in Brain Disorders

While epigenetic mechanisms control gene accessibility, transcriptomic changes reflect the downstream functional output.[199] In brain disorders, altered transcriptional profiles can disrupt neurotransmission, neurodevelopment, and synaptic plasticity.[199] With the rise of high-throughput sequencing technologies, transcriptomic profiling has uncovered both disease-specific and shared gene expression signatures across a wide range of neurological and psychiatric conditions.[200,201]

Disease-associated transcriptome changes can include changes in mRNA quantity, alternative splicing patterns, RNA editing, and non-coding RNA expression, all of which can have a significant impact on neuronal function.[202] These alterations in the epitranscriptome, the complete set of RNA transcripts, can disrupt critical cellular processes like development, protein synthesis, and DNA repair, ultimately contributing to neurodevelopmental, neurodegenerative, and other brain disorders.[203] In neurodevelopmental diseases, dysregulated transcriptional processes can affect neuronal migration, axon guidance, and synapse formation, resulting in long-term circuit connectivity deficiency.[204] Transcriptomic changes in psychiatric illnesses frequently affect genes that regulate neurotransmitter synthesis, receptor composition, and intracellular signalling cascades, changing synaptic efficacy and network dynamics.[116]

Neurodegenerative illnesses commonly show transcriptional suppression of genes involved in mitochondrial function, protein quality control, and antioxidant defence, while also activating inflammatory and apoptotic pathways.[205,206] Non-coding RNA dysregulation, particularly of microRNAs and long non-coding RNAs, has the potential to magnify these changes by post-transcriptionally altering numerous neuronal and glial communication networks.[123]

Advances in single-cell and spatial transcriptomics now enable these patterns to be resolved at a cell-type level, exposing variability that bulk RNA sequencing cannot.[207] High-resolution profiling has revealed early molecular fingerprints that predict disease progression, as well as shared transcriptional modules that could serve as convergent therapeutic targets.[208] These findings are increasingly being incorporated into systems biology frameworks in order to better understand the molecular foundations of brain illnesses and guide precision medicine approaches.[209]

7.5.1 Neurodevelopmental Disorders

7.5.1.1 Autism Spectrum Disorder (ASD)

Transcriptomic analyses of ASD brain tissue and induced pluripotent stem cell (iPSC)-derived neurons reveal widespread dysregulation of genes involved in synaptic development, cell adhesion, and chromatin remodelling.[210,211] Key networks affected include:

- Synaptic genes: such as SHANK3, NRXN1, NLGN3[211–215]
- Transcriptional regulators: including CHD8, TBR1[111,161,216]

- Neuroimmune components: dysfunction of microglia will ultimately affect the neural function in a variety of ways, including the formation of synapses and alteration of excitatory–inhibitory balance, microglial involvement.[217–219]

Single-cell RNA sequencing (scRNA-seq) studies show that excitatory and inhibitory neuron subtypes in ASD brains display unique expression changes, pointing to circuit-level imbalances that underlie core symptoms.[220]

7.5.1.2 Rett Syndrome

Rett syndrome, caused by MeCP2 dysfunction, exhibits global transcriptomic repression.[221] Neurons from MeCP2-null models show decreased expression of genes critical for synaptic maintenance, metabolism, and ion channel function. These shifts impair excitability and lead to synaptic silencing.[27,222] Moreover, the downstream transcriptional targets of MeCP2,[223] including BDNF,[224] are consistently downregulated.[29,225]

7.5.2 Neuropsychiatric Disorders

7.5.2.1 Schizophrenia

Transcriptomic studies in schizophrenia consistently reveal altered expressions of genes involved in:

- Synaptic transmission: Transcriptomic studies in schizophrenia repeatedly show altered expression of genes involved in synaptic transmission, which supports the "synaptic hypothesis" of the disorder.[27] This includes genes for both inhibitory and excitatory neurotransmission, like *GAD1* and *SYN1*, respectively.[226–228]
- Mitochondrial function: Studies frequently report decreased activity in the oxidative phosphorylation system, which is responsible for cellular energy production.[229] Specifically, reduced activity has been observed in electron transport chain complexes, particularly complex I and complex IV.[229] A related change is the lower adenosine triphosphate (ATP) levels found in the temporal lobe of schizophrenia patients, suggesting that mitochondrial dysfunction is a key component of the disorder.[230] This includes altered gene expression in areas like the prefrontal cortex, which is critical for cognitive function.[231]
- Myelination and oligodendrocyte signalling: The dysfunction of these genes can disrupt myelination, the process by which oligodendrocytes wrap nerve fibres in a fatty substance called myelin, and is a significant feature in the pathophysiology of the disease.[232–234]

Alternative splicing is significantly dysregulated in schizophrenia, particularly in genes related to synaptic vesicle cycling and calcium signalling.[235]

Notably, isoform shifts in DISC1 and NRG1 impact protein–protein interactions critical for neural connectivity.[236,237] Transcriptomic signatures also differ between early- and late-onset cases, suggesting temporal dynamics in disease development.[235]

7.5.2.2 Major Depressive Disorder (MDD)

In MDD, transcriptomic changes are prominent in the prefrontal cortex, hippocampus, and amygdala. Key observations include:

- Downregulation of neurotrophic genes like brain-derived neurotrophic factor (BDNF) and cAMP response element-binding protein (CREB) is linked to reduced neurogenesis, synaptic dysfunction, and neuronal atrophy in key brain regions.[238,239] The specific effects are primarily observed in the prefrontal cortex (PFC), hippocampus, and amygdala, and they contribute to the neurotrophic hypothesis of depression.[238]
- Upregulation of inflammatory mediators: Studies show that major MDD is associated with an upregulation of inflammatory mediators, including the pro-inflammatory cytokines interleukin-6 (IL-6) and tumour necrosis factor-alpha (TNF-α).[240] This link between inflammation and depression has led to the "inflammatory hypothesis of depression," which suggests that inflammation may contribute to the development and symptoms of MDD.[241]
- Altered monoamine signalling pathways: Alterations in the genes SLC6A4 and MAOA affect the monoamine signalling pathways that regulate neurotransmitters like serotonin, norepinephrine, and dopamine, a process linked to MDD.[242] However, the monoamine hypothesis, which suggests that low monoamine levels cause depression, is now considered an oversimplification, as alterations in these genes are often intertwined with environmental factors and have complex effects.[243,244]

Chronic stress models show suppression of activity-dependent transcription, implicating the CREB signalling pathway.[245] Non-coding RNA changes, such as increased miR-1202 and reduced miR-135a, modulate serotonin transporter expression and are being explored as potential biomarkers.[246]

7.5.3 Neurodegenerative Diseases

7.5.3.1 Alzheimer's Disease (AD)

Transcriptomic profiles in AD reveal:

- Suppression of genes involved in synaptic function (SYN1, DLG4, CAMK2A): Transcriptomic studies in AD's confirm the widespread downregulation of synaptic function genes, including SYN1, DLG4, and

CAMK2A, which are essential for neuronal communication and plasticity.[247,248] This suppression contributes directly to the cognitive deficits seen in AD.[249]

- Upregulation of innate immune pathways: A significant upregulation of innate immune pathways, particularly genes associated with microglial activation, such as TREM2 and TYROBP.[250] TREM2, a receptor on microglia, and its associated signalling adapter TYROBP, are key drivers of microglial responses to AD pathology, influencing their function in clearing amyloid plaques and their transition to a disease-associated state.[251,252]
- Alternative splicing in APP and MAPT (tau) transcripts: Studies have revealed that altered alternative splicing of the amyloid precursor protein (APP) and microtubule-associated protein (MAPT) genes produces dysfunctional protein isoforms implicated in AD pathogenesis.[253] Unlike normal regulation, this altered splicing leads to a pathological balance of isoforms, contributing to the hallmark amyloid plaques and neurofibrillary tangles of AD.[254]

Longitudinal studies show that these changes precede cognitive decline, indicating their value in early diagnosis.[255] Integration with methylome data has revealed coordinated changes in promoter methylation and transcript levels, particularly in genes governing amyloid metabolism.[256]

7.5.3.2 Parkinson's Disease (PD)

In PD, dopaminergic neurons exhibit downregulation of genes linked to mitochondrial respiration, vesicle trafficking, and dopamine synthesis (TH, VMAT2, PINK1).[257,258] Upregulation of stress-response genes and small heat shock proteins suggests cellular attempts at neuroprotection.[259]

Distinct miRNA signatures, such as reduced miR-133b, which normally promotes dopaminergic differentiation, highlight transcriptomic instability.[260] Splicing changes in SNCA and LRRK2 may exacerbate protein aggregation and cellular toxicity.[261,262]

7.5.3.3 Huntington's Disease (HD)

HD transcriptomes show broad downregulation of genes essential for neuronal identity and synaptic signalling.[263] Transcriptional repression is driven by mutant huntingtin's interference with co-activators like CBP, leading to reduced expression of BDNF, PENK, and DARPP-32, all essential for striatal neuron function.[264]

Additionally, RNA toxicity may arise from expanded CAG repeat RNA, which forms nuclear foci and sequesters splicing factors.[265] Mis-spliced isoforms of genes like MBD1 and ATXN2 further contribute to disease progression.[266]

7.6 Epigenetic and Transcriptomic Crosstalk

Epigenetic and transcriptomic regulation are not independent layers of control, they operate in concert, forming intricate feedback and feedforward loops that govern neural identity, function, and plasticity.[267,268] In the brain, where cells require precise, context-specific gene expression, the interaction between these systems ensures both stability and adaptability.[269] When this crosstalk is disrupted, it can amplify molecular dysfunction and accelerate disease progression.[270]

The combination of epigenetic and transcriptomic pathways creates a dynamic framework within which neurons and glial cells respond to developmental signals, environmental stimuli, and metabolic conditions.[268,271] Epigenetic modifications, such as DNA methylation, histone post-translational changes, and chromatin remodelling, have a direct impact on the accessibility of transcriptional machinery to certain genomic areas, altering the transcriptome profile.[148,272] Transcriptomic outputs, particularly non-coding RNAs, can influence epigenetic regulators by recruiting chromatin-modifying complexes to target loci or changing their enzymatic activity.[268] This bidirectional link allows for fine-tuning of gene expression patterns, which is necessary for activities like synaptic plasticity, memory consolidation, and stress adaptation.[178,273]

Perturbations in this interaction can cause maladaptive transcriptional processes, which contribute to neuropathology. Dysregulated non-coding RNAs, for example, in neurodegenerative and neurodevelopmental disorders, can reinforce aberrant epigenetic states, causing cells to become dysfunctional.[274] Similarly, long-term epigenetic regulation of activity-dependent genes can reduce neuronal reactivity and circuit remodelling.[12,275] The confluence of these flaws frequently initiates a self-sustaining cycle of molecular instability, oxidative stress, and metabolic compromise, further decreasing brain resilience.[12] Understanding the interdependence of these regulatory layers not only reveals illness causes, but also identifies prospective therapeutic options for restoring coordinated epigenetic-transcriptomic signalling and preserving brain function.[175] As shown in Figure 7.3, epigenetic and transcriptomic mechanisms operate within an integrated feedback framework, wherein transcription factors recruit histone acetyltransferases to activate gene expression, while the resulting non-coding RNAs direct DNA methyltransferases to silence other genomic regions, thereby maintaining transcriptional homeostasis.

7.6.1 Epigenetic Control of Transcription

Epigenetic marks directly influence the transcriptional landscape:[276,277]

- **DNA methylation** at gene promoters usually suppresses transcription by preventing transcription factor binding or attracting repressors like MeCP2.[278–280] Adding a methyl group to cytosine bases at gene promoters can stop transcription factors from binding or attract proteins

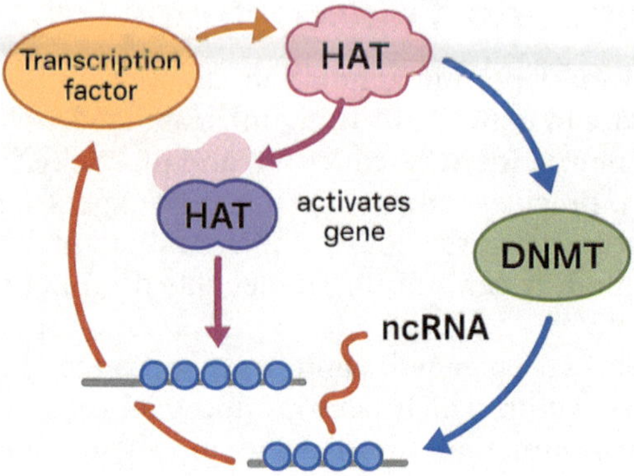

Figure 7.3 Integrated epigenetic–transcriptomic feedback loop. Schematic representation of the bidirectional interplay between epigenetic modifications and transcriptional regulation in neural cells. The diagram illustrates a feedback loop in which a transcription factor (TF) recruits a histone acetyltransferase (HAT) complex to specific promoter regions, resulting in histone acetylation and transcriptional activation of a target gene. The activated gene encodes a non-coding RNA (ncRNA) that, once expressed, interacts with DNA methyltransferase (DNMT) to promote DNA methylation and silencing of another gene, thereby closing the regulatory loop. Arrows depict the circular, dynamic flow of molecular interactions, while color coding distinguishes between proteins (TFs, HATs, DNMTs), RNA species (ncRNAs), and chromatin components. This integrated model highlights how epigenetic and transcriptomic layers cooperatively maintain neural gene expression homeostasis and mediate adaptive or maladaptive responses in brain disorders.

like MeCP2, which recruit other proteins to create repressive chromatin structures, ultimately silencing gene expression.[278] This process is a crucial mechanism for turning genes "off" and is vital for processes such as cell differentiation, development, and maintaining proper cellular function, as well as contributing to disease states when it becomes dysregulated.[276,279]

- **Histone modifications** such as H3K4me3 (activation) or H3K27me3 (repression) determine whether chromatin is in an open or closed conformation.[281] Histone modifications are key epigenetic mechanisms that regulate gene transcription by altering the accessibility of DNA packaged in chromatin.[282] The chemical modifications to histone proteins, particularly their protruding tails, act as signals that dictate whether a region of DNA is in an "open" (transcriptionally active) or "closed" (transcriptionally repressed) conformation. Trimethylation of the fourth lysine residue on histone H3 (H3K4me3) is a hallmark of active transcription. It is predominantly found at the promoters and transcription start sites of actively expressed genes.[282] Conversely, trimethylation of the 27th lysine residue on histone H3 (H3K27me3) is a canonical mark of transcriptional repression.[276,281]

- **Chromatin remodelling complexes** reposition nucleosomes to expose or conceal regulatory regions.[276,277] Epigenetic marks do not directly change the DNA sequence but act as signals that influence the transcriptional landscape, largely by guiding ATP-dependent chromatin remodelling complexes.[283] Various protein complexes, including SWI/SNF, ISWI, and CHD, act as "readers" that are recruited to specific regions of the genome based on the epigenetic marks present.[284] Chromatin remodels are essential because they use energy from ATP hydrolysis to physically move, eject, or restructure the nucleosomes themselves, while most epigenetic modifiers only add or remove marks.[283]

These epigenetic states define transcriptional competence. For example, activity-dependent genes like FOS and ARC are kept in a "poised" state with bivalent histone marks, ready to be rapidly activated during synaptic activity.[276,285,286]

7.6.2 Transcription Factors as Epigenetic Modulators

Conversely, transcription factors can shape the epigenetic landscape: TFs act as epigenetic modulators by binding to DNA and recruiting epigenetic machinery to alter chromatin structure and regulate gene expression.[287] While TFs traditionally influence gene expression by binding to specific DNA sequences, they also directly influence epigenetic modifiers like histone-modifying enzymes and chromatin remodelling complexes.[288] This interplay allows TFs to create a specific nuclear environment, establishing epigenetic "on–off" switches on the chromatin to control gene accessibility and activity, a mechanism crucial in processes like cell differentiation, cancer progression, and immune responses.[287]

- **Pioneer transcription factors**, such as ASCL1 and PU.1, are unique proteins that can bind to and initiate changes in the structure of condensed chromatin.[289] This ability to interact with nucleosomal DNA in "closed" chromatin allows them to recruit other proteins, like chromatin remodels, to open up the DNA. This process creates accessibility for other factors, ultimately leading to cell-type-specific gene expression and establishing specific cell fates during development.[290,291]
- TFs can act as **epigenetic modulators by recruiting co-activators** like CBP/p300 to activate genes or co-repressors like HDACs and DNMTs to silence them.[292,293] By interacting with these enzymes, TFs direct chemical modifications to histones and DNA, controlling which genes are turned on or off. This allows TFs to regulate gene expression by either making the DNA more accessible for transcription or promoting a condensed, silent chromatin state.[292] For example, TFs recruit co-activators like CBP/p300, which acetylate histones, or co-repressors such as HDACs and DNMTs, which silence gene expression.[292,293]

This interaction creates a self-reinforcing circuit. In neurodevelopment, TFs not only activate target genes but also epigenetically prime them for future expression. Disruption in this feedback—such as mutations in chromatin readers or TFs—leads to long-term transcriptional instability, seen in disorders like Rett syndrome and schizophrenia.[294,295]

7.6.3 Non-coding RNAs as Integrators of Epigenetic and Transcriptomic Regulation

Non-coding RNAs bridge the two regulatory layers: ncRNAs integrate epigenetic and transcriptomic regulation by recruiting epigenetic machinery to specific DNA sites, mediating changes in DNA methylation and histone modification, and thereby controlling gene expression at the transcriptional level.[296] Long non-coding RNAs (lncRNAs) are particularly known for this role, recruiting epigenetic regulators to chromatin to influence gene expression, which can have significant implications in diseases like cancer and in response to environmental stressors.[296,297]

- **miRNAs** are a key class of small non-coding RNAs that regulate gene expression by targeting mRNAs for degradation or translational repression.[298] Acting as integrators, they connect epigenetic mechanisms with the regulation of the transcriptome, influencing processes like development, cell proliferation, and disease.[299]
- **lncRNAs** act as crucial integrators of epigenetic and transcriptomic regulation by serving as scaffolds or guides to recruit epigenetic modifiers, such as histone-modifying enzymes and DNA methyltransferases, to specific genomic loci.[296] This recruitment leads to targeted DNA methylation, histone modifications, and chromatin remodelling at those sites, thereby altering chromatin structure and fine-tuning gene expression.[300] These precise epigenetic changes, guided by lncRNAs, establish patterns of epigenetic memory and control transcriptional activity, playing significant roles in normal physiological processes and the development of diseases.[301]

For example, HOTAIR recruits the Polycomb complex to silence gene clusters in trans,[302] while GOMAFU, implicated in schizophrenia, regulates alternative splicing and chromatin compaction simultaneously.[303] These RNAs can be both products and regulators of epigenetic states; many are themselves controlled by methylation or histone marks, creating tight regulatory loops that respond to environmental cues or developmental signals.[304] As outlined in Table 7.2, the interplay between epigenetic and transcriptomic alterations underlies the molecular pathology of major brain disorders, with characteristic patterns such as ANK1 hypermethylation in Alzheimer's disease, reduced H3K9 acetylation in schizophrenia, and dysregulated microRNA expression in autism spectrum and Parkinson's disease.

Table 7.2 Epigenetic–transcriptomic interplay in major brain disorders: summary of representative disorders illustrating how epigenetic modifications interface with transcriptomic alterations to influence neural function and pathology. The table highlights disease-specific epigenetic changes, including DNA methylation, histone modifications, and non-coding RNA dysregulation, and their corresponding transcriptomic outcomes, such as altered gene expression or pathway activity. These molecular interactions collectively shape the functional consequences observed in major neurological and psychiatric disorders, including Alzheimer's disease (AD), schizophrenia (SCZ), autism spectrum disorder (ASD), and Parkinson's disease (PD).

Disorder	Epigenetic alteration	Transcriptomic outcome	Functional consequence
AD	*ANK1* hypermethylation (glia); ↓5hmC BACE1-AS↑	↑ Neuroinflammation ↓ Synaptic genes	Aβ accumulation; synaptic loss
SCZ	↓ H3K9ac at *GAD1* miR-137 dysregulation	↓GABAergic transcripts ↑Immune genes	E/I imbalance; cortical dysfunction
ASD	*CHD8* mutations ↑ miR-132 in the cortex	Dysregulated neurodevelopmental pathways	Altered synaptogenesis; circuit defects
PD	*SNCA* hypomethylation ↓ miR-7	↑ α-Synuclein ↓ Mitochondrial genes	Proteotoxicity; oxidative stress

7.7 Technological Advances and Methodologies

Rapid advances in high-throughput sequencing, genome editing, and computational biology have transformed our ability to investigate the epigenetic and transcriptomic landscapes of the brain. These technologies allow researchers to identify cell-type-specific regulatory mechanisms, track disease progression at the molecular level, and uncover therapeutic targets with unprecedented resolution. This section outlines the key technology and tools that have propelled the field forward.[305]

Cutting-edge experimental platforms such as single-cell RNA sequencing (scRNA-seq) and single-cell ATAC sequencing (scATAC-seq) have enabled unprecedented resolution in dissecting transcriptional and chromatin accessibility profiles, revealing cellular heterogeneity and dynamic state transitions within the brain.[306] Multi-omics integration techniques now allow for simultaneous assessment of gene expression, DNA methylation, histone changes, and chromatin architecture inside the same cell, offering a comprehensive picture of regulatory networks.[307] Spatial transcriptomics provides additional anatomical context by mapping molecular markers directly onto tissue architecture to reveal region-specific gene regulation.[178,308]

On the computational front, machine learning and network analysis methods are increasingly being used to forecast regulatory interactions, uncover disease-associated modules, and infer causal links between epigenetic changes and transcriptome outcomes.[309,310] Predictive modelling simulates treatments at the molecular level, which aids in drug target prioritization.[311,312] CRISPR-based functional genomics screens support these efforts by allowing for high-throughput alteration of regulatory elements to validate their involvement in neuronal function and disease.[313] Collectively, these technological improvements are bridging the gap between descriptive molecular mapping and mechanistic knowledge, hastening the translation of fundamental discoveries into targeted, precision-based therapy for neurological and psychiatric illnesses.[184]

7.7.1 Epigenomic and Transcriptomic Profiling

Advances in sequencing technologies have provided unprecedented resolution in mapping the molecular landscape of the brain. For epigenetics, bisulfite sequencing (BS-seq, WGBS) enables base-pair resolution of DNA methylation, while reduced representation bisulfite sequencing (RRBS) offers a cost-efficient focus on CpG-rich regions.[314] Histone modifications and transcription factor binding can be mapped using ChIP-seq, with newer low-input methods such as CUT&RUN and CUT&Tag particularly suitable for scarce neuronal samples. Chromatin accessibility profiling has been transformed by ATAC-seq, which requires minimal input and captures active enhancer, promoter regions. On the transcriptomic side, bulk RNA-seq remains a standard for differential expression and splicing analysis, but single-cell RNA sequencing (scRNA-seq) now dominates in revealing cellular heterogeneity.[314] Spatial transcriptomics adds another dimension by retaining anatomical context, mapping expression patterns directly onto brain architecture.[315,316]

7.7.2 Multi-omics Integration

A major advance has been the rise of multi-omics platforms, which combine different molecular layers from the same cells. Tools such as 10× Multiome allow simultaneous profiling of gene expression and chromatin accessibility, while computational frameworks like Seurat, ArchR, and MOFA integrate transcriptomic, epigenomic, and proteomic data.[317] These methods uncover causal links between regulatory elements and downstream gene activity, revealing disease-specific circuits and stage-dependent changes in brain disorders.[317] By aligning methylation, histone state, chromatin accessibility, and transcriptomic output, multi-omics strategies provide a systems-level understanding that single-assay approaches cannot capture.[317,318]

7.7.3 Functional Genomics and Epigenome Editing

Beyond descriptive profiling, CRISPR-based technologies have enabled functional testing of regulatory elements.[319,320] Catalytically inactive Cas9 (dCas9) fused to epigenetic modifiers such as DNMT3A, TET1, or HDACs can selectively alter methylation or histone states at specific loci, offering a means of directly testing causal relationships between chromatin marks and transcriptional outcomes.[319,321] Similarly, CRISPR activation (CRISPRa) and interference (CRISPRi) platforms modulate transcription without DNA cleavage, expanding the therapeutic potential for neurological disorders.[320] These functional tools are complemented by high-throughput CRISPR screens that identify genes and enhancers critical for neuronal survival, plasticity, or disease progression.[320]

7.7.4 Experimental Models and Computational Resources

The emergence of brain organoids derived from induced pluripotent stem cells (iPSCs) provides a powerful model for studying human cortical development and disease in controlled settings.[322,323] Combined with lineage-tracing approaches such as scGESTALT, researchers can reconstruct developmental trajectories at cellular resolution.[322] Integration with optogenetics and chemogenetics further links activity-dependent regulation with epigenetic and transcriptomic states.[324] Complementing these experimental systems are large-scale resources such as ENCODE, Roadmap Epigenomics, Allen Brain Atlas, and PsychENCODE, which provide reference datasets across brain regions and disease contexts.[325] Computational pipelines including DESeq2, edgeR, and MAESTRO make it feasible to analyze and interpret the vast quantities of omics data, accelerating discovery and therapeutic translation.[326–328] As summarized in Table 7.3, multi-omics technologies such as RNA-seq, ATAC-seq, ChIP-seq, and single-cell RNA-seq provide complementary insight into gene expression, chromatin accessibility, and regulatory landscapes, offering a comprehensive framework for dissecting molecular complexity in the brain.

Table 7.3 Multi-omics technologies in brain research. This table summarizes major omics platforms applied to the study of brain function and disease, highlighting their molecular targets, resolution, data output, and primary applications. RNA-seq and single-cell RNA-seq (scRNA-seq) enable quantitative profiling of transcriptomes across bulk tissues or individual cells, revealing differential gene expression and cellular heterogeneity. ATAC-seq identifies regions of open chromatin to map accessible regulatory elements, while ChIP-seq characterizes histone modifications and transcription factors binding to define chromatin states. Together, these multi-omics approaches provide an integrated framework for understanding gene regulation, cellular diversity, and network dysfunction in the brain.

Method	Target	Resolution	Output	Application
RNA-seq	mRNA	Cell/tissue	Gene expression	DEGs, splicing
ATAC-seq	Open chromatin	Single cell	Accessibility	Enhancers
ChIP-seq	Histone marks	Region specific	Binding maps	Regulatory state
scRNA-seq	mRNA	Single cell	Cell types, states	Brain heterogeneity

7.8 Therapeutic Implications

Understanding the epigenetic and transcriptomic underpinnings of brain disorders has opened new therapeutic frontiers.[20] Unlike genetic mutations, many epigenetic and transcriptomic alterations are reversible, making them attractive targets for disease intervention. Current strategies aim to correct aberrant gene expression, restore neuronal function, and even reprogram disease trajectories.[148] Therapeutic development in this sector includes both pharmacological and molecular therapies that aim to reset gene regulatory networks.[329] Epigenetic-targeting drugs, including histone deacetylase inhibitors, histone methyltransferase inhibitors, and DNA methyltransferase modulators, seek to restore chromatin accessibility and reactivate suppressed neuroprotective genes.[148] In parallel, RNA-based techniques like antisense oligonucleotides, small interfering RNAs, and microRNA mimics or inhibitors are being used to regulate transcript abundance, repair splicing errors, and suppress harmful gain-of-function transcripts.[193] Advances in genome-editing technologies, particularly CRISPR-based epigenome editors, enable precise control of gene expression without affecting DNA sequence, allowing for locus-specific restoration of normal regulatory states.[330] Combination therapies that combine epigenetic medicines and transcriptome-modifying medications show potential for synergistic benefits, particularly in multifactorial conditions like Alzheimer's, Parkinson's, and autism spectrum disorders. Importantly, these therapies can be adjusted to target specific cell types or brain areas, reducing off-target effects while increasing therapeutic efficacy.[220] The combination of single-cell multi-omics and patient-derived cellular models is accelerating the identification of personalized treatment regimens, thereby facilitating the shift toward precision neurotherapeutics.[331] By harnessing the reversibility and plasticity of these regulatory mechanisms, such techniques attempt to improve functional recovery in afflicted brain circuits while also halting disease progression.[332]

7.8.1 Epigenetic Drugs

Epigenetic drugs represent a key category of these therapeutic approaches. Histone deacetylase (HDAC) inhibitors, such as valproic acid, vorinostat, and romidepsin, increase histone acetylation, promoting a more open chromatin state and enhancing gene expression.[333] In preclinical models of Rett syndrome, HDAC inhibitors restore BDNF expression and partially rescue neuronal deficits.[334] Clinical trials are ongoing for HDAC inhibitors in Alzheimer's disease, Huntington's disease, and depression, though challenges remain regarding blood–brain barrier (BBB) penetration and off-target effects.[333] DNA methylation inhibitors, including DNMT inhibitors like 5-azacytidine and decitabine, are approved for cancer treatment and under investigation for neurological uses. These compounds demethylate gene promoters, potentially reactivating silenced neuroprotective genes. However, their global demethylation effects raise safety concerns in the CNS.[335]

Bromodomain inhibitors, such as BET inhibitors like JQ1, disrupt the reading of acetylated histones and can suppress the expression of pro-inflammatory and oncogenic genes.[336] Early studies suggest potential roles in dampening neuroinflammation and aberrant gene activation in glioblastoma and AD models.[336]

7.8.2 RNA-based Therapeutics

RNA-based therapeutics offer another promising avenue.[337] Antisense oligonucleotides (ASOs) are short, synthetic nucleotides that bind specific RNA sequences to degrade or modify their function.[337] Notable successes include nusinersen for spinal muscular atrophy, which alters SMN2 splicing; preclinical ASOs targeting HTT mRNA in Huntington's disease; and experimental ASOs aimed at reducing SNCA expression in Parkinson's disease.[338] Small interfering RNAs (siRNAs) and miRNA mimics silence target mRNAs *via* RNA interference pathways.[339] For example, miR-124 mimics have shown promise in restoring neuronal gene expression in schizophrenia and stroke models.[340] Circular RNA modulators, through therapeutic strategies that involve synthetic constructs or interference with endogenous circRNAs, are being explored in gliomas, epilepsy, and AD, though still in the early stages.[339,341]

7.8.3 Epigenome Editing

Epigenome editing provides precise tools for intervention. CRISPR/dCas9-based epigenome editing systems offer locus-specific precision with reduced off-target effects.[342,343] Targeted demethylation of promoters for neuroprotective genes like BDNF and GDNF has improved outcomes in rodent models of depression and Parkinson's disease.[238,344] Editing enhancers or silencers to modulate long-range gene regulation is being explored for autism-related genes such as SHANK3 and CHD8.[345,346] These strategies are not yet in clinical use but represent a transformative direction for long-term correction of transcriptional dysfunction.[347]

7.8.4 Personalized Medicine and Biomarkers

Personalized medicine and biomarkers are advancing through multi-omics profiling, allowing for individualized molecular diagnosis and treatment stratification.[348,349] Transcriptomic signatures from blood or cerebrospinal fluid (CSF) are being developed as early diagnostic biomarkers in Alzheimer's and Parkinson's diseases.[350] Epigenetic markers, such as methylation of NR3C1 or BDNF promoters, have been proposed as predictors of antidepressant response or PTSD risk.[351] Pharmacogenomic screening also leverages transcriptomic data to anticipate drug efficacy and side effects, enabling personalized regimens for psychiatric medications and neuroprotective agents.[348,352,353]

7.8.5 Neuroprotective and Reprogramming Strategies

Neuroprotective and reprogramming strategies further expand therapeutic options.[354] HDAC inhibitors and CREB activators can protect neurons by maintaining gene expression programs essential for survival.[196,355] Reprogramming glial cells into neurons using transcription factor cocktails, such as NeuroD1 and Ascl1, along with epigenetic modulation, is being explored as a regenerative strategy in stroke and trauma models.[356] While these therapies are promising, translating them to clinical use requires overcoming challenges in delivery, specificity, and long-term safety.[356] Still, the integration of epigenomic and transcriptomic knowledge has fundamentally shifted the therapeutic landscape for brain disorders.[357]

7.9 Challenges and Future Directions

Despite major progress in characterizing epigenetic and transcriptomic regulation in the brain, significant obstacles remain.[358] These challenges stem from the biological complexity of neural systems, the limitations of current technologies, and the slow pace of translation into clinical practice.[358] Addressing them will be critical for realizing the full therapeutic potential of this field.[359]

7.9.1 Biological and Cellular Complexity

The brain contains an extraordinary diversity of cell types, each with distinct regulatory landscapes. Bulk tissue analysis often masks cell-type-specific signatures, limiting our ability to detect subtle but disease-relevant changes.[360] Single-cell and spatial omics approaches are beginning to resolve this heterogeneity, enabling analysis of transcriptional activity, chromatin accessibility, and epigenetic states at single-cell resolution.[361] These methods have revealed rare or transient cell states and lineage trajectories that bulk methods miss, while spatial approaches preserve anatomical context and cell–cell interactions.[362] However, the resulting datasets are vast and complex, requiring sophisticated computational frameworks and raising concerns about capture efficiency, batch effects, and reproducibility.[363] As costs fall and analytical tools improve, the integration of single-cell and spatial omics is expected to become a standard in brain research, offering unprecedented precision in mapping disease mechanisms.[235]

7.9.2 Temporal and Spatial Dynamics of Regulation

Epigenetic and transcriptomic profiles are highly dynamic, evolving across development, in response to stimuli, and throughout disease progression.[364] Yet most studies provide only static snapshots of molecular states, especially in human brain tissue.[363] Longitudinal sampling is difficult, creating major gaps in understanding how regulatory programs unfold over time.[365]

To address this, researchers increasingly rely on iPSC-derived neurons, brain organoids, and animal models, which allow dynamic tracking of molecular changes.[366] Time-course multi-omics approaches are beginning to shed light on developmental trajectories and disease progression, though scaling these efforts to human populations remains a challenge.[367]

7.9.3 Annotation, Integration, and Interpretation

Non-coding regions of the genome, including enhancers, silencers, and non-coding RNAs, play critical roles in brain function, yet their annotation remains incomplete.[368,369] Many disease-associated transcriptomic changes occur in poorly characterized genomic regions. Large-scale initiatives such as ENCODE and PsychENCODE are expanding the catalogue of regulatory elements, while machine learning approaches are being used to predict their functions.[370] At the same time, integrating information across omics layers, transcriptomics, epigenomics, proteomics, and metabolomics, poses computational hurdles.[371] High dimensionality, noise, and lack of standardized pipelines complicate interpretation.[367] Emerging integration tools such as $MOFA^{+}$, Harmony, and graph-based models are improving the construction of disease-specific regulatory networks, and greater emphasis on reproducibility and open data sharing is expected to accelerate progress.[372,373]

7.9.4 Translational and Ethical Barriers

Translating molecular insights into diagnostics or therapies remains slow.[374] Epigenetic drugs often lack specificity and can trigger off-target effects, while many transcriptomic signatures remain correlative rather than causal.[342] Emerging strategies, including precision epigenetic editing tools, biomarker-guided patient stratification, and testing therapies in patient-derived organoids or brain-on-a-chip models, are beginning to bridge this gap.[375] Equally important are ethical and societal considerations.[374,376] As epigenome profiling and editing enter neuropsychiatric research, issues of privacy, consent, and equity must be addressed.[374] Neuroethics guidelines, transparent governance, and public engagement will be essential to ensure responsible application of these technologies.[368]

7.9.5 Future Outlook

Looking ahead, several trends are likely to define the next decade. Integrated single-cell multi-omics will become routine for dissecting complex brain states, while artificial intelligence will accelerate biomarker discovery and therapeutic target prediction. Gene and epigenome editing therapies are poised to enter clinical trials, especially for monogenic neurodevelopmental disorders. At the same time, personalized psychiatry and neurology will increasingly incorporate molecular signatures into diagnostics and treatment decisions. The convergence of high-resolution technologies, computational biology, and translational models promises not only a deeper mechanistic understanding of brain disorders but also new avenues for early intervention, precision therapies, and ultimately prevention.

7.10 Conclusion

Epigenetic and transcriptomic regulation form the core of the brain's molecular control system, governing everything from neurodevelopment and synaptic plasticity to aging and disease. In healthy brains, these layers of regulation operate in concert to produce the diversity, adaptability, and precision required for complex cognitive function. In brain disorders, however, this balance is disrupted, leading to altered gene expression, impaired cellular function, and progressive neurological or psychiatric symptoms.

Across neurodevelopmental, neuropsychiatric, and neurodegenerative diseases, common patterns emerge such as aberrant DNA methylation, dysfunctional histone modifications, misregulated non-coding RNAs, and widespread transcriptional shifts. These changes are not merely downstream consequences but are often early contributors to pathophysiology. Importantly, many of these molecular changes are reversible, offering hope for intervention.

Advances in sequencing technologies, single-cell analysis, spatial omics, and genome editing have revolutionized our ability to study and manipulate the epigenome and transcriptome. These tools are not only reshaping research but are now being translated into diagnostic platforms, therapeutic strategies, and precision medicine approaches.

Still, challenges remain. The brain's cellular complexity, dynamic regulation, and sensitivity to perturbation make therapeutic translation difficult. Future efforts must focus on refining our understanding of causality, improving specificity in molecular therapies, and building ethical frameworks for clinical applications.

As the field moves forward, the integration of epigenetic and transcriptomic insights will be key to unlocking the mysteries of brain disorders. Whether through early diagnosis, targeted treatment, or prevention, these molecular pathways are becoming central to how we understand and manage brain health in the 21st century.

Research focuses on linking molecular changes to functional and behavioural outcomes, bridging the gap between mechanistic understanding and therapeutic relevance. Integrative multi-omics techniques give a systems-level knowledge of brain illnesses, while emerging approaches such as CRISPR-based editing have the potential to modify disease-relevant pathways.

Abbreviations

AD	Alzheimer's disease
ADAR	Adenosine deaminase acting on RNA

ALS	Amyotrophic lateral sclerosis
APP	Amyloid precursor protein
ASD	Autism spectrum disorder
ATAC-seq	Assay for transposase-accessible chromatin using sequencing
BDNF	Brain-derived neurotrophic factor
BS-seq	Bisulfite sequencing
CBP	CREB-binding protein
ChIP-seq	Chromatin immunoprecipitation sequencing
CHD	Chromodomain helicase DNA-binding protein
circRNA	Circular RNA
CRISPR	Clustered regularly interspaced short palindromic repeats
CRISPRa/i	CRISPR activation/CRISPR interference
CUT&RUN/ CUT&Tag	Cleavage under targets & release using nuclease/cleavage under targets & tagmentation
dCas9	Dead Cas9 (catalytically inactive Cas9)
DEG	Differentially expressed gene
DNMT	DNA methyltransferase
eRNA	Enhancer RNA
FAIRE-seq	Formaldehyde-assisted isolation of regulatory elements sequencing
FOXP2	Forkhead box protein P2
GAD1	Glutamate decarboxylase 1
HAT	Histone acetyltransferase
HD	Huntington's disease
HDAC	Histone deacetylase
LINE-1	Long interspersed nuclear element-1
lncRNA	Long non-coding RNA
MAPT	Microtubule-associated protein tau
MeCP2	Methyl-CpG-binding protein 2
MDD	Major depressive disorder
MERFISH	Multiplexed error-robust fluorescence *in situ* hybridization
miRNA	MicroRNA
MOFA	Multi-omics factor analysis
ncRNA	Non-coding RNA
NR3C1	Nuclear receptor subfamily 3 group C member 1 (glucocorticoid receptor)
NRXN	Neurexin
OXTR	Oxytocin receptor
PD	Parkinson's disease
PRC2	Polycomb repressive complex 2
PTBP	Polypyrimidine tract-binding protein
RBFOX	RNA binding Fox protein
REST	RE1-silencing transcription factor

RRBS	Reduced representation bisulfite sequencing
scATAC-seq	Single-cell ATAC sequencing
scGESTALT	Single-cell genome editing of synthetic target arrays for lineage tracing
scRNA-seq	Single-cell RNA sequencing
SNCA	Alpha-synuclein gene
SYN1	Synapsin I
TET	Ten-eleven translocation enzyme
TF	Transcription factor
TH	Tyrosine hydroxylase
UTR	Untranslated region
VMAT2	Vesicular monoamine transporter 2
WGBS	Whole-genome bisulfite sequencing

References

1. H. J. Kang, *et al.*, Spatio-temporal transcriptome of the human brain, *Nature*, 2011, **478**, 483–489.
2. M. Li, *et al.*, Integrative functional genomic analysis of human brain development and neuropsychiatric risks, *Science*, 2018, **362**, eaat7615.
3. Y. Zhu, *et al.*, Spatiotemporal transcriptomic divergence across human and macaque brain development, *Science*, 2018, **362**, eaat8077.
4. A. E. West and M. E. Greenberg, Neuronal activity-regulated gene transcription in synapse development and cognitive function, *Cold Spring Harb. Perspect. Biol.*, 2011, **3**, a005744.
5. G. Hermey, N. Blüthgen and D. Kuhl, Neuronal activity-regulated alternative mRNA splicing, *Int. J. Biochem. Cell Biol.*, 2017, **91**, 184–193.
6. B. J. Lukasak and E. Korb, Histone variants: expanding the epigenetic potential of neurons one amino acid at a time, *Trends Biochem. Sci.*, 2025, **50**, 532–543.
7. T. L. Bale, Epigenetic and transgenerational reprogramming of brain development, *Nat. Rev. Neurosci.*, 2015, **16**, 332–344.
8. D. Kofink, M. P. M. Boks, H. T. M. Timmers and M. J. Kas, Epigenetic dynamics in psychiatric disorders: environmental programming of neurodevelopmental processes, *Neurosci. Biobehav. Rev.*, 2013, **37**, 831–845.
9. A. P. Monaco, An epigenetic, transgenerational model of increased mental health disorders in children, adolescents and young adults, *Eur. J. Hum. Genet.*, 2021, **29**, 387–395.
10. J. P. Curley, C. L. Jensen, R. Mashoodh and F. A. Champagne, Social influences on neurobiology and behavior: epigenetic effects during development, *Psychoneuroendocrinology*, 2011, **36**, 352–371.
11. P. L. Sokolov, N. V. Chebanenko and D. M. Mednaya, Epigenetic influences and brain development, *Zh. Nevrol. Psikhiatr. Im. S. S. Korsakova*, 2023, **123**, 12–19.

12. M. Fagiolini, C. L. Jensen and F. A. Champagne, Epigenetic influences on brain development and plasticity, *Curr. Opin. Neurobiol.*, 2009, **19**, 207–212.
13. W. Sun, *et al.*, Histone Acetylome-wide Association Study of Autism Spectrum Disorder, *Cell*, 2016, **167**, 1385–1397.e11.
14. S. Loebrich and E. Nedivi, The function of activity-regulated genes in the nervous system, *Physiol. Rev.*, 2009, **89**, 1079–1103.
15. S. Akbarian and H.-S. Huang, Epigenetic regulation in human brain-focus on histone lysine methylation, *Biol. Psychiatry*, 2009, **65**, 198–203.
16. S. Kim, *et al.*, An integrative single-cell atlas for exploring the cellular and temporal specificity of genes related to neurological disorders during human brain development, *Exp. Mol. Med.*, 2024, **56**, 2271–2282.
17. D. Zhang, *et al.*, Spatial epigenome-transcriptome co-profiling of mammalian tissues, *Nature*, 2023, **616**, 113–122.
18. F. M. Vassoler and M. E. Wimmer, Consequences of parental opioid exposure on neurophysiology, behavior, and health in the next generations, *Cold Spring Harb. Perspect. Med.*, 2021, **11**, a040436.
19. L. Li, *et al.*, Epigenomic programming in early fetal brain development, *Epigenomics*, 2020, **12**, 1053–1070.
20. L. Lossi, C. Castagna and A. Merighi, An Overview of the Epigenetic Modifications in the Brain under Normal and Pathological Conditions, *Int. J. Mol. Sci.*, 2024, **25**, 3881.
21. S. Maity, K. Farrell, S. Navabpour, S. N. Narayanan and T. J. Jarome, Epigenetic Mechanisms in Memory and Cognitive Decline Associated with Aging and Alzheimer's Disease, *Int. J. Mol. Sci.*, 2021, **22**, 12280.
22. B. S. Basavarajappa and S. Subbanna, Unlocking the epigenetic symphony: histone acetylation's impact on neurobehavioral change in neurodegenerative disorders, *Epigenomics*, 2024, **16**, 331–358.
23. L. Li, *et al.*, The epigenetic modification of DNA methylation in neurological diseases, *Front. Immunol.*, 2024, **15**, 1401962.
24. W. Qu, Y. Zhuang and X. Li, The roles of epigenetic modifications in neurodegenerative diseases, *Zhejiang Da Xue Xue Bao Yi Xue Ban*, 2021, **50**, 642–650.
25. L. Wang, B. Wang, C. Wu, J. Wang and M. Sun, Autism spectrum disorder: neurodevelopmental risk factors, biological mechanism, and precision therapy, *Int. J. Mol. Sci.*, 2023, **24**, 1819.
26. G. Mastrototaro, M. Zaghi and A. Sessa, Epigenetic mistakes in neurodevelopmental disorders, *J. Mol. Neurosci.*, 2017, **61**, 590–602.
27. D. Marano, S. Fioriniello, M. D'Esposito and F. Della Ragione, Transcriptomic and epigenomic landscape in rett syndrome, *Biomolecules*, 2021, **11**, 967.
28. C.-T. Roberts, K. S. Arezoumand, A. Kadar Shahib, J. R. Davie and M. Rastegar, Epigenetics in rare neurological diseases, *Front. Cell Dev. Biol.*, 2024, **12**, 1413248.
29. J. N. Kuehner, E. C. Bruggeman, Z. Wen and B. Yao, Epigenetic regulations in neuropsychiatric disorders, *Front. Genet.*, 2019, **10**, 268.

30. P. Ghosh and A. Saadat, Neurodegeneration and epigenetics: A review, *Neurologia (Engl. Ed.)*, 2023, **38**, e62–e68.
31. S. Younesian, A.-M. Yousefi, M. Momeny, S. H. Ghaffari and D. Bashash, The DNA methylation in neurological diseases, *Cells*, 2022, **11**, 3439.
32. J. van de Leemput, J. L. Hess, S. J. Glatt and M. T. Tsuang, Genetics of schizophrenia: historical insights and prevailing evidence, *Adv. Genet.*, 2016, **96**, 99–141.
33. E. A. Thomas, Histone posttranslational modifications in schizophrenia, *Adv. Exp. Med. Biol.*, 2017, **978**, 237–254.
34. D. A. Chistiakov, Y. V. Bobryshev and V. P. Chekhonin, Epigenetic Alterations in DNA and Histone Modifications Caused by Depression and Antidepressant Drugs: Lessons from the Rodent Models, *Curr. Pharm. Des.*, 2017, **23**, 6828–6840.
35. B. S. Basavarajappa and S. Subbanna, Histone methylation regulation in neurodegenerative disorders, *Int. J. Mol. Sci.*, 2021, **22**, 4654.
36. A. S. Rathore, *et al.*, Epigenetic modulation in parkinson's disease and potential treatment therapies, *Neurochem. Res.*, 2021, **46**, 1618–1626.
37. J. Men, *et al.*, Neurodegenerative diseases: Epigenetic regulatory mechanisms and therapeutic potential, *Cell. Signalling*, 2025, **131**, 111715.
38. M. A. T. S. Hashmi, H. Fatima, S. Ahmad, A. Rehman and F. Safdar, The interplay between epitranscriptomic RNA modifications and neurodegenerative disorders: Mechanistic insights and potential therapeutic strategies, *iBrain*, 2024, **10**, 395–426.
39. R. Holliday, Epigenetics: a historical overview, *Epigenetics*, 2006, **1**, 76–80.
40. N. Ingram, Waddington, holmyard and alchemy: perspectives on the epigenetic landscape, *Endeavour*, 2019, **43**, 100690.
41. J. D. Sweatt, The emerging field of neuroepigenetics, *Neuron*, 2013, **80**, 624–632.
42. A. Mitchell, P. Roussos, C. Peter, N. Tsankova and S. Akbarian, The future of neuroepigenetics in the human brain, *Prog. Mol. Biol. Transl. Sci.*, 2014, **128**, 199–228.
43. E. Borrelli, E. J. Nestler, C. D. Allis and P. Sassone-Corsi, Decoding the epigenetic language of neuronal plasticity, *Neuron*, 2008, **60**, 961–974.
44. F. A. Champagne and J. P. Curley, Epigenetic mechanisms mediating the long-term effects of maternal care on development, *Neurosci. Biobehav. Rev.*, 2009, **33**, 593–600.
45. E. W. Fish, *et al.*, Epigenetic programming of stress responses through variations in maternal care, *Ann. N. Y. Acad. Sci.*, 2004, **1036**, 167–180.
46. I. C. G. Weaver, Epigenetic programming by maternal behavior and pharmacological intervention. Nature *versus* nurture: let's call the whole thing off, *Epigenetics*, 2007, **2**, 22–28.
47. J. J. Day and J. D. Sweatt, DNA methylation and memory formation, *Nat. Neurosci.*, 2010, **13**, 1319–1323.

48. C. Dulac, Brain function and chromatin plasticity, *Nature*, 2010, **465**, 728–735.
49. K. L. Brida and J. J. Day, Molecular and genetic mechanisms of plasticity in addiction, *Curr. Opin. Neurobiol.*, 2025, **93**, 103032.
50. E. J. Nestler, Epigenetic mechanisms of drug addiction, *Neuropharmacology*, 2014, **76**(Pt B), 259–268.
51. M. A. Christopher, S. M. Kyle and D. J. Katz, Neuroepigenetic mechanisms in disease, *Epigenetics Chromatin*, 2017, **10**, 47.
52. S. L. Berger, T. Kouzarides, R. Shiekhattar and A. Shilatifard, An operational definition of epigenetics, *Genes Dev.*, 2009, **23**, 781–783.
53. F. Crick, Memory and molecular turnover, *Nature*, 1984, **312**, 101.
54. J. E. Lisman, A mechanism for memory storage insensitive to molecular turnover: a bistable autophosphorylating kinase, *Proc. Natl. Acad. Sci. U. S. A.*, 1985, **82**, 3055–3057.
55. R. Holliday, Is there an epigenetic component in long-term memory?, *J. Theor. Biol.*, 1999, **200**, 339–341.
56. A. Bird, DNA methylation patterns and epigenetic memory, *Genes Dev.*, 2002, **16**, 6–21.
57. L. D. Moore, T. Le and G. Fan, DNA Methylation and Its Basic Function, *Neuropsychopharmacology*, 2013, **38**, 23–38.
58. D. Schübeler, Function and information content of DNA methylation, *Nature*, 2015, **517**, 321–326.
59. K. F. Santos, T. N. Mazzola and H. F. Carvalho, The prima donna of epigenetics: the regulation of gene expression by DNA methylation, *Braz. J. Med. Biol. Res.*, 2005, **38**, 1531–1541.
60. J. A. Cain, B. Montibus and R. J. Oakey, Intragenic cpg islands and their impact on gene regulation, *Front. Cell Dev. Biol.*, 2022, **10**, 832348.
61. S. Ghosh, *et al.*, Tissue specific DNA methylation of CpG islands in normal human adult somatic tissues distinguishes neural from non-neural tissues, *Epigenetics*, 2010, **5**, 527–538.
62. D. Zilberman, The evolving functions of DNA methylation, *Curr. Opin. Plant Biol.*, 2008, **11**, 554–559.
63. L. M. De Plano, A. Saitta, S. Oddo and A. Caccamo, Epigenetic changes in alzheimer's disease: DNA methylation and histone modification, *Cells*, 2024, **13**, 719.
64. M. Kim and J. Costello, DNA methylation: an epigenetic mark of cellular memory, *Exp. Mol. Med.*, 2017, **49**, e322.
65. L. Xie, *et al.*, Emerging roles for DNA 6ma and RNA m6a methylation in mammalian genome, *Int. J. Mol. Sci.*, 2023, **24**, 13897.
66. Y. Zhang, *et al.*, Overview of histone modification, *Adv. Exp. Med. Biol.*, 2021, **1283**, 1–16.
67. C. L. Peterson and M.-A. Laniel, Histones and histone modifications, *Curr. Biol.*, 2004, **14**, R546–R551.
68. E. L. Greer and Y. Shi, Histone methylation: a dynamic mark in health, disease and inheritance, *Nat. Rev. Genet.*, 2012, **13**, 343–357.

69. L. Verdone, M. Caserta and E. Di Mauro, Role of histone acetylation in the control of gene expression, *Biochem. Cell Biol.*, 2005, **83**, 344–353.
70. T. Banerjee and D. Chakravarti, A peek into the complex realm of histone phosphorylation, *Mol. Cell. Biol.*, 2011, **31**, 4858–4873.
71. S. Lagger, *et al.*, MeCP2 recognizes cytosine methylated tri-nucleotide and di-nucleotide sequences to tune transcription in the mammalian brain, *PLoS Genet.*, 2017, **13**, e1006793.
72. F. Mattiroli and L. Penengo, Histone ubiquitination: an integrative signaling platform in genome stability, *Trends Genet.*, 2021, **37**, 566–581.
73. H.-Y. Ryu and M. Hochstrasser, Histone sumoylation and chromatin dynamics, *Nucleic Acids Res.*, 2021, **49**, 6043–6052.
74. D. Zhang, *et al.*, Metabolic regulation of gene expression by histone lactylation, *Nature*, 2019, **574**, 575–580.
75. P. Gong, Z. Guo, S. Wang, S. Gao and Q. Cao, Histone phosphorylation in DNA damage response, *Int. J. Mol. Sci.*, 2025, **26**, 2405.
76. E. Seto, Histone modifications, *Methods*, 2003, **31**, 1–2.
77. T. G. Kutateladze, O. Gozani, M. Bienz and M. Ostankovitch, Histone modifications for chromatin dynamics and cellular plasticity, *J. Mol. Biol.*, 2017, **429**, 1921–1923.
78. M. Esteller, Non-coding RNAs in human disease, *Nat. Rev. Genet.*, 2011, **12**, 861–874.
79. K. Nemeth, R. Bayraktar, M. Ferracin and G. A. Calin, Non-coding RNAs in disease: from mechanisms to therapeutics, *Nat. Rev. Genet.*, 2024, **25**, 211–232.
80. G. Catanzaro, Non-Coding RNAs in Health and Disease: Editorial, *Biomedicines*, 2023, **11**, 14.
81. J. S. Mattick and I. V. Makunin, Non-coding RNA, *Hum. Mol. Genet.*, 2006, **15**(Spec No 1), R17–R29.
82. J.-W. Wei, K. Huang, C. Yang and C.-S. Kang, Non-coding RNAs as regulators in epigenetics (Review), *Oncol. Rep.*, 2017, **37**, 3–9.
83. S. Uchida and J. C. Adams, Physiological roles of non-coding RNAs, *Am. J. Physiol., Cell Physiol.*, 2019, **317**, C1–C2.
84. T. Loganathan and C. G. P. Doss, Non-coding RNAs in human health and disease: potential function as biomarkers and therapeutic targets, *Funct. Integr. Genomics*, 2023, **23**, 33.
85. J.-H. Nie, T.-X. Li, X.-Q. Zhang and J. Liu, Roles of Non-Coding RNAs in Normal Human Brain Development, Brain Tumor, and Neuropsychiatric Disorders, *Noncoding RNA*, 2019, **5**, 36.
86. S. Trivedi and G. Ramakrishna, miRNA and neurons, *Int. J. Neurosci.*, 2009, **119**, 1995–2016.
87. A. X. Sun, G. R. Crabtree and A. S. Yoo, MicroRNAs: regulators of neuronal fate, *Curr. Opin. Cell Biol.*, 2013, **25**, 215–221.
88. D. Tardito, A. Mallei and M. Popoli, Lost in translation. New unexplored avenues for neuropsychopharmacology: epigenetics and microRNAs, *Expert Opin. Investig. Drugs*, 2013, **22**, 217–233.

89. W. Filipowicz, L. Jaskiewicz, F. A. Kolb and R. S. Pillai, Post-transcriptional gene silencing by siRNAs and miRNAs, *Curr. Opin. Struct. Biol.*, 2005, **15**, 331–341.
90. K. S. Kosik, The neuronal microRNA system, *Nat. Rev. Neurosci.*, 2006, **7**, 911–920.
91. L. Zuo, Z. Wang, Y. Tan, X. Chen and X. Luo, piRNAs and Their Functions in the Brain, *Int. J. Hum. Genet.*, 2016, **16**, 53–60.
92. V. Aliperti and A. Donizetti, Long Non-coding RNA in Neurons: New Players in Early Response to BDNF Stimulation, *Front. Mol. Neurosci.*, 2016, **9**, 15.
93. S. Memczak, *et al.*, Circular RNAs are a large class of animal RNAs with regulatory potency, *Nature*, 2013, **495**, 333–338.
94. S. Gokul and G. K. Rajanikant, Circular rnas in brain physiology and disease, *Adv. Exp. Med. Biol.*, 2018, **1087**, 231–237.
95. H. Gan, *et al.*, Circular RNAs in depression: Biogenesis, function, expression, and therapeutic potential, *Biomed. Pharmacother.*, 2021, **137**, 111244.
96. J. L. Ronan, W. Wu and G. R. Crabtree, From neural development to cognition: unexpected roles for chromatin, *Nat. Rev. Genet.*, 2013, **14**, 347–359.
97. M. Xylaki, B. Atzler and T. F. Outeiro, Epigenetics of the synapse in neurodegeneration, *Curr. Neurol. Neurosci. Rep.*, 2019, **19**, 72.
98. N. Ballas and G. Mandel, The many faces of REST oversee epigenetic programming of neuronal genes, *Curr. Opin. Neurobiol.*, 2005, **15**, 500–506.
99. D. S. Samuels, *et al.*, Gene regulation and transcriptomics, *Curr. Issues Mol. Biol.*, 2021, **42**, 223–266.
100. M. H. Flight, Gene expression: the dynamics of the brain transcriptome revealed, *Nat. Rev. Neurosci.*, 2011, **12**, 703.
101. S. M. Fernández-Moya, A. J. Ganesh and M. Plass, Neural cell diversity in the light of single-cell transcriptomics, *Transcription*, 2023, **14**, 158–176.
102. C. Gao, J. Jiang, Y. Tan and S. Chen, Microglia in neurodegenerative diseases: mechanism and potential therapeutic targets, *Signal Transduct. Target. Ther.*, 2023, **8**, 359.
103. F. Telese, A. Gamliel, D. Skowronska-Krawczyk, I. Garcia-Bassets and M. G. Rosenfeld, Seq-ing" insights into the epigenetics of neuronal gene regulation, *Neuron*, 2013, **77**, 606–623.
104. C. J. Pena, Epigenetic regulation of brain development, plasticity, and response to early-life stress, *Neuropsychopharmacology*, 2026, **51**, 5–15.
105. C.-W. Wei, T. Luo, S.-S. Zou and A.-S. Wu, The role of long noncoding rnas in central nervous system and neurodegenerative diseases, *Front. Behav. Neurosci.*, 2018, **12**, 175.
106. S. Brummelte, E. Mc Glanaghy, A. Bonnin and T. F. Oberlander, Developmental changes in serotonin signaling: Implications for early

brain function, behavior and adaptation, *Neuroscience*, 2017, **342**, 212–231.

107. D. Saha, S. Animireddy and B. Bartholomew, The SWI/SNF ATP dependent chromatin remodeling complex in cell lineage priming and early development, *Biochem. Soc. Trans.*, 2024, **52**, 603–616.
108. S. S. Purushotham and Y. Buskila, Astrocytic modulation of neuronal signalling, *Front. Netw. Physiol.*, 2023, **3**, 1205544.
109. Y. Luo and Z. Wang, The impact of microglia on neurodevelopment and brain function in autism, *Biomedicines*, 2024, **12**, 210.
110. R. Singh, *et al.*, Autism Spectrum Disorders: A Recent Update on Targeting Inflammatory Pathways with Natural Anti-Inflammatory Agents, *Biomedicines*, 2023, **11**, 115.
111. A. Hoffmann and D. Spengler, Chromatin remodeler CHD8 in autism and brain development, *J. Clin. Med.*, 2021, **10**, 366.
112. M. Hou, *et al.*, Advances in DNA-based proteolysis-targeting chimeras technology: Novel strategies for targeting previously undruggable proteins in cancer therapy, *Bioorg. Med. Chem.*, 2025, **129**, 118297.
113. O. Oksuz, *et al.*, Transcription factors interact with RNA to regulate genes, *Mol. Cell*, 2023, **83**, 2449–2463.e13.
114. S. Maksour, L. Ooi and M. Dottori, More than a Corepressor: The Role of CoREST Proteins in Neurodevelopment, *eNeuro*, 2020, 7(2), ENEURO.0337-19.2020.
115. D. Zheng, K. Zhao and M. F. Mehler, Profiling RE1/REST-mediated histone modifications in the human genome, *Genome Biol.*, 2009, **10**, R9.
116. R. Butler-Ryan and I. C. Wood, The functions of repressor element 1-silencing transcription factor in models of epileptogenesis and post-ischemia, *Metab. Brain Dis.*, 2021, **36**, 1135–1150.
117. E. Battaglioli, *et al.*, REST repression of neuronal genes requires components of the hSWI.SNF complex, *J. Biol. Chem.*, 2002, **277**, 41038–41045.
118. Y. Wang, *et al.*, Mechanism of alternative splicing and its regulation, *Biomed. Rep.*, 2015, **3**, 152–158.
119. Z.-X. Lu, P. Jiang and Y. Xing, Genetic variation of pre-mRNA alternative splicing in human populations, *Wiley Interdiscip. Rev.: RNA*, 2012, **3**, 581–592.
120. E. Park, Z. Pan, Z. Zhang, L. Lin and Y. Xing, The expanding landscape of alternative splicing variation in human populations, *Am. J. Hum. Genet.*, 2018, **102**, 11–26.
121. E. T. Wang, *et al.*, Alternative isoform regulation in human tissue transcriptomes, *Nature*, 2008, **456**, 470–476.
122. Q. Pan, O. Shai, L. J. Lee, B. J. Frey and B. J. Blencowe, Deep surveying of alternative splicing complexity in the human transcriptome by high-throughput sequencing, *Nat. Genet.*, 2008, **40**, 1413–1415.
123. W. Jiang and L. Chen, Alternative splicing: Human disease and quantitative analysis from high-throughput sequencing, *Comput. Struct. Biotechnol. J.*, 2021, **19**, 183–195.

124. Y. Koike, Abnormal Splicing Events due to Loss of Nuclear Function of TDP-43: Pathophysiology and Perspectives, *JMA J.*, 2024, **7**, 313–318.
125. G. E. Aranda-Abreu, *et al.*, The role of tau in neuronal function and neurodegeneration, *Neurol. Int.*, 2025, **17**, 75.
126. A. Chongtham, *et al.*, Neocortical tau propagation is a mediator of clinical heterogeneity in Alzheimer's disease, *Mol. Psychiatry*, 2025, **30**, 4194–4213.
127. I. D'Souza and G. D. Schellenberg, Regulation of tau isoform expression and dementia, *Biochim. Biophys. Acta*, 2005, **1739**, 104–115.
128. J. S. Satterlee, *et al.*, Noncoding RNAs in the brain, *J. Neurosci.*, 2007, **27**, 11856–11859.
129. H. K. Karnati, M. K. Panigrahi, R. K. Gutti, N. H. Greig and I. A. Tamargo, miRNAs: Key Players in Neurodegenerative Disorders and Epilepsy, *J. Alzheimers Dis.*, 2015, **48**, 563–580.
130. S. Aten, K. F. Hansen, K. R. Hoyt and K. Obrietan, The miR-132/212 locus: a complex regulator of neuronal plasticity, gene expression and cognition, *RNA Dis.*, 2016, **3**, e1375.
131. Y. Statsenko, N. V. Kuznetsov and M. Ljubisaljevich, Hallmarks of brain plasticity, *Biomedicines*, 2025, **13**, 460.
132. L.-C. Cheng, E. Pastrana, M. Tavazoie and F. Doetsch, miR-124 regulates adult neurogenesis in the subventricular zone stem cell niche, *Nat. Neurosci.*, 2009, **12**, 399–408.
133. M. Coolen, S. Katz and L. Bally-Cuif, miR-9: a versatile regulator of neurogenesis, *Front. Cell. Neurosci.*, 2013, **7**, 220.
134. S. Zhou, *et al.*, Long Non-coding RNAs in Pathogenesis of Neurodegenerative Diseases, *Front. Cell Dev. Biol.*, 2021, **9**, 719247.
135. I. Cajigas, *et al.*, The Evf2 Ultraconserved Enhancer lncRNA Functionally and Spatially Organizes Megabase Distant Genes in the Developing Forebrain, *Mol. Cell*, 2018, **71**, 956–972.e9.
136. J. P. Bohnsack, T. Teppen, E. J. Kyzar, S. Dzitoyeva and S. C. Pandey, The lncRNA BDNF-AS is an epigenetic regulator in the human amygdala in early onset alcohol use disorders, *Transl. Psychiatry*, 2019, **9**, 34.
137. D. Bernard, *et al.*, A long nuclear-retained non-coding RNA regulates synaptogenesis by modulating gene expression, *EMBO J.*, 2010, **29**, 3082–3093.
138. W. Xiao, *et al.*, The lncRNA Malat1 is trafficked to the cytoplasm as a localized mRNA encoding a small peptide in neurons, *Genes Dev.*, 2024, **38**, 294–307.
139. R. Ben-Tov Perry, M. Tsoory, M. Tolmasov and I. Ulitsky, Silc1 long noncoding RNA is an immediate-early gene promoting efficient memory formation, *Cell Rep.*, 2023, **42**, 113168.
140. I. Espadas, *et al.*, Synaptically-targeted long non-coding RNA SLAMR promotes structural plasticity by increasing translation and CaMKII activity, *Nat. Commun.*, 2024, **15**, 2694.
141. V. Galli, S. Vinciguerra, M. Biagioli and J. Morandell, Circular RNAs as disease modifiers of complex neurologic disorders, *Front. Pharmacol.*, 2025, **16**, 1577496.

142. S. L. Mehta, R. J. Dempsey and R. Vemuganti, Role of circular RNAs in brain development and CNS diseases, *Prog. Neurobiol.*, 2020, **186**, 101746.
143. A. J. Zimmerman, *et al.*, A psychiatric disease-related circular RNA controls synaptic gene expression and cognition, *Mol. Psychiatry*, 2020, **25**, 2712–2727.
144. D. Kelly, *et al.*, A functional screen uncovers circular RNAs regulating excitatory synaptogenesis in hippocampal neurons, *Nat. Commun.*, 2025, **16**, 3040.
145. H. Gruner, M. Cortés-López, D. A. Cooper, M. Bauer and P. Miura, CircRNA accumulation in the aging mouse brain, *Sci. Rep.*, 2016, **6**, 38907.
146. K. Kirio, *et al.*, Circular RNAs exhibit exceptional stability in the aging brain and serve as reliable age and experience indicators, *Cell Rep.*, 2025, **44**, 115485.
147. R. G. Urdinguio, J. V. Sanchez-Mut and M. Esteller, Epigenetic mechanisms in neurological diseases: genes, syndromes, and therapies, *Lancet Neurol.*, 2009, **8**, 1056–1072.
148. D. B. Olawade, *et al.*, Reversing epigenetic dysregulation in neurodegenerative diseases: mechanistic and therapeutic considerations, *Int. J. Mol. Sci.*, 2025, **26**, 4929.
149. R. R. Ratan, Epigenetics and the nervous system: epiphenomenon or missing piece of the neurotherapeutic puzzle?, *Lancet Neurol.*, 2009, **8**, 975–977.
150. T. Wang, J. Zhang and Y. Xu, Epigenetic Basis of Lead-Induced Neurological Disorders, *Int. J. Environ. Res. Public Health*, 2020, **17**, 4878.
151. B. M. Gruber, Epigenetics and etiology of neurodegenerative diseases, *Postepy Hig. Med. Dosw.*, 2011, **65**, 542–551.
152. P. Petazzi, *et al.*, Dysregulation of the long non-coding RNA transcriptome in a Rett syndrome mouse model, *RNA Biol.*, 2013, **10**, 1197–1203.
153. M. Jakovcevski and S. Akbarian, Epigenetic mechanisms in neurological disease, *Nat. Med.*, 2012, **18**, 1194–1204.
154. T. Lu, *et al.*, Gene regulation and DNA damage in the ageing human brain, *Nature*, 2004, **429**, 883–891.
155. V. K. Ma, *et al.*, Prader-Willi and Angelman Syndromes: Mechanisms and Management, *Appl. Clin. Genet.*, 2023, **16**, 41–52.
156. K. Conboy, D. C. Henshall and G. P. Brennan, Epigenetic principles underlying epileptogenesis and epilepsy syndromes, *Neurobiol. Dis.*, 2021, **148**, 105179.
157. M. W. Tremblay and Y.-H. Jiang, DNA methylation and susceptibility to autism spectrum disorder, *Annu. Rev. Med.*, 2019, **70**, 151–166.
158. A. Vogel Ciernia and J. LaSalle, The landscape of DNA methylation amid a perfect storm of autism aetiologies, *Nat. Rev. Neurosci.*, 2016, **17**, 411–423.
159. R. Kumsta, E. Hummel, F. S. Chen and M. Heinrichs, Epigenetic regulation of the oxytocin receptor gene: implications for behavioral neuroscience, *Front. Neurosci.*, 2013, **7**, 83.

160. C. Kao, K. K. Lim and E. S. Chen, Emerging epigenetic therapeutics and diagnostics for autism spectrum disorder, *Curr. Issues Mol. Biol.*, 2025, **47**, 491.
161. O. Weissberg and E. Elliott, The mechanisms of CHD8 in neurodevelopment and autism spectrum disorders, *Genes*, 2021, **12**, 1133.
162. U. Sorrentino, *et al.*, CHD8-related disorders redefined: an expanding spectrum of dystonic phenotypes, *J. Neurol.*, 2024, **271**, 2859–2865.
163. M. W. Mitchel *et al.*, CHD8-Related Neurodevelopmental Disorder with Overgrowth, in *GeneReviews*®, ed. M. P. Adam *et al.*, University of Washington, Seattle, 1993.
164. M. L. Herrera, J. Paraíso-Luna, I. Bustos-Martínez and Á. Barco, Targeting epigenetic dysregulation in autism spectrum disorders, *Trends Mol. Med.*, 2024, **30**, 1028–1046.
165. M. Al-Beltagi, N. K. Saeed, A. S. Bediwy, E. A. Bediwy and R. Elbeltagi, Decoding the genetic landscape of autism: A comprehensive review, *World J. Clin. Pediatr.*, 2024, **13**, 98468.
166. S. Pejhan and M. Rastegar, Role of DNA Methyl-CpG-Binding Protein $MeCP_2$ in Rett Syndrome Pathobiology and Mechanism of Disease, *Biomolecules*, 2021, **11**, 75.
167. V. R. B. Liyanage and M. Rastegar, Rett syndrome and MeCP2, *Neuromolecular Med.*, 2014, **16**, 231–264.
168. A. K. Percy, Rett syndrome: clinical correlates of the newly discovered gene, *Brain Dev.*, 2001, **23**(Suppl 1), S202–S205.
169. O. Rittinger, Clinical aspects and genetics of Prader-Willi syndrome, *Klin. Padiatr.*, 2001, **213**, 91–98.
170. O. Davis, Abnormal Chromatin Folding in the Molecular Pathogenesis of Epilepsy and Autism Spectrum Disorder: a Meta-synthesis with Systematic Searching, *Mol. Neurobiol.*, 2023, **60**, 768–779.
171. H. Yang, W. Sun, J. Li and X. Zhang, Epigenetics factors in schizophrenia: future directions for etiologic and therapeutic study approaches, *Ann. Gen. Psychiatry*, 2025, **24**, 21.
172. Ç. Gürel, G. C. Kuşçu, A. Yavaşoğlu and Ç. Biray Avcı, The clues in solving the mystery of major psychosis: The epigenetic basis of schizophrenia and bipolar disorder, *Neurosci. Biobehav. Rev.*, 2020, **113**, 51–61.
173. P. Li, *et al.*, Decoding disease-specific ageing mechanisms through pathway-level epigenetic clock: insights from multi-cohort validation, *EBioMedicine*, 2025, **118**, 105829.
174. J. Mill, *et al.*, Epigenomic profiling reveals DNA-methylation changes associated with major psychosis, *Am. J. Hum. Genet.*, 2008, **82**, 696–711.
175. J. Park, K. Lee, K. Kim and S.-J. Yi, The role of histone modifications: from neurodevelopment to neurodiseases, *Signal Transduct. Target. Ther.*, 2022, **7**, 217.
176. S. Chakrapani, N. Eskander, L. A. De Los Santos, B. A. Omisore and J. A. Mostafa, Neuroplasticity and the biological role of brain derived neurotrophic factor in the pathophysiology and management of depression, *Cureus*, 2020, **12**, e11396.

177. M. Yuan, *et al.*, Epigenetic regulation in major depression and other stress-related disorders: molecular mechanisms, clinical relevance and therapeutic potential, *Signal Transduct. Target. Ther.*, 2023, **8**, 309.
178. D. Cicchetti and E. D. Handley, Methylation of the glucocorticoid receptor gene, nuclear receptor subfamily 3, group C, member 1 (NR3C1), in maltreated and nonmaltreated children: Associations with behavioral undercontrol, emotional lability/negativity, and externalizing and internalizing symptoms, *Dev. Psychopathol.*, 2017, **29**, 1795–1806.
179. E. Raffetti, *et al.*, DNA methylation of the glucocorticoid receptor gene predicts substance use in adolescence: longitudinal data from over 1000 young individuals, *Transl. Psychiatry*, 2021, **11**, 477.
180. A. Menke, T. Klengel and E. B. Binder, Epigenetics, depression and antidepressant treatment, *Curr. Pharm. Des.*, 2012, **18**, 5879–5889.
181. M. Shaikh and G. Doshi, Epigenetic aging in major depressive disorder: Clocks, mechanisms and therapeutic perspectives, *Eur. J. Pharmacol.*, 2024, **978**, 176757.
182. B. Feng, *et al.*, An epigenetic manifestation of alzheimer's disease: DNA methylation, *Actas Esp. Psiquiatr.*, 2024, **52**, 365–374.
183. Y. Rasmi, *et al.*, The role of DNA methylation in progression of neurological disorders and neurodegenerative diseases as well as the prospect of using DNA methylation inhibitors as therapeutic agents for such disorders, *IBRO Neurosci. Rep.*, 2023, **14**, 28–37.
184. J. Z. Li, N. Ramalingam and S. Li, Targeting epigenetic mechanisms in amyloid-β-mediated Alzheimer's pathophysiology: unveiling therapeutic potential, *Neural Regen. Res.*, 2025, **20**, 54–66.
185. J.-P. Issa, Aging and epigenetic drift: a vicious cycle, *J. Clin. Invest.*, 2014, **124**, 24–29.
186. S.-C. Wang, B. Oelze and A. Schumacher, Age-specific epigenetic drift in late-onset Alzheimer's disease, *PLoS One*, 2008, **3**, e2698.
187. H. Song, *et al.*, Epigenetic modification in Parkinson's disease, *Front. Cell Dev. Biol.*, 2023, **11**, 1123621.
188. D. Zhang, *et al.*, Targeting epigenetic modifications in Parkinson's disease therapy, *Med. Res. Rev.*, 2023, **43**, 1748–1777.
189. S. P. Sardi, J. M. Cedarbaum and P. Brundin, Targeted therapies for parkinson's disease: from genetics to the clinic, *Mov. Disord*, 2018, **33**, 684–696.
190. G. P. Bates, History of genetic disease: the molecular genetics of Huntington disease – a history, *Nat. Rev. Genet.*, 2005, **6**, 766–773.
191. R. Lardenoije, E. Pishva, K. Lunnon and D. L. van den Hove, Neuroepigenetics of Aging and Age-Related Neurodegenerative Disorders, *Prog. Mol. Biol. Transl. Sci.*, 2018, **158**, 49–82.
192. L. Francelle, C. Lotz, T. Outeiro, E. Brouillet and K. Merienne, Contribution of neuroepigenetics to huntington's disease, *Front. Hum. Neurosci.*, 2017, **11**, 17.

193. D. C. Joshi, *et al.*, The role of mitochondrial dysfunction in Huntington's disease: Implications for therapeutic targeting, *Biomed. Pharmacother.*, 2025, **183**, 117827.
194. S. Bassi, T. Tripathi, A. Monziani, F. Di Leva and M. Biagioli, Epigenetics of huntington's disease, *Adv. Exp. Med. Biol.*, 2017, **978**, 277–299.
195. A. Didonna and P. Opal, The promise and perils of HDAC inhibitors in neurodegeneration, *Ann. Clin. Transl. Neurol.*, 2015, **2**, 79–101.
196. L. P. Diniz, P. D. S. M. Pinheiro, L. S. Franco and F. C. A. Gomes, Hdacs in the brain: from chromatin remodeling to neurodegenerative disease, *Cells*, 2025, **14**, 1338.
197. H. A. Mai, C. M. Thomas, G. G. Nge and F. Elefant, Modulating Cognition-Linked Histone Acetyltransferases (HATs) as a Therapeutic Strategy for Neurodegenerative Diseases: Recent Advances and Future Trends, *Cells*, 2025, **14**, 873.
198. G. Sadri-Vakili, *et al.*, Histones associated with downregulated genes are hypo-acetylated in Huntington's disease models, *Hum. Mol. Genet.*, 2007, **16**, 1293–1306.
199. I. Sadeghi, *et al.*, Brain transcriptomic profiling reveals common alterations across neurodegenerative and psychiatric disorders, *Comput. Struct. Biotechnol. J.*, 2022, **20**, 4549–4561.
200. A. Arnatkeviciute, B. D. Fulcher, M. A. Bellgrove and A. Fornito, Imaging transcriptomics of brain disorders, *Biol. Psychiatry Glob. Open Sci.*, 2022, **2**, 319–331.
201. W.-T. Chen, *et al.*, Spatial transcriptomics and *in situ* sequencing to study alzheimer's disease, *Cell*, 2020, **182**, 976–991.e19.
202. Y. Tao, Q. Zhang, H. Wang, X. Yang and H. Mu, Alternative splicing and related RNA binding proteins in human health and disease, *Signal Transduct. Target. Ther.*, 2024, **9**, 26.
203. C.-W. Park, S.-M. Lee and K.-J. Yoon, Epitranscriptomic regulation of transcriptome plasticity in development and diseases of the brain, *BMB Rep.*, 2020, **53**, 551–564.
204. J. Mathoux, D. C. Henshall and G. P. Brennan, Regulatory mechanisms of the RNA modification m6a and significance in brain function in health and disease, *Front. Cell. Neurosci.*, 2021, **15**, 671932.
205. A. Adamu, S. Li, F. Gao and G. Xue, The role of neuroinflammation in neurodegenerative diseases: current understanding and future therapeutic targets, *Front. Aging Neurosci.*, 2024, **16**, 1347987.
206. P. Ciryam, *et al.*, A transcriptional signature of Alzheimer's disease is associated with a metastable subproteome at risk for aggregation, *Proc. Natl. Acad. Sci. U. S. A.*, 2016, **113**, 4753–4758.
207. M. Piwecka, N. Rajewsky and A. Rybak-Wolf, Single-cell and spatial transcriptomics: deciphering brain complexity in health and disease, *Nat. Rev. Neurol.*, 2023, **19**, 346–362.
208. S. Wei, *et al.*, Charting the spatial transcriptome of the human cerebral cortex at single-cell resolution, *Nat. Commun.*, 2025, **16**, 7702.

209. N. Schonrock and J. Götz, Decoding the non-coding RNAs in Alzheimer's disease, *Cell. Mol. Life Sci.*, 2012, **69**, 3543–3559.
210. A. J. AlShawaf, *et al.*, Autism spectrum disorder induced pluripotent stem cells display dysregulated calcium signaling during neural differentiation, *Cells*, 2025, **14**, 1402.
211. A. K. Sauer, J. E. Stanton, S. Hans and A. M. Grabrucker, Autism spectrum disorders: etiology and pathology, in *Autism Spectrum Disorders*, ed. A. M. Grabrucker, Exon Publications, 2021, DOI: 10.36255/exonpublications.autismspectrumdisorders.2021.etiology.
212. S. Guang, *et al.*, Synaptopathology involved in autism spectrum disorder, *Front. Cell. Neurosci.*, 2018, **12**, 470.
213. I. W. Riemersma, R. Havekes and M. J. H. Kas, Spatial and Temporal Gene Function Studies in Rodents: Towards Gene-Based Therapies for Autism Spectrum Disorder, *Genes*, 2021, **13**, 28.
214. M. Daghsni, *et al.*, Autism throughout genetics: Perusal of the implication of ion channels, *Brain Behav.*, 2018, **8**, e00978.
215. R. Moessner, *et al.*, Contribution of SHANK3 mutations to autism spectrum disorder, *Am. J. Hum. Genet.*, 2007, **81**, 1289–1297.
216. H.-C. Chuang, T.-N. Huang and Y.-P. Hsueh, T-Brain-1–A Potential Master Regulator in Autism Spectrum Disorders, *Autism Res.*, 2015, **8**, 412–426.
217. C. Hu, H. Li, J. Li, X. Luo and Y. Hao, Microglia: Synaptic modulator in autism spectrum disorder, *Front. Psychiatry*, 2022, **13**, 958661.
218. G. Fan, *et al.*, Microglia modulate neurodevelopment in autism spectrum disorder and schizophrenia, *Int. J. Mol. Sci.*, 2023, **24**, 17297.
219. A. Inamdar, B. Gurupadayya and H. Sharma, The role of glial cells in autism spectrum disorder: molecular mechanisms and therapeutic approaches, *CNS Neurol. Disord.: Drug Targets*, 2025, DOI: 10.2174/0118715273337007241115102118.
220. J. Nomura, *et al.*, ESC models of autism with copy-number variations reveal cell-type-specific translational vulnerability, *Cell Genomics*, 2025, **5**, 100877.
221. Y. Li, *et al.*, Global transcriptional and translational repression in human-embryonic-stem-cell-derived Rett syndrome neurons, *Cell Stem Cell*, 2013, **13**, 446–458.
222. N. E. Ali, *et al.*, Rett syndrome: advances in Understanding MeCP2 function, potential gene therapies, and public health implications, *Mol. Biol. Rep.*, 2025, **52**, 687.
223. F. Ehrhart, *et al.*, Rett syndrome - biological pathways leading from MECP2 to disorder phenotypes, *Orphanet J. Rare Dis.*, 2016, **11**, 158.
224. L. Abuhatzira, K. Makedonski, Y. Kaufman, A. Razin and R. Shemer, MeCP2 deficiency in the brain decreases BDNF levels by REST/CoREST-mediated repression and increases TRKB production, *Epigenetics*, 2007, **2**, 214–222.
225. Y. Liu, *et al.*, Exploring the complexity of MECP2 function in Rett syndrome, *Nat. Rev. Neurosci.*, 2025, **26**, 379–398.

226. S. Akbarian and H.-S. Huang, Molecular and cellular mechanisms of altered GAD1/GAD67 expression in schizophrenia and related disorders, *Brain Res. Rev.*, 2006, **52**, 293–304.
227. J. Hall and N. J. Bray, Schizophrenia genomics: convergence on synaptic development, adult synaptic plasticity, or both?, *Biol. Psychiatry*, 2022, **91**, 709–717.
228. O. D. Howes and E. C. Onwordi, The synaptic hypothesis of schizophrenia version III: a master mechanism, *Mol. Psychiatry*, 2023, **28**, 1843–1856.
229. D. Arion, *et al.*, Distinctive transcriptome alterations of prefrontal pyramidal neurons in schizophrenia and schizoaffective disorder, *Mol. Psychiatry*, 2015, **20**, 1397–1405.
230. X. Song, *et al.*, Transcriptomics analysis reveals shared pathways in peripheral blood mononuclear cells and brain tissues of patients with schizophrenia, *Front. Psychiatry*, 2021, **12**, 716722.
231. L. Bast *et al.*, Transcriptomic and genetic analysis suggests a role for mitochondrial dysregulation in schizophrenia, medRxiv, 2025, DOI: 10.1101/2025.03.14.25323827.
232. J. J. D. Bergstrom and M.-M. Fu, Dysregulation of myelination-related genes in schizophrenia, *J. Neurochem.*, 2024, **168**, 2227–2242.
233. Y. Hakak, *et al.*, Genome-wide expression analysis reveals dysregulation of myelination-related genes in chronic schizophrenia, *Proc. Natl. Acad. Sci. U. S. A.*, 2001, **98**, 4746–4751.
234. I. Chavarria-Siles, *et al.*, Myelination-related genes are associated with decreased white matter integrity in schizophrenia, *Eur. J. Hum. Genet.*, 2016, **24**, 381–386.
235. A. de Bartolomeis, *et al.*, Dysregulated signaling at postsynaptic density: A systematic review and translational appraisal for the pathophysiology, clinics, and antipsychotics' treatment of schizophrenia, *Cells*, 2023, **12**, 574.
236. W. J. Muir, B. S. Pickard and D. H. R. Blackwood, Disrupted-in-Schizophrenia-1, *Curr. Psychiatry Rep.*, 2008, **10**, 140–147.
237. S. J. Park, *et al.*, Disrupted-in-schizophrenia-1 (DISC1) Regulates Endoplasmic Reticulum Calcium Dynamics, *Sci. Rep.*, 2015, **5**, 8694.
238. A. S. Correia, A. Cardoso and N. Vale, BDNF unveiled: exploring its role in major depression disorder serotonergic imbalance and associated stress conditions, *Pharmaceutics*, 2023, **15**, 2081.
239. R. S. Duman and N. Li, A neurotrophic hypothesis of depression: role of synaptogenesis in the actions of NMDA receptor antagonists, *Philos. Trans. R. Soc., B*, 2012, **367**, 2475–2484.
240. B. R. Kouba, L. de Araujo Borba, P. Borges de Souza, J. Gil-Mohapel and A. L. S. Rodrigues, Role of inflammatory mechanisms in major depressive disorder: from etiology to potential pharmacological targets, *Cells*, 2024, **13**, 423.
241. Y. Zhan, *et al.*, Alterations of multiple peripheral inflammatory cytokine levels after repeated ketamine infusions in major depressive disorder, *Transl. Psychiatry*, 2020, **10**, 246.

242. A. Caspi, *et al.*, Influence of life stress on depression: moderation by a polymorphism in the 5-HTT gene, *Science*, 2003, **301**, 386–389.
243. R. Li, *et al.*, Genetically predicted circulating levels of glycine, glutamate, and serotonin in relation to the risks of three major neurodegenerative diseases: A Mendelian randomization analysis, *Front. Aging Neurosci.*, 2022, **14**, 938408.
244. C. Gao, *et al.*, Combination of spontaneous regional brain activity and HTR1A/1B DNA methylation to predict early responses to antidepressant treatments in MDD, *J. Affect. Disord.*, 2022, **302**, 249–257.
245. Y. Dwivedi, Emerging role of microRNAs in major depressive disorder: diagnosis and therapeutic implications, *Dialogues Clin. Neurosci.*, 2014, **16**, 43–61.
246. L. Zhou, Y. Zhu, W. Chen and Y. Tang, Emerging role of microRNAs in major depressive disorder and its implication on diagnosis and therapeutic response, *J. Affect. Disord.*, 2021, **286**, 80–86.
247. A. Ghosh and K. P. Giese, Calcium/calmodulin-dependent kinase II and Alzheimer's disease, *Mol. Brain*, 2015, **8**, 78.
248. S. Kumar, *et al.*, Synaptosome microRNAs regulate synapse functions in Alzheimer's disease, *npj Genom. Med.*, 2022, **7**, 47.
249. J. B. Williams, Q. Cao and Z. Yan, Transcriptomic analysis of human brains with Alzheimer's disease reveals the altered expression of synaptic genes linked to cognitive deficits, *Brain Commun.*, 2021, **3**, fcab123.
250. J.-V. Haure-Mirande, M. Audrain, M. E. Ehrlich and S. Gandy, Microglial TYROBP/DAP12 in Alzheimer's disease: Transduction of physiological and pathological signals across TREM2, *Mol. Neurodegener.*, 2022, **17**, 55.
251. L. Zhang, X. Xiang, Y. Li, G. Bu and X.-F. Chen, TREM2 and sTREM2 in Alzheimer's disease: from mechanisms to therapies, *Mol. Neurodegener.*, 2025, **20**, 43.
252. Z. Valiukas, K. Tangalakis, V. Apostolopoulos and J. Feehan, Microglial activation states and their implications for Alzheimer's Disease, *J. Prev. Alzheimers Dis.*, 2025, **12**, 100013.
253. J. E. Love, E. J. Hayden and T. T. Rohn, Alternative splicing in alzheimer's disease, *J. Parkinsons Dis. Alzheimers Dis.*, 2015, **2**, 6.
254. S. Buchholz and H. Zempel, The six brain-specific TAU isoforms and their role in Alzheimer's disease and related neurodegenerative dementia syndromes, *Alzheimers Dement.*, 2024, **20**, 3606–3628.
255. H. Mathys, *et al.*, Single-cell transcriptomic analysis of Alzheimer's disease, *Nature*, 2019, **570**, 332–337.
256. L. Chouliaras, *et al.*, Peripheral DNA methylation, cognitive decline and brain aging: pilot findings from the Whitehall II imaging study, *Epigenomics*, 2018, **10**, 585–595.
257. C. Bus, *et al.*, Human Dopaminergic Neurons Lacking PINK1 Exhibit Disrupted Dopamine Metabolism Related to Vitamin B6 Co-Factors, *iScience*, 2020, **23**, 101797.

258. G.-R. Luo, S. Chen and W.-D. Le, Are heat shock proteins therapeutic target for Parkinson's disease?, *Int. J. Biol. Sci.*, 2006, **3**, 20–26.
259. H. E. Moon and S. H. Paek, Mitochondrial dysfunction in parkinson's disease, *Exp. Neurobiol.*, 2015, **24**, 103–116.
260. L.-M. Zhang, *et al.*, Dopaminergic neuron injury in Parkinson's disease is mitigated by interfering lncRNA SNHG14 expression to regulate the miR-133b/ α-synuclein pathway, *Aging*, 2019, **11**, 9264–9279.
261. X. Zhang, *et al.*, Reduced Circulating Levels of miR-433 and miR-133b Are Potential Biomarkers for Parkinson's Disease, *Front. Cell. Neurosci.*, 2017, **11**, 170.
262. D. Li, C. S. McIntosh, F. L. Mastaglia, S. D. Wilton and M. T. Aung-Htut, Neurodegenerative diseases: a hotbed for splicing defects and the potential therapies, *Transl. Neurodegener.*, 2021, **10**, 16.
263. L. Lin, *et al.*, Transcriptome sequencing reveals aberrant alternative splicing in Huntington's disease, *Hum. Mol. Genet.*, 2016, **25**, 3454–3466.
264. A. Kumar, M. Vaish and R. R. Ratan, Transcriptional dysregulation in Huntington's disease: a failure of adaptive transcriptional homeostasis, *Drug Discovery Today*, 2014, **19**, 956–962.
265. J. Schilling, *et al.*, Deregulated Splicing Is a Major Mechanism of RNA-Induced Toxicity in Huntington's Disease, *J. Mol. Biol.*, 2019, **431**, 1869–1877.
266. J. W. Hyeon, A. H. Kim and H. Yano, Epigenetic regulation in Huntington's disease, *Neurochem. Int.*, 2021, **148**, 105074.
267. M. Tomljanović, C. H. Muflihah, D. Rajkovski and P. Mikulski, The epigenetic circle: feedback loops in the maintenance of cellular memory, *Epigenetics Chromatin*, 2025, **18**, 56.
268. R. L. Kan, J. Chen and T. Sallam, Crosstalk between epitranscriptomic and epigenetic mechanisms in gene regulation, *Trends Genet.*, 2022, **38**, 182–193.
269. J. Kim-Ha and Y.-J. Kim, Age-related epigenetic regulation in the brain and its role in neuronal diseases, *BMB Rep.*, 2016, **49**, 671–680.
270. J. Feng, S. Fouse and G. Fan, Epigenetic regulation of neural gene expression and neuronal function, *Pediatr. Res.*, 2007, **61**, 58R–63R.
271. M. F. Mehler, Epigenetic principles and mechanisms underlying nervous system functions in health and disease, *Prog. Neurobiol.*, 2008, **86**, 305–341.
272. B. G. Dias, S. Maddox, T. Klengel and K. J. Ressler, Epigenetic mechanisms underlying learning and the inheritance of learned behaviors, *Trends Neurosci.*, 2015, **38**, 96–107.
273. C. Li, K. Chen, X. Li and X. Xiong, Epitranscriptome-epigenome interactions in development and disease mechanisms, *Trends Genet.*, 2025, **41**, 691–705.
274. V. Miguel, S. Lamas and C. Espinosa-Diez, Role of non-coding-RNAs in response to environmental stressors and consequences on human health, *Redox Biol.*, 2020, **37**, 101580.

275. K. V. Good, J. B. Vincent and J. Ausió, Mecp2: the genetic driver of rett syndrome epigenetics, *Front. Genet.*, 2021, **12**, 620859.
276. D. Gökbuget and R. Blelloch, Epigenetic control of transcriptional regulation in pluripotency and early differentiation, *Development*, 2019, **146**, dev164772.
277. M. J. Friedman, H. Lee, J.-Y. Lee and S. Oh, Transcriptional and Epigenetic Regulation of Context-Dependent Plasticity in T-Helper Lineages, *Immune Netw.*, 2023, **23**, e5.
278. Y. M. Vuu, C.-T. Roberts and M. Rastegar, MeCP2 Is an Epigenetic Factor That Links DNA Methylation with Brain Metabolism, *Int. J. Mol. Sci.*, 2023, **24**, 4218.
279. S. Kaluscha, *et al.*, Evidence that direct inhibition of transcription factor binding is the prevailing mode of gene and repeat repression by DNA methylation, *Nat. Genet.*, 2022, **54**, 1895–1906.
280. H. Cedar, O. Sabag and Y. Reizel, The role of DNA methylation in genome-wide gene regulation during development, *Development*, 2022, **149**, dev200118.
281. H. Wang and K. Helin, Roles of H3K4 methylation in biology and disease, *Trends Cell Biol.*, 2025, **35**, 115–128.
282. E. Shen, H. Shulha, Z. Weng and S. Akbarian, Regulation of histone H3K4 methylation in brain development and disease, *Philos. Trans. R. Soc., B*, 2014, **369**, 20130514.
283. S. S. Nair and R. Kumar, Chromatin remodeling in cancer: a gateway to regulate gene transcription, *Mol. Oncol.*, 2012, **6**, 611–619.
284. N. R. Stone, *et al.*, Context-Specific Transcription Factor Functions Regulate Epigenomic and Transcriptional Dynamics during Cardiac Reprogramming, *Cell Stem Cell*, 2019, **25**, 87–102.e9.
285. D. Puri, H. Gala, R. Mishra and J. Dhawan, High-wire act: the poised genome and cellular memory, *FEBS J.*, 2015, **282**, 1675–1691.
286. J. Keifer, Primetime for learning genes, *Genes*, 2017, **8**, genes8020069.
287. M. Aziz, K. A. Jandeleit-Dahm and A. W. Khan, Interplay between epigenetic mechanisms and transcription factors in atherosclerosis, *Atherosclerosis*, 2024, **395**, 117615.
288. H. Jiang and G. Li, Transcription factors direct epigenetic reprogramming at specific loci in human cancers, *Front. Genet.*, 2023, **14**, 1234515.
289. M. Iwafuchi-Doi and K. S. Zaret, Pioneer transcription factors in cell reprogramming, *Genes Dev.*, 2014, **28**, 2679–2692.
290. G. Mirizio, S. Sampson and M. Iwafuchi, Interplay between pioneer transcription factors and epigenetic modifiers in cell reprogramming, *Regener. Ther.*, 2025, **28**, 246–252.
291. K. S. Zaret and J. S. Carroll, Pioneer transcription factors: establishing competence for gene expression, *Genes Dev.*, 2011, **25**, 2227–2241.
292. C. C. Bell, *et al.*, Comparative cofactor screens show the influence of transactivation domains and core promoters on the mechanisms of transcription, *Nat. Genet.*, 2024, **56**, 1181–1192.

293. T. Ahsendorf, F.-J. Müller, V. Topkar, J. Gunawardena and R. Eils, Transcription factors, coregulators, and epigenetic marks are linearly correlated and highly redundant, *PLoS One*, 2017, **12**, e0186324.
294. A. Mayran and J. Drouin, Pioneer transcription factors shape the epigenetic landscape, *J. Biol. Chem.*, 2018, **293**, 13795–13804.
295. V. Sartorelli and P. L. Puri, Shaping Gene Expression by Landscaping Chromatin Architecture: Lessons from a Master, *Mol. Cell*, 2018, **71**, 375–388.
296. C. Dai, *et al.*, Epigenetic and epitranscriptomic role of lncRNA in carcinogenesis (Review), *Int. J. Oncol.*, 2025, **66**, 29.
297. E. Larriba and J. del Mazo, Role of Non-Coding RNAs in the Transgenerational Epigenetic Transmission of the Effects of Reprotoxicants, *Int. J. Mol. Sci.*, 2016, **17**, 452.
298. J. O'Brien, H. Hayder, Y. Zayed and C. Peng, Overview of microrna biogenesis, mechanisms of actions, and circulation, *Front. Endocrinol.*, 2018, **9**, 402.
299. K. M. T. Arif, E. K. Elliott, L. M. Haupt and L. R. Griffiths, Regulatory Mechanisms of Epigenetic miRNA Relationships in Human Cancer and Potential as Therapeutic Targets, *Cancers*, 2020, **12**, 2922.
300. D. Segal and J. Dostie, The Talented LncRNAs: Meshing into Transcriptional Regulatory Networks in Cancer, *Cancers*, 2023, **15**, 3433.
301. X. Chen, *et al.*, Long noncoding RNA: multiple players in gene expression, *BMB Rep.*, 2018, **51**, 280–289.
302. J. S. Dasen, Long noncoding RNAs in development: solidifying the Lncs to Hox gene regulation, *Cell Rep.*, 2013, **5**, 1–2.
303. P. M. Zakutansky and Y. Feng, The Long Non-Coding RNA GOMAFU in Schizophrenia: Function, Disease Risk, and Beyond, *Cells*, 2022, **11**, 1949.
304. X. Zhang, *et al.*, Mechanisms and Functions of Long Non-Coding RNAs at Multiple Regulatory Levels, *Int. J. Mol. Sci.*, 2019, **20**, 5573.
305. H. C. Beird, *et al.*, Epigenomic and Transcriptomic Profiling of Solitary Fibrous Tumors Identifies Site-Specific Patterns and Candidate Genes Regulated by DNA Methylation, *Lab. Invest.*, 2024, **104**, 102146.
306. D. Deshpande, *et al.*, RNA-seq data science: From raw data to effective interpretation, *Front. Genet.*, 2023, **14**, 997383.
307. Z.-J. Cao and G. Gao, Multi-omics single-cell data integration and regulatory inference with graph-linked embedding, *Nat. Biotechnol.*, 2022, **40**, 1458–1466.
308. E. Athieniti and G. M. Spyrou, A guide to multi-omics data collection and integration for translational medicine, *Comput. Struct. Biotechnol. J.*, 2023, **21**, 134–149.
309. K. Yang, *et al.*, A network-based machine-learning framework to identify both functional modules and disease genes, *Hum. Genet.*, 2021, **140**, 897–913.
310. D. Kim, *et al.*, Gene regulatory network reconstruction: harnessing the power of single-cell multi-omic data, *NPJ Syst. Biol. Appl.*, 2023, **9**, 51.

311. M. De Vivo, M. Masetti, G. Bottegoni and A. Cavalli, Role of molecular dynamics and related methods in drug discovery, *J. Med. Chem.*, 2016, **59**, 4035–4061.
312. P. Srivastava and A. Tiwari, Critical role of computer simulations in drug discovery and development, *Curr. Top. Med. Chem.*, 2017, **17**, 2422–2432.
313. K. Li, M. Ouyang, J. Zhan and R. Tian, CRISPR-based functional genomics screening in human-pluripotent-stem-cell-derived cell types, *Cell Genomics*, 2023, **3**, 100300.
314. C. Couldrey and V. Cave, Assessing DNA methylation levels in animals: choosing the right tool for the job, *Anim. Genet.*, 2014, **45**(Suppl 1), 15–24.
315. X. Qian, *et al.*, Spatial transcriptomics reveals human cortical layer and area specification, *Nature*, 2025, **644**, 153–163.
316. X. Qian *et al.,* Spatial Single-cell Analysis Decodes Cortical Layer and Area Specification, *bioRxiv*, 2024, preprint, DOI: 10.1101/2024.06.05.597673.
317. D. B. Gutierrez, *et al.*, An Integrated, High-Throughput Strategy for Multiomic Systems Level Analysis, *J. Proteome Res.*, 2018, **17**, 3396–3408.
318. C. Wang, *et al.*, Microfluidic Biochips for Single-Cell Isolation and Single-Cell Analysis of Multiomics and Exosomes, *Adv. Sci.*, 2024, **11**, e2401263.
319. X. S. Liu, *et al.*, Editing DNA methylation in the mammalian genome, *Cell*, 2016, **167**, 233–247.e17.
320. J. Pulecio, N. Verma, E. Mejía-Ramírez, D. Huangfu and A. Raya, CRISPR/Cas9-Based Engineering of the Epigenome, *Cell Stem Cell*, 2017, **21**, 431–447.
321. L. Villiger, *et al.*, CRISPR technologies for genome, epigenome and transcriptome editing, *Nat. Rev. Mol. Cell Biol.*, 2024, **25**, 464–487.
322. M. Swingler, M. Donadoni, A. Bellizzi, S. Cakir and I. K. Sariyer, iPSC-derived three-dimensional brain organoid models and neurotropic viral infections, *J. Neurovirol.*, 2023, **29**, 121–134.
323. J. Cerneckis, H. Cai and Y. Shi, Induced pluripotent stem cells (iPSCs): molecular mechanisms of induction and applications, *Signal Transduct. Target. Ther.*, 2024, **9**, 112.
324. C.-L. Lo, S. R. Choudhury, J. Irudayaraj and F. C. Zhou, Epigenetic editing of ascl1 gene in neural stem cells by optogenetics, *Sci. Rep.*, 2017, **7**, 42047.
325. I. Faravelli, N. Antón-Bolaños, J. R. Brown and P. Arlotta, Brain organoids: tools for understanding the uniqueness and individual variability of the human brain, *Annu. Rev. Genomics Hum. Genet.*, 2025, **26**, 299–320.
326. G. Jin and B. Lim, Epigenome editing and epigenetic gene regulation in disease phenotypes, *Korean J. Chem. Eng.*, 2022, **39**, 1361–1367.
327. Y. Chen, L. Chen, A. T. L. Lun, P. L. Baldoni and G. K. Smyth, edgeR v4: powerful differential analysis of sequencing data with expanded

functionality and improved support for small counts and larger datasets, *Nucleic Acids Res.*, 2025, **53**, gkaf018.
328. V. Paton, *et al.*, Assessing the impact of transcriptomics data analysis pipelines on downstream functional enrichment results, *Nucleic Acids Res.*, 2024, **52**, 8100–8111.
329. Z. Liu and S.-Y. Song, Genomic and transcriptomic approaches advance the diagnosis and prognosis of neurodegenerative diseases, *Genes*, 2025, **16**, 135.
330. R. Mallick and A. K. Duttaroy, Epigenetic modification impacting brain functions: Effects of physical activity, micronutrients, caffeine, toxins, and addictive substances, *Neurochem. Int.*, 2023, **171**, 105627.
331. A. G. Kazantsev and S. M. Hersch, Drug targeting of dysregulated transcription in Huntington's disease, *Prog. Neurobiol.*, 2007, **83**(4), 249–259.
332. A. A. A. Aljabali, *et al.*, Deciphering depression: epigenetic mechanisms and treatment strategies, *Biology*, 2024, **13**, 638.
333. C. Zhou, *et al.*, Role of histone deacetylase inhibitors in non-neoplastic diseases, *Heliyon*, 2024, **10**, e33997.
334. K. Ranganna, C. Selvam, A. Shivachar and Z. Yousefipour, Histone Deacetylase Inhibitors as Multitarget-Directed Epi-Drugs in Blocking PI3K Oncogenic Signaling: A Polypharmacology Approach, *Int. J. Mol. Sci.*, 2020, **21**, 8198.
335. A. K. Giri and T. Aittokallio, DNMT inhibitors increase methylation in the cancer genome, *Front. Pharmacol.*, 2019, **10**, 385.
336. S. Meng, *et al.*, BET inhibitor JQ1 blocks inflammation and bone destruction, *J. Dent. Res.*, 2014, **93**, 657–662.
337. K. Anthony, RNA-based therapeutics for neurological diseases, *RNA Biol.*, 2022, **19**, 176–190.
338. S. McDowall, M. Aung-Htut, S. Wilton and D. Li, Antisense oligonucleotides and their applications in rare neurological diseases, *Front. Neurosci.*, 2024, **18**, 1414658.
339. J.-B. Brunet de Courssou, A. Durr, D. Adams, J.-C. Corvol and L.-L. Mariani, Antisense therapies in neurological diseases, *Brain*, 2022, **145**, 816–831.
340. R. Mejzini, *et al.*, Allele-Selective Thiomorpholino Antisense Oligonucleotides as a Therapeutic Approach for Fused-in-Sarcoma Amyotrophic Lateral Sclerosis, *Int. J. Mol. Sci.*, 2024, **25**, 8495.
341. X. Liu, *et al.*, Circular RNA: An emerging frontier in RNA therapeutic targets, RNA therapeutics, and mRNA vaccines, *J. Controlled Release*, 2022, **348**, 84–94.
342. G. V. Roth, I. R. Gengaro and L. S. Qi, Precision epigenetic editing: Technological advances, enduring challenges, and therapeutic applications, *Cell Chem. Biol.*, 2024, DOI: 10.1016/j.chembiol.2024.07.007.
343. V. Tadić, G. Josipović, V. Zoldoš and A. Vojta, CRISPR/Cas9-based epigenome editing: An overview of dCas9-based tools with special emphasis on off-target activity, *Methods*, 2019, **164–165**, 109–119.

344. H. Grezenko, *et al.*, Epigenetics in neurological and psychiatric disorders: A comprehensive review of current understanding and future perspectives, *Cureus*, 2023, **15**, e43960.
345. J. Cotney, *et al.*, The autism-associated chromatin modifier CHD8 regulates other autism risk genes during human neurodevelopment, *Nat. Commun.*, 2015, **6**, 6404.
346. M. Apte and A. Kumar, Correlation of mutated gene and signalling pathways in ASD, *IBRO Neurosci. Rep.*, 2023, **14**, 384–392.
347. Y. Zhu, L. Zhu, X. Wang and H. Jin, RNA-based therapeutics: an overview and prospectus, *Cell Death Dis.*, 2022, **13**, 644.
348. G. Molla and M. Bitew, Revolutionizing Personalized Medicine: Synergy with Multi-Omics Data Generation, Main Hurdles, and Future Perspectives, *Biomedicines*, 2024, **12**, 2750.
349. R. C. Wang and Z. Wang, Precision medicine: disease subtyping and tailored treatment, *Cancers*, 2023, **15**, 3837.
350. M. Jamalinia and R. Weiskirchen, Advances in personalized medicine: translating genomic insights into targeted therapies for cancer treatment, *Ann. Transl. Med.*, 2025, **13**, 18.
351. S. Jiang, L. Postovit, A. Cattaneo, E. B. Binder and K. J. Aitchison, Epigenetic modifications in stress response genes associated with childhood trauma, *Front. Psychiatry*, 2019, **10**, 808.
352. C. Young and D. MacDougall, *An overview of pharmacogenomic testing for psychiatric disorders: CADTH horizon scan*, Canadian Agency for Drugs and Technologies in Health, 2023.
353. C. A. Wall, P. E. Croarkin, C. Swintak and B. A. Koplin, Psychiatric pharmacogenomics in pediatric psychopharmacology, *Child Adolesc. Psychiatr. Clin. N. Am.*, 2012, **21**, 773–788.
354. M. Lisek, N. Bochenska, J. Tomczak, J. Duraj and T. Boczek, Epigenetic regulation in ischemic neuroprotection: the dual role of hdacs and hats in neuroinflammation and recovery, *Antioxidants*, 2025, **14**, 1015.
355. H. Wang, J. Xu, P. Lazarovici, R. Quirion and W. Zheng, cAMP Response Element-Binding Protein (CREB): A Possible Signaling Molecule Link in the Pathophysiology of Schizophrenia, *Front. Mol. Neurosci.*, 2018, **11**, 255.
356. F. Wang, L. Cheng and X. Zhang, Reprogramming Glial Cells into Functional Neurons for Neuro-regeneration: Challenges and Promise, *Neurosci. Bull.*, 2021, **37**, 1625–1636.
357. L. Vasan, E. Park, L. A. David, T. Fleming and C. Schuurmans, Direct neuronal reprogramming: bridging the gap between basic science and clinical application, *Front. Cell Dev. Biol.*, 2021, **9**, 681087.
358. M. Fallet, *et al.*, Present and future challenges for the investigation of transgenerational epigenetic inheritance, *Environ. Int.*, 2023, **172**, 107776.
359. V. Labrie, S. Pai and A. Petronis, Epigenetics of major psychosis: progress, problems and perspectives, *Trends Genet.*, 2012, **28**, 427–435.
360. L. Prokopuk, P. S. Western and J. M. Stringer, Transgenerational epigenetic inheritance: adaptation through the germline epigenome?, *Epigenomics*, 2015, **7**, 829–846.

361. Y.-L. Tan, Y. Yuan and L. Tian, Microglial regional heterogeneity and its role in the brain, *Mol. Psychiatry*, 2020, **25**, 351–367.
362. A. Rao, D. Barkley, G. S. França and I. Yanai, Exploring tissue architecture using spatial transcriptomics, *Nature*, 2021, **596**, 211–220.
363. S. M. Sheen-Chen, Laparoscopic antegrade spinchterotomy for the management of complex choledocholithiasis, *Ann. Surg.*, 1996, **224**, 236.
364. C. Tiffon, The impact of nutrition and environmental epigenetics on human health and disease, *Int. J. Mol. Sci.*, 2018, **19**, 3425.
365. M. Machnik and U. Oleksiewicz, Dynamic signatures of the epigenome: friend or foe?, *Cells*, 2020, **9**, 653.
366. C. E. McKinney, Using induced pluripotent stem cells derived neurons to model brain diseases, *Neural Regen. Res.*, 2017, **12**, 1062–1067.
367. A. E. Mohr, C. P. Ortega-Santos, C. M. Whisner, J. Klein-Seetharaman and P. Jasbi, Navigating Challenges and Opportunities in Multi-Omics Integration for Personalized Healthcare, *Biomedicines*, 2024, **12**, 1496.
368. J. S. Mattick, *et al.*, Long non-coding RNAs: definitions, functions, challenges and recommendations, *Nat. Rev. Mol. Cell Biol.*, 2023, **24**, 430–447.
369. A. Frydas, E. Wauters, J. van der Zee and C. Van Broeckhoven, Uncovering the impact of noncoding variants in neurodegenerative brain diseases, *Trends Genet.*, 2022, **38**, 258–272.
370. P. Lu, *et al.*, Systematic characterization of full-length RNA isoforms in human colorectal cancer at single-cell resolution, *Protein Cell*, 2025, **16**(10), 873–895.
371. P. Aramillo Irizar, *et al.*, Transcriptomic alterations during ageing reflect the shift from cancer to degenerative diseases in the elderly, *Nat. Commun.*, 2018, **9**, 327.
372. R. Argelaguet, *et al.*, MOFA+: a statistical framework for comprehensive integration of multi-modal single-cell data, *Genome Biol.*, 2020, **21**, 111.
373. I. Korsunsky, *et al.*, Fast, sensitive and accurate integration of single-cell data with Harmony, *Nat. Methods*, 2019, **16**, 1289–1296.
374. T. Feehley, C. W. O'Donnell, J. Mendlein, M. Karande and T. McCauley, Drugging the epigenome in the age of precision medicine, *Clin. Epigenetics*, 2023, **15**, 6.
375. R. Prabhakaran, *et al.*, Epigenetic frontiers: miRNAs, long non-coding RNAs and nanomaterials are pioneering to cancer therapy, *Epigenetics Chromatin*, 2024, **17**, 31.
376. O. Pich, *et al.*, The translational challenges of precision oncology, *Cancer Cell*, 2022, **40**, 458–478.

CHAPTER 8

Biomarker and Precision Medicine in Neurology

SYED MOHAMMAD NASAR ATA,*[a] NITISH KUMAR SINGH,[b] MAYANK GANGWAR,[c] JAIKEE KUMAR SINGH[d] AND MUKESH KUMAR[e]

[a] Department of Neurology, Henry Ford Health System, USA; [b] Department of Biosciences, Manipal University Jaipur, Rajasthan, India; [c] Clinical Studies and Trials Unit, Division of Development/Indian Council of Medical Research, Department of Health Research (Min. of Health & FW), India; [d] Department of Bioinformatics, Faculty of Engineering and Technology, Marwadi University, Rajkot Gujarat, India; [e] Integrative Neurochemistry Laboratory, Harvard Medical School, McLean Hospital, USA

*Email: sata1@hfhs.org

8.1 Introduction

Biomarkers have emerged as a cornerstone in the advancement of precision medicine, especially within the field of neurology, where complex and multifactorial disorders such as Alzheimer's disease, Parkinson's disease, multiple sclerosis, amyotrophic lateral sclerosis, and various neuropsychiatric and cerebrovascular conditions present significant diagnostic and therapeutic challenges. In the simplest terms, a biomarker, or biological marker, is a measurable indicator of a normal biological process, a pathogenic process, or a pharmacologic response to a therapeutic intervention. Biomarkers can take various forms, including molecular (such as proteins, genes, or metabolites), cellular, imaging, or physiological indicators that provide objective data on the state of a disease or health condition.[1] In neurology, biomarkers are of

Chemical Biology No. 26
Chemical Signatures of the Brain: Bridging Biochemistry, Neuroanatomy, and Brain Disorders
Edited by Sachchida Nand Rai and Abhishek Kumar Singh

Published by the Royal Society of Chemistry, www.rsc.org

paramount importance because they enable clinicians and researchers to transcend traditional symptom-based diagnosis by offering molecular and structural insights into disease mechanisms. The importance of biomarkers lies in their ability to facilitate early and accurate diagnosis, monitor disease progression, predict therapeutic responses, and evaluate treatment efficacy, thereby enabling a shift from generalized to personalized and predictive medical care. Neurological disorders are often progressive and characterized by overlapping clinical manifestations, making timely and precise identification of their underlying pathology difficult.[2] Therefore, validated biomarkers act as vital tools in differentiating between disease subtypes, identifying pre-symptomatic stages, and improving the stratification of patients for targeted interventions. The role of biomarkers in neurology extends across the entire spectrum of disease management from early detection and differential diagnosis to prognosis, monitoring, and treatment optimization.[3] In neurodegenerative diseases such as Alzheimer's disease (AD), biomarkers have transformed understanding and clinical evaluation. Core AD biomarkers include amyloid-beta (Aβ) peptides, total tau (t-tau), and phosphorylated tau (p-tau) measured in cerebrospinal fluid (CSF) or detected through positron emission tomography (PET) imaging.[4] These biomarkers not only reflect the fundamental neuropathological processes of amyloid deposition and neurofibrillary tangle formation but also allow for the identification of at-risk individuals decades before clinical symptoms emerge. Similarly, in Parkinson's disease (PD), biomarkers such as alpha-synuclein levels, dopamine transporter imaging, and various neuroinflammatory and mitochondrial markers are being explored to define disease stages and predict motor and non-motor progression.[5] In multiple sclerosis (MS), neurofilament light chain (NfL) serves as a sensitive biomarker for axonal injury and disease activity, correlating with both magnetic resonance imaging (MRI) findings and clinical relapses. Beyond these classical examples, biomarkers are integral in stroke diagnosis and management, where imaging markers like diffusion-weighted MRI aid in determining infarct core and penumbra, while molecular biomarkers like matrix metalloproteinases or inflammatory cytokines assist in assessing prognosis and therapeutic windows.[6] In epilepsy, biomarkers may help distinguish between epileptic and non-epileptic seizures, while in psychiatric and neurodevelopmental conditions, such as schizophrenia and autism spectrum disorder, emerging genomic, transcriptomic, and metabolomic biomarkers are paving the way for molecular classification of previously heterogeneous disorders.[7] Furthermore, biomarkers serve as valuable pharmacodynamic and predictive tools, guiding the use of disease-modifying therapies and evaluating drug responses in clinical trials, thus reducing time, cost, and uncertainty in drug development.

8.2 Evolution of Neurological Biomarkers in Clinical Practice

The evolution of biomarkers in neurology reflects a paradigm shift from conventional, symptom-driven diagnostics toward a precision medicine framework grounded in molecular characterization, technological innovation, and data integration.[8] Early neurological practice relied heavily on clinical observation and neuroimaging to define disorders, but with the advent of molecular biology, proteomics, genomics, metabolomics, and advanced imaging modalities, biomarker discovery and validation have accelerated dramatically. In the early 2000s, CSF-based biomarkers for Alzheimer's disease marked the first breakthrough in neurodiagnostic applications, leading to the establishment of diagnostic criteria incorporating biomarker evidence alongside cognitive assessment.[9] Subsequently, neuroimaging biomarkers such as PET tracers for amyloid and tau, as well as structural and functional MRI, enhanced the *in vivo* visualization of neuropathology, allowing researchers to map disease progression with unprecedented accuracy. Advances in high-throughput omics technologies have further expanded the biomarker landscape, enabling the simultaneous analysis of thousands of molecules to identify specific patterns associated with neurological diseases.[10] For example, transcriptomic profiling of blood and brain tissues has revealed gene expression signatures predictive of neuroinflammation, while metabolomic studies have uncovered metabolic alterations preceding neurodegeneration. In parallel, the integration of artificial intelligence (AI) and machine learning algorithms into biomarker research has revolutionized data analysis, enabling the synthesis of multimodal data—from imaging, genetics, electrophysiology, and clinical metrics —into predictive models that can forecast disease onset and progression. The rise of liquid biopsy techniques, particularly those detecting circulating microRNAs, exosomes, and cell-free DNA in blood, offers a non-invasive and scalable alternative to traditional CSF sampling, thus enhancing accessibility in clinical settings. Moreover, large-scale collaborative initiatives, such as the Alzheimer's Disease Neuroimaging Initiative (ADNI), Parkinson's Progression Markers Initiative (PPMI), and the Human Brain Project, have standardized biomarker protocols and data sharing, fostered reproducibility and accelerated translation from bench to bedside. In clinical practice today, neurological biomarkers play a pivotal role in implementing precision medicine—a patient-centered approach that tailor's diagnosis and treatment based on individual molecular and phenotypic profiles.[11] Precision neurology integrates biomarker data with clinical, genetic, and environmental factors to predict disease trajectories and select personalized therapeutic strategies. For instance, stratifying Alzheimer's patients based on amyloid and tau status enables targeted therapeutic trials for anti-amyloid or anti-tau agents, while identifying specific genetic mutations such as leucine-rich repeat kinase 2 (LRRK2) or glucocerebrosidase (GBA) in Parkinson's disease patients informs both prognosis and therapeutic selection. In multiple sclerosis, monitoring serum NfL levels allows real-time assessment of neuronal injury and guides therapeutic adjustments.[12] Furthermore, in stroke management, biomarkers combined with imaging criteria enhance decision-making for thrombolytic and endovascular interventions. Beyond diagnostics, biomarkers are instrumental in understanding disease heterogeneity, elucidating mechanisms of neurodegeneration, and identifying novel therapeutic targets. They are also critical in clinical trials, where they serve as surrogate endpoints, reducing dependence on long-term clinical outcomes. The ultimate vision of biomarker-guided precision neurology is to shift from a reactive to a proactive model, where diseases are detected before symptoms appear, interventions are personalized, and outcomes are continuously monitored through measurable biological indicators. However, challenges remain in the validation, standardization, and translation of biomarkers into routine care.[13] Variability in sample collection, assay sensitivity, population diversity, and ethical considerations in genetic testing pose ongoing barriers. Furthermore, neurological diseases are often multifactorial, requiring composite biomarker panels rather than single markers for accurate representation of complex pathophysiology. Nevertheless, continuous advancements in systems biology, neuroinformatics, and longitudinal cohort studies are addressing these limitations, promising a future where biomarker-driven precision medicine becomes the foundation of neurology.[14] In summary, biomarkers are redefining the landscape of neurological diagnosis and treatment. Their integration with precision medicine offers unparalleled potential to unravel disease mechanisms, enhance diagnostic accuracy, optimize therapeutic efficacy, and ultimately improve patient outcomes. As the field evolves, the synergy between technological innovation, molecular neuroscience, and clinical practice will continue to propel biomarker research from discovery to direct patient benefit, heralding a new era in the personalized management of neurological disorders.[15]

8.3 Types of Biomarkers in Neurology

Biomarkers in neurology are broadly classified based on their function and clinical application. These include diagnostic biomarkers, which help in disease detection and differentiation (*e.g.*, amyloid-beta and tau in Alzheimer's disease); prognostic biomarkers, which predict disease progression; predictive biomarkers, indicating response to specific therapies; pharmacodynamic or response biomarkers, which assess treatment effects; and monitor biomarkers, which track disease activity over time.[16] Additionally, susceptibility or risk biomarkers identify individuals predisposed to developing neurological disorders. Biomarkers may also be categorized by modality, such as molecular, imaging, genetic, proteomic, or metabolomic biomarkers, each offering distinct insights into neurological function and pathology.[17]

8.3.1 Genetic Biomarkers

Genetic biomarkers have become indispensable tools in modern neurology, enabling the identification, diagnosis, and management of a wide array of neurological disorders that are influenced by genetic alterations. They are defined as specific DNA or RNA sequences, mutations, polymorphisms, or expression patterns that indicate the presence or risk of a disease, its progression, or therapeutic responsiveness.[18] In the context of neurology, genetic biomarkers provide molecular insights into the hereditary basis and pathogenic mechanisms underlying complex disorders of the brain and nervous system. With the advent of next-generation sequencing, genome-wide association studies (GWAS), and transcriptomic profiling, researchers can now detect subtle variations and rare mutations that contribute to neurodegeneration, neurodevelopmental anomalies, and neuroinflammatory diseases.[19] These biomarkers not only facilitate early and accurate diagnosis but also guide personalized therapeutic approaches by predicting individual responses to drugs, thereby embodying the principles of precision medicine in neurology.[20] Gene mutations are among the most studied genetic biomarkers in neurology, as they directly influence the onset, severity, and progression of many neurological disorders. A mutation may involve a single nucleotide substitution (point mutation), deletion, insertion, duplication, or chromosomal rearrangement that disrupts normal gene function. In several neurodegenerative diseases, mutations lead to protein misfolding, aggregation, and neuronal toxicity. For instance, in Alzheimer's disease (AD), mutations in the APP (amyloid precursor protein), PSEN1, and PSEN2 (presenilin 1 and 2) genes are associated with early-onset familial AD. These mutations enhance amyloid-beta (Aβ) peptide production and aggregation, a hallmark of AD pathology. Similarly, in Parkinson's disease (PD), mutations in the SNCA gene encoding alpha-synuclein, LRRK2 (leucine-rich repeat kinase 2), PINK1, and DJ-1 genes contribute to mitochondrial dysfunction, oxidative stress, and neuronal degeneration.[21] The LRRK2 G2019S mutation, for example, is one of the most common genetic causes of autosomal dominant PD, while PTEN-induced kinase 1 (PINK1) and parkin RBR E3 ubiquitin protein ligase (PARK2) mutations are linked to early-onset forms. In Huntington's disease (HD), a classic example of a trinucleotide repeat disorder, an expansion of CAG repeats in the HTT gene leads to abnormal huntingtin protein aggregation, resulting in progressive motor dysfunction, cognitive decline, and psychiatric disturbances. Similarly, amyotrophic lateral sclerosis (ALS), a fatal motor neuron disease, is associated with mutations in genes like Superoxide Dismutase 1 (SOD1), Chromosome 9 Open Reading Frame 72 (C9ORF72), FUsed in Sarcoma (FUS), and TAR DNA-Binding Protein (TARDBP), which cause protein misfolding, RNA metabolism defects, and neurotoxicity.[22] The C9ORF72 hexanucleotide repeat expansion is the most frequent genetic cause of familial and sporadic ALS, also implicated in frontotemporal dementia (FTD), highlighting the overlapping genetic basis of neurodegenerative diseases. Beyond these, mutations in the Methyl-CpG binding protein 2 (MECP2) gene cause Rett syndrome, a neurodevelopmental disorder primarily affecting females, characterized by loss of acquired motor and communication skills. In multiple sclerosis (MS), although not caused by a single mutation, polymorphisms in genes like Human leukocyte antigen-DR beta 1 (HLA-DRB1) have been identified as strong genetic risk factors. Similarly, mutations in ion channel genes such as sodium channel alpha subunit 1 (SCN1A) are linked to epileptic encephalopathies, whereas mutations in Calcium voltage-gated channel subunit alpha1 A (CACNA1A) and ATPase Na^+/K^+ transporting subunit alpha 2 (ATP1A2) genes are associated with familial hemiplegic migraine.[23] Mitochondrial gene mutations also play a crucial role in neurological disorders such as mitochondrial encephalomyopathy, lactic acidosis, and stroke-like episodes (MELAS), where mutations in mitochondrial DNA compromise energy metabolism.[24] These discoveries underscore how genetic biomarkers not only clarify disease mechanisms but also offer diagnostic and prognostic value. For example, identifying PSEN1 mutations can confirm familial Alzheimer's diagnosis before symptoms appear, while genetic testing for C9ORF72 or SOD1 mutations aids in counseling ALS patients and their families.[25] Moreover, the development of gene-targeted therapies—such as antisense oligonucleotides (ASOs) for SOD1 or HTT mutations—demonstrates how genetic biomarkers can guide the design of disease-modifying interventions, marking a major step toward precision neurology. Genetic predisposition refers to the inherited susceptibility or increased likelihood of developing a particular neurological condition due to specific genetic variants. Unlike monogenic disorders driven by a single mutation, most common neurological diseases such as Alzheimer's, Parkinson's, multiple sclerosis, and stroke have polygenic origins, meaning they result from the combined effects of multiple genes interacting with environmental factors.[26] Genetic biomarkers associated with predisposition often involve single-nucleotide polymorphisms (SNPs) that subtly influence gene expression, protein function, or metabolic pathways. In Alzheimer's disease, the Apolipoprotein E (APOE) gene stands as the most significant genetic risk factor for late-onset forms. The APOE ε4 allele increases amyloid-beta deposition and neuroinflammation, substantially elevating the risk and lowering the age of onset compared to ε2 or ε3 alleles. Similarly, in Parkinson's disease, carriers of the LRRK2 G2019S or GBA mutations exhibit higher susceptibility, with GBA mutations also linking PD to lysosomal dysfunction.[27] Genome-wide studies have identified numerous PD susceptibility loci, including microtubule-associated protein tau (MAPT) and Bone marrow stromal cell antigen 1 (BST1), emphasizing the polygenic nature of disease risk. In multiple sclerosis, genetic predisposition is strongly associated with specific human leukocyte antigen (HLA) alleles, particularly Human leukocyte antigen-DR beta 1 (HLA-DRB1) and Human leukocyte antigen-DQ beta 1 (HLA-DQB1), which modulate immune system responses. These genetic variants increase the risk of autoimmune demyelination in response to environmental triggers like viral infections or vitamin D deficiency. Similarly, migraine and epilepsy display hereditary patterns where ion channel and neurotransmitter-related gene polymorphisms contribute to neuronal hyperexcitability. For instance, CACNA1A mutations in calcium channels predispose individuals to familial hemiplegic migraine, while SCN1A and GABRG2 variants are implicated in generalized epilepsy syndromes.[28] Genetic predisposition also increases stroke risk through polymorphisms in Apolipoprotein E Angiotensinogen Angiotensin-Converting Enzyme Coagulation Factor V and Methylenetetrahydrofolate Reductase genes which affect lipid metabolism, blood pressure and coagulation. In neurodevelopmental disorders, variants in synaptic genes such as SH3 and Multiple Ankyrin Repeat Domains Protein 3 (SHANK3), Neurexin 1 (NRXN1), and Contactin-associated protein-like 2 (CNTNAP2) predispose individuals to autism spectrum disorders and intellectual disabilities by disrupting synaptic connectivity. Schizophrenia and bipolar disorder are similarly influenced by complex genetic networks involving neurotransmitter regulation, calcium signaling, and immune pathways. The clinical utility of identifying genetic predispositions lies in risk stratification, preventive strategies, and personalized interventions. For example, individuals carrying the APOE ε4 allele may benefit from early lifestyle modifications, cognitive monitoring, and participation in preventive clinical trials. In Parkinson's, identifying GBA or LRRK2 mutation carriers enables targeted surveillance and inclusion in genotype-specific therapeutic studies.[29] Moreover, integrating genetic predisposition data with other biomarkers—such as imaging, proteomic, or metabolomic signatures—enhances predictive accuracy and helps define disease subtypes. Recent advances in polygenic risk scoring (PRS) and machine learning models have made it possible to estimate an individual's cumulative genetic risk by aggregating data from thousands of SNPs. This approach is particularly promising in predicting complex neurological diseases where no single mutation predominates. Combined with environmental and lifestyle data, PRS supports a holistic precision medicine framework that identifies at-risk populations long before clinical onset. In conclusion, genetic biomarkers, encompassing both disease-causing mutations and predispositional variants, are revolutionizing our understanding of neurological diseases.[30] They not only uncover molecular mechanisms of pathogenesis but also enable earlier diagnosis, patient stratification, and individualized treatment approaches. As genomic technologies continue to advance, the integration of genetic biomarkers with multi-omic and clinical data will refine predictive models, improve therapeutic targeting, and ultimately transform the landscape of neurological care into one that is predictive, preventive, and personalized.

8.3.2 Proteomic Biomarkers

Proteomic biomarkers are among the most promising tools for advancing precision medicine in neurology, offering a dynamic and functional perspective on disease processes. Unlike genetic biomarkers that provide static information about inherited susceptibility, proteomic biomarkers reflect real-time changes in protein expression, structure, post-translational modifications, and interactions occurring in response to physiological or pathological states. The proteome represents the complete set of proteins expressed by a cell, tissue, or organism at a given time, and it varies significantly depending on age, environmental factors, and disease stage. In neurological disorders, proteomic biomarkers play a crucial role in understanding molecular pathogenesis, identifying disease stages, and predicting therapeutic responses. With the advent of high-throughput technologies such as mass spectrometry (MS), two-dimensional gel electrophoresis, and protein microarrays, researchers can now quantify and characterize thousands of proteins in biological fluids like cerebrospinal fluid (CSF), plasma, serum, and even extracellular vesicles such as exosomes.[31] These advancements have made it possible to identify specific protein signatures associated with neurodegeneration, neuroinflammation, synaptic dysfunction, and oxidative stress processes central to diseases like Alzheimer's disease (AD), Parkinson's disease (PD), amyotrophic lateral sclerosis (ALS), multiple sclerosis (MS), and traumatic brain injury (TBI). Proteins serve as direct functional effectors in cells, and their altered expression or modification often mirrors the underlying molecular mechanisms driving neurological disease progression. In neurodegenerative disorders, specific proteins accumulate abnormally or undergo pathological modifications that contribute to neuronal dysfunction and cell death.[32] In Alzheimer's disease, for instance, amyloid-beta (Aβ) peptides and tau proteins are the most established proteomic biomarkers. The aggregation of Aβ into extracellular plaques and the hyperphosphorylation of tau leading to neurofibrillary tangles are central to AD pathology. Measuring CSF levels of decreased Aβ42 and increased phosphorylated tau (p-tau) and total tau (t-tau) allows for early and differential diagnosis, even before clinical symptoms appear. Longitudinal monitoring of these proteins helps assess disease progression and therapeutic response, as increasing tau levels often correlate with ongoing neuronal damage. Similarly, in Parkinson's disease, misfolded α-synuclein accumulates to form Lewy bodies, a hallmark of the disease.[33] Quantification of total, oligomeric, or phosphorylated α-synuclein in CSF and plasma serves as a biomarker for disease severity and progression. Proteins like DJ-1 and neurofilament light chain (NfL) have also been linked to PD progression, with elevated NfL levels indicating axonal injury and neurodegeneration. In amyotrophic lateral sclerosis (ALS), elevated concentrations of NfL and phosphorylated neurofilament heavy chain (pNfH) in CSF and blood are reliable indicators of motor neuron damage and correlate strongly with disease progression and prognosis. In multiple sclerosis, proteomic biomarkers provide insight into immune-mediated demyelination and neurodegeneration. Elevated CSF levels of NfL, glial fibrillary acidic protein (GFAP), and chitinase-3-like protein 1 (CHI3L1) reflect active axonal damage and gliosis. These biomarkers not only differentiate between relapsing–remitting and progressive MS but also help evaluate the efficacy of disease-modifying therapies.[34] In stroke and traumatic brain injury, proteins such as S100B, neuron-specific enolase (NSE), and glial-derived proteins like GFAP serve as acute injury markers, helping clinicians assess the extent of neuronal and glial damage. Elevated plasma S100B or GFAP levels post-stroke or trauma correlate with poor outcomes and can guide clinical management.

Beyond individual markers, proteomic changes in pathways involving inflammation, oxidative stress, and synaptic function provide deeper insights into disease mechanisms. For example, inflammatory cytokines such as interleukin-6 (IL-6), tumor necrosis factor-alpha (TNF-α), and C-reactive protein (CRP) are elevated in many neurological diseases, linking systemic inflammation to neurodegeneration. Oxidative stress-related proteins like peroxiredoxins and superoxide dismutase are also dysregulated, reflecting mitochondrial dysfunction and energy imbalance.[35] Collectively, these protein alterations not only indicate disease presence but also reveal temporal dynamics, helping clinicians monitor disease trajectory and evaluate therapeutic impact. While single-protein biomarkers have proven valuable, the complexity and heterogeneity of neurological disorders necessitate the use of biomarker panel combinations of multiple proteins that collectively enhance diagnostic accuracy and predictive power.[36] Such panels capture the multifactorial nature of disease processes by integrating data from multiple molecular pathways. In Alzheimer's disease, combined proteomic panels have demonstrated superior diagnostic performance compared to individual markers. For example, a panel including Aβ42, p-tau, t-tau, neurogranin, and YKL-40 has been shown to improve differentiation between AD and other dementias. Neurogranin, a postsynaptic protein involved in synaptic plasticity, reflects synaptic loss, while YKL-40, secreted by astrocytes and microglia, indicates neuroinflammation.[37] Together, these markers provide a multidimensional view of amyloid pathology, tau-mediated neurodegeneration, and inflammatory responses. Similarly, in Parkinson's disease, multi-protein panels combining α-synuclein, DJ-1, urate, and inflammatory cytokines enhance early diagnosis and classification of disease subtypes. Proteomic profiling of CSF and serum has revealed distinct signatures distinguishing idiopathic PD from atypical parkinsonian disorders such as multiple system atrophy and progressive supranuclear palsy.[38] These panels also help predict therapeutic outcomes, for instance, urate levels inversely correlate with disease progression, suggesting its neuroprotective potential. In multiple sclerosis, proteomic panels incorporating NfL, CHI3L1, Chemokine (C-X-C motif) ligand 13 (CXCL13), and osteopontin have shown high predictive value for disease activity, relapse rate, and transition to progressive stages. These biomarkers reflect distinct aspects of MS pathology, including axonal damage, astrocytosis, and immune activation. Similarly, in stroke, combinations of GFAP, NSE, S100B, and matrix metalloproteinases (MMPs) enhance differentiation between ischemic and hemorrhagic events, aiding in rapid clinical decision-making. Emerging studies also highlight the diagnostic value of plasma- and exosome-derived protein panels. Blood-based biomarkers, though less specific than CSF markers, offer non-invasive and scalable options for large-scale screening.[39] Exosomes—small extracellular vesicles secreted by neurons and glia—carry disease-specific proteins and reflect CNS pathology even in peripheral fluids. Proteomic analysis of exosomes isolated from plasma or saliva can detect early alterations in AD-related proteins, making them attractive for future clinical applications.

The use of machine learning and bioinformatics has further enhanced biomarker discovery and panel optimization. Algorithms can analyze high-dimensional proteomic datasets to identify combinations of proteins that best discriminate disease states, predict progression, or forecast therapeutic responses. These computational approaches help overcome biological variability and improve reproducibility, paving the way for clinically validated proteomic signatures. Clinically, biomarker panels are invaluable for prognosis and treatment monitoring. For example, longitudinal changes in CSF tau or plasma NfL levels can indicate therapeutic efficacy in AD or MS trials. In ALS, sustained elevation of NfL predicts rapid progression, guiding treatment planning and patient counseling. Moreover, integrating proteomic data with genomic, metabolomic, and imaging biomarkers enhances diagnostic precision and provides a holistic view of neurological disorders, an approach known as multi-omic integration. Despite these advances, challenges remain in standardization, assay sensitivity, and validation across diverse populations. Differences in sample collection, storage, and analytical methods can influence results, highlighting the need for international standards. Nonetheless, the future of proteomic biomarkers is promising, driven by continuous technological improvements, automation, and clinical collaborations.[40] In summary, proteomic biomarkers and multi-protein panels have revolutionized neurological research and clinical practice by enabling earlier diagnosis, accurate prognosis, and personalized treatment strategies. They bridge the gap between molecular pathology and clinical manifestation, providing actionable insights into disease mechanisms. As the field progresses, integration of proteomic data with other biomarker modalities and artificial intelligence-driven analytics will further refine diagnostic and therapeutic precision, marking a significant step toward truly individualized neurology.

8.3.3 Metabolomic Biomarkers

Metabolomic biomarkers represent a rapidly advancing area of research in neurology, offering powerful insights into the biochemical alterations that accompany and drive neurological diseases. Metabolomics is the comprehensive study of small molecules, or metabolites, produced during cellular metabolism, providing a direct reflection of physiological and pathological states. Since metabolites are the end products of gene and protein activity, their profiling offers a highly sensitive snapshot of dynamic biological processes, making metabolomic biomarkers valuable for early diagnosis, monitoring disease progression, and evaluating therapeutic responses. In the context of neurological diseases, metabolic disturbances often precede overt clinical symptoms, as the brain's high energy demand and dependence on precise metabolic regulation make it particularly vulnerable to biochemical imbalance. Metabolic changes in neurological diseases are diverse and disease-specific, involving disruptions in energy metabolism, oxidative stress pathways, neurotransmitter synthesis, and lipid homeostasis.[41] In Alzheimer's disease (AD), for example, abnormal glucose utilization and mitochondrial dysfunction are among the earliest detectable changes. Decreased glucose uptake observed through fluorodeoxyglucose-positron emission tomography (FDG-PET) imaging and altered levels of metabolites like lactate, pyruvate, and tricarboxylic acid (TCA) cycle intermediates reflect impaired neuronal energy metabolism. Additionally, dysregulation of lipid metabolism, including reduced phosphatidylcholine and sphingomyelin levels, correlates with amyloid-beta accumulation and synaptic dysfunction. Elevated homocysteine and decreased folate and choline levels further indicate impaired one-carbon metabolism, contributing to neuroinflammation and oxidative damage. In Parkinson's disease (PD), alterations in dopamine metabolism, oxidative stress markers, and mitochondrial metabolites are prominent.[42] Reduced levels of urate and glutathione (GSH) indicate increased oxidative burden, while elevated kynurenine and quinolinic acid levels from disrupted tryptophan metabolism contribute to neurotoxicity. Moreover, changes in fatty acid oxidation and carnitine metabolism reflect defective mitochondrial bioenergetics in dopaminergic neurons. Similarly, amyotrophic lateral sclerosis (ALS) exhibits altered amino acid and energy metabolism, with elevated glutamate (indicating excitotoxicity), decreased creatine and pyruvate, and imbalances in lipid species such as ceramides and lysophospholipids that drive neuronal degeneration. In multiple sclerosis (MS), metabolomic studies reveal dysregulation in lipid and energy pathways due to demyelination and chronic inflammation.[43] Increased lactate and decreased *N*-acetylaspartate (NAA) reflect mitochondrial dysfunction, while alterations in sphingolipid and phospholipid metabolism indicate active demyelination and neuroinflammatory responses. Even in psychiatric and neurodevelopmental disorders such as schizophrenia, autism spectrum disorder, and depression, metabolic imbalances involving amino acids (glutamate, GABA, tryptophan), lipids, and oxidative stress markers play key roles in altered neurotransmission and neuronal connectivity.[44] These observations highlight that neurological disorders share common metabolic disturbances, particularly those affecting mitochondrial energy production, oxidative balance, and neurotransmitter metabolism, though each disease exhibits distinct metabolomic signatures. Identification of metabolite profiles for early diagnosis has become an important goal of precision neurology. Advances in analytical technologies like nuclear magnetic resonance (NMR) spectroscopy, liquid chromatography–mass spectrometry (LC–MS), and gas chromatography–mass spectrometry (GC–MS) have enabled the detection of hundreds of metabolites in biofluids such as cerebrospinal fluid, plasma, serum, and urine, allowing researchers to identify disease-specific profiles or "metabolic fingerprints." For example, in AD, reduced levels of glucose and amino acids combined with elevated lipid peroxidation products have been proposed as early diagnostic indicators even before cognitive decline manifests. In PD, specific metabolite combinations such as decreased urate and GSH with increased kynurenine and acylcarnitines can help distinguish early PD from other movement disorders.[45] In MS, metabolite panels including NAA, choline derivatives, and sphingomyelins provide diagnostic and prognostic insights. Similarly, in ALS, elevated glutamate and decreased creatine can help predict disease progression and therapeutic response. Recent integration of metabolomics with other "omics" approaches, such as genomics, proteomics, and lipidomics, has enhanced diagnostic precision by linking metabolic alterations to genetic and proteomic pathways.[46] Moreover, the application of artificial intelligence and machine learning to metabolomic data analysis allows for the identification of subtle, multidimensional biomarker patterns that are otherwise undetectable through traditional statistical methods. These computational models can predict disease onset, differentiate subtypes, and personalize treatment strategies based on individual metabolic signatures. Importantly, metabolomic biomarkers also serve as pharmacodynamic indicators, reflecting biochemical responses to drug therapy and guiding dosage optimization.[47] Despite these advances, challenges remain in standardizing sample collection, addressing variability due to diet, age, and comorbidities, and validating biomarkers across diverse populations. Nonetheless, metabolomic profiling holds immense promise for transforming neurological diagnostics by providing a non-invasive, cost-effective, and highly sensitive approach for detecting early metabolic perturbations, monitoring disease trajectories, and tailoring therapeutic interventions.[48] As the field progresses, integrating metabolomic biomarkers into clinical practice will enable earlier intervention, improved disease management, and ultimately, a more personalized approach to neurological care.

8.3.4 Neuroimaging Biomarkers

Neuroimaging biomarkers have revolutionized the field of neurology by enabling the non-invasive visualization of brain structure, function, and molecular processes associated with neurological diseases. These biomarkers provide crucial insights into disease mechanisms, progression, and treatment response, bridging the gap between molecular pathology and clinical manifestation. Advances in imaging technologies such as magnetic resonance imaging (MRI), positron emission tomography (PET), functional MRI (fMRI), diffusion tensor imaging (DTI), and single-photon emission computed tomography (SPECT) have allowed researchers and clinicians to detect early pathological alterations long before symptoms become evident. The integration of imaging biomarkers with molecular, genetic, and biochemical data underpins the foundation of precision neurology, enhancing diagnostic accuracy and enabling patient-specific therapeutic strategies. MRI and PET imaging for brain diseases represent two of the most powerful and widely utilized modalities in neuroimaging biomarker research.[49] MRI provides detailed anatomical and physiological information using strong magnetic fields and radiofrequency pulses, allowing the assessment of gray and white matter integrity, volumetric changes, and functional connectivity. Structural MRI is invaluable in detecting cortical atrophy, white matter lesions, and ventricular enlargement, which are hallmark features of neurodegenerative disorders such as Alzheimer's disease (AD), Parkinson's disease (PD), and multiple sclerosis (MS).[50] Functional MRI (fMRI), based on blood oxygenation level-dependent (BOLD) contrast, measures brain activity and connectivity patterns, identifying altered neural networks in conditions like mild cognitive impairment, schizophrenia, and depression. Diffusion tensor imaging (DTI), a specialized MRI technique, quantifies white matter microstructural integrity by measuring the diffusion of water molecules, revealing axonal damage in traumatic brain injury, stroke, and neurodegeneration.[51] MRI also includes magnetic resonance spectroscopy (MRS), which measures brain metabolites such as *N*-acetylaspartate (NAA), choline, and creatine to assess neuronal viability and energy metabolism. Decreased NAA levels, for example, indicate neuronal loss in AD and MS, while increased choline suggests demyelination or membrane turnover. On the other hand, positron emission tomography (PET) provides metabolic and molecular information by detecting radiolabeled tracers that bind to specific targets in the brain.[52] The most used PET tracer, fluorodeoxyglucose (FDG), measures cerebral glucose metabolism, serving as a marker of neuronal activity. Hypometabolism in temporoparietal and posterior cingulate regions detected by FDG-PET is a diagnostic hallmark of Alzheimer's disease, even in early stages. PET imaging with amyloid-binding tracers such as [11C]PiB, [18F] florbetapir, and [18F] florbetaben has revolutionized AD diagnosis by allowing direct visualization of amyloid-beta deposition *in vivo*.[53] Similarly, tau-specific PET tracers like [18F] flortaucipir enable the mapping of tau pathology, providing complementary information for staging disease progression. In Parkinson's disease and other parkinsonian syndromes, PET tracers targeting the dopamine transporter (DAT), DOPA decarboxylase, and dopamine receptors assess nigrostriatal integrity, differentiating PD from atypical movement disorders.[54] Moreover, PET imaging for neuroinflammation using TSPO (translocator protein) ligands has been applied in MS, AD, and traumatic brain injury, highlighting microglial activation and glial pathology. These MRI- and PET-based biomarkers not only improve diagnostic precision but also facilitate clinical trial design, as they serve as surrogate endpoints for assessing therapeutic efficacy and disease modification. Biomarkers identified through neuroimaging modalities encompass a broad spectrum of structural, functional, and molecular parameters that provide objective measures of disease states.[55] Structural imaging biomarkers include brain volume reduction, cortical thinning, and white matter lesion burden, all of which correlate with cognitive decline and disease severity. In Alzheimer's disease, MRI-based hippocampal atrophy is one of the most validated biomarkers, reflecting neuronal loss in memory-related regions.[56] In multiple sclerosis, MRI detects white matter lesions, gadolinium enhancement (signifying active inflammation), and brain atrophy, all of which are used to monitor disease activity and therapeutic response. Similarly, in stroke, diffusion-weighted imaging (DWI) identifies ischemic areas within minutes of onset, while perfusion imaging assesses blood flow deficits, guiding timely interventions.[57] Functional imaging biomarkers, such as altered connectivity patterns on fMRI, reveal network disruptions underlying cognitive and emotional symptoms in neurodegenerative and psychiatric conditions. For example, reduced connectivity in the default mode network (DMN) is consistently observed in AD and mild cognitive impairment, while hyperconnectivity in motor circuits is reported in early PD. Molecular imaging biomarkers derived from PET and SPECT provide even greater specificity, detecting neurotransmitter dysregulation, protein aggregation, and neuroinflammation.[58] Dopaminergic imaging with [123I] FP-CIT SPECT (DaTSCAN) identifies dopaminergic neuron loss in PD, while PET tracers for amyloid and tau distinguish between AD, frontotemporal dementia (FTD), and vascular dementia. Moreover, multimodal integration of MRI, PET, and CSF data enhances biomarker reliability and supports personalized diagnosis. Cerebrospinal fluid (CSF) biomarkers represent another vital class of indicators in neurology, providing a direct biochemical window into central nervous system (CNS) function and pathology.[59] CSF circulates around the brain and spinal cord, maintaining homeostasis and reflecting dynamic molecular changes in the CNS. It is an ideal source for biomarkers because it is in direct contact with brain extracellular fluid, allowing the detection of neuronal, glial, and metabolic alterations that may not be apparent in blood or imaging. Advances in proteomics, metabolomics, and immunoassays have enabled the identification of numerous CSF biomarkers that inform diagnosis, prognosis, and therapeutic monitoring across neurological disorders. The role of CSF in neurological disorder diagnosis lies in its ability to detect disease-specific molecular changes long before structural or functional symptoms emerge. In neurodegenerative disorders, CSF biomarkers provide molecular evidence of neuronal injury, synaptic dysfunction, protein aggregation, and neuroinflammation. For example, in Alzheimer's disease, characteristic alterations in CSF levels of amyloid-beta and tau preceded cognitive impairment by many years, allowing early identification of at-risk individuals.[60] Similarly, in Parkinson's disease and other synucleinopathies, measurement of α-synuclein levels in CSF aids in diagnosis and differentiation from atypical parkinsonism. In multiple sclerosis, CSF analysis revealing the presence of oligoclonal bands (OCBs) indicative of intrathecal IgG synthesis has long been a diagnostic criterion, reflecting immune activation within the CNS.[61] Elevated levels of inflammatory cytokines, chemokines, and neurofilament light chain (NfL) in CSF also provide evidence of active neurodegeneration and axonal injury, supporting disease monitoring and prognosis. In infectious and autoimmune neurological disorders, such as meningitis, encephalitis, or neuromyelitis optica spectrum disorder (NMOSD), CSF biomarkers help distinguish between inflammatory subtypes, detect pathogens, and guide immunotherapy. Thus, CSF analysis remains indispensable for understanding CNS pathology at a molecular level. Key CSF biomarkers in Alzheimer's, Parkinson's, and other disorders have transformed both research and clinical practice by enabling biochemical diagnosis and disease staging.

In Alzheimer's disease, the three core CSF biomarkers—amyloid-beta 42 (Aβ42), total tau (t-tau), and phosphorylated tau (p-tau)—form the foundation of modern diagnostic criteria.[62] Aβ42 levels decrease due to plaque deposition in the brain, while t-tau and p-tau increase because of neuronal injury and tangle formation. The ratio of Aβ42/Aβ40 improves diagnostic specificity, distinguishing AD from other dementias. Recent research has also identified neurogranin (a synaptic protein) and YKL-40 (a marker of astrocytic activation) as emerging CSF biomarkers reflecting synaptic dysfunction and neuroinflammation, respectively.[63] In Parkinson's disease, decreased total α-synuclein levels are commonly observed in CSF, consistent with its aggregation in Lewy bodies, while increased oligomeric α-synuclein correlates with disease severity. Additionally, elevated NfL levels in PD patients indicate axonal degeneration and are associated with faster progression and cognitive decline. In multiple sclerosis, beyond OCBs, biomarkers like CHI3L1, CXCL13, and GFAP provide valuable information on inflammatory activity and glial pathology. In amyotrophic lateral sclerosis (ALS), CSF concentrations of NfL and phosphorylated neurofilament heavy chain (pNfH) serve as robust indicators of motor neuron degeneration and correlate strongly with disease progression.[64] Likewise, in Huntington's disease, increased CSF levels of mutant huntingtin protein directly reflect pathogenic processes, providing a target for therapeutic monitoring. The diagnostic utility of CSF biomarkers has also extended to psychiatric and neurovascular conditions, where metabolic and inflammatory profiles offer potential indicators of disease state and treatment response.[65] Collectively, the integration of neuroimaging and CSF biomarkers represents a powerful approach for understanding and managing neurological diseases. While neuroimaging provides spatial and functional insights, CSF biomarkers offer molecular specificity, and together they create a multidimensional view of CNS pathology. This convergence facilitates earlier diagnosis, precise disease stratification, and evaluation of therapeutic efficacy, driving the evolution of precision neurology.[66] Continuous advancements in imaging resolution, biomarker quantification, and data analytics combined with artificial intelligence and machine learning promise to refine biomarker-based diagnosis further, paving the way toward individualized management of neurological disorders and improved patient outcomes.

8.4 Precision Medicine in Neurology

The sub-classification of diseases is based on unique molecular and biological etiologies rather than solely on clinical manifestations. Alzheimer's disease is currently categorized as having multiple biological etiologies alongside amyloid-β and tau pathologies. Precision medicine (PM) represents a conceptual change in health care to shift from the "one-size-fits-all" treatment approach. It is a new style of treating and preventing disease that considers the variability among individuals in genes, environment, and lifestyle for every individual. The objective of PM is to deliver optimally

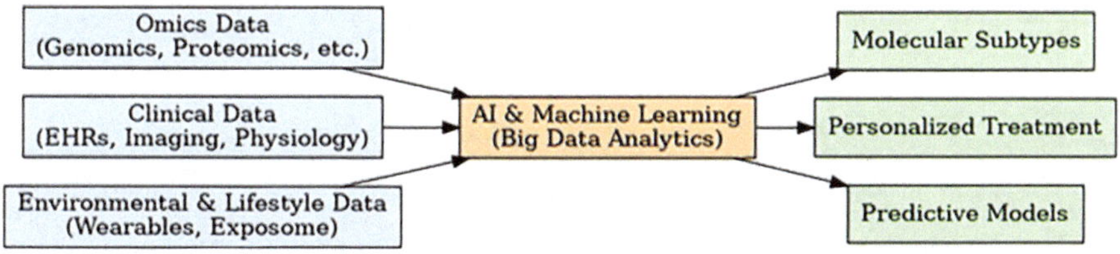

Figure 8.1 Precision medicine data integration workflow: a schematic representation depicting how AI and machine learning use omics data, clinical data, and environmental/lifestyle data to generate molecular subtypes, predictive models, and personalized treatment strategies.

targeted and scheduled interventions to a subpopulation of patients with a common susceptibility or molecular disease driver, so that the right patient gets the right drug in the right phase.[67] Precision medicine builds upon the integration of enormous datasets, also known as big data, coming from diverse sources, such as (Figure 8.1):

- Omics data: Genomics, transcriptomics, proteomics, metabolomics, and epigenomics.
- Clinical data: Electronic health records (EHRs), imaging (*e.g.*, PET, MRI), and physiological measurements.
- Environmental and lifestyle data: Data from wearables, patient-reported outcomes, and exposome mapping.

All these data are analyzed with the aid of sophisticated computational methods, including artificial intelligence (AI) and machine learning (ML), to define molecular subtypes of disease and anticipate treatment response.[68]

8.4.1 The Role of Precision Medicine in Neurological Diseases

The use of precision medicine is especially important in neurology because the majority of important neurological and neuropsychiatric diseases such as Alzheimer's disease (AD), Parkinson's disease (PD), and multiple sclerosis (MS) are extremely heterogeneous. Patients with a shared clinical diagnosis have essentially different molecular and mechanistic bases for their disease.[69] Precision neurology aims to surpass the repetitive failure in conventional drug trials, which tend to address a clinically characterized syndrome as one disease process (Figure 8.2)

1. Mechanistic stratification: Sub-classification of diseases according to distinctive molecular and biological causes, as opposed to mere clinical signs. For instance, Alzheimer's disease is now classified as having various biological causes in addition to amyloid-β and tau pathology.

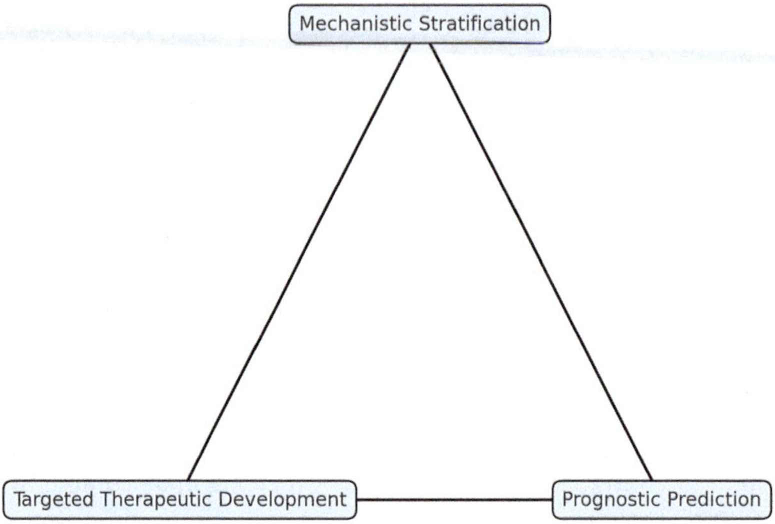

Figure 8.2 Roles of precision neurology: a triangle model depicting the three primary functions of precision medicine in neurology: mechanistic stratification, tailored therapy development, and prognosis prediction, with examples from Alzheimer's disease, Parkinson's disease, and ALS.

2. Targeted therapeutic development: Validation and targeting specific, actionable molecular targets for treatment. Success with therapies such as tofersen for individuals with SOD1 mutations in amyotrophic lateral sclerosis (ALS) represents a proof-of-concept for this strategy.[70]
3. Prognostic prediction: Utilizing a person's molecular profile to predict disease course, enabling earlier and more intense treatment of rapid-progressing subtypes.

8.4.2 Personalized Treatment Plans Based on Biomarkers

Biomarkers are key precision medicine tools that exist as measurable surrogates for normal biological function, disease process, or pharmacologic response to a therapeutic intervention. They lie at the foundation for developing truly personalized treatment plans (Figure 8.3).

- Diagnostic and subtyping biomarkers: Biomarkers are employed to establish a diagnosis, stage illness, and, most importantly, position the patient in a therapeutically relevant molecular subtype. The ratio of amyloid-β and phosphorylated tau (p-tau) in cerebrospinal fluid (CSF) or blood for AD is an example.[71]
- Prognostic biomarkers: Markers predicting a patient's long-term status, *e.g.*, Neurofilament Light Chain (NfL), which is a nonspecific marker of neuro-axonal injury and can forecast disease severity and progression in a few disorders.

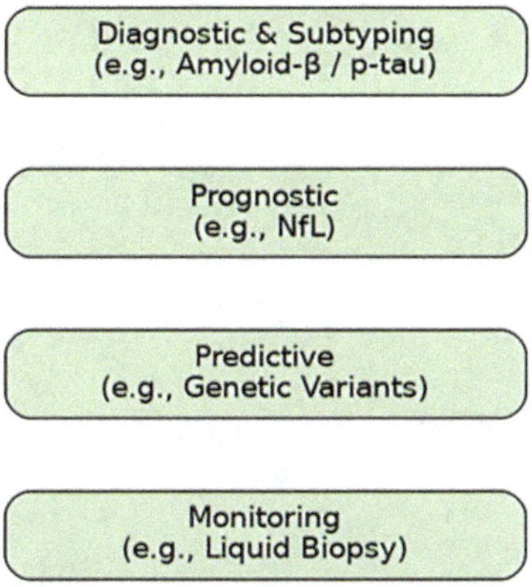

Figure 8.3 Biomarker classification: a hierarchical chart showing the four main types of biomarkers: diagnostic and subtyping, prognostic, predictive, and monitoring. Examples include amyloid-β/p-tau, neurofilament light chain, genetic variations, and liquid biopsy.

- Predictive biomarkers (pharmacogenomics): These biomarkers, frequently genetic variants, are used to forecast an individual's response to a given drug. Pharmacogenomics employs a patient's genomic profile to rationalize drug selection and dosage, reducing side effects and enhancing efficacy.
- Monitoring biomarkers (liquid biopsy): Liquid biopsy improvements (studying blood for markers such as NfL or exosomal proteins) enable minimally invasive, serial monitoring of treatment response and disease activity longitudinally.[72]

8.4.3 Genetic and Environmental Considerations in Precision Medicine

Complex neurological disorders are a consequence of a non-linear interaction between several genetic risk factors and environmental agents. Precision medicine requires the merger of both these elements for a comprehensive patient profile (Figure 8.4).

8.4.3.1 Genetic Considerations

The emergence of next-generation sequencing (NGS) and genome-wide association studies (GWASs) has revealed hundreds of genetic variants implicated in neurological risk.[73]

- Monogenic disorders: In diseases such as Huntington's disease or certain types of epilepsy, a single gene mutation determines the disease, and the genetic constitution becomes a direct target for therapy (*e.g.*, antisense oligonucleotides).

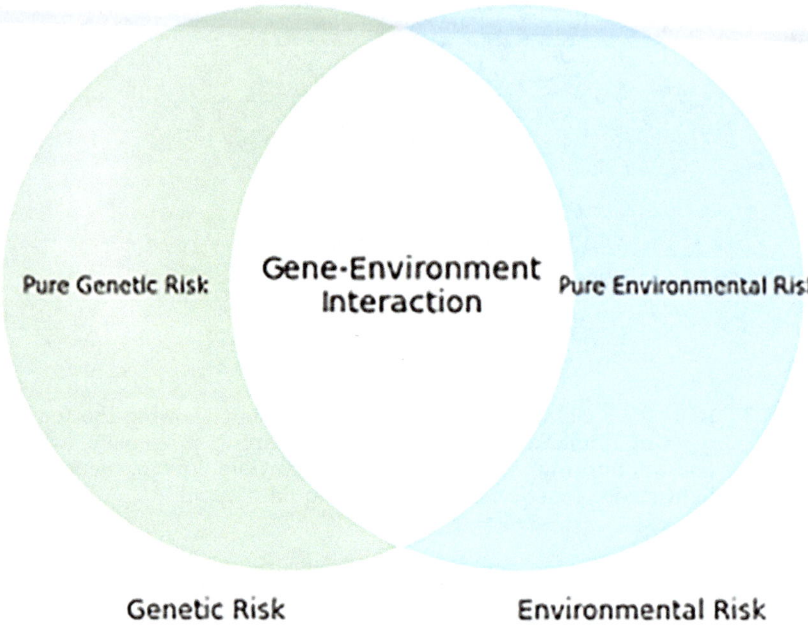

Figure 8.4 Gene–environment interaction model: a Venn diagram illustrates the interaction between genetic variables (NGS, SNPs, monogenic disorders) and environmental factors (lifestyle, poisons, exposome), with epigenetics representing the intersection that influences individual risk profiles.

- Complex disorders: In AD and PD, hundreds of single-nucleotide polymorphisms (SNPs) account for risk. Precision medicine combines this polygenic risk score with additional information to recognize individuals at elevated risk and inform early preventative measures
- Epigenetics: Gene expression is modified by environmental factors *via* mechanisms such as DNA methylation and histone modification without altering the DNA sequence itself. The epigenome shows how lifestyle and the environment (*e.g.*, stress, diet) affect genetic risk.

8.4.3.2 Environmental and Lifestyle Considerations

The exposome is the sum of all environmental exposures and the resulting biological responses during a lifetime. Because the genome only determines inherited risk, this combined metric is increasingly acknowledged as having similar relevance in predicting health outcomes, explaining a significant percentage of chronic disease and phenotypic diversity.[73]

- Lifestyle and comorbidities: Diet, physical activity, sleep, and vascular status are strongly incorporated into PM models for intervention development and risk stratification. For example, an elevated step count has been associated inversely with chronic disease risk, including neurological disease.[74]

- Toxin exposure: Environmental toxin or infectious agent exposure may initiate or hasten neurodegenerative processes, particularly in subjects with genetic susceptibility.
- Data integration: PM frameworks rely on sophisticated computational models to decipher the complex gene–environment interactions, thereby determining a specific patient's risk profile (their disease subtype) and resultantly, derive a tailored prevention or treatment plan which considers both their inborn biology as well as their life setting.

8.5 Applications for Biomarkers in Neurological Disorders

Biomarkers play a transformative role in neurology, enabling early diagnosis, disease classification, progression monitoring, and therapeutic assessment. They support differential diagnosis of overlapping disorders like Alzheimer's, Parkinson's, and multiple sclerosis, while aiding patient stratification in clinical trials. By integrating genetic, proteomic, metabolomic, and neuroimaging data, biomarkers enhance precision medicine, predict disease risk, guide personalized therapies, and improve treatment efficacy—ultimately advancing individualized care and patient outcomes (Table 8.1).

8.6 Challenges in Implementing Biomarkers and Precision Medicine

The potential of precision medicine (PM) in neurology for providing targeted treatments based on a person's specific biological profile is vast. Translating sophisticated "omics data" and new biomarkers into standard clinical practice is, nonetheless, faced with enormous scientific, ethical, economic, and logistic barriers to overcome, which need a collective global effort. For a biological marker to be incorporated into clinical decision-making, it should display high reliability and reproducibility in a wide range of settings. This is a main bottleneck for neurological biomarkers:

- Validation complexity: In contrast to most other disciplines, direct sampling of the central nervous system (CNS) is extremely invasive. This means that there must be dependence on surrogate biomarkers from cerebrospinal fluid (CSF) or blood, which must accurately describe the pathological processes in the brain. Clinical validation involves large, longitudinal, and heterogeneous patient populations to establish that a biomarker is reproducibly linked with disease presence, severity, and treatment response.[74]
- Lack of standardization: Even for prospective markers such as Neurofilament Light Chain (NfL) or blood-based p-tau for Alzheimer's disease (AD), measurement units, sample handling protocols, and assay

Table 8.1 Overview of biomarkers in major neurological disorders and their clinical relevance.[71–74]

Neurological disorder	Key biomarkers/ indicators	Clinical applications	Purpose/outcome
Alzheimer's disease and dementia	Aβ42, total tau (t-tau), phosphorylated tau (p-tau), APOE ε4, neurogranin, YKL-40, PET (amyloid/tau imaging). neurofilament light chain (NfL), inflammatory cytokines	Early diagnosis and disease monitoring, potential disease-modifying therapies	Detects amyloid deposition and tau pathology before cognitive symptoms, monitors progression and treatment response. Evaluates neuronal injury and therapeutic efficacy; aids in clinical trials of anti-amyloid and anti-tau agents
Parkinson's disease	α-Synuclein (total/ oligomeric), DJ-1, LRRK2, GBA mutations, urate, DAT-SPECT imaging. neurofilament light chain (NfL), mitochondrial proteins, PET (dopamine transporter imaging)	Identifying biomarkers for early detection. Monitoring disease progression	Enables differentiation between idiopathic PD and atypical parkinsonism; detects dopaminergic dysfunction pre-symptomatically. Tracks axonal degeneration, mitochondrial impairment, and treatment response over time
Multiple sclerosis (MS)	Neurofilament light chain (NfL), CHI3L1, CXCL13, oligoclonal bands (OCBs), GFAP. T2 lesion load, gadolinium enhancement, brain atrophy metrics	Disease activity assessment. Role in diagnosis and monitoring	Indicates axonal damage, immune activation, and glial pathology; supports diagnosis and prognosis. Detects demyelination and inflammation; measures therapeutic effectiveness and disease progression
Epilepsy	EEG patterns, SCN1A, GABRG2, CACNA1A mutations, inflammatory cytokines. mTOR pathway mutations (TSC1, TSC2), ion channel genes	Biomarkers for seizure prediction. Genetic markers for epileptic syndromes	Identifies preictal changes and neuronal excitability patterns; enables timely intervention and seizure forecasting. Defines syndrome subtypes and guides use of targeted antiepileptic therapies

Table 8.1 (*Continued*)

Neurological disorder	Key biomarkers/ indicators	Clinical applications	Purpose/outcome
Stroke	S100B, GFAP, NSE, matrix metalloproteinases (MMPs), D-dimer, CRP. Lactate, TNF-α, interleukins, NfL	Biomarkers for stroke subtypes. Early diagnosis and prognostic indicators	Distinguishes between ischemic and hemorrhagic strokes, aids in rapid classification and treatment decision-making. Evaluates extent of neuronal injury and predicts recovery outcomes; supports personalized rehabilitation strategies

platforms differ significantly across laboratories. The absence of standardization invalidates cross-study or cross-institutional comparison of results, making widespread clinical use problematic and obstructing multi-center clinical trials.[75]

- Multimodal data integration: Most CNS diseases need multimodal biomarkers (*e.g.*, genetic, imaging, and fluid-based data together). Testing the clinical utility of such integrated complex models is much more difficult than testing a single marker.[76]

8.6.1 Ethical and Privacy Issues with Genetic Data

The aggregation and utilization of large volumes of highly sensitive individual and family data in precision neurology pose profound ethics and legal challenges:

- Data security and re-identification: Genomic and multi-omics data are permanently identifying, unique, and transmissible between members of the same family. With increasing expansion of large-scale biobanks, there is always a chance of data leak or re-identification of anonymized data, potentially with an unprecedented impact on the individuals concerned.[77]
- Genetic discrimination: Notwithstanding legislative protection (*e.g.*, GINA in the US), there remain concerns about the use of genetic information by life insurance, long-term care companies, or employers to discriminate against someone because of a susceptibility to a future neurological condition (*e.g.*, Huntington's disease or late-onset AD risk).[78]
- Complex informed consent: The range of genetic testing is broad, frequently yielding results that are incidental to the major purpose of the study (*e.g.*, risk for an entirely unrelated condition). Securing truly informed consent for current and future use of such complex, lifetime data continues to be a huge challenge.[79]

8.6.2 Cost and Accessibility of Precision Medicine in Neurology

Precision medicine relies on modern technology that costs a lot and is difficult to obtain, which could make healthcare inequality worse:

- High initial costs: Next-generation sequencing (NGS), advanced neuroimaging (such amyloid/tau PET scans), and novel targeted gene therapies are all very expensive. These costs usually mean that only large academic medical centers and specialty clinics can use them.
- Problems with reimbursement: Health systems and payers are sometimes hesitant to adopt and pay for novel, expensive tests and treatments until they can show that they work in the real world. This delay slows down the integration of clinical data.
- Fairness and access: The costs and specialized infrastructure required render patients in disadvantaged communities, low-income nations, and even many primary care offices unable to access the diagnostic and therapeutic benefits of precision medicine, leading to inequitable health outcomes.[79]

8.6.3 Integrating Multi-omics Data for Comprehensive Disease Profiling

Precision medicine seeks an all-around understanding through the integration of information from the genome, transcriptome, proteome, metabolome, and microbiome (multi-omics). Yet, this technical issue of their integration is formidable:

- Data sparsity and heterogeneity: Every 'omics' layer produces data with varying formats, scales, and completeness. Also, having all 'omics' data for the same patient sample is a rare occurrence, resulting in missing modalities.[79]
- Computational and algorithmic needs: Conventional statistical tools do not work with high-dimensional, multi-omics data. There is a need for specialized artificial intelligence (AI) and machine learning (ML) algorithms to integrate these data streams, recognize meaningful biological patterns, and handle missing data robustly.[79,80]
- Biological interpretation: After integrating the data, the challenge then becomes converting the intricate patterns (*e.g.*, a particular combination of gene expression and protein phosphorylation) into biologically reasonable mechanisms and treatable clinical targets (Figure 8.5).

8.6.4 The Need for Global Cooperation in Neurological Research

To overcome the challenges of limited data availability, heterogeneity, and validation, neurological research must evolve into a truly global and collaborative initiative.

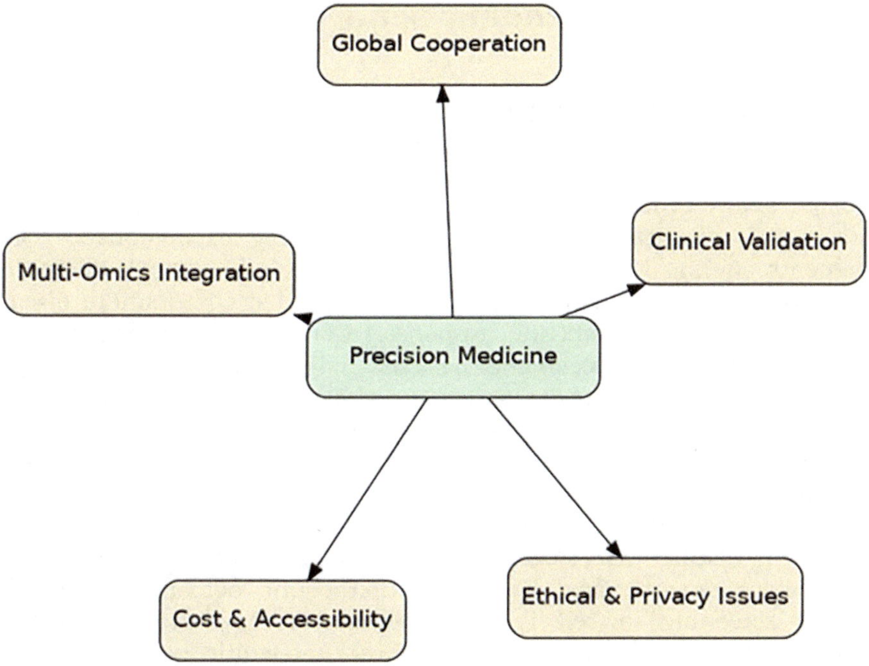

Figure 8.5 Challenges in precision medicine implementation: a hub-and-spoke graphic that shows the five main problems that make it hard for neurologists to use precision medicine in their work: clinical validation, ethical/privacy issues, cost/accessibility, multi-omics data integration, and worldwide cooperation.

Augmented data heterogeneity: Neurological disorders exhibit racial, ethnic, and sex-based variation in incidence, prevalence, and response to treatment. To prepare precision treatments that work for all groups, studies must involve large, heterogeneous, global cohorts representing global human genetic diversity.[80]

Standardization of protocols: International consortia (*e.g.*, the ENIGMA Consortium) are needed to create common best practice for data collection, quality control, and analysis procedures (*e.g.*, imaging and meta-data standards). Standardization allows international data to be directly compared and significantly speeds up discovery.

Resource and technology sharing: Cooperation makes it possible to share costly resources, sophisticated technologies, and computational skills beyond borders, which is especially crucial for institutions in resource-constrained environments. The International Brain Initiative and other global projects want to get people from all around the world to work together to map, monitor, and control brain activity to have the biggest scientific impact.[81]

8.7 Advances in Genomic Sequencing and CRISPR Technology

The foundation of precision medicine lies in the ability to understand an individual's genetic architecture. The rapid evolution of genomic sequencing technologies, particularly next-generation sequencing (NGS), has revolutionized our ability to map entire genomes with unprecedented speed and accuracy. NGS enables the simultaneous sequencing of millions of DNA fragments, offering detailed insights into genetic mutations, single-nucleotide polymorphisms (SNPs), and structural variations that contribute to disease susceptibility and therapeutic response. Complementing genomic sequencing, CRISPR-Cas9 technology has emerged as a precise genome-editing tool that enables scientists to alter genetic sequences with surgical accuracy. CRISPR works by utilizing a guide RNA to target specific DNA sequences, allowing for the addition, deletion, or modification of genetic material.[82] This innovation has opened the door to correcting disease-causing mutations directly within the genome. For instance, CRISPR applications in monogenic diseases like sickle cell anemia and Duchenne muscular dystrophy have demonstrated promising therapeutic outcomes. Moreover, CRISPR screening in cell lines has enhanced our understanding of drug–gene interactions and identified novel therapeutic targets. Together, genomic sequencing and CRISPR technology form a powerful synergy—sequencing identifies the molecular defects, while CRISPR offers the tools to correct them. Their integration supports drug discovery, precision oncology, and regenerative medicine, creating a landscape where treatments are tailored at the molecular level. Future developments, such as base editing and prime editing, promise even higher precision with reduced off-target effects, moving humanity closer to a new era of curative genetic medicine. The complexity of biological data necessitates advanced computational systems to extract clinically meaningful insights.[83] Artificial intelligence (AI) and machine learning (ML) have become indispensable in the discovery and validation of biomarker molecules that signal disease onset, progression, or therapeutic response. Traditional biomarker discovery relies on hypothesis-driven research, which can be slow and limited by human bias. In contrast, AI and ML leverage vast multi-omic datasets (genomic, proteomic, metabolomic) to uncover hidden patterns that might otherwise remain undetected. Machine learning algorithms—such as random forests, support vector machines, and deep neural networks—enable the analysis of complex, high-dimensional biological data. These algorithms identify correlations between molecular signatures and disease phenotypes, assisting in early diagnosis and risk prediction. In oncology, AI-based models can differentiate tumor subtypes and predict drug responsiveness based on gene expression profiles. Similarly, in neurodegenerative diseases like Alzheimer's, ML has been used to analyze neuroimaging and cerebrospinal fluid biomarkers to predict disease progression. Moreover, AI-driven biomarker discovery accelerates clinical trials by identifying patient subgroups most likely to respond to specific therapies, enhancing trial efficiency and reducing costs.[84] Integrating AI with natural language processing (NLP) further allows the mining of biomedical literature and clinical records, revealing new associations between genes, drugs, and diseases. As datasets continue to expand, AI models become more robust, driving precision diagnostics and therapeutics. However, challenges remain in ensuring data standardization, transparency, and interpretability. Addressing these will ensure AI serves as a trustworthy partner in personalized medicine, transforming biomarker discovery into a data-driven science. In traditional diagnostics, tissue biopsies are invasive, risky, and often unfeasible for continuous disease monitoring. Liquid biopsy, a groundbreaking non-invasive diagnostic tool, addresses these limitations by analyzing circulating biomarkers such as circulating tumor DNA (ctDNA), cell-free RNA, exosomes, and circulating tumor cells (CTCs) from blood or other body fluids. This approach enables the detection of genetic and epigenetic alterations associated with various diseases, especially cancers, without surgical intervention. Liquid biopsy provides real-time insights into tumor heterogeneity and evolution, making it invaluable for monitoring disease progression and treatment response.[85] For instance, ctDNA analysis can detect minimal residual disease (MRD) after treatment, guiding timely therapeutic interventions. In addition to oncology, liquid biopsies are being applied in prenatal testing, organ transplant monitoring, and neurodegenerative disease diagnostics, where traditional biopsies are impractical. Recent advancements in microfluidics, nanotechnology, and digital PCR have increased the sensitivity and specificity of liquid biopsy assays, enabling the detection of rare molecular events at single-molecule resolution. Furthermore, combining liquid biopsy with AI-based data analysis enhances diagnostic accuracy and predictive capability. Despite challenges such as standardization and cost, liquid biopsy holds immense promise as a cornerstone of precision medicine, enabling continuous and personalized disease management through minimally invasive sampling. Neurological disorders such as Alzheimer's disease, Parkinson's disease, and multiple sclerosis are characterized by multifactorial and heterogeneous pathologies. The integration of big data and bioinformatics is transforming neurology by enabling the analysis of complex biological networks underlying these conditions.[86] Modern research generates vast datasets from genomic, transcriptomic, proteomic, and neuroimaging studies. Bioinformatics provides computational frameworks to integrate, visualize, and interpret multilayered datasets to reveal molecular mechanisms and identify novel therapeutic targets. Big data analytics in neurology involve large-scale collection of clinical, imaging, and molecular information from diverse populations.[87] Machine learning models trained on such data can identify subtle disease patterns and predict progression trajectories. For instance, deep learning applied to MRI and PET scans enables early detection of structural and functional brain changes before clinical symptoms appear. Similarly, integrating omics data helps in mapping molecular pathways implicated in neurodegeneration and discovering predictive biomarkers. Bioinformatics pipelines facilitate the annotation and comparison of genetic variants, the reconstruction of neural pathways, and the simulation of drug–target interactions. Public databases like the Allen Brain Atlas and Human Connectome Project provide reference frameworks for multi-dimensional brain data analysis.[88] Moreover, the fusion of big data with wearable sensors and real-time monitoring expands precision neurology into everyday clinical practice, allowing continuous assessment of patient status. However, the success of big data in neurology depends on addressing challenges such as data heterogeneity, privacy, and computational complexity. Cloud computing, federated learning and data harmonization initiatives are helping overcome these barriers.[89] Ultimately, big data and bioinformatics are driving a paradigm shift from descriptive neuroscience to predictive and personalized neurology, where data-driven insights guide targeted interventions for brain health.

8.8 Future Directions and Innovations

The landscape of neurology is undergoing a profound transformation, driven by the convergence of molecular biology, computational science, and personalized medicine. Future directions in neurological research emphasize precision diagnostics, novel biomarkers, regenerative therapeutics, and the integration of holistic biological systems such as the gut–brain axis. Emerging technologies—ranging from stem cell therapies to AI-driven predictive modeling—are redefining how neurological diseases are understood, diagnosed, and treated. As neuroscience transitions from symptom-based management to mechanism-based interventions, innovations in molecular medicine, bioinformatics, and patient-specific care promise to improve prognosis and enhance quality of life.[90] The future of neurology lies in embracing these multidisciplinary approaches that unite genomics, systems biology, and clinical neuroscience to achieve truly individualized treatment strategies. The identification of emerging biomarkers is reshaping diagnostic and therapeutic paradigms in neurology. Biomarkers—biological indicators of physiological or pathological processes—are vital for early diagnosis, disease monitoring, and therapeutic stratification. Traditionally, neurological diagnosis relied on symptomatic presentation and imaging, often identifying disease only after irreversible damage occurred. However, advances in multi-omics profiling (genomics, proteomics, metabolomics, and transcriptomics) have unveiled novel molecular signatures that precede clinical manifestation. For example, neurofilament light chain (NfL) in cerebrospinal fluid and plasma has emerged as a reliable marker for neuronal injury, applicable across multiple neurodegenerative diseases such as Alzheimer's, ALS, and multiple sclerosis. Similarly, plasma phosphorylated tau (p-tau181, p-tau217) serves as a non-invasive biomarker for Alzheimer's pathology, correlating strongly with amyloid and tau PET imaging.[91] Emerging exosomal biomarkers, derived from extracellular vesicles released by neurons and glial cells, offer minimally invasive windows into central nervous system (CNS) activity. The integration of AI and bioinformatics further accelerates biomarker discovery by mining vast datasets for predictive patterns. For instance, deep learning models analyzing gene expression profiles can distinguish between overlapping neurodegenerative phenotypes. Additionally, epigenetic markers—such as DNA methylation and microRNA expression—are revealing the dynamic regulatory mechanisms underlying disease susceptibility and progression.[92] As biomarker validation progresses through multicentric clinical trials, personalized biomarker panels tailored to individual pathophysiological signatures will guide targeted therapeutic interventions. Ultimately, the discovery of emerging biomarkers will enable earlier diagnosis, monitor therapeutic response in real time, and transform neurological care into a predictive, preventive, and personalized practice. Regenerative medicine offers one of the most promising frontiers in neurological research, aiming to restore or replace damaged neurons and neural circuits. The brain and spinal cord possess limited intrinsic regenerative capacity, making degenerative and traumatic neurological disorders particularly devastating. However, stem cell-based therapies—including embryonic stem cells (ESCs), induced pluripotent stem cells (iPSCs), and mesenchymal stem cells (MSCs)—hold the potential to repopulate lost neural tissue and restore function. Preclinical studies have demonstrated that iPSC-derived dopaminergic neurons can ameliorate symptoms in Parkinson's disease models, while oligodendrocyte progenitor cell transplants have improved myelination in spinal cord injuries.[93] Beyond cell replacement, stem cells also secrete neurotrophic factors that support endogenous repair, modulate inflammation, and enhance synaptic plasticity. In neurodegenerative diseases like ALS and Huntington's, MSCs are being investigated for their ability to create neuroprotective microenvironments. Advancements in biomaterial scaffolds, 3D bioprinting, and gene-editing tools like CRISPR are refining stem cell delivery and integration into neural tissue.[94] These technologies improve cell survival, differentiation, and connectivity, enhancing the efficacy of regenerative interventions. Furthermore, patient-specific iPSCs minimize immunogenicity and allow for personalized modeling of disease mechanisms and drug testing. Despite the promise, challenges persist—ethical concerns over ESCs, risk of tumorigenicity, and ensuring functional synaptic integration. Nonetheless, ongoing clinical trials and translational research continue to advance regenerative medicine toward safe, effective clinical application. The future may witness a convergence of stem cell therapy, gene editing, and nanotechnology, heralding an era where neurological restoration becomes a clinical reality. The gut–brain axis (GBA)—a complex bidirectional communication network linking the enteric and central nervous systems—has emerged as a crucial player in neurological health. The human gut microbiome, consisting of trillions of microorganisms, influences neurodevelopment, neurotransmission, and neuroinflammation through metabolic, immune, and neural pathways. Disruptions in microbial composition, termed dysbiosis, have been associated with numerous neurological disorders, including Parkinson's disease (PD), Alzheimer's disease (AD), autism spectrum disorder (ASD), and multiple sclerosis (MS). Microbes modulate the production of neurotransmitters such as serotonin, dopamine, and GABA, and produce short-chain fatty acids (SCFAs) that influence blood–brain barrier integrity and neuroinflammation.[95] In PD, specific microbial taxa alterations correlate with motor and non-motor symptom severity, suggesting diagnostic potential. In AD, microbial metabolites and systemic inflammation may exacerbate amyloid deposition and cognitive decline. Emerging research explores probiotic, prebiotic, and dietary interventions to restore microbial balance and improve neurological outcomes. Fecal microbiota transplantation (FMT) and postbiotics represent next-generation therapeutic strategies that modulate the microbiome–brain axis. Integration of metagenomics and metabolomics provides mechanistic insights into how microbial metabolites affect neural signaling and disease progression. Furthermore, microbiome-based biomarkers are being developed for early disease detection and treatment monitoring. As precision medicine embraces microbiomics, the gut–brain axis will become a therapeutic target for modulating neuroinflammation, neurodegeneration, and even mental health disorders.[96] The growing recognition of microbiomes as a modifiable factor highlights its transformative potential in the future of personalized neurology. The future of neurology envisions a fully personalized treatment paradigm where interventions are tailored to the molecular, genetic, and environmental profile of each patient. Traditional "one-size-fits-all" treatments often fail to address interindividual variability in disease mechanisms and drug responses. Precision neurology integrates genomic sequencing, digital phenotyping, AI-based predictive analytics, and real-time monitoring to deliver individualized care. Advances in pharmacogenomics enable the identification of genetic polymorphisms influencing drug metabolism and efficacy, optimizing medication choices and dosages. Wearable sensors and neuroinformatics platforms continuously collect physiological and behavioral data, allowing dynamic adjustment of therapies. AI models can predict disease trajectories and suggest personalized intervention strategies based on multimodal datasets, including imaging, omics, and clinical parameters. Furthermore, digital twins—virtual replicas of patients integrating biological and lifestyle data—are emerging as tools to simulate treatment responses before real-world implementation. Integrating psychological and social determinants ensures holistic patient care beyond molecular precision. The success of personalized neurology will depend on ethical frameworks for data privacy, equitable access, and clinician training in digital medicine.[97] Ultimately, personalized treatment plans will not only improve efficacy but also empower patients through participatory care, marking the transition from reactive to proactive neurological medicine.

8.9 Conclusion

The convergence of molecular science, computational analytics, and clinical innovation has propelled precision medicine to the forefront of modern neurology. This evolution is redefining how we conceptualize, diagnose, and manage neurological disorders. As we move from population-based generalizations toward individualized interventions, biomarkers, AI, and regenerative therapies serve as the foundation of transformative care. Yet, with innovation comes responsibility—to ensure equity, ethics, and accessibility in the deployment of advanced technologies. The concluding insights of this discourse synthesize key developments, reflect on existing challenges and opportunities, and project a visionary outlook for the future of neurology and patient-centered care. Despite remarkable advancements, the translation of precision neurology into routine clinical practice faces numerous challenges. Chief among them are issues of data integration, cost, ethical governance, and equitable access. The heterogeneity of neurological disorders complicates biomarker validation, while data silos across institutions hinder interoperability and reproducibility. Ethical dilemmas surrounding genetic privacy, consent, and algorithmic bias require robust policy frameworks. Financial and infrastructural limitations also impede the widespread adoption of advanced diagnostics and therapeutics, particularly in low-resource settings. However, these challenges coexist with unprecedented opportunities. The rise of open-access biobanks, federated data networks, and global research collaborations is democratizing scientific discovery. AI-driven clinical decision-support systems can bridge gaps in expertise, especially in regions with limited neurological care. Furthermore, public–private partnerships and patient advocacy movements are accelerating innovation and regulatory adaptation. Addressing challenges through interdisciplinary collaboration, standardization, and policy innovation will ensure that the promise of precision neurology benefits all of humanity. In this vision, data-driven precision medicine converges with compassionate clinical care to deliver truly personalized outcomes. Neuroinformatics platforms, powered by AI, will continuously learn from global patient datasets, refining diagnostic models and predicting disease trajectories with unprecedented accuracy. Patients will play active roles in their care, empowered by digital tools, wearable sensors, and personalized feedback systems. Regenerative therapies, neuroprosthetics, and brain–computer interfaces will enable the restoration of lost neurological functions, while microbiome modulation and gene editing will address root causes at molecular levels. The boundaries between neurology, psychiatry, and behavioral science will blur, promoting integrated brain health management. This vision also demands a commitment to ethical innovation, emphasizing data privacy, inclusivity, and human oversight. The goal is a neuroscience of empathy and precision—where science not only deciphers the brain but also heals it holistically. The next generation of neurology will not merely treat disease but will nurture human potential.

Abbreviations

Aβ	Amyloid-beta
ACE	Angiotensin-converting enzyme

AD	Alzheimer's disease
ADNI	Alzheimer's Disease Neuroimaging Initiative
AGT	Angiotensinogen
AI	Artificial intelligence
ALS	Amyotrophic lateral sclerosis
APP	Amyloid precursor protein
ASOs	Antisense oligonucleotides
ATP1A2	ATPase Na^+/K^+ transporting subunit alpha 2
BOLD	Blood oxygenation level-dependent
BST1	Bone marrow stromal cell antigen 1
C9ORF72	Chromosome 9 open reading frame 72
CACNA1A	Calcium voltage-gated channel subunit alpha1 A
CHI3L1	Chitinase-3-like protein 1
CNS	Central nervous system
CNTNAP2	Contactin-associated protein-like 2
CRP	C-reactive protein
CSF	Cerebrospinal fluid
CTCs	Circulating tumor cells
DAT	Dopamine transporter
DMN	Default mode network
DTI	Diffusion tensor imaging
DWI	Diffusion-weighted imaging
EHRs	Electronic health records
ESCs	Embryonic stem cells
F5	Coagulation factor V
FDG-PET	Fluorodeoxyglucose-positron emission tomography
fMRI	Functional MRI
FMT	Fecal microbiota transplantation
FTD	Frontotemporal dementia
FUS	FUsed in Sarcoma
GBA	Glucocerebrosidase
GC–MS	Gas chromatography–mass spectrometry
GFAP	Glial fibrillary acidic protein
GWAS	Genome-wide association studies
HD	Huntington's disease
HLA	Human leukocyte antigen
HLA-DQB1	Human leukocyte antigen-DQ beta 1
HLA-DRB1	Human leukocyte antigen-DR beta 1
IL-6	Interleukin-6
iPSCs	Induced pluripotent stem cells
LC–MS	Liquid chromatography–mass spectrometry
LRRK2	Leucine-rich repeat kinase 2
MAPT	Microtubule-associated protein tau
MECP2	Methyl-CpG binding protein 2
MELAS	Mitochondrial encephalomyopathy, lactic acidosis, and stroke-like episodes

MMPs	Matrix metalloproteinases
MRD	Minimal residual disease
MRI	Magnetic resonance imaging
MS	Multiple sclerosis
MS	Mass spectrometry
MSCs	Mesenchymal stem cells
MTHFR	Methylenetetrahydrofolate reductase
NAA	*N*-Acetylaspartate
NGS	Next-generation sequencing
NMOSD	Neuromyelitis optica spectrum disorder
NMR	Nuclear magnetic resonance spectroscopy
NRXN1	Neurexin
1NfL	Neurofilament light chain
NSE	Neuron-specific enolase
OCBs	Oligoclonal bands
PARK2	Parkin RBR E3 ubiquitin protein ligase
PD	Parkinson's disease
PET	Positron emission tomography
PINK1	PTEN-induced kinase 1
PM	Precision medicine
pNfH	Phosphorylated neurofilament heavy chain
PPMI	Parkinson's Progression Markers Initiative
p-tau	Phosphorylated tau
PRS	Polygenic risk scoring
SCFAs	Short-chain fatty acids
SCN1A	Sodium channel alpha subunit 1
SNPs	Single-nucleotide polymorphisms
SOD1	Superoxide dismutase 1
SPECT	Single-photon emission computed tomography
TARDBP	TAR DNA-binding protein
TBI	Traumatic brain injury
TNF-α	Tumor necrosis factor-alpha
t-tau	Total tau

References

1. C. R. Jack Jr, D. A. Bennett, K. Blennow, M. C. Carrillo, B. Dunn and S. B. Haeberlein, *et al.*, NIA-AA Research Framework: Toward a biological definition of Alzheimer's disease, *Alzheimers Dement.*, 2018, **14**(4), 535–562.
2. C. R. Jack Jr, D. S. Knopman, W. J. Jagust, R. C. Petersen, M. W. Weiner and P. S. Aisen, *et al.*, Tracking pathophysiological processes in Alzheimer's disease: an updated AT(N) framework, *Neurology*, 2016, **87**(5), 539–547.
3. S. Palmqvist, S. Janelidze, Y. T. Quiroz, H. Zetterberg, F. Lopera and E. Stomrud, *et al.*, Discriminative accuracy of plasma phospho-tau217

for Alzheimer disease *vs* other neurodegenerative disorders, *JAMA*, 2020, **324**(8), 772–781.
4. E. H. Thijssen, R. La Joie, A. Wolf, A. Strom, P. Wang and L. Iaccarino, *et al.*, Associations of plasma phospho-tau217 levels with tau PET in early Alzheimer disease, *JAMA Neurol.*, 2020, **77**(10), 1332–1341.
5. S. Janelidze, N. Mattsson-Carlgren, S. Palmqvist, H. Vanderstichele, O. Lindberg and D. van Westen, *et al.*, Plasma P-tau181 in Alzheimer's disease: relationship to other biomarkers, differential diagnosis, neuropathology and longitudinal progression to Alzheimer's dementia, *Nat. Med.*, 2020, **26**(3), 379–386.
6. M. Khalil, C. E. Teunissen, M. Otto, F. Piehl, M. P. Sormani and T. Gattringer, *et al.*, Neurofilaments as biomarkers in neurological disorders, *Nat. Rev. Neurol.*, 2018, **14**(10), 577–589.
7. G. Disanto, C. Barro, P. Benkert, Y. Naegelin, S. Schädelin and A. Giardiello, *et al.*, Swiss Multiple Sclerosis Cohort Study Group. Serum neurofilament light: a biomarker of neuronal damage in multiple sclerosis, *Ann. Neurol.*, 2017, **81**(6), 857–870.
8. J. Kuhle, H. Kropshofer, D. A. Haering, U. Kundu, R. Meinert and C. Barro, *et al.*, Blood neurofilament light chain as a biomarker of MS disease activity and treatment response, *Neurology*, 2019, **92**(10), e1007–e1015.
9. N. Mattsson, N. C. Cullen, U. Andreasson, H. Zetterberg and K. Blennow, Association between longitudinal plasma neurofilament light and neurodegeneration in Alzheimer disease, *JAMA Neurol.*, 2019, **76**(7), 791–799.
10. A. Leuzy, R. Smith, R. Ossenkoppele, A. Santillo, E. Borroni and G. Klein, *et al.*, Diagnostic performance of [18F]flortaucipir tau PET for Alzheimer's disease *vs* other neurodegenerative disorders, *JAMA Neurol.*, 2020, **77**(8), 955–965.
11. K. Marek, D. Jennings, S. Lasch, A. Siderowf, C. Tanner and T. Simuni, *et al.*, The Parkinson Progression Marker Initiative (PPMI), *Prog. Neurobiol.*, 2011, **95**(4), 629–635.
12. M. W. Weiner, D. P. Veitch, P. S. Aisen, L. A. Beckett, N. J. Cairns and R. C. Green, *et al.*, Impact of the Alzheimer's Disease Neuroimaging Initiative, 2004 to 2014, *Alzheimers Dement.*, 2015, **11**(7), 865–884.
13. N. J. Ashton, A. Leuzy, T. K. Karikari, N. Mattsson-Carlgren, A. Dodich and M. Boccardi, *et al.*, A multicentre validation study of the diagnostic value of plasma neurofilament light, *Nat. Commun.*, 2021, **12**, 3400.
14. H. Hampel, J. Cummings, K. Blennow, P. Gao, C. R. Jack Jr and A. Vergallo, Blood-based biomarkers for Alzheimer's disease: current state and future use in a transformed global healthcare landscape, *Neuron*, 2023, **111**(15), 2781–2799.
15. E. J. Topol, High-performance medicine: the convergence of human and artificial intelligence, *Nat. Med.*, 2019, **25**(1), 44–56.
16. J. L. Cummings, Biomarkers in Alzheimer's disease drug development, *Alzheimers Dement.*, 2011, **7**(3), e13–e44.

17. H. Hampel, S. E. O'Bryant, J. L. Molinuevo, H. Zetterberg, C. L. Masters and S. Lista, *et al.*, Blood-based biomarkers for Alzheimer disease: mapping the road to the clinic, *Nat. Rev. Neurol.*, 2018, **14**(11), 639–652.
18. M. Khalil, C. E. Teunissen, M. Otto, F. Piehl, M. P. Sormani and T. Gattringer, *et al.*, Neurofilaments as biomarkers in neurological disorders, *Nat. Rev. Neurol.*, 2018, **14**(10), 577–589.
19. M. W. Weiner, D. P. Veitch, P. S. Aisen, L. A. Beckett, N. J. Cairns and R. C. Green, *et al.*, Impact of the Alzheimer's Disease Neuroimaging Initiative, 2004 to 2014, *Alzheimers Dement.*, 2015, **11**(7), 865–884.
20. C. R. Jack Jr, D. A. Bennett, K. Blennow, M. C. Carrillo, B. Dunn and S. B. Haeberlein, *et al.*, NIA-AA Research Framework: Toward a biological definition of Alzheimer's disease, *Alzheimers Dement.*, 2018, **14**(4), 535–562.
21. R. Guerreiro and J. Hardy, Genetic variants and their role in Alzheimer's disease, *Neurobiol. Aging*, 2014, **35**(4), 872–878.
22. M. A. Nalls, C. Blauwendraat, C. L. Vallerga, K. Heilbron, S. Bandres-Ciga and D. Chang, *et al.*, Identification of novel risk loci, causal insights, and heritable risk for Parkinson's disease: a meta-analysis, *Nat. Genet.*, 2019, **51**(2), 208–218.
23. A. E. Renton, A. Chio and B. J. Traynor, State of play in amyotrophic lateral sclerosis genetics, *Nat. Neurosci.*, 2014, **17**(1), 17–23.
24. International Multiple Sclerosis Genetics Consortium, Multiple sclerosis genomic map implicates peripheral immune cells and microglia in susceptibility, *Science*, 2019, **365**(6460), eaav7188.
25. C. Bellenguez, F. Küçükali, I. E. Jansen, L. Kleineidam, S. Moreno-Grau and N. Amin, *et al.*, New insights into the genetic etiology of Alzheimer's disease and related dementias, *Nat. Genet.*, 2022, **54**(4), 412–436.
26. B. De Strooper and E. Karran, The cellular phase of Alzheimer's disease, *Cell*, 2016, **164**(4), 603–615.
27. A. B. Singleton, M. J. Farrer and V. Bonifati, The genetics of Parkinson's disease: progress and therapeutic implications, *Mov. Disord*, 2013, **28**(1), 14–23.
28. R. Chia, A. Chiò and B. J. Traynor, Novel genes associated with amyotrophic lateral sclerosis: diagnostic and clinical implications, *Lancet Neurol.*, 2018, **17**(1), 94–102.
29. A. Goris, I. Pauwels, M. W. Gustavsen, B. van Son, K. Hilven and S. D. Bos, *et al.*, Genetic variants are major determinants of CSF antibody levels in multiple sclerosis, *Brain*, 2015, **138**(3), 632–643.
30. B. W. Kunkle, B. Grenier-Boley, R. Sims, J. C. Bis, V. Damotte and A. C. Naj, *et al.*, Alzheimer Disease Genetics Consortium. Genetic meta-analysis of diagnosed Alzheimer's disease identifies new risk loci and implicates Aβ, tau, immunity, and lipid processing, *Nat. Genet.*, 2019, **51**(3), 414–430.
31. M. Thambisetty and S. Lovestone, Blood-based protein biomarkers for Alzheimer's disease: challenging but feasible, *Biomark. Med.*, 2010, **4**(1), 65–79.

32. M. Sattlecker, S. J. Kiddle, S. Newhouse, P. Proitsi, S. Nelson and S. Williams, *et al.*, Alzheimer's disease biomarker discovery using SOMAscan multiplexed protein technology, *Alzheimers Dement.*, 2014, **10**(6), 724–734.
33. A. Keshavan, J. Pannee, T. K. Karikari, J. L. Rodriguez, N. J. Ashton and K. Blennow, *et al.*, Plasma neurofilament light levels in Alzheimer's disease: a cross-cohort study, *Nat. Med.*, 2021, **27**(8), 1436–1443.
34. J. L. Dage, A. M. V. Wennberg, D. C. Airey, C. E. Hagen, D. S. Knopman and M. M. Machulda, *et al.*, Levels of tau protein in plasma are associated with neurodegeneration and cognitive decline, *Nat. Med.*, 2016, **22**(4), 449–457.
35. B. Olsson, R. Lautner, U. Andreasson, A. Öhrfelt, E. Portelius and M. Bjerke, *et al.*, CSF and blood biomarkers for the diagnosis of Alzheimer's disease: a systematic review and meta-analysis, *Lancet Neurol.*, 2016, **15**(7), 673–684.
36. O. Hansson, H. Zetterberg, P. Buchhave, U. Andreasson, E. Londos and L. Minthon, *et al.*, Association between CSF biomarkers and incipient Alzheimer's disease in patients with mild cognitive impairment: a follow-up study, *Lancet Neurol.*, 2006, **5**(3), 228–234.
37. A. L. Baird, S. Westwood and S. Lovestone, Blood-based proteomic biomarkers of Alzheimer's disease pathology, *Front. Neurol.*, 2015, **6**, 236.
38. R. C. Barber, Biomarkers for early detection of Alzheimer's disease: current and future perspectives, *Front. Aging Neurosci.*, 2018, **10**, 391.
39. P. Oeckl, P. Weydt, P. Steinacker, S. Anderl-Straub, F. Nordin and A. E. Volk, *et al.*, Proteomic fingerprinting in cerebrospinal fluid distinguishes ALS from controls and other neurodegenerative diseases, *Mol. Neurodegener.*, 2020, **15**(1), 36.
40. D. S. Spellman, K. R. Wildsmith, L. A. Honigberg, M. Tuefferd, D. Baker and N. Raghavan, *et al.*, Development and evaluation of a multiplexed mass-spectrometry-based assay for measuring candidate peptide biomarkers in Alzheimer's disease neuroimaging initiative (ADNI) CSF, *Proteomics Clin Appl*, 2015, **9**(7–8), 715–731.
41. E. Trushina, T. Dutta, X. M. Persson, M. M. Mielke and R. C. Petersen, Identification of altered metabolic pathways in plasma and CSF in mild cognitive impairment and Alzheimer's disease using metabolomics, *PLoS One*, 2013, **8**(5), e63644.
42. S. F. Graham, O. P. Chevallier, C. T. Elliott, C. Hölscher, J. Johnston and B. McGuinness, *et al.*, Untargeted metabolomic analysis of human plasma indicates differentially affected polyunsaturated fatty acid metabolism in mild cognitive impairment, *J. Proteome Res.*, 2013, **12**(9), 3562–3570.
43. P. A. LeWitt, J. Li, M. Lu, L. Guo and P. Auinger, Parkinson Study Group–DATATOP Investigators. Metabolomic biomarkers as strong correlates of Parkinson disease progression, *Neurology*, 2017, **88**(9), 862–869.
44. M. Orešič, T. Hyötyläinen, S. K. Herukka, M. Sysi-Aho, I. Mattila and T. Seppänan-Laakso, *et al.*, Metabolome in progression to Alzheimer's disease, *Transl. Psychiatry*, 2011, **1**(12), e57.

45. R. Kaddurah-Daouk, H. Zhu, S. Sharma, M. B. Bogdanov, S. G. Rozen and W. Matson, *et al.*, Alterations in metabolic pathways and networks in Alzheimer's disease, *Transl. Psychiatry*, 2013, **3**(4), e244.
46. J. M. Wilkins and E. Trushina, Application of metabolomics in Alzheimer's disease, *Front. Neurol.*, 2018, **8**, 719.
47. M. Cuperlovic-Culf, D. Ferguson, A. Culf and P. Morin Jr, 1H NMR metabolomics analysis of glioblastoma subtypes: correlation between metabolomics and gene expression characteristics, *J. Biol. Chem.*, 2012, **287**(24), 20164–20175.
48. X. Pan, M. B. Nasaruddin, C. T. Elliott, B. McGuinness, A. P. Passmore and P. G. Kehoe, *et al.*, Alzheimer's disease-like pathology has transient effects on the brain and plasma metabolome, *Neurobiol. Aging*, 2016, **38**, 151–163.
49. A. Leuzy, R. Smith, R. Ossenkoppele, A. Santillo, E. Borroni and G. Klein, *et al.*, Diagnostic performance of [18F]flortaucipir tau PET for Alzheimer's disease *vs* other neurodegenerative disorders, *JAMA Neurol.*, 2020, **77**(8), 955–965.
50. C. C. Rowe and V. L. Villemagne, Brain amyloid imaging, *J. Nucl. Med.*, 2011, **52**(11), 1733–1740.
51. M. Filippi, M. A. Rocca, O. Ciccarelli, N. De Stefano, N. Evangelou and L. Kappos, *et al.*, MRI criteria for the diagnosis of multiple sclerosis: MAGNIMS consensus guidelines, *Lancet Neurol.*, 2016, **15**(3), 292–303.
52. J. Cummings, G. Lee, K. Zhong, J. Fonseca and K. Taghva, Alzheimer's disease drug development pipeline: 2021, *Alzheimers Dement.*, 2021, **7**(1), e12179.
53. E. R. de Natale, F. Niccolini, H. Wilson and M. Politis, Molecular imaging of the dopaminergic system in idiopathic Parkinson's disease, *Int Rev Neurobiol*, 2018, **141**, 131–172.
54. N. Villain and G. Chételat, Alzheimer disease: Biomarkers of structural and functional brain changes, *Radiology*, 2020, **295**(3), 636–639.
55. K. A. Johnson, S. Minoshima, N. I. Bohnen, K. J. Donohoe, N. L. Foster and P. Herscovitch, *et al.*, Appropriate use criteria for amyloid PET: A report of the Amyloid Imaging Task Force, the Society of Nuclear Medicine and Molecular Imaging, and the Alzheimer's Association, *J. Nucl. Med.*, 2013, **54**(3), 476–490.
56. M. Bergamino, Y. Su, B. J. Hanseeuw, H. Cho, J. N. Grace and J. Sepulcre, *et al.*, Early tau PET signal in nucleus basalis of Meynert in preclinical Alzheimer's disease, *Brain*, 2021, **144**(4), 1168–1179.
57. M. Schöll, S. N. Lockhart, D. R. Schonhaut, J. P. O'Neil, M. Janabi and R. Ossenkoppele, *et al.*, PET imaging of tau deposition in the aging human brain, *Neuron*, 2016, **89**(5), 971–982.
58. J. Oh, L. Airas, D. Harrison, E. Järvinen, T. Livingston and S. Lanker, *et al.*, Neuroimaging to monitor worsening of multiple sclerosis: advances supported by the Grant for Multiple Sclerosis Innovation, *Front. Neurol.*, 2023, **14**, 1319869.
59. T. Rudroff, O. Rainio and R. Klén, AI for the prediction of early stages of Alzheimer's disease from neuroimaging biomarkers — A narrative review of a growing field, *Alzheimers Res. Ther.*, 2024, **16**(1), 85.

60. S. D. Reddy, I. Younus, V. Sridhar and D. Samba Reddy, Neuroimaging Biomarkers of Experimental Epileptogenesis and Refractory Epilepsy, *Int. J. Mol. Sci.*, 2019, **20**(1), 220.
61. A. M. Landau, Novel Imaging Biomarkers for Brain PET Imaging, *Biomolecules*, 2025, **15**(4), 517.
62. N. Mattsson, N. C. Cullen, U. Andreasson, H. Zetterberg and K. Blennow, Association between longitudinal plasma neurofilament light and neurodegeneration in Alzheimer disease, *JAMA Neurol.*, 2019, **76**(7), 791–799.
63. T. Mitchell, S. Lehéricy and S. Y. Chiu, *et al.*, Emerging neuroimaging biomarkers across disease stages in Parkinson disease: a review, *JAMA Neurol.*, 2021, **78**(10), 1262–1272.
64. R. Ossenkoppele, O. Hansson and P. Insel, *et al.*, Imaging biomarkers in neurodegeneration: current and future practices, *Alzheimers Res. Ther.*, 2020, **12**(1), 49.
65. M. Walterfang, D. Velakoulis and S. Barton, *et al.*, A neuroimaging biomarker for individual brain-related abnormalities in early Alzheimer's disease progression, *eClinicalMedicine*, 2023, **49**, 101453.
66. A. J. Peltsch, Y. Su and O. Sabri, *et al.*, Neuroimaging biomarkers in the biological definition of Parkinson's and Lewy body dementia, *Eur. J. Nucl. Med. Mol. Imaging*, 2024, **51**(4), 758–771.
67. E. A. Ashley, The precision medicine initiative: a new national effort, *JAMA*, 2015, **313**(21), 2119–2120.
68. A. Rodriguez-Martinez, D. Kothalawala, R. M. Carrillo-Larco and A. Poulakakis-Daktylidis, Artificial intelligence in precision medicine: transforming disease subtyping, medical imaging, and pharmacogenomics, *Emerg. Top. Life Sci.*, 2025, **8**(2), 73–82.
69. T. J. Montine and K. S. Montine, Precision medicine: Clarity for the clinical and biological complexity of Alzheimer's and Parkinson's diseases, *J. Exp. Med.*, 2015, **212**(5), 601–605.
70. K. J. Friston, Precision psychiatry, *Biol. Psychiatry Cogn. Neurosci. Neuroimaging*, 2017, **2**(8), 640–643.
71. J. Deng, L. Hua, L. Bian, H. Chen, L. Chen, H. Cheng, C. Dou, D. Geng, T. Hong, H. Ji and Y. Jiang, Molecular diagnosis and treatment of meningiomas: an expert consensus, *Chin. Med. J.*, 20222022, **135**(16), 1894–1912.
72. M. Zia Ul Haq, X. Zhao, S. Obeng Apori, B. Singh and F. Tian, Molecular Biomarkers for Early Detection of Alzheimer's Disease and the Complementary Role of Engineered Nanomaterials: A Systematic Review, *Int. J. Mol. Sci.*, 2025, **26**(19), 9282.
73. J. Milic, M. Vucurovic, D. Jovic, V. Stankovic, E. Grego, S. Jankovic and R. Sapic, Exploring the Potential of Precision Medicine in Neuropsychiatry: A Commentary on New Insights for Tailored Treatments Based on Genetic, Environmental, and Lifestyle Factors, *Genes*, 2025, **16**(4), 371.
74. O. Strianese, F. Rizzo, M. Ciccarelli, G. Galasso, Y. D'Agostino, A. Salvati, C. Del Giudice, P. Tesorio and M. R. Rusciano, Precision and

personalized medicine: how genomic approach improves the management of cardiovascular and neurodegenerative disease, *Genes*, 2020, **11**(7), 747.
75. J. Deng, L. Hua, L. Bian, H. Chen, L. Chen, H. Cheng, C. Dou, D. Geng, T. Hong, H. Ji and Y. Jiang, Molecular diagnosis and treatment of meningiomas: an expert consensus, *Chin. Med. J.*, 20222022, **135**(16), 1894–1912.
76. S. Velmurugan, S. Waheeda, L. Kulanthaivel and G. K. Subbaraj, Applications of machine learning and multimodal integration for the early diagnosis of neurodegenerative diseases, *World Acad. Sci. J.*, 2025, 7(6), 1–8.
77. *Realizing the Promise and Minimizing the Perils of AI for Science and the Scientific Community*, ed. K. H. Jamieson, W. Kearney and A. M. Mazza, University of Pennsylvania Press, 2024.
78. S. L. Rossi, P. Subramanian and D. E. Bovenkamp, The future is precision medicine-guided diagnoses, preventions and treatments for neurodegenerative diseases, *Front. Aging Neurosci.*, 2023, **15**, 1128619.
79. S. Park, K. Lee, S. Yim, D. Hwang, D. Kim, S. Lee, A. Dunn, D. Gatti, E. Chesler, K. O'Connell and K. Kim, Robust Multi-Omics Integration from Incomplete Modalities Significantly Improves Prediction of Alzheimer's Disease, *arXiv*, 2025, preprint, arXiv:2509.20842.
80. Y. H. Chan, C. Wang, W. K. Soh and J. C. Rajapakse, Combining neuroimaging and omics datasets for disease classification using graph neural networks, *Front. Neurosci.*, 2022, **16**, 866666.
81. K. T. Motlana, Cortical Thickness Changes as a Potential Biomarker for Epilepsy in People With First-Time Unprovoked Seizure, Master's thesis, The University of Western Ontario, Canada.
82. M. Kolmogorov, K. J. Billingsley, M. Mastoras, M. Meredith, J. Monlong, R. Lorig-Roach, M. Asri, P. Alvarez Jerez, L. Malik, R. Dewan and X. Reed, Scalable Nanopore sequencing of human genomes provides a comprehensive view of haplotype-resolved variation and methylation, *Nat. Methods*, 2023, **20**(10), 1483–1492.
83. P. Gupta, H. O'Neill, E. J. Wolvetang, A. Chatterjee and I. Gupta, Advances in single-cell long-read sequencing technologies, *NAR Genom. Bioinform.*, 2024, **6**(2), lqae047.
84. P. J. Chen and D. R. Liu, Prime editing for precise and highly versatile genome manipulation, *Nat. Rev. Genet.*, DOI: 10.1038/s41576-022-00541-1.
85. S. Zou, Y. Sun and W. Tang, Charting the development and engineering of CRISPR base editors: lessons and inspirations, *Cell Chem. Biol.*, 2025, DOI: 10.1016/j.chembiol.2025.05.003.
86. D. K. Sahel, G. Giriprasad, R. Jatyan, S. Guha, A. Korde, A. Mittal, S. Bhand and D. Chitkara, Next-generation CRISPR/Cas-based ultrasensitive diagnostic tools: current progress and prospects, *RSC Adv.*, 2024, **14**(44), 32411–32435.
87. F. O. Bagger, L. Borgwardt, A. S. Jespersen, A. R. Hansen, B. Bertelsen, M. Kodama and F. C. Nielsen, Whole genome sequencing in clinical practice, *BMC Med. Genomics*, 2024, **17**(1), 39.

88. S. S. Azeez, R. S. Hamad, B. K. Hamad, M. S. Shekha and P. Bergsten, Advances in CRISPR-Cas technology and its applications: Revolutionising precision medicine, *Front. Genome Ed.*, 2024, **6**, 1509924.
89. M. E. Tarozzi, Next generation sequencing technologies, bioinformatics and artificial intelligence: A shared timeline, *Sci. Rev. Biol.*, 2024, **3**, 13–21.
90. A. Myrou, K. Barmpagiannos, A. Ioakimidou and C. Savopoulos, Molecular Biomarkers in Neurological Diseases: Advances in Diagnosis and Prognosis, *Int. J. Mol. Sci.*, 2025, **26**(5), 2231.
91. M.-Y. Noh, H. S. Kwon, M.-S. Kwon, M. Nahm, H.-K. Jin, J.-s. Bae and S. H. Kim, Biomarkers and therapeutic strategies targeting microglia in neurodegenerative diseases: current status and future directions, *Mol. Neurodegener.*, 2025, **20**, 82.
92. M. A. AbuAlrob and B. Mesraoua, Harnessing artificial intelligence for the diagnosis and treatment of neurological emergencies: A comprehensive review of recent advances and future directions, *Front. Neurol.*, 2024, **15**, 1485799.
93. Y. Zhang, L. Yu, Y. Lv, T. Yang and Q. Guo, Artificial intelligence in neurodegenerative diseases research: a bibliometric analysis since 2000, *Front. Neurol.*, 2025, **16**, 1607924.
94. Z. Meng, Y. He, K. Guo and L. Li, Imaging biomarkers in optic neuritis: current tools and future directions, *Front. Neurol.*, 2025, **16**, 1666835.
95. L. Liu, K. Guo and D. Yang, Advances in biomarkers for optic neuritis and neuromyelitis optica spectrum disorders: A multi-omics perspective, *Front. Neurol.*, 2025, **16**, 1559172.
96. P. Li, Y. Xin, C. Li, L. Yao and Y. Su, Blood-based biomarkers and neuroimaging for early detection of post-stroke cognitive impairment: current evidence and synergistic prospects, *Front. Neurol.*, 2025, **16**, 1596940.
97. A. Varesi, A. Carrara, V. G. Pires, V. Floris, E. Pierella, G. Savioli, S. Prasad, C. Esposito, G. Ricevuti, S. Chirumbolo and A. Pascale, Blood-based biomarkers for Alzheimer's disease diagnosis and progression: an overview, *Cells*, 2022, **11**(8), 1367.

CHAPTER 9

Environmental Influence on Neurochemistry

ANUSHKA SHARMA* AND DAKSH KUMAR

Department of Pharmaceutical Technology, Meerut Institute of Engineering and Technology, Meerut, India
*Email: anushkash2709@gmail.com

9.1 Introduction

Environmental factors are paramount in neurochemistry across the human lifespan, and, therefore, brain functioning and behavior.[1] The relationship between internal neural functions and external environmental input has been of much interest to neuroscience research over the last several decades.[2] This multi-factorial relationship exemplifies many environmental factors, including early life development, stress, diet, chemical exposure, and social interaction.[3] Neuroplasticity allows for continual adaptation to environmental changes through changes in gene expression, synaptic connections, and neurotransmitter systems. Neurochemical changes play a significant role in emotional regulation, cognitive function, and general mental health.[4] Understanding the contribution of environmental factors in neurochemical functioning is key to pinpointing the etiology of many neurological and psychiatric disorders and designing individualized prevention and treatment plans.[5] The issue most commonly investigated in this area is the influence of experiences in early life on subsequent brain development and neurochemical function. It has been indicated that maternal care, early life stress, and sensory input during sensitive developmental periods produce lasting impacts on the brain with respect to the serotonergic, dopaminergic, and noradrenergic neurotransmitter systems.[6]

Chemical Biology No. 26
Chemical Signatures of the Brain: Bridging Biochemistry, Neuroanatomy, and Brain Disorders
Edited by Sachchida Nand Rai and Abhishek Kumar Singh

Published by the Royal Society of Chemistry, www.rsc.org

Those effects (which are important themselves) related to childhood experiences can create downstream effects in terms of stress reactivity, emotion regulation, and cognitive functioning as an adult.[7] Additionally, hypothalamic–pituitary–adrenal (HPA) axis dysregulation to hypercortisolemia resulting in neurochemical imbalance has been implicated in a lifetime of chronic stress.[8] Neurochemical dysregulation, in turn, as a consequence of stress, has been implicated in several mood disorders, including anxiety and depression. The HPA axis and chronic stress have been implicated in several neurochemical systems, including dopamine, serotonin, and norepinephrine.[9] Alterations in neurogenesis, synaptic plasticity, and neuronal circuit function may explain stress-related mental illness. Effects of nutrition may also be significantly related to the modulation of neurochemistry.[10]

Omega-3 fatty acids, vitamins, and minerals are all important for the appropriate production and functioning of neurotransmitters. On the contrary, alterations in neurotransmitter composition and receptor sensitivity related to nutrient deficiency or imbalance have been implicated in mood disorders and cognitive dysfunction.[11] For example, omega-3 fatty acids are a part of the structural composition of neuronal membranes, and have been reported to affect neurotransmitter release and receptor activity.[12] B vitamin deficiencies, particularly folate and vitamin B12, have been shown to affect the biosynthesis of serotonin and dopamine neurotransmitters, which are involved in mood disorders and cognitive dysregulation.[13] Environmental exposure to persistent pollutants, like pesticides, heavy metals, and airborne toxins, has been demonstrated to affect neurochemical processes *via* a number of mechanisms, notably oxidative stress, inflammation, or by modulating a direct effect on neurochemical systems.[14] Long-term cumulative effects of toxin exposure may cause mental incapacitation and neurodegenerative illnesses. For example, exposure to lead has been claimed to disturb dopamine and glutamate neurotransmission and result in behavioral abnormalities and intellectual impairment.[15] In addition, cholinergic neurotransmission that is crucial in learning and memory has been proposed to be disturbed by organophosphate herbicides. Social relationships and culture play an important role in neurochemistry.[16] Both social contact quantity and quality are associated with variations in the oxytocin and dopamine systems that control social bonding, reward system processing, and emotional.[17] Coping strategies and stress responses, which may also lead to neurochemical changes, can also be influenced by cultural practices and beliefs. An illustration of how social contexts can affect the expression of genes involved in neurotransmitter synthesis, receptor functioning, and synaptic plasticity in both development and ongoing life experiences is provided by the pervasive role of social experience as a potential influencer of brain chemistry and neurobiology.[18] Through epigenetics, we now have a new understanding that has successfully outlined the molecular influence on environmental risks impacting neurochemistry. Underlying environmental change can affect gene expression patterns through epigenetic changes such as DNA methylation and histone modification without altering the original DNA sequence.[19] The resiliency of the environment influences neurotransmitter functioning, as changes that occurred due to the environment persist and can even become inherited by future generations.[20] By targeting gene-related changes in neurotransmitter synthesis, receptor expression, and synaptic plasticity, epigenetic mechanisms have established a molecular link between environmental experience and neurochemical alteration.[21] Public policy, social policy, and education all have an impact on the wide-ranging implications of research on how the environment influences neurochemistry. By describing the dynamic interaction between brain function and environment, researchers can identify specific tactics to achieve optimal neurochemical control and psychological well-being in a range of groups.[22]

For instance, since it was determined that early adversity affected neurochemistry, therapies that encourage joyful parenting and lessen child adversity have been created.[23] In order to maintain mental function and brain integrity, neurochemical monitoring of nutritional effect has simultaneously guided dietary prescriptions and supplement regimens.[24] Also, increased efforts are being made to limit and prevent exposure to toxic chemicals, particularly during critical periods of brain development, against the backdrop of awareness of the sweeping and striking influence of environmental poisons on neurochemistry change. Efforts at reducing air and water pollution and promoting environmentally friendly agricultural practices may trickle into the population's neurochemical health outcomes.[25]

Because of knowledge gained about the neurochemical foundations of learning and memory, teachers have been able to establish evidence-informed teaching strategies and interventions for learning disability. Teachers can create better learning environments that optimize brain function and cognitive performance by being aware of the impacts of stress, diet, and social pressure on neurochemistry.[26] Incorporating neurochemical science into social policy will help resolve even more deeply rooted social issues, such as addiction, mental disorder, and cognitive impairment.[27] For instance, neurochemical health would be greatly improved by diminishing social isolation and encouraging community participation programs, especially among higher-risk populations such as the elderly and mentally ill.[28] The need for a complete understanding of brain activity and function is emphasized by the actual and observable effects of environmental influences on neurochemistry. The interplay between genetic vulnerability and environmental influences underscores the importance of individualizing interventions to sensitivity for environmental influences.[29] Future research in this area will give us more insight into the complex mechanisms by which our environment modifies our neurochemical landscape, opening the door to novel strategies for maintaining optimal brain well-being and health throughout life.[30]

The complicated ways in which our surroundings affect brain chemistry and function are becoming increasingly clear through the emerging science of environmental neuroscience. Perhaps most interesting is emerging science around how the built environment affects neurochemistry. Urban, environmental, and architectural designs can significantly influence stress, social interaction, and use of natural environments, all of which massively affect neurochemical functioning.[31] Green space, for example, has been correlated with better mood and lower stress level, perhaps because of the influence of alterations to cortisol and serotonin metabolism. In contrast, chronic stress and inflammation resulting from exposure to excess air pollutants and environmental noise have been shown to disrupt neurochemical regulatory equilibrium and promote susceptibility to mental disease.[32] Research on the effect of exercise on neurochemistry has been another area of research focus. Exercise has been shown to have significant effects on brain function which include increased production and release of neurotransmitters, increased synthesis of neurotrophic factors, like brain-derived neurotrophic factor (BDNF), and increased synaptic plasticity.[33] Exercise-induced changes in neurochemistry have been connected to enhanced cognitive function, a decreased risk of neurodegenerative illness, and better mood regulation. Understanding the most effective types and durations of exercise to maximize neurochemical health can help build more customized exercise regimens for different individuals.[34] The effect of technology on neurochemistry has emerged as a new area of study in the age of increasing daytime electronic device use. Dopamine signaling changes linked to social media and excessive screen time have been linked to addiction and attention issues.[35] Additionally, blue light from digital devices has been shown to interfere with circadian rhythm and melatonin production, which leads to sleep disruptions and the associated neurochemical abnormalities.[36]

Designing computer interfaces with fewer detrimental neurochemical effects and creating guidelines for the use of healthy technology should be the main goals of future research and policy-making initiatives. The gut–brain axis has been found to be a significant regulator of environmental effects on neurochemistry.[37] Dietary and environmental factors that impact the gut microbiota may also govern the creation of neurotransmitters, inflammatory mediators, and other signaling molecules that impact brain function.[38] The effect of probiotics and prebiotics on mood and cognition, *via* altering gut microbiota composition and secondary neurochemical modulation, is supporting a range of new microbial- and dietary-based interventions aimed at sustaining neurochemical health and a growing understanding of the enteric–brain axis.[39] Climate change and the resulting changes to the environment threaten the world's neurochemicals. Increased temperature, changing air composition (such as CO_2), and increasing extreme weather changes will pose similar adverse effects on neurochemical systems, possibly causing oxidative stress, inflammation, and circadian rhythm desynchronization.[40] Starvation due to dehydration (thirst-induced) and environmental stress can also lead to nutritional shortcomings that influence neurotransmitter production and activity. Addressing the problems from the perspective of brain health requires coordinated efforts to simultaneously understand and innovate neurological, environmental, and public health.[41]

Pharmacogenomics examines how human variation (both genetic and environmental) modifies the way neurochemistry responds to therapeutic agents. Personalized pharmacology has the potential to upend the treatment of neurological and psychiatric disorders by allowing therapeutic agents to be approached according to the individual's genetic and environmental constitution.[42] Increasingly, it is becoming apparent that supporting brain health requires a deliberate, comprehensive, lifelong process as we begin to learn more about how environments affect neurochemistry.[43] This means a supportive environment during developmental-sensitive periods, a healthy lifestyle complimented with a healthy and nutritious diet, limiting exposure to environmental pollutants, maintaining healthy, uplifting social connections, and regularly engaging in physical and mental training.[44] The combination of findings from neuroscience, environmental science, public health and other related fields enables researchers and practitioners to develop life stage-specific approaches for promoting brain health and well-being among diverse populations.[45] To conclude, environmental hazards and neurochemistry are an active and ongoing science that continues to have significant relevance to public and personal health.[46] As we further untangle the complex interdependencies of our world and brain chemistry, we are given a science that can be applied to develop more effective interventions, policies, and practices to enable optimal neurochemical functioning and mental health. Our evolving world's challenges highlight the necessity for continued efforts and multidisciplinary collaboration in this vital field of neuroscience.[47]

9.2 Heavy Metals, Pollutants and Neurotoxicity

9.2.1 Heavy Metals

The environment contains high levels of lead (Pb), mercury (Hg), cadmium (Cd), aluminum (Al), and arsenic (As), all of which can seriously harm the neurological system and pose neurotoxicological risks.[48] All of these metals cause neurotoxicity through similar processes, such as oxidative stress, disruption of calcium homeostasis, and changes to neurotransmitter systems. However, the effects of these metals on brain function and neurochemistry vary.[49] Exposure to lead (Pb) still poses a serious neurotoxicological risk since it alters synapse formation and interferes with neurotransmitter systems, particularly glutamatergic and dopaminergic transmission. By competitively inhibiting calcium ions, lead interferes with calcium-dependent functions such as neurotransmitter release, synaptic plasticity, and neuronal excitation.[50]

Over time, low levels of exposure result in cognitive damage, behavioral abnormalities, learning impairments, and a decline in IQ. Adults who are exposed to lead neurotoxicity are more likely to experience neurological problems, cognitive decline, and motor disability.[51] A highly deadly neurotoxin, mercury (Hg) easily penetrates the blood–brain barrier and can accumulate in neural tissue. This is particularly true for methylmercury, or organic mercury, which causes oxidative stress, interferes with microtubule polymerization, and inhibits neural migration.[52] Mercury interferes with neurotransmission *via* altering glutamate signaling pathways and blocking dopaminergic and serotonergic transmission. These actions result in cognitive and motor dysfunction, cerebellar damage, and neuronal cell death. Mercury exposure during pregnancy and early childhood is linked to autism spectrum disorder symptoms and developmental delays.[53] Cadmium (Cd) has become increasingly implicated in neurological dysfunctions. It causes oxidative stress, mitochondrial injury, and interferes with calcium homeostasis within neurons.[54] Cadmium is detrimental to cholinergic neurotransmission and GABAergic inhibitory neurotransmission and plays a role in excitotoxicity. Epidemiology studies have identified Cd exposure with reduced attentiveness, memory loss, and vulnerability to Alzheimer's disease and Parkinsonian signs in old age.[55] Aluminum (Al) has been extensively researched in relation to neurodegenerative illnesses, despite the fact that it is not biologically required. It can be absorbed from a number of sources, including food, drinking water, and antiperspirants, and is stored in brain tissue.[56] Two key characteristics that distinguish Alzheimer's disease are tau hyperphosphorylation and β-amyloid aggregation, both of which are facilitated by aluminum. Aluminum also disrupts iron homeostasis, induces oxidative stress, and promotes neuroinflammation.[57] Long-term exposure to aluminum has been associated with cognitive deterioration, learning impairments, and memory loss.[58] Exposure to arsenic (As), especially through contaminated water, is one unique neurotoxic risk. Inorganic arsenic crosses the blood–brain barrier and causes chronic neuroinflammation.[59] By interacting with monoamine neurotransmitters, including serotonin and dopamine, arsenic changes moods and induces feelings of worry and despair. Arsenic also inhibits neurogenesis and synaptic plasticity, which leads to long-term cognitive damage.[60] Epidemiological evidence has linked childhood exposure to arsenic to a higher incidence of neurological disease and poor mental development.[61] Although trace amounts of copper (Cu), manganese (Mn), and iron (Fe) are essential for brain function, their imbalance causes serious neurotoxicity. This apparent paradox captures the necessity to attain complete homeostatic regulation in the central nervous system.[62]

Iron is necessary for brain metabolism and the production of neurotransmitters, but if its balance is off, it can be harmful. Through Fenton reactions, excess iron catalytically produces hydroxyl radicals that oxidatively damage proteins, DNA, and lipids.[63] Oxidative stress has an impact on the pathogenesis of neurodegenerative diseases, such as Parkinson's and Alzheimer's. β-Amyloid deposition in Alzheimer's disease and α-synuclein fibrillation in Parkinson's disease are both brought on by iron excess. Iron-induced oxidative stress causes dopaminergic neuron degeneration, which leads to motor impairment and cognitive failure.[64]

Another necessary element for normal brain function, manganese, can be hazardous in excess. Frequent exposure can result in "manganism", a type of Parkinson's disease that is most commonly observed in work-related environments like welding and mining. When manganese is preferentially accumulated in the basal ganglia, it interferes with mitochondrial activity and dopaminergic neurotransmission.[65] The presentation is also linked to motor symptoms of bradykinesia and rigidity, as well as neurobehavioral symptoms of irritability and memory impairment. Prolonged manganese exposure has a comparable effect on glutamate absorption and GABAergic neurotransmission, potentially leading to excitotoxic neuronal damage.[66]

Copper has several roles in maintaining the health of neurons, including promoting the synthesis of neurotransmitters and providing antioxidant defense. Conversely, a copper deficiency can cause oxidative stress and protein misfolding, as well as a devastating effect on neuronal redox equilibrium.[67] Increased copper has been linked to β-amyloid and α-synuclein aggregation, which supports neurodegenerative processes in Parkinson's and Alzheimer's diseases. Conversely, cognitive impairment and neurodevelopmental delay may result from copper-deficiency-induced synapse malfunction and myelination problems.[68] Therefore, maintaining neurochemical balance and general brain health requires adequate copper control.[69]

Among the heavy metals that share neurotoxic mechanisms involving the creation of oxidative stress and the disruption of calcium homeostasis are lead, mercury, cadmium, aluminum, and arsenic. Every metal affects neurochemistry and neurologic function differently.[70] While mercury interferes with microtubule polymerization and hinders neuronal migration, lead has a major impact on neurotransmitter function and synaptic formation. Cadmium affects both cholinergic neurotransmission and GABAergic transmission blocking.[71] Alzheimer's disease pathogeneses, such as tau hyperphosphorylation and the development of β-amyloid fibrils, are influenced by aluminum. Arsenic inhibits monoamine neurotransmitter pathways, alters neurogenesis, and changes synaptic plasticity.[72] The complex interaction between these metals and neuronal function emphasizes the importance of metal-specific neuroprotection treatments as well as the necessity to build specific therapeutic treatments against metal-induced neurotoxicity (Table 9.1).[73]

9.2.2 Neurotoxicity and Contaminants

A class of environmental contaminants known as "persistent organic pollutants" (POPs) share three characteristics: toxicity, bioaccumulation, and persistence. Polychlorinated biphenyls (PCBs), dioxins, and polybrominated diphenyl ethers (PBDEs) are a few examples.[79] POPs can accumulate in adipose tissue, cross the blood–brain barrier, and infiltrate brain tissue due to their lipid solubility. POPs cause complex neurotoxicity by interacting with several signal pathways and neuronal components.[80]

POPs interact with membrane receptors and transporters to disturb neurotransmitter balance at the cellular level. Through a variety of mechanisms, they were found to specifically decrease dopamine and γ-aminobutyric acid (GABA)

Table 9.1 Heavy metals and their toxicity.

Metal	Source	Neurochemical impact	Neurological effects	Ref.
Lead (Pb)	Old paints, batteries, plumbing, dust	Blocks NMDA receptors, disrupts dopamine system	Learning deficits, attention issues, IQ decline	74
Mercury (Hg)	Fish (methylmercury), industrial waste	Disrupts glutamate and GABA signaling, mitochondrial dysfunction	Tremors, memory loss, autism-like symptoms	75
Cadmium (Cd)	Cigarette smoke, industrial emissions	Promotes ROS, interferes with neurotransmitter storage	Mood disorders, neurodegeneration	76
Arsenic (As)	Contaminated water, pesticides	Impairs acetylcholine and dopamine release	Cognitive impairment, motor dysfunction	77
Aluminum (Al)	Food additives, cookware, antacids	Promotes β-amyloid aggregation, oxidative stress	Linked to Alzheimer's-like symptoms	78

levels. Because POPs have been shown to damage dopaminergic neurons, they can reduce dopamine synthesis and release.[81] They also affect the function of GABAergic interneurons and interfere with inhibitory neurotransmission. The disturbance of thyroid hormone transmission is a significant component of POP-induced neurotoxicity. Thyroid hormones are essential for the development of the nervous system because they control myelination, synaptogenesis, and neuronal migration.[82]

POPs can interfere with thyroid hormone production, transport, and receptor binding, which can lead to developing neurological defects in the central nervous system. The neurotoxicity of POPs is measured using a variety of behavioral and cognitive deficiencies, such as executive, attentional, and cognitive impairments.[83] The severity and kind of these anomalies are most severe when exposure takes place during crucial periods of neurodevelopment, like pregnancy or early childhood. Because of this heightened vulnerability during developmental windows, exposure to POP during infancy may have long-term neurological effects.[84]

Toluene, benzene, and xylenes are examples of solvents and volatile organic compounds (VOCs). These chemical solvents can readily penetrate the central nervous system (CNS) and cause acute and chronic neurotoxicity because they are lipophilic.[85] Toluene, for instance, inhibits *N*-methyl-D-aspartate (NMDA) and γ-aminobutyric acid type A (GABA_A) receptors, upsetting the relative balance of excitation and inhibition and leading to mental health problems and cognitive impairment.[86]

Diffuse white matter degradation brought on by long-term solvent usage leads to "chronic toxic encephalopathy", a condition marked by ataxia, mood swings, memory loss, and symptoms similar to dementia. Solvent and volatile organic compound (VOC) neurotoxicity includes more than just acute intoxication. Repeated exposure may result in mild brain changes that accumulate over time, even at low levels.[87] They could manifest as a disruption in executive function, a reduction in psychomotor speed, or a change in how sensory data are processed. Particularly in the paint, printing, and chemical sectors, occupational solvent exposure has been linked to an increased risk of neurological conditions such as multiple sclerosis and Parkinson's disease.[88]

Polycyclic aromatic hydrocarbons (PAHs) and polychlorinated diphenyl ethers (PBDEs) are flame retardants and fire byproducts. Two crucial indicators of neurodevelopment are the aryl hydrocarbon receptor (AhR) pathway and thyroid hormone homeostasis, which they disrupt.[89] They also encourage oxidative stress and disrupt calcium homeostasis in neurons.[90] Concentration issues, IQ decrease, childhood developmental delays, and heightened vulnerability to neurobehavioral illnesses have all been connected with them. The neurotoxicity of flame retardants is particularly significant because they are widely utilized in consumer items and do not decompose easily in the environment.[91]

PBDEs can, however, bioaccumulate in human tissues and breast milk, which could impact fetuses' and babies' neurodevelopment. According to recent studies, pregnant women who are exposed to PBDEs have altered brain connections and an increased risk of attention deficit hyperactivity disorder (ADHD) in their unborn children. PAHs, which are usually generated during incomplete combustion processes, have been associated with reduced white matter integrity and cognitive impairment in affected persons.[92] Pesticides: organophosphates, like parathion and chlorpyrifos, and organochlorines, such as dichlorodiphenyltrichloroethane, or DDT, are extremely neurotoxic. Acetylcholine accumulation in synapses and abnormal cholinergic hyperstimulation can cause seizures, neuroinflammation, and neuronal death because organophosphates block AChE.[93]

Organochlorines disrupt GABA receptors and voltage-gated sodium channels, which results in hyperexcitability in the central nervous system. By increasing oxidative stress and dopaminergic damage, frequent exposure to these chemicals increases the risk of neurodegenerative and neurodevelopmental issues.[94] Acute poisoning is not the only instance of pesticide neurotoxicity. Long-term low-dose exposure to organophosphates has been linked to cognitive deficits, increased susceptibility to depression, and changes in the structure and function of the brain. Rural residents and agricultural workers who live in areas with significant pesticide use are the most vulnerable to these effects.[95]

Recent research has connected prenatal exposure to some pesticides to a higher risk of autism spectrum disorder and other neurological problems. The potential contribution of pesticide exposure to the pathogenesis of Parkinson's disease has drawn special attention, and epidemiologic evidence has consistently shown a connection between pesticide exposure and the risk of developing Parkinson's disease.[96] Air pollutants: sulfur dioxide (SO_2), nitrogen oxides (NOx), ozone (O_3), and fine and ultrafine particulate matter (PM2.5/PM0.1) are among the many air pollutants that are neurotoxic. These contaminants travel directly to the brain through the olfactory nerve, along with metals and organic toxins. They cause neuroinflammation by activating glia, inducing pro-inflammatory cytokines, and producing more reactive oxygen species (ROS).[97]

Oxidative stress causes mitochondrial damage and synaptic dysfunction. Epidemiological studies have linked prolonged exposure to air pollution to a higher risk of Alzheimer's disease, depression, brain aging, and diminished cognitive performance.[98] The neurotoxic consequences of air pollution are carefully nuanced since it is so prevalent in urban environments. Air pollution may affect the brain even during fetal development and may even reroute neurodevelopmental processes, according to recent studies.[99]

According to scientific studies employing state-of-the-art neuroimaging techniques, high levels of exposure to air pollution have been associated with reduced brain sizes, especially in regions linked to cognitive functioning and emotional control. There are lifelong neurologic effects of air pollution, including the ability to trigger the beginning of neurodegenerative disease and to hasten the decline of cognitive function in the elderly.[100] Industrial chemicals and plasticizers: Endocrine disruptors, such as phthalates and bisphenol A (BPA), induce the brain to reroute androgenic and estrogenic signals. Through its binding to estrogen receptors in the prefrontal cortex and hippocampus, BPA modifies dendritic spine development and synaptic plasticity.[101]

Phthalates interfere with mood and reward systems by inhibiting serotonin receptors and dopamine transporters. Behavioral abnormalities, anxiety, changes in mood and social interactions, and developmental neurobehavioral impairment are examples of neurotoxic consequences.[102] Human exposure to plasticizers is almost universal due to their widespread use in consumer goods, and long-term neurologic effects are expected. For instance, studies on animals have shown that BPA affects behavior and brain development, which may have consequences for human health.[103] Exposure to phthalates during pregnancy has been linked to behavioral changes in children, including increased aggression and difficulty focusing. Furthermore, by causing epigenetic modifications, these endocrine-disrupting compounds may potentially exhibit transgenerational effects, which could result in neurological abnormalities in future generations.[104] Environmental neurotoxicants work through a variety of multifaceted processes, including endocrine disruption, oxidative stress induction, and disruption of neurotransmitter systems. Understanding and preventing these pathways' effects on brain health is extremely difficult due to their polyvalent nature, sometimes low-dose exposure, and chronic nature.[105] Estimating neurotoxic hazards is further complicated by the potential for synergy between several types of neurotoxicants and the role of genetic susceptibility factors.[106] In order to protect risk groups and lessen the global burden of environmentally induced neurological disease, more extensive research, stronger regulatory frameworks, and public health initiatives are obviously required as our understanding of these environmental insults to neurologic function grows (Table 9.2).[107]

9.3 Diet, Gut-Microbiota, Neurochemical Interactions

9.3.1 Dietary Components and Neurochemical Modulation

Because they directly affect neurotransmitter production, neuronal function, and brain health, nutritional components are essential for neurochemical modulation. There are several mechanisms and routes involved in the intricate relationship between nutrition and neurochemistry: the precursors of neurotransmitters: the amino acids found in food proteins serve as building blocks for the production of neurotransmitters. The primary building block for the production of serotonin is tryptophan, which is found in diets high in protein. Likewise, dopamine production requires tyrosine, another amino acid found in food proteins.[113]

Neurotransmitter levels and, consequently, mood, cognition, and behavior can be directly impacted by the presence or absence of these precursors in the diet. Co-factors for enzymes: as enzymatic cofactors in the production and metabolism of neurotransmitters, micronutrients are essential. The synthesis of several neurotransmitters, including serotonin, dopamine, and GABA, depends on B-vitamins B6, B9 (folate), and B12.[114] Tyrosine hydroxylase, the rate-limiting enzyme involved in catecholamine production, requires iron as a cofactor. Numerous enzymatic processes related to neurotransmitter function and synaptic plasticity involve zinc and magnesium.[115]

Micronutrient deficiencies can lead to neurological and mental disorders by disrupting the development and function of neurotransmitters. Protection from antioxidants: neuroprotection depends on nutrient antioxidants like vitamins C and E as well as plant-based polyphenols found in fruits and vegetables.[116] These compounds inhibit oxidative stress, which otherwise destroys proteins, DNA, and neuronal membranes, and combat reactive oxygen species (ROS). Antioxidants can be used to prevent neurodegenerative events by preventing oxidative damage, which in turn prevents the breakdown of neuronal function. To further support neuroprotection, some antioxidants have also been shown to control neuroinflammatory responses.[117]

Neurotrophic factors and neuroplasticity: by influencing neurotrophic factors like brain-derived neurotrophic factor (BDNF), nutritional control might affect neuroplasticity. It has been discovered that omega-3 fatty acids, which are present in fatty fish and certain plant diets, increase the expression of BDNF and promote neuroplasticity.[118] Conversely, "Western" diets that include a lot of refined sugar and saturated fats have been

Table 9.2 Pollutants causing neurotoxicity.

Category	Examples	Primary neurochemical targets	Neurotoxic consequences	Ref.
Pesticides	Organophosphates (chlorpyrifos, parathion), organochlorines (DDT), pyrethroids	Acetylcholinesterase inhibition, GABA_A receptor blockade, altered dopamine signaling, oxidative stress	Seizures, impaired attention, mood disorders, developmental delays, increased Parkinson's risk	108
Persistent organic pollutants (POPs)	PCBs, dioxins, PBDEs	Thyroid hormone disruption, dopamine depletion, reduced GABAergic signaling, oxidative stress	Learning deficits, executive dysfunction, behavioral abnormalities, developmental neurotoxicity	109
Air pollutants	Particulate matter (PM2.5, PM0.1), nitrogen oxides, ozone, sulfur dioxide	Neuroinflammation, ROS generation, impaired monoamine neurotransmission, HPA axis activation	Cognitive impairment, depression, accelerated brain aging, neurodegeneration (Alzheimer's, Parkinson's)	110
Industrial and household chemicals	Bisphenol A (BPA), phthalates, volatile organic compounds (toluene, benzene, xylene)	Estrogen receptor disruption, dopamine transporter interference, NMDA/GABA receptor inhibition	Mood disturbances, anxiety, impaired memory, neurodevelopmental deficits, toxic encephalopathy	111
Combustion byproducts	Polycyclic aromatic hydrocarbons (PAHs), flame retardants	AhR activation, disruption of calcium signaling, thyroid imbalance	Reduced IQ, developmental delay, attention and behavioral problems	112

associated with lower BDNF levels and problems with hippocampus development. Learning, memory, and cognition are all impacted by these dietary neuroplasticity effects.[119] Metabolic dysregulation and neuroinflammation: the "Western" dietary pattern of high-fat and high-sugar meals may cause both metabolic dysregulation and neuroinflammatory cascades in the brain. This includes an increased release of pro-inflammatory cytokines by activated microglia and astrocytes brought on by these diets.[120] Chronic low-grade inflammation may lead to desynchronization of neurotransmitter activity, synaptic plasticity, and the pathophysiology of mood disorders and cognitive failure. Furthermore, oxidative stress and neuroinflammation in the brain may be exacerbated by metabolic diseases associated with such diets, such as insulin resistance.[121] Gut–brain axis: through the gut–brain axis, the intestinal microbiota plays a crucial role in regulating neurochemistry when it is strictly regulated by food. Dietary components, particularly fermented foods and fiber, may promote the growth of the good colonic bacteria that generate SCFAs.[122] These SCFAs are also said to regulate blood–brain barrier integrity, neuroinflammation, and neurotransmitter synthesis. Second, the gut flora also metabolizes tryptophan, which impacts the brain and gut serotonin levels.[123] Epigenetic modulation: gene expression related to neurotransmitter function and neuroplasticity can be changed by specific food items. Methyl donor nutrients like folate, vitamin B12, and choline, for instance, may change DNA methylation patterns, which may impact the expression of genes related to neurotransmitter synthesis and receptor function. These epigenetic processes open a line of communication between dietary intake and neurochemical activity alterations that last longer than an axis of moderation.[124] In summary, the dietary components and neurochemical modulation are complex and diverse. The antioxidant effect provides neuroprotection, neuroplasticity is promoted, healthy neuroinflammation is preserved, and good nutrition supports the synthesis of neurotransmitters.[125] Conversely, a poor diet causes persistent neuroinflammation, elevated oxidative stress, neurochemical imbalance, mood disorders, and even cognitive impairment.[126] In addition to supporting dietary treatments for the prevention and treatment of neurological and mental illnesses, an understanding of these complex interactions highlights the vital role nutrition plays in preserving brain health.[127] Gut microbiota: the gut microbiota is responsible for a majority of the neuroactive substances that can directly modulate processes in the brain through a variety of mechanisms. There has been a growing interest in how gut microbes interact with the central nervous system, referred to as the gut–brain axis, in the neuroscientific literature. The gut microbiome, or the complex gut microbiota comprising the gastrointestinal tract, contains trillions of species of all types of microbes, which include bacteria, fungi, viruses, and archaea.[128]

These microbes have a synergetic effect on multiple physiological functions such as digestion, immunity, and nervous function. These microbes have also been documented to secrete various neurotransmitters and neuromodulators, which include short-chain fatty acids (SCFAs) such as acetate, propionate, and butyrate that are produced by different gut microbiota, serotonin, acetylcholine, and dopamine from the gut microbiota, and γ-aminobutyric acid (GABA) produced by *Bifidobacterium* and *Lactobacillus* species in the gut microbiome.[129] The production of neuroactive molecules by the gut microbiota exemplifies the highly interactive nature of microbial metabolism with host neurochemistry. For example, the primary inhibitory neurotransmitter in the central nervous system, GABA, plays an important role in mood control, anxiety, and cognition. Similarly, serotonin, the "feel good" neurotransmitter, influences mood, appetite, and sleep.[130]

These metabolites from gut microbiota modulate enteroendocrine cell function, activate vagal sensory nerve fibers, and influence the blood–brain barrier, among other neural influences. The vagus nerve, a branch of the autonomic nervous system that is distributed throughout the body, provides a unique pathway for communication between the gut and the brain.[131] Microbial metabolites can activate vagal afferent neurons, which are sensory neurons that send sensory signals from the gut to the brain and drive many neural processes. Enteroendocrine cells in the gut will secrete hormones and peptides that act on the brain and influence behavior after exposure to metabolites.[132] Reports state that SCFAs, butyrate in particular, have an inhibitory effect on histone deacetylase, which stimulates the production of BDNF, neurogenesis, and synaptic plasticity. Butyrate, a result of bacterial fermentation of dietary fiber, can directly change the gene expression of brain cells and cross the blood–brain barrier.[133] Butyrate promotes histone acetylation by inhibiting histone deacetylases, which leads to an open chromatin shape and increased gene transcription. This epigenetic regulation may lead to an increase in the expression of neuroprotective and neuroplasticity genes, including BDNF, which is crucial for synaptic plasticity, neuronal survival, and neuronal growth.[134]

When dysbiosis presents as an abnormal gut microbial community, neurological function is negatively impacted. Reduced levels of beneficial metabolites, increased bacterial production of lipopolysaccharide (LPS) and other toxic substances, systemic inflammation, microglia activation, and impairment of synaptic integrity and neurotransmission are some of these effects.[135] Dysbiosis can result for a variety of reasons, such as a poor diet, antibiotic use, stress, and environmental pollutants. The altered microbiota may suppress the production of SCFAs, the advantageous metabolites, and increase potentially dangerous substances like LPS, a component of Gram-negative bacterial cell walls that can induce inflammation.[136]

Dysbiosis is not limited to the gut; it also impacts the brain through a number of routes. Systemic inflammation brought on by dysbiosis might weaken the blood–brain barrier and let potentially dangerous substances and pro-inflammatory mediators into the brain. Neuroinflammation may result from this activation of microglia, the brain's inherent immune cells.[137] Long-term microglial activation has been linked to a variety of neurological illnesses, such as mood disorders and neurodegenerative diseases. Moreover, dysbiosis may disrupt the brain's delicate neurotransmitter balance. For example, tryptophan, the precursor molecule of serotonin, is influenced in its bioavailability by changes in gut flora, thus inducing mood and cognitive problems.[138]

Preclinical disruption of synaptic integrity, required for normal brain communication, may also result in neurological dysfunction. The gut microbiota–neurochemistry bidirectional relationship demonstrates how essential healthy gut flora is for optimal brain function. The gut–brain axis is bidirectional; the gut microbiome modulates brain activity and *vice versa*, brain function is modulated by the gut microbiome.[139] This relationship underscores the overall connection of gut and CNS functions. There is growing evidence that it is possible to alter gut microbiota through dietary means, pre- and probiotics, to treat various neurological and psychiatric disorders.[140] Probiotic supplementation, for instance, has been shown to be of benefit in the treatment of depression and anxiety symptomology. Reversing neuroinflammation and improvement of cognitive function have also been established as outcomes of dietary therapies that boost gut bacteria activity. Further studies in this field will be needed to better understand the processes of the gut–brain axis and work towards potential therapeutic interventions on the microbiome in neurological and mental illness.[141] Increasing understanding of the gut–brain axis offers exciting opportunities for creating new therapies and averting a number of neurological disorders. Future studies should focus on identifying microbial strains or metabolites that show neuromodulation or neuroprotection and evaluating the possibility of microbiome-based biomarkers for the early detection and tracking of neurological disorders (Figure 9.1).[142]

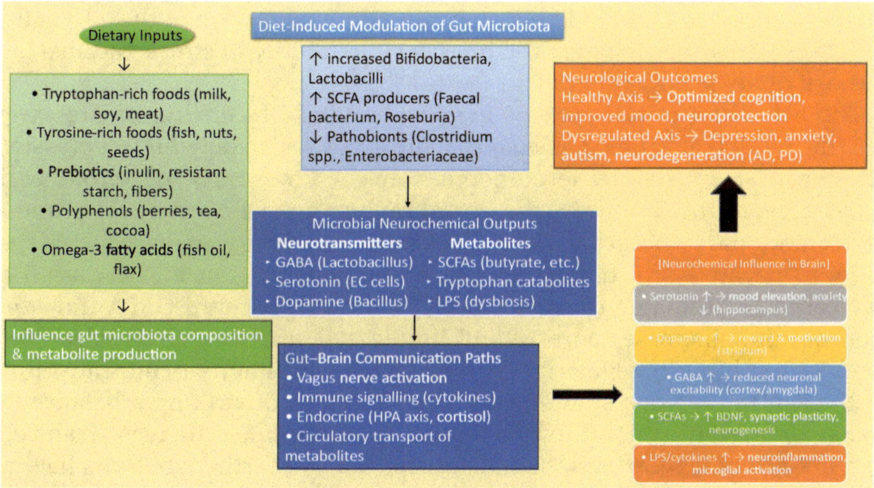

Figure 9.1 Flow chart of dietary and gut microbial influence on neurological interactions.

9.4 Environmental Stressors and Neurochemical Consequences

External stressors, such as physical, pharmacological, and psychological agents that disrupt homeostasis and result in either maladaptive or adaptive physiological reactions, are especially dangerous to human brain systems. Even though acute stress reactions are necessary for survival, long-term exposure to stressors typically has detrimental consequences on neurochemical systems that change cognitive processing, emotional regulation, and neurological integrity overall.[143] Environmental stressors can be categorized as physical (noise pollution, circadian disturbance), chemical/toxicological (pesticides, heavy metals), or psychosocial (isolation, job stress). Stress activates many neurochemical pathways, including the hypothalamic–pituitary–adrenal (HPA) axis and the sympatho–adrenal–medullary (SAM) system.[144]

Long-term HPA axis stimulation can lead to chronic cortisol secretion, hippocampus atrophy, prefrontal brain dysfunction, and impaired feedback inhibition. It plays a part in sadness, anxiety, and neurodegeneration. The hippocampus, one of the main brain areas involved in memory consolidation and emotional processing, has been shown to be most affected by chronic stress.[145] Prolonged elevations in cortisol levels can cause dendritic atrophy, decreased neurogenesis, and aberrant long-term potentiation in hippocampal neurons. Structural and functional alterations lead to cognitive impairments, especially in declarative memory and spatial navigation.[146]

Maladaptive remodeling in dopaminergic and noradrenergic circuits caused by repeated activation of the SAM system raises susceptibility to addiction and stress. Repetitive activation of the SAM system increases the release of catecholamines, especially norepinephrine and epinephrine, from the sympathetic nerve terminals and adrenal medulla.[147] This continuous increase in catecholamine levels may cause desensitization of the adrenaline receptors which may change the brain's vulnerability to stress and increase the risk of anxiety disorders and PTSD.[148]

The neurotransmitter systems are also impacted by environmental stresses. In the dorsal raphe nucleus, the substrate for anxiety and depression, prolonged stress lowers serotonergic tone. A key source of serotonergic projections to the forebrain, the dorsal raphe nucleus plays a role in mood and emotional processing. Sleep–wake irregularities, increased susceptibility to negative affect, and poor emotional control can all be caused by decreased serotonergic transmission.[149] Similarly, reduced GABA transmission not only increases cortical and limbic hyperexcitability but also interferes with inhibitory regulation.[150] The primary inhibitory neurotransmitter in the central nervous system, GABA, is in charge of striking a balance between excitation and inhibition in neuronal circuits. Increased anxiety, disturbance of fear extinction, and susceptibility to further stressors have all been linked to stress-repeated reductions in GABA transmission.[151] Repeated stress exposure increases extracellular glutamate levels, especially in the hippocampus and prefrontal cortex. This disrupts synaptic plasticity and elevates the risk of neurodegenerative disease. Glutamate overrelease can result in excitotoxicity, a disorder where repeated activation of glutamate receptors results in neurodegeneration or death. Glutamate-evoked excitotoxicity has been implicated in a number of neurodegenerative diseases, such as Parkinson's and Alzheimer's diseases.[152] Environmental stresses have an impact on neurochemical systems in addition to these extensive networks of neurotransmitters. Stress also affects the endocannabinoid system, which is essential for emotional regulation and stress adaptation. Endocannabinoid signaling may be dysregulated by prolonged stress exposure, which is likely connected to the pathophysiology of stress-related mental illnesses.[153]

Environmental stressors can also change neuroinflammatory processes by triggering astrocytes and microglia and generating pro-inflammatory cytokines. This neuroinflammation will therefore increase the negative effects of stress on brain function and contribute to the pathogenesis of many neurological and psychiatric illnesses. This complex interaction between environmental stressors and neurochemical pathways also involves epigenetic mechanisms. Chronic stress may result in long-lasting alterations in gene expression through DNA methylation, histone modification, and modifications in non-coding RNA expression. These persistent epigenetic alterations may even be inherited by subsequent generations, making offspring more vulnerable to stress.[154]

Pharmacological manipulations of particular neurotransmitter systems, stress-reduction strategies to alter HPA axis activity, and lifestyle modifications to improve resistance to environmental stressors are examples of effective treatments.[155] To comprehend how environmental adversity results in mental and neurological disorders, one must have a solid understanding of the neurochemical mechanisms underlying stress responses. In order to stop the negative consequences of ongoing stress on neurological health, it can also be used to propel the creation of specific therapies.[156]

Environmental stresses induce neuroinflammation and oxidative stress in addition to directly causing neurotransmitter dysregulation. Neuronal function is hampered by the overproduction of reactive oxygen species brought on by stress-induced mitochondrial damage.[157] Proteins, lipids, DNA, and other cell components can be broken down by oxidative stress, which can further impair neuronal function and result in cell death. The brain is particularly vulnerable due to its high lipid content and fast rate of oxygen consumption.[158] Additionally, stress causes pro-inflammatory cytokines like TNF-α and IL-6 to be released as well as microglia to become activated. Activated microglia, the resident immune cells of the central nervous system, emit these pro-inflammatory substances. The cascade of neuroinflammation exacerbates neurotransmitter dysregulation, which increases long-term mental vulnerability.[159]

Chronic inflammation is the underlying cause of several neuropsychiatric disorders, such as anxiety, depression, and post-traumatic stress disorder (PTSD), and it also changes the metabolism of neurotransmitters and may harm the blood–brain barrier.[160] The cumulative impacts of environmental stressors cause structural and functional modification of neural networks. While amygdala activation exacerbates emotional dysregulation and fear conditioning, impaired hippocampus function decreases learning and memory. The hippocampus area, which is involved in stress management and memory consolidation, is significantly impacted by chronic stress.[161]

Long-term stimulation of stress hormones such as cortisol can lead to dendritic withdrawal, reduced neurogenesis, and eventually volume loss in this region. However, long-term stress hyperactivates the amygdala, which controls fear and emotional stimulation, leading to an increase in emotionality and anxiety.[162] Stress tolerance, executive function, and decision-making are all hampered by prefrontal brain damage. The prefrontal cortex, which controls higher-order thinking, undergoes significant change as a result of repeated stress.[163] This can show up as altered dendritic form, decreased synaptic density, and poor functional connections with other parts of the brain. Working memory, impulse control, and cognitive flexibility—all essential for handling and adapting to stressful situations—may be affected as a result.[164]

Reduced gray matter volume in brain regions that control stress is associated with the severity of mental symptoms, according to neuroimaging studies. These structural changes are not limited to the hippocampus, amygdala, and prefrontal cortex; they also affect the insula and anterior cingulate cortex, which are also involved in stress management. The extent of these volumetric declines is often associated with the length and severity of stress exposure, as well as the severity of mental symptoms.[165]

For the purpose of treatment and prevention, it is essential to investigate the neurochemical effects of environmental stressors. Knowledge enables the development of targeted therapies that may undo the harm that long-term stress does to the structure and functionality of the brain.[166] Pharmacotherapy uses selective serotonin reuptake inhibitors (SSRIs), which can restore neurotransmitter balance and enhance neuroplasticity. Benzodiazepines and other GABA modulators that enhance inhibitory neurotransmission will control anxiety. Novel anti-inflammatory medications are being used to treat the neuroinflammatory component of stress disorders, potentially opening up a new therapeutic option.[167]

It may also be beneficial to adopt other lifestyle habits including regular exercise, mindfulness-based stress reduction, and proper sleep hygiene. Exercise has been shown to increase neurotrophin levels, promote neurogenesis, and reduce inflammation, all of which can help to lessen the negative consequences of chronic stress.[168] Engaging in mindfulness practices can improve prefrontal brain function, reduce amygdala reactivity, and influence emotional responses. Sleep hygiene is particularly crucial since sleep disruptions can exacerbate stress symptoms and reduce cognitive function.[169]

Another method that policy initiatives could use to mitigate the detrimental effects of stressors on public health is by reducing environmental

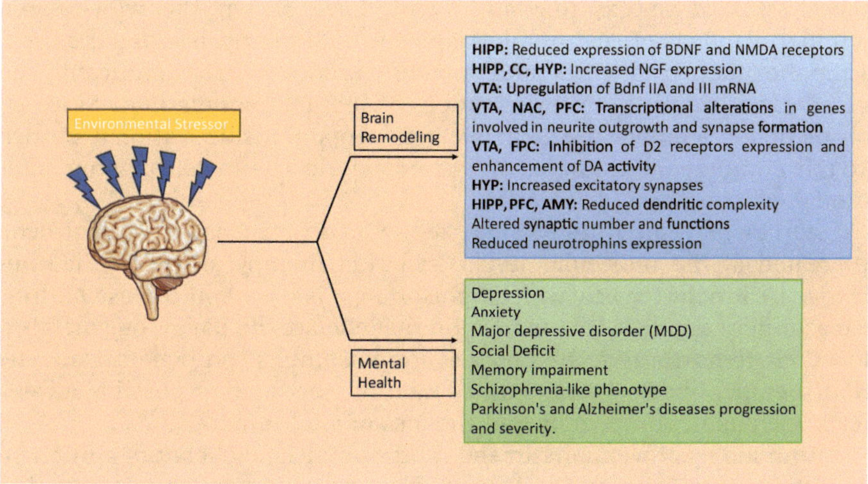

Figure 9.2 Effects of environmental stressors on neurochemistry.

toxins. This includes stricter control of air and water quality, a reduction in urban noise pollution, and more green space to promote stress reduction and physical activity.[170] Additionally, organizational initiatives that reduce chronic stress, such as promoting work–life balance and providing mental health services, may help avoid stress-related disorders. Public education programs that raise awareness of the neurobiologic effects of chronic stress and healthy coping strategies may also be necessary to create more stress-resistant populations (Figure 9.2).[171]

9.5 Prevention Strategies for Environmental Neurotoxicity

Notwithstanding the unbridled expansion of business, lifestyle, and urbanization, environmental neurotoxicity is increasing. Humans are increasingly exposed to hazardous environmental substances, including heavy metals, pesticides, industrial solvents, air pollutants, and endocrine-disrupting chemicals.[172] By disrupting neurotransmitter systems, producing oxidative stress, reducing synaptic plasticity, and initiating neuroinflammation, these neurotoxic chemicals collectively increase the risk of cognitive impairment, mental disorders, and neurodegenerative illnesses. To lessen environmental neurotoxicity, a comprehensive approach including dietary modifications, clinical care, molecular therapy, and broad public health campaigns is required. The best way to reduce exposure to neurotoxics is to address the underlying cause.[173]

This can be accomplished through strict regulatory control of industrial discharges, enhanced enforcement of exposure limits, and the phase-out of neurotoxic pesticides and industrial chemicals. Previous efforts, such as the removal of lead in gasoline and paint, have shown the wide-ranging neurological effects of environmental policy.[174] Similarly, limiting the use of organophosphate pesticides and mercury products has significantly decreased exposure to these toxins among vulnerable populations. Since environmental pollutants are less likely to remain inside national borders through commercial, aquatic, and atmospheric channels, ongoing international collaboration is essential.[175]

To stop or fix cellular damage caused by neurotoxic assaults, treatments are created at the molecular level. Chelation therapy, a well-studied and effective treatment for heavy metal poisonings, is based on the use of drugs such as dimercaprol, EDTA, and DMSA that chelate the dangerous metals to allow their removal from the body. Acute poisonings can benefit from chelation therapy, but its use should be cautious because of possible adverse effects such as nephrotoxicity and the loss of vital minerals.[176]

Continuous improvements in the synthesis of new chelators aim to improve their specificity and safety profiles, maybe even broadening their therapeutic use to treat chronic low-level exposure. Even though control measures and the phase-down of toxic chemicals have been successful in limiting exposures to neurotoxins, such as successful efforts to reduce fuel and paint lead content, the global distribution of environmental toxins necessitates ongoing international cooperation to maximize the prevention of neurological risks.[177]

Treating neurotoxicity with antioxidants is another way to reduce it. Oxidative stress from many environmental toxicants results in lipid peroxidation, protein misfolding, and DNA damage. The neuroprotective benefits of dietary and pharmaceutical antioxidants, including vitamins C and E, coenzyme Q10, *N*-acetylcysteine, and polyphenolic compounds curcumin and resveratrol, have been shown in experimental models.[178] These drugs improve endogenous antioxidant defense, restore mitochondrial function, and block pro-inflammatory signals. Despite continued difficulties with clinic translation due to worries about bioavailability and result heterogeneity, recent developments in targeted formulations and nanoparticle administration provide fresh hope for an efficient antioxidant-based treatment.[179]

Anti-inflammatory techniques have significant therapeutic utility since microglial activation and cytokine production are important contributors to neurotoxicity. Furthermore, pharmacologic drugs that work by inducing neuroinflammation, such as minocycline, new cytokine inhibitors, and non-steroidal anti-inflammatory drugs, have shown promise in preclinical models.[180] Furthermore, nutritional therapy containing large amounts of flavonoids, omega-3 fatty acids, and other anti-inflammatory substances is necessary to reduce neuroinflammatory responses. Changing the gut microbiota has become increasingly recognized as an indirect but very effective way to influence environmental neurotoxicity, since microbial metabolites regulate central nervous system function and systemic inflammation.[181]

To increase resistance to environmental neurotoxins, lifestyle modifications are crucial in addition to medication treatments. Most of the adverse effects of toxic exposure can be reversed by exercise, which has been shown to increase neurogenesis, activate defensive antioxidant mechanisms, and improve mitochondrial function.[182] Proper sleep and circadian control are also essential to allow for optimal detoxification processes and prevent neuroinflammation, as chronic sleep abnormalities accelerate the rate of neuronal deterioration and increase oxidative stress. Dietary approaches that prioritize whole grains, fruits, and vegetables along with healthy fats supply the essential micronutrients and phytochemicals that support antioxidant and anti-inflammatory pathways.[183]

Together, these lifestyle modifications at the population level provide a low-cost and convenient means of reversing the risks associated with exposure to environmental neurotoxins. One emerging area in this field is the application of precision medicine. Both genetic and epigenetic variability have a substantial impact on an individual's vulnerability to environmental neurotoxicants. Gene polymorphisms that impact detoxification enzymes, antioxidant activities, or neurotransmitter pathways regulate the level of vulnerability to specific toxins.[184]

Environmental exposure-induced epigenetic modifications, such as histone acetylation and DNA methylation, affect gene expression over time and are occasionally inherited by subsequent generations. Focused prevention, early intervention, and customized therapy intervention are made possible by the identification of biomarkers for certain genetic vulnerabilities. Thus, genomes, metabolomics, and exposomics must be integrated to create personalized neuroprotection models.[185] Public health interventions are still required to prevent environmental neurotoxicity. Education programs that increase the understanding of exposure sources and safe behaviors empower communities and the general public to reduce risk. Routine screening for neurotoxic exposures in high-risk groups, including children, pregnant women, and employees exposed at work, enables early identification and intervention.[186]

By providing green space, clean air, and noise reduction as a top priority in urban development, the cumulative impact of environmental stressors on neurological health can be reduced. Particularly in the rapidly industrializing globe, unilateral deployment of neuroprotective measures is prevented from weakening them by international collaboration on environmental control.[187] Finally, toxicology, neuroscience, medicine, and politics must work together to combat environmental neurotoxicity. In addition to being a clinical issue, mitigation calls for government action, community engagement, and technological advancement. Technological developments in green chemistry, sustainable agriculture, and renewable energy reduce reliance on neurotoxic chemicals at the point of production, while research into neuroprotective medications and therapies continues to enhance therapeutic response capacity.[188]

Overall, reducing exposure, pharmaceutical and nutritional therapies, lifestyle modification, and macro-policy change are all important goals of a comprehensive strategy to mitigate environmental neurotoxicity. Although each strategy focuses on a different aspect of the neurotoxic issue, their combined use has the greatest potential to protect brain health.[189] Even though exposure to the environment is a necessary part of modern life, proactive steps that increase resilience and reduce risk should take precedence over reactive approaches to controlling neurotoxic effects. It is feasible to considerably lessen the neurological effects of environmental pollutants and support cognitive health throughout one's lifespan by combining advancements in biomedicine with preventive legislation and public awareness.[190]

9.6 Future Perspective

To enable greater understanding and better intervention techniques, future research in environmental neurotoxicology should be directed in a number of broad directions. It is possible to identify the ways in which specific stressors alter neurochemical pathways by using exposomics technology to measure complex lifetime environmental exposures in conjunction with metabolomics, epigenetics, and proteomics.[191] This can offer a more comprehensive view of the intricate relationships between environmental factors and neurological health, shedding light on the long-term cumulative consequences of joint exposures.[192]

In order to collect data on the impacts of early exposure and developmental stages that are sensitive, longitudinal cohort studies of human subjects that follow participants from birth to old age are crucial. The late-onset effects of environmental exposure, windows of neurodevelopmental risk, and the causal relationship between early exposures and later neurologic outcomes can all be clarified by longitudinal cohort studies. The possible transgenerational effects of environmental pollutants on brain health can also be determined by longitudinal studies.[193]

The effects of environmental agent combinations on neurotransmitters, synaptic signaling, and glial function are transmitted by sophisticated human-source models such as brain organoids, organ-on-chip platforms, and gut–brain axis platforms.[194] Compared to conventional animal models, these novel models more closely mimic human neurophysiology and enable researchers to examine the complex mechanism of neurotoxicity in an experimental setting. They also make it possible to study the intricate relationships between many organ systems, including the gut–brain axis, neurological illness, and the focus on neurodevelopment.[195]

Vulnerable brain cells and their responses can be described using spatially resolved techniques and single-cell sequencing. These advanced techniques offer previously unheard-of clarity for examining the effects of environmental toxins on specific brain cell types, describing susceptibilities and adaptive reactions at the cell-type level. This resolution can help uncover the cellular basis of neurotoxicity and guide the creation of focused therapies.[196]

Establishing the long-term impacts of brief exposure to behavior, mental illness, and cognition requires an understanding of the brain's epigenetic markers. After exposure, epigenetic changes like DNA methylation and histone modification can persist for a few years and control gene expression models all the way through life. Studies on these markings can demonstrate the long-lasting effects of environmental exposure on behavior and brain function.[197]

Because real-world environmental exposures include multipollutant mixtures of chemicals, heavy metals, noise, heat, and chronic life stress that interact with diet, sleep, socioeconomic position, and lifestyle, it is necessary to look at both chemical and psychosocial stressors simultaneously. The multifactorial approach aims to identify the synergy or antagonistic relationship between many stressors while taking into account the multifaceted nature of environmental impacts on brain health. Additionally, it takes into account how lifestyle choices and genetic vulnerability may alter the consequences of environmental exposure.[198]

To comprehend the impact of environmental stress on the brain, more accurate models of these coupled stresses must be created. They must allow for the dynamic character of long-term environmental exposures and be multifactorial. They must allow for the presence of resilience mechanisms and the potential for adaptive responses to long-term environmental stress.[199] One of the top priorities is to bridge the gap between solution-oriented and mechanistic science. It entails creating preventative measures such as microbiota modification, diet-based therapies, antioxidant supplements, and legislation reducing exposure to neurotoxic chemicals. A multidisciplinary strategy is necessary to translate research findings into action-oriented solutions, involving not only scientists but also healthcare professionals, policymakers, and the wider community.[200]

Early-life prevention and targeted intervention can be guided by customized risk-assessment instruments that record exposure history and neurochemical indicators. To inform customized risk assessments and preventive measures, these tools may include exposure and effect biomarkers, genetic data, and epigenetic information. Individuals with a history of major environmental exposures or high-risk groups may benefit most from these customized solutions.[201] The combined efforts of neuroscientists, physicians, toxicologists, environmentalists, and public health policymakers are needed to advance this field. The complicated problems brought up by environmental neurotoxicology necessitate interdisciplinary cooperation. It can support the translation of scientific knowledge into practical public health policy, the creation of new research strategies, and the integration of various data.[202]

We will be better equipped to safeguard the brain throughout life in a world that is becoming more complicated if we can connect basic research to a clearer link to prevention and policy. This strategy entails developing and putting into practice evidence-based strategies for reducing environmental insults to brain health in addition to our advancements in the understanding of neurotoxic mechanisms. Maintaining neurological health at the societal and individual levels requires constant communication between researchers, decision-makers, and the general public in an effort to connect scientific data with practical action.[203]

9.7 Conclusion

Neurochemistry, which governs behavior and brain function throughout an organism's life, is greatly influenced by environmental circumstances. Early experiences, stress, diet, exposure to toxins, and social interaction are examples of heterogeneous environmental effects that illustrate the dynamic interaction between external stimulation and internal and exterior brain activities. Because of its remarkable plasticity, the brain frequently reorganizes itself in response to changes in the environment by modifying gene expression, synaptic pathways, and neurotransmitter systems. These neurochemical changes may significantly affect cognition, regulation, and mental health. The complex mechanisms *via* which the environment affects neurochemistry are central to the pathophysiology of most neurological and psychiatric illnesses.

The complex mechanisms by which the environment influences neurochemistry are a target for the pathophysiology of most neurological and psychiatric illnesses. Stress early in life, for instance, has been implicated in alterations of long-term HPA axis function, along with effects on susceptibility to mood disorder and stress responsiveness. Omega-3 fatty acid intake is one of the dietary elements discovered to affect neurotransmitter synthesis and synaptic plasticity that can determine the course of mental illness as well as mental status. Environmental toxins like air pollution and heavy metals can drive neurodegenerative processes and interfere with neurotransmitter function. The environment and social interaction significantly influence neurochemistry. The stability of social connections and social support networks can affect the release of neurotransmitters like oxytocin and serotonin, which are important in emotional regulation and social bonding. Cultural values and beliefs can affect the neurochemical processes that govern stress management and are associated with resilience and well-being.

Environmental neuroscience has made significant progress in mapping how the environment affects brain function and physiology through specific mechanisms. With technologies such as positron emission tomography (PET) and functional magnetic resonance imaging (fMRI), scientists are now able to capture neurotransmitter changes and live brain responses to external stimuli. Evidence from epigenetic studies suggests that environmental stimuli influence gene expression over time, and thus neurochemical transmission, to later generations. The interaction of environmental factors and neurochemistry is the foundation for effective individualized prevention and treatment approaches. The therapy can even inform individualized treatment approaches, as it is individualized around each patient's unique risk factor and environmental exposure to that factor. To avoid a negative effect of environmental exposure on neurological well-being, first, public health policies could pertain to restricting air pollution or giving consumers information about food. Second, since environmental neuroscience calls for, in some way, interdisciplinary approaches, it will be important to achieve maximum neurochemical balance and address critic issues we face in modern society. Future research will need to investigate the cumulative impact on brain function across the lifespan and the dynamic interactions of exposures to more than one environmental factor. This knowledge will guide evidence-based policy, public health programs, and individualized treatment options to alleviate the negative inequities related to ongoing environmental toxic exposures as they relate to our neurohealth. The integration of consumer exposomics, epigenetics, and advancements in neuroimaging will provide insights into how the environment shapes our brain chemistry and consequent function across the life course. We will gain new insights into how environmental exposures predict variations in neurochemical function and the lasting implications long-term environmental exposures can have on brain functioning across longitudinal cohorts. If we broaden our area of research to include examinations of the concept that environmental enrichment and healthier lifestyle choices foster neuroplasticity and cognitive resilience, we will be able to further develop new approaches to preserve or evolve cognition. By approaching this work without discipline constraints and with an integrated worldview, we can more effectively explore how to support neurologic health within an increasingly complex environmental context. Importantly, the depth of knowledge on the influence the environment has on neurochemistry allows us to contribute to new frameworks to develop prevention, treatment, and public health actions. Such developments will promote optimal brain function across a range of contexts and populations, improve quality of life, and minimize or relieve the global burden of neurologic and mental health disorders.

Abbreviations

BDNF	Brain-derived neurotrophic factor
CRH	Corticotropin-releasing hormone
EDTA	Ethylenediaminetetraacetic acid
GABA	Gamma-aminobutyric acid
HPA	Hypothalamic–pituitary–adrenal
5-HT	5-Hydroxytryptamine
IL-6	Interleukin-6
NMDA	*N*-Methyl-D-aspartate
PTSD	Post-traumatic stress disorder
ROS	Reactive oxygen species
SAM	Sympatho–adrenal–medullary
SSRI	Selective serotonin reuptake inhibitor
TNF-α	Tumor necrosis factor alpha

References

1. S. Chourbaji, H. Hörtnagl, R. Molteni, M. A. Riva, P. Gass and R. Hellweg, The impact of environmental enrichment on sex-specific

neurochemical circuitries–effects on brain-derived neurotrophic factor and the serotonergic system, *Neuroscience*, 2012, **220**, 267–276.
2. D. J. Amit and N. Brunel, Learning internal representations in an attractor neural network with analogue neurons, *Network Comput. Neural Syst.*, 1995, **6**(3), 359–388.
3. M. A. Scattolin, R. M. Resegue and M. C. Rosário, The impact of the environment on neurodevelopmental disorders in early childhood, *J. Pediatr.*, 2022, **98**, 66–72.
4. C. J. Axelrod, S. P. Gordon and B. A. Carlson, Integrating neuroplasticity and evolution, *Curr. Biol.*, 2023, **33**(8), R288–R293.
5. J. Reis, A. Buguet, G. C. Román and P. S. Spencer, Environmental neurology: concepts and short history of an interdisciplinary approach to etiology, treatment and prevention, *J. Neurol. Sci.*, 2023, **454**, 120861.
6. P. Onu, A. Pradhan and C. Mbohwa, Potential to use metaverse for future teaching and learning, *Educ. Inf. Technol.*, 2024, **29**(7), 8893–8924.
7. K. Langer, V. L. Jentsch and O. T. Wolf, Rapid effects of acute stress on cognitive emotion regulation, *Psychoneuroendocrinology*, 2023, **151**, 106054.
8. N. V. Gulyaeva, Neuroendocrine control of hyperglutamatergic states in brain pathologies: the effects of glucocorticoids, *J. Evol. Biochem. Physiol.*, 2022, **58**(5), 1425–1438.
9. C. Marcolongo-Pereira, F. C. Castro, R. M. Barcelos, K. C. Chiepe, J. V. Rossoni Junior, R. P. Ambrosio, O. Chiarelli-Neto and A. P. Pesarico, Neurobiological mechanisms of mood disorders: Stress vulnerability and resilience, *Front. Behav. Neurosci.*, 2022, **16**, 1006836.
10. X. Zhang, X. Wei, Y. Mei, D. Wang, J. Wang, Y. Zhang, X. Li, Y. Gu, G. Peng and B. Sun, Modulating adult neurogenesis affects synaptic plasticity and cognitive functions in mouse models of Alzheimer's disease, *Stem Cell Rep.*, 2021, **16**(12), 3005–3019.
11. M. L. Raza, S. T. Hassan, S. Jamil, W. Fatima and M. Fatima, Nutritional interventions in depression: The role of vitamin D and omega-3 fatty acids in neuropsychiatric health, *Clin. Nutr.*, 2025, **45**, 270–280.
12. M. Pilecky, L. Závorka, M. T. Arts and M. J. Kainz, Omega-3 PUFA profoundly affect neural, physiological, and behavioural competences–Implications for systemic changes in trophic interactions, *Biol. Rev.*, 2021, **96**(5), 2127–2145.
13. A. Sobczyńska-Malefora, E. Delvin, A. McCaddon, K. R. Ahmadi and D. J. Harrington, Vitamin B12 status in health and disease: a critical review. Diagnosis of deficiency and insufficiency–clinical and laboratory pitfalls, *Crit. Rev. Clin. Lab. Sci.*, 2021, **58**(6), 399–429.
14. S. S. Shetty, D. Deepthi, S. Harshitha, S. Sonkusare, P. B. Naik and H. Madhyastha, Environmental pollutants and their effects on human health, *Heliyon*, 2023, **9**(9), e19496.
15. M. Nabi and N. Tabassum, Role of environmental toxicants on neurodegenerative disorders, *Front. Toxicol.*, 2022, **4**, 837579.

16. D. Jerusalinsky, E. Kornisiuk and I. Izquierdo, Cholinergic neurotransmission and synaptic plasticity concerning memory processing, *Neurochem. Res.*, 1997, **22**(4), 507–515.
17. D. Fiorillo and F. Sabatini, Quality and quantity: the role of social interactions in self-reported individual health, *Soc. Sci. Med.*, 2011, **73**(11), 1644–1652.
18. S. Choudhuri, Y. Cui and C. D. Klaassen, Molecular targets of epigenetic regulation and effectors of environmental influences, *Toxicol. Appl. Pharmacol.*, 2010, **245**(3), 378–393.
19. E. Paizanis, S. Kelaï, T. Renoir, M. Hamon and L. Lanfumey, Life-long hippocampal neurogenesis: environmental, pharmacological and neurochemical modulations, *Neurochem. Res.*, 2007, **32**(10), 1762–1771.
20. T. Wu, W. Cai and X. Chen, Epigenetic regulation of neurotransmitter signaling in neurological disorders, *Neurobiol. Dis.*, 2023, **184**, 106232.
21. R. I. Teleanu, A. G. Niculescu, E. Roza, O. Vladâcenco, A. M. Grumezescu and D. M. Teleanu, Neurotransmitters—key factors in neurological and neurodegenerative disorders of the central nervous system, *Int. J. Mol. Sci.*, 2022, **23**(11), 5954.
22. L. Green, K. Ashton, S. Azam, M. Dyakova, T. Clemens and M. A. Bellis, Using health impact assessment (HIA) to understand the wider health and well-being implications of policy decisions: the COVID-19 'staying at home and social distancing policy'in Wales, *BMC Public Health*, 2021, **21**(1), 1456.
23. B. J. Ellis, M. A. Sheridan, J. Belsky and K. A. McLaughlin, Why and how does early adversity influence development? Toward an integrated model of dimensions of environmental experience, *Dev. Psychopathol.*, 2022, **34**(2), 447–471.
24. A. Wajnerman-Paz, F. Aboitiz, F. Álamos and P. R. Vergara, A healthcare approach to mental integrity, *J. Med. Ethics*, 2024, **50**(10), 664–669.
25. A. M. Ali and N. A. Fetyan, Biocontrol and Remediation Help in a Safe Environment Production, in *Sustainable Remediation for Pollution and Climate Resilience*, Springer Nature Singapore, Singapore, 2025, pp. 467–505.
26. C. Basckin, I. Strnadova and T. M. Cumming, Teacher beliefs about evidence-based practice: A systematic review, *Int. J. Educ. Res.*, 2021, **106**, 101727.
27. A. T. Mohammadi, E. Ghanbarzadeh, R. Zamiri, M. K. Dana, Y. K. Far, M. A. Avarani, R. Asgari, S. Keshtgar, A. Sanaei and M. Shamshiri, Advancements in Neuroscience: Modern Tools and Techniques Fueling Thrilling Discoveries in the 21st Century, *Nobel Sciences.*, 2024.
28. Y. Xiong, H. Hong, C. Liu and Y. Q. Zhang, Social isolation and the brain: effects and mechanisms, *Mol. Psychiatry*, 2023, **28**(1), 191–201.
29. J. Lin, L. Li, N. Pan, X. Liu, X. Zhang, X. Suo, G. J. Kemp, S. Wang and Q. Gong, Neural correlates of neuroticism: a coordinate-based meta-analysis of resting-state functional brain imaging studies, *Neurosci. Biobehav. Rev.*, 2023, **146**, 105055.

30. K. Gallagher, *Deeper reading: Comprehending challenging texts*, Routledge, 2023, pp. 4–12.
31. S. E. Manahan, *Environmental chemistry*, CRC Press, 2022.
32. C. E. Reid, E. S. Rieves and K. Carlson, Perceptions of green space usage, abundance, and quality of green space were associated with better mental health during the COVID-19 pandemic among residents of Denver, *PLoS One*, 2022, **17**(3), e0263779.
33. P. Bhattacharya, S. Chatterjee and D. Roy, Impact of exercise on brain neurochemicals: a comprehensive review, *Sport Sci. Health*, 2023, **19**(2), 405–452.
34. M. Augusto-Oliveira, G. P. Arrifano, C. G. Leal-Nazaré, L. Santos-Sacramento, A. Lopes-Araujo, L. F. Royes and M. E. Crespo-Lopez, Exercise reshapes the brain: molecular, cellular, and structural changes associated with cognitive improvements, *Mol. Neurobiol.*, 2023, **60**(12), 6950–6974.
35. J. O. Owolabi, O. S. Ilesanmi and V. Luximon-Ramma, A Study of RFR Effects on the Brain Neurochemistry during Pre-and Postnatal Brain Development, *J. Adv. Med. Med. Res.*, 2021, 53–63.
36. N. A. Wong and H. Bahmani, A review of the current state of research on artificial blue light safety as it applies to digital devices, *Heliyon*, 2022, **8**(8), e10282.
37. D. Chen, Z. Zhao, J. Shi, S. Li, X. Xu, Z. Wu, Y. Tang, N. Liu, W. Zhou, C. Ni and B. Ma, Harnessing the sensing and stimulation function of deep brain-machine interfaces: a new dawn for overcoming substance use disorders, *Transl. Psychiatry*, 2024, **14**(1), 440.
38. L. M. Dicks, Gut bacteria and neurotransmitters, *Microorganisms*, 2022, **10**(9), 1838.
39. T. M. Freijy, L. Cribb, G. Oliver, N. J. Metri, R. S. Opie, F. N. Jacka, J. A. Hawrelak, J. J. Rucklidge, C. H. Ng and J. Sarris, Effects of a high-prebiotic diet *versus* probiotic supplements *versus* synbiotics on adult mental health: The "Gut Feelings" randomised controlled trial, *Front. Neurosci.*, 2023, **16**, 1097278.
40. E. A. Ayeni, A. M. Aldossary, D. A. Ayejoto, L. A. Gbadegesin, A. A. Alshehri, H. A. Alfassam, H. K. Afewerky, F. A. Almughem, S. M. Bello and E. A. Tawfik, Neurodegenerative diseases: implications of environmental and climatic influences on neurotransmitters and neuronal hormones activities, *Int. J. Environ. Res. Public Health*, 2022, **19**(19), 12495.
41. S. Mansoor, T. Khan, I. Farooq, L. R. Shah, V. Sharma, C. Sonne, J. Rinklebe and P. Ahmad, Drought and global hunger: biotechnological interventions in sustainability and management, *Planta*, 2022, **256**(5), 97.
42. A. H. TumkurSattar, F. Abushoumi, Z. Al Dobayan and F. Al-Ghamdi, Investigate how genetic variations impact drug response and explore the integration of pharmacogenomics into clinical practice, *Eur. J. Cardiovasc. Med.*, 2023, **13**(4), 785.

43. A. Modak, V. Ronghe, K. P. Gomase and K. P. Dukare, The psychological benefits of breastfeeding: fostering maternal well-being and child development, *Cureus*, 2023, **15**(10), 1007–1020.
44. H. Skouteris, R. Green, A. Chung, H. Bergmeier, L. H. Amir, S. K. Baidwan, A. M. Chater, C. Chamberlain, R. Emond, K. Gibbons and M. Gooey, Nurturing children's development through healthy eating and active living: Time for policies to support effective interventions in the context of responsive emotional support and early learning, *Health Soc. Care Community*, 2022, **30**(6), e6719–e6729.
45. G. Schumann, O. A. Andreassen, T. Banaschewski, V. D. Calhoun, N. Clinton, S. Desrivieres, R. E. Brandlistuen, J. Feng, S. Hese, E. Hitchen and P. Hoffmann, Addressing global environmental challenges to mental health using population neuroscience: a review, *JAMA Psychiatry*, 2023, **80**(10), 1066–1074.
46. M. Haripriyaa and K. Suthindhiran, The important role of systems biology in neuroscience and neurology and its associated diseases, in *Systems Biology Approaches: Prevention, Diagnosis, and Understanding Mechanisms of Complex Diseases*, Springer Nature Singapore, Singapore, 2024, pp. 343–392.
47. A. Korade, Exploring the Human Mind: A Multidisciplinary Perspective, *Int. J. LifeSci. Biotechnol. Pharma. Res.*, 2024, **13**, 2250–3137.
48. N. Dağ and Ö. K. Arıcı, Heavy metals in soils Pb (lead), Hg (mercury), Cd (cadmium), As (arsenic) effects on human health, *Ulus. Çevresel Eğilimler Derg.*, 2021, **5**(2), 48–59.
49. F. Islam, S. Shohag, S. Akhter, M. R. Islam, S. Sultana, S. Mitra, D. Chandran, M. U. Khandaker, G. M. Ashraf, A. M. Idris and T. B. Emran, Exposure of metal toxicity in Alzheimer's disease: An extensive review, *Front. Pharmacol.*, 2022, **13**, 903099.
50. D. Ramírez Ortega, D. F. González Esquivel, T. Blanco Ayala, B. Pineda, S. Gómez Manzo, J. Marcial Quino, P. Carrillo Mora and V. Pérez de la Cruz, Cognitive impairment induced by lead exposure during lifespan: mechanisms of lead neurotoxicity, *Toxics*, 2021, **9**(2), 23.
51. Y. S. Wu, A. I. Osman, M. Hosny, A. M. Elgarahy, A. S. Eltaweil, D. W. Rooney, Z. Chen, N. S. Rahim, M. Sekar, S. C. Gopinath and N. N. Mat Rani, The toxicity of mercury and its chemical compounds: molecular mechanisms and environmental and human health implications: a comprehensive review, *ACS Omega*, 2024, **9**(5), 5100–5126.
52. A. Chamoli and S. K. Karn, The effects of mercury exposure on neurological and cognitive dysfunction in human: a review, *Mercury toxicity mitigation: sustainable nexus approach*, 2024, 117–135.
53. A. E. Charkiewicz, W. J. Omeljaniuk, K. Nowak, M. Garley and J. Nikliński, Cadmium toxicity and health effects—a brief summary, *Molecules*, 2023, **28**(18), 6620.
54. F. Qu and W. Zheng, Cadmium exposure: mechanisms and pathways of toxicity and implications for human health, *Toxics*, 2024, **12**(6), 388.

55. I. Suhani, S. Sahab, V. Srivastava and R. P. Singh, Impact of cadmium pollution on food safety and human health. Current opinion in toxicology, 2021, **27**, 1–7.
56. Q. Niu, Overview of the relationship between aluminum exposure and human health, *Neurotoxicity of Aluminum*, 2023, 1–32.
57. B. Dubois, von, C. A. F. Arnim, N. Burnie, S. Bozeat and J. Cummings, Biomarkers in Alzheimer's disease: role in early and differential diagnosis and recognition of atypical variants, *Alzheimers Res. Ther.*, 2023, **15**(1), 175.
58. N. W. Andersson, I. Bech Svalgaard, S. S. Hoffmann and A. Hviid, Aluminum-Adsorbed Vaccines and Chronic Diseases in Childhood: A Nationwide Cohort Study, *Ann. Intern. Med.*, 2025, 1369–1377.
59. K. Khosravi-Darani, Y. Rehman, I. A. Katsoyiannis, E. Kokkinos and A. I. Zouboulis, Arsenic exposure *via* contaminated water and food sources, *Water*, 2022, **14**(12), 1884.
60. Y. Jiang, D. Zou, Y. Li, S. Gu, J. Dong, X. Ma, S. Xu, F. Wang and J. H. Huang, Monoamine neurotransmitters control basic emotions and affect major depressive disorders, *Pharmaceuticals*, 2022, **15**(10), 1203.
61. E. Haverinen, M. F. Fernandez, V. Mustieles and H. Tolonen, Metabolic syndrome and endocrine disrupting chemicals: an overview of exposure and health effects, *Int. J. Environ. Res. Public Health*, 2021, **18**(24), 13047.
62. B. Tarnacka, A. Jopowicz and M. Maślińska, Copper, iron, and manganese toxicity in neuropsychiatric conditions, *Int. J. Mol. Sci.*, 2021, **22**(15), 7820.
63. C. Berthou, J. P. Iliou and D. Barba, Iron, neuro-bioavailability and depression, *EJHaem*, 2022, **3**(1), 263–275.
64. A. T. Aborode, M. Pustake, W. A. Awuah, M. Alwerdani, P. Shah, R. Yarlagadda, S. Ahmad, I. F. Silva Correia, A. Chandra, E. P. Nansubuga and T. Abdul-Rahman, Targeting oxidative stress mechanisms to treat Alzheimer's and Parkinson's disease: a critical review, *Oxid. Med. Cell. Longevity*, 2022, **2022**(1), 7934442.
65. J. Baj, W. Flieger, A. Barbachowska, B. Kowalska, M. Flieger, A. Forma, G. Teresiński, P. Portincasa, G. Buszewicz, E. Radzikowska-Büchner and J. Flieger, Consequences of disturbing manganese homeostasis, *Int. J. Mol. Sci.*, 2023, **24**(19), 14959.
66. J. Jankovic and A. E. Lang, Diagnosis and assessment of Parkinson disease and other movement disorders, *Bradley's Neurology in Clinical Practice E-Book*, 2021, vol. 24, pp. 310–333.
67. K. Jomova, S. Y. Alomar, R. Valko, E. Nepovimova, K. Kuca and M. Valko, The role of redox-active iron, copper, manganese, and redox-inactive zinc in toxicity, oxidative stress, and human diseases, *EXCLI J.*, 2025, **24**, 880.
68. G. Walke, R. Kumar and P. Wittung-Stafshede, Copper ion incorporation in α-synuclein amyloids, *Protein Sci.*, 2024, **33**(4), e4956.

69. A. Górska, A. Markiewicz-Gospodarek, R. Markiewicz, Z. Chilimoniuk, B. Borowski, M. Trubalski and K. Czarnek, Distribution of iron, copper, zinc and cadmium in glia, their influence on glial cells and relationship with neurodegenerative diseases, *Brain Sci.*, 2023, **13**(6), 911.
70. S. Pyatha, H. Kim, D. Lee and K. Kim, Association between heavy metal exposure and Parkinson's disease: a review of the mechanisms related to oxidative stress, *Antioxidants*, 2022, **11**(12), 2467.
71. R. D. Graff and K. R. Reuhl, Cytoskeletal toxicity of heavy metals, in *Toxicology of Metals*, CRC Press, 2023, vol. I, pp. 639–658.
72. H. Zhang, W. Wei, M. Zhao, L. Ma, X. Jiang, H. Pei, Y. Cao and H. Li, Interaction between Aβ and tau in the pathogenesis of Alzheimer's disease, *Int. J. Biol. Sci.*, 2021, **17**(9), 2181.
73. A. Carmona, S. Roudeau and R. Ortega, Molecular mechanisms of environmental metal neurotoxicity: a focus on the interactions of metals with synapse structure and function, *Toxics*, 2021, **9**(9), 198.
74. S. Gudadhe, S. K. Singh and J. Ahsan, Cellular and neurological effects of lead (Pb) toxicity, in *Lead toxicity mitigation: Sustainable nexus approaches*, Springer Nature Switzerland, Cham, 2024, pp. 125–145.
75. A. Loan, *Prenatal low-dose methylmercury (MeHg) exposure causes premature neuronal differentiation and autism spectrum disorder (ASD)-like behaviours in a rodent model*, Doctoral dissertation, Université d'Ottawa/University of Ottawa, 2023.
76. M. Nabi and N. Tabassum, Role of environmental toxicants on neurodegenerative disorders, *Front. Toxicol.*, 2022, **4**, 837579.
77. P. Kumar, Arsenic exposure and neurotoxic effects, in *Heavy Metal Toxicity and Neurodegeneration*, Academic Press, 2025, pp. 77–101.
78. M. Hayat, N. U. Khola and T. Ahmed, A Systematic Review of Preclinical Studies Investigating the Effects of Pharmacological Agents on Learning and Memory in Prolonged Aluminum-Exposure-Induced Neurotoxicity, *Brain Sci.*, 2025, **15**(8), 849.
79. A. B. Akhtar, S. Naseem, A. Yasar and Z. Naseem, Persistent organic pollutants (POPs): sources, types, impacts, and their remediation, in *Environmental pollution and remediation*, Springer Singapore, Singapore, 2021, pp. 213–246.
80. J. Aaseth, D. Javorac, A. B. Djordjevic, Z. Bulat, A. V. Skalny, I. P. Zaitseva, M. Aschner and A. A. Tinkov, The role of persistent organic pollutants in obesity: a review of laboratory and epidemiological studies, *Toxics*, 2022, **10**(2), 65.
81. S. E. Latchney and A. K. Majewska, Persistent organic pollutants at the synapse: Shared phenotypes and converging mechanisms of developmental neurotoxicity, *Dev. Neurobiol.*, 2021, **81**(5), 623–652.
82. J. R. Marafiga, M. V. Pasquetti and M. E. Calcagnotto, GABAergic interneurons in epilepsy: More than a simple change in inhibition, *Epilepsy Behav.*, 2021, **121**, 106935.
83. W. K. Jasim and A. U. Mosa, Effects of Environmental Pollutants on Thyroid Hormone Functions, *Karbala J. Pharm. Sci.*, 2023, **1**(21), 42.

84. M. M. Haghighi, C. Y. Wright, J. Ayer, M. F. Urban, M. D. Pham, M. Boeckmann, A. Areal, B. Wernecke, C. P. Swift, M. Robinson and R. S. Hetem, Impacts of high environmental temperatures on congenital anomalies: a systematic review, *Int. J. Environ. Res. Public Health*, 2021, **18**(9), 4910.
85. H. Huang, Z. Wang, C. Dai, H. Wu, J. Guo, C. Wang and X. Zhang, Species profile and reactivity of volatile organic compounds emission in solvent uses, industry activities and from vehicular tunnels, *J. Environ. Sci.*, 2024, **135**, 546–559.
86. H. Ghouas, B. Haddou, J. P. Canselier and C. Gourdon, Wastewater Pollution Prevention for volatile organic compounds (Benzene, Toluene, Ethylbenzene, and Xylene) using cloud point extraction and regeneration of surfactant by evaporation, *Euro-Mediterr. J. Environ. Integr.*, 2022, **7**(1), 1–2.
87. Y. Wang and P. Du, Acute organic solvent toxic encephalopathy: A case report and literature review, *Biomed. Rep.*, 2024, **21**(5), 163.
88. S. Moosavi, M. Ghalenoei, M. Amerzadeh and A. S. Variani, The relationship between shift work, circadian rhythms, and cognitive function in ICU nursing, *BMC Nurs.*, 2025, **24**(1), 324.
89. O. D. Ekpe, W. Sim, S. Choi, G. Choo and J. E. Oh, Assessment of exposure of Korean firefighters to polybrominated diphenyl ethers and polycyclic aromatic hydrocarbons *via* their measurement in serum and polycyclic aromatic hydrocarbon metabolites in urine, *Environ. Sci. Technol.*, 2021, **55**(20), 14015–14025.
90. Q. Yan, S. Liu, Y. Sun, C. Chen, S. Yang, M. Lin, J. Long, J. Yao, Y. Lin, F. Yi and L. Meng, Targeting oxidative stress as a preventive and therapeutic approach for cardiovascular disease, *J. Transl. Med.*, 2023, **21**(1), 519.
91. Z. Jia, H. Zhang, L. Yu, F. Qiu, Y. Lv, J. Guan, H. Gang, J. Zuo, T. Zheng, H. Liu and W. Xia, Prenatal lead exposure, genetic factors, and cognitive developmental delay, *JAMA Netw. Open*, 2023, **6**(10), e2339108.
92. J. T. Wallenborn, C. Hyland, S. K. Sagiv, K. R. Kogut, A. Bradman and B. Eskenazi, Prenatal exposure to polybrominated diphenyl ether (PBDE) and child neurodevelopment: The role of breastfeeding duration, *Sci. Total Environ.*, 2024, **921**, 171202.
93. J. Horák, L. Kuboňová, F. Hopan, J. Kremer, M. Dej, T. Tomšej, K. Krpec, J. Ryšavý, O. Molchanov, M. Garba and T. Ochodek, Influence of co-combustion of unsuitable fuels with standardized fuels in households on CO, OGC, PM, and PAH emissions, *Environ. Sci. Pollut. Res.*, 2022, **29**(29), 44297–44307.
94. M. Bhattu, M. Verma and D. Kathuria, Recent advancements in the detection of organophosphate pesticides: a review, *Anal. Methods*, 2021, **13**(38), 4390–4428.
95. N. Xie and A. D. Gross, Muscarinic acetylcholine receptor activation synergizes the knockdown and toxicity of GABA-gated chloride channel insecticides, *Pest Manage. Sci.*, 2022, **78**(11), 4599–4607.
96. L. Pulkrabkova, B. Svobodova, J. Konecny, T. Kobrlova, L. Muckova, J. Janousek, J. Pejchal, J. Korabecny and O. Soukup, Neurotoxicity

evoked by organophosphates and available countermeasures, *Arch. Toxicol.*, 2023, **97**(1), 39–72.

97. K. Felisbino, S. da Silva Milhorini, N. Kirsten, K. Bernert, R. Schiessl and I. C. Guiloski, Exposure to pesticides during pregnancy and the risk of neural tube defects: A systematic review, *Sci. Total Environ.*, 2024, **913**, 169317.
98. S. A. Meo, M. A. Salih, J. M. Alkhalifah, A. H. Alsomali and A. A. Almushawah, Environmental pollutants particulate matter (PM2. 5, PM10), Carbon Monoxide (CO), Nitrogen dioxide (NO2), Sulfur dioxide (SO2), and Ozone (O3) impact on lung functions, *J. King Saud Univ., Sci.*, 2024, **36**(7), 103280.
99. G. Plascencia-Villa and G. Perry, Roles of oxidative stress in synaptic dysfunction and neuronal cell death in Alzheimer's disease, *Antioxidants*, 2023, **12**(8), 1628.
100. K. M. Rentschler and U. P. Kodavanti, Mechanistic insights regarding neuropsychiatric and neuropathologic impacts of air pollution, *Crit. Rev. Toxicol.*, 2024, **54**(10), 953–980.
101. K. Tenório, E. Pereira, S. Remigio, D. Costa, W. Oliveira, D. Dermeval, A. P. da Silva, I. I. Bittencourt and L. B. Marques, Brain-imaging techniques in educational technologies: A systematic literature review, *Educ. Inf. Technol.*, 2022, **27**(1), 1183–1212.
102. K. Tumu, K. Vorst and G. Curtzwiler, Endocrine modulating chemicals in food packaging: A review of phthalates and bisphenols, *Compr. Rev. Food Sci. Food Saf.*, 2023, **22**(2), 1337–1359.
103. H. Hlisníková, I. Petrovičová, B. Kolena, M. Šidlovská and A. Sirotkin, Effects and mechanisms of phthalates' action on neurological processes and neural health: a literature review, *Pharmacol. Rep.*, 2021, **73**(2), 386–404.
104. J. Jung, Y. Cho, Y. Lee and K. Choi, Uses and occurrences of five major alternative plasticizers, and their exposure and related endocrine outcomes in humans: A systematic review, *Crit. Rev. Environ. Sci. Technol.*, 2024, **54**(16), 1165–1194.
105. L. Cohen-Eliraz, A. Ornoy, E. Ein-Mor, M. Bar-Nitsan, T. P. Peleg and R. Calderon-Margalit, Prenatal exposure to phthalates and emotional/behavioral development in young children, *Neurotoxicology*, 2023, **98**, 39–47.
106. H. Gouri and G. Bhalla, Neurotoxic chemistry: Unraveling the chemical mechanisms connecting environmental toxin exposure to neurological disorders, *J. Integr. Sci. Technol.*, 2025, **13**(3), 1064.
107. S. V. M. de Sá, C. S. Monteiro, J. O. Fernandes, E. Pinto, M. A. Faria and S. C. Cunha, Evaluating the human neurotoxicity and toxicological interactions impact of co-occurring regulated and emerging mycotoxins, *Food Res. Int.*, 2024, **184**, 114239.
108. L. Recupero, *Environmental-induced migration: gaps and challenges in the current legal system of international protection*, Mater's degree thesis, Ca' Foscari University of Venice., 2021.

109. Y. O. Ayipo, C. F. Chong, A. A. Jimoh, L. B. Abdulra'uf and W. A. Osunniran, The Fate of Pesticides in Wastewater: A Review, *Pesticide Removal Methods from Wastewater*, 2025, pp. 1–39.
110. M. P. Hoyeck, G. Matteo, E. M. MacFarlane, I. Perera and J. E. Bruin, Persistent organic pollutants and β-cell toxicity: a comprehensive review, *Am. J. Physiol., Endocrinol. Metab.*, 2022, **322**(5), E383–E413.
111. World Health Organization, *WHO global air quality guidelines: particulate matter (PM2. 5 and PM10), ozone, nitrogen dioxide, sulfur dioxide and carbon monoxide*, World Health Organization, 2021.
112. B. C. Lee, H. Yoon, B. Lee, P. Kim, H. B. Moon and Y. Kim, Occurrence of bisphenols and phthalates in indoor dust collected from Korean homes, *J. Ind. Eng. Chem.*, 2021, **99**, 68–73.
113. K. Kim, The role of endocrine disruption chemical-regulated aryl hydrocarbon receptor activity in the pathogenesis of pancreatic diseases and cancer, *Int. J. Mol. Sci.*, 2024, **25**(7), 3818.
114. A. Dhailappan and S. Samiappan, Impact of diet on neurotransmitters, in *Role of nutrients in neurological disorders*, Springer Singapore, Singapore, 2022, pp. 363–383.
115. A. Gasmi, A. Nasreen, A. Menzel, A. Gasmi Benahmed, L. Pivina, S. Noor, M. Peana, S. Chirumbolo and G. Bjørklund, Neurotransmitters regulation and food intake: The role of dietary sources in neurotransmission, *Molecules*, 2022, **28**(1), 210.
116. M. Electricwala, Neurotransmitters: Roles, Functions, and Impact on Mental Health, *Mave Health*, 2024, **18**, 2024.
117. L. M. Vecchio, P. Sullivan, A. R. Dunn, M. K. Bermejo, R. Fu, S. T. Masoud, E. Gregersen, N. M. Urs, R. Nazari, P. H. Jensen and A. Ramsey, Enhanced tyrosine hydroxylase activity induces oxidative stress, causes accumulation of autotoxic catecholamine metabolites, and augments amphetamine effects *in vivo*, *J. Neurochem.*, 2021, **158**(4), 960–979.
118. P. Sethi, A. Prajapati, T. Mishra, T. Chaudhary and S. Kumar, Effects of malnutrition on brain development, in *Nutrition and psychiatric disorders*, Springer Nature Singapore, Singapore, 2022, pp. 75–88.
119. A. Gupta and J. A. Imlay, How a natural antibiotic uses oxidative stress to kill oxidant-resistant bacteria, *Proc. Natl. Acad. Sci. U. S. A.*, 2023, **120**(52), e2312110120.
120. V. Miranda, S. Castiglioni and J. A. Maier, Neuroplasticity and tinnitus: the role of Brain-Derived Neurotrophic Factor in pathogenesis and treatment, *Front. Neurosci.*, 2025, **19**, 1620894.
121. L. Tsan, L. Décarie-Spain, E. E. Noble and S. E. Kanoski, Western diet consumption during development: setting the stage for neurocognitive dysfunction, *Front. Neurosci.*, 2021, **15**, 632312.
122. S. Zhang, Y. Ning, Y. Yang, G. Mu, Y. Yang, C. Ren, C. Liao, C. Ou and Y. Zhang, Decoding pain chronification: mechanisms of the acute-to-chronic transition, *Front. Mol. Neurosci.*, 2025, **18**, 1596367.
123. J. Maiuolo, M. Gliozzi, V. Musolino, C. Carresi, F. Scarano, S. Nucera, M. Scicchitano, F. Oppedisano, F. Bosco, S. Ruga and M. C. Zito, The

contribution of gut microbiota–brain axis in the development of brain disorders, *Front. Neurosci.*, 2021, **15**, 616883.
124. E. Fock and R. Parnova, Mechanisms of blood–brain barrier protection by microbiota-derived short-chain fatty acids, *Cells*, 2023, **12**(4), 657.
125. T. Wu, W. Cai and X. Chen, Epigenetic regulation of neurotransmitter signaling in neurological disorders, *Neurobiol. Dis.*, 2023, **184**, 106232.
126. Y. K. Ghallab and O. S. Elassal, Biochemical and neuropharmacology of psychiatric disorders, in *Nutrition and Psychiatric Disorders: An Evidence-Based Approach to Understanding the Diet-Brain Connection*, Springer Nature Singapore, Singapore, 2024, pp. 25–47.
127. S. Jiang, H. Liu and C. Li, Dietary regulation of oxidative stress in chronic metabolic diseases, *Foods*, 2021, **10**(8), 1854.
128. D. Ambroselli, F. Masciulli, E. Romano, G. Catanzaro, Z. M. Besharat, M. C. Massari, E. Ferretti, S. Migliaccio, L. Izzo, A. Ritieni and M. Grosso, New advances in metabolic syndrome, from prevention to treatment: the role of diet and food, *Nutrients*, 2023, **15**(3), 640.
129. V. Ronan, R. Yeasin and E. C. Claud, Childhood development and the microbiome—the intestinal microbiota in maintenance of health and development of disease during childhood development, *Gastroenterology*, 2021, **160**(2), 495–506.
130. D. Medina-Félix, E. Garibay-Valdez, F. Vargas-Albores and M. Martínez-Porchas, Fish disease and intestinal microbiota: A close and indivisible relationship, *Rev. Aquacult.*, 2023, **15**(2), 820–839.
131. P. D. Cani, A. Everard and T. Duparc, Gut microbiota, enteroendocrine functions and metabolism, *Curr. Opin. Pharmacol.*, 2013, **13**(6), 935–940.
132. D. Y. Chuyue, Q. J. Xu and R. B. Chang, Vagal sensory neurons and gut-brain signaling, *Curr. Opin. Neurobiol.*, 2020, **62**, 133–140.
133. J. R. Davie, Inhibition of histone deacetylase activity by butyrate, *J. Nutr.*, 2003, **133**(7), 2485S–2493S.
134. K. Steliou, M. S. Boosalis, S. P. Perrine, J. Sangerman and D. V. Faller, Butyrate histone deacetylase inhibitors, *BioRes. Open Access*, 2012, **1**(4), 192–198.
135. H. Brüssow, Problems with the concept of gut microbiota dysbiosis, *Microb. Biotechnol.*, 2020, **13**(2), 423–434.
136. D. Dahiya and P. S. Nigam, Antibiotic-therapy-induced gut dysbiosis affecting gut microbiota—brain axis and cognition: restoration by intake of probiotics and synbiotics, *Int. J. Mol. Sci.*, 2023, **24**(4), 3074.
137. G. B. Rogers, D. J. Keating, R. L. Young, M. L. Wong, J. Licinio and S. Wesselingh, From gut dysbiosis to altered brain function and mental illness: mechanisms and pathways, *Mol. Psychiatry*, 2016, **21**(6), 738–748.
138. V. Mondelli, A. C. Vernon, F. Turkheimer, P. Dazzan and C. M. Pariante, Brain microglia in psychiatric disorders, *Lancet Psychiatry*, 2017, **4**(7), 563–572.
139. M. Colom-Cadena, T. Spires-Jones, H. Zetterberg, K. Blennow, A. Caggiano, S. T. DeKosky, H. Fillit, J. E. Harrison, L. S. Schneider, P. Scheltens and W. de Haan, The clinical promise of biomarkers of

synapse damage or loss in Alzheimer's disease, *Alzheimers Res. Ther.*, 2020, **12**(1), 21.
140. S. M. Collins and P. Bercik, The relationship between intestinal microbiota and the central nervous system in normal gastrointestinal function and disease, *Gastroenterology*, 2009, **136**(6), 2003–2014.
141. M. Pirbaglou, J. Katz, R. J. de Souza, J. C. Stearns, M. Motamed and P. Ritvo, Probiotic supplementation can positively affect anxiety and depressive symptoms: a systematic review of randomized controlled trials, *Nutr. Res.*, 2016, **36**(9), 889–898.
142. A. Chakrabarti, L. Geurts, L. Hoyles, P. Iozzo, A. D. Kraneveld, G. La Fata, M. Miani, E. Patterson, B. Pot, C. Shortt and D. Vauzour, The microbiota–gut–brain axis: pathways to better brain health. Perspectives on what we know, what we need to investigate and how to put knowledge into practice, *Cell. Mol. Life Sci.*, 2022, **79**(2), 80.
143. J. T. Cacioppo and G. G. Berntson, The brain, homeostasis, and health: Balancing demands of the internal and external milieu, *The Oxford handbook of health psychology*, 2011, pp. 121–137.
144. O. Hahad, J. H. Prochaska, A. Daiber and T. Muenzel, Environmental noise-induced effects on stress hormones, oxidative stress, and vascular dysfunction: Key factors in the relationship between cerebrocardiovascular and psychological disorders, *Oxid. Med. Cell. Longevity*, 2019, **2019**(1), 4623109.
145. O. T. Wolf, HPA axis and memory, *Best Pract. Res., Clin. Endocrinol. Metab.*, 2003, **17**(2), 287–299.
146. E. Knezevic, K. Nenic, V. Milanovic and N. N. Knezevic, The role of cortisol in chronic stress, neurodegenerative diseases, and psychological disorders, *Cells*, 2023, **12**(23), 2726.
147. T. Furuyashiki and S. Kitaoka, Neural mechanisms underlying adaptive and maladaptive consequences of stress: Roles of dopaminergic and inflammatory responses, *Psychiatry Clin. Neurosci.*, 2019, **73**(11), 669–675.
148. J. Motiejunaite, L. Amar and E. Vidal-Petiot, Adrenergic receptors and cardiovascular effects of catecholamines, in *Annales d'endocrinologie*, Elsevier Masson, 2021, vol. 82, no. 3–4, pp. 193–197.
149. G. Tsujimoto, W. M. Manger and B. B. Hoffman, Desensitization of β-adrenergic receptors by pheochromocytoma, *Endocrinology*, 1984, **114**(4), 1272–1278.
150. Z. Yan and B. Rein, Mechanisms of synaptic transmission dysregulation in the prefrontal cortex: pathophysiological implications, *Mol. Psychiatry*, 2022, **27**(1), 445–465.
151. R. Gulias-Cañizo, E. D. Ruíz-Leyja, H. Sánchez-Castillo, J. Parodí and L. D. Ochoa-de la Paz, The role of GABA neurotransmitter in the human central nervous system, physiology, and pathophysiology, *Rev. Mex. Neuroci.*, 2021, **22**(2), 67–76.
152. M. Popoli, Z. Yan, B. S. McEwen and G. Sanacora, The stressed synapse: the impact of stress and glucocorticoids on glutamate transmission, *Nat. Rev. Neurosci.*, 2012, **13**(1), 22–37.

153. M. A. Olude, A. Mouihate, O. A. Mustapha, C. Farina, F. J. Quintana and J. O. Olopade, Astrocytes and microglia in stress-induced neuroinflammation: the African perspective, *Front. Immunol.*, 2022, **13**, 795089.
154. A. S. Correia and N. Vale, Advancements exploring major depressive disorder: insights on oxidative stress, serotonin metabolism, BDNF, HPA axis dysfunction, and pharmacotherapy advances, *Int. J. Transl. Med.*, 2024, **4**(1), 176–196.
155. A. K. Short and T. Z. Baram, Early-life adversity and neurological disease: age-old questions and novel answers, *Nat. Rev. Neurol.*, 2019, **15**(11), 657–669.
156. E. Taylor and J. W. Rogers, Practitioner review: early adversity and developmental disorders, *J. Child Psychol. Psychiatry*, 2005, **46**(5), 451–467.
157. P. Agostinho, A. Cunha and R. Oliveira, C. Neuroinflammation, oxidative stress and the pathogenesis of Alzheimer's disease, *Curr. Pharm. Des.*, 2010, **16**(25), 2766–2778.
158. R. E. Pacifici and K. J. Davies, Protein, lipid and DNA repair systems in oxidative stress: the free-radical theory of aging revisited, *Gerontology*, 1991, **37**(1–3), 166–180.
159. W. Y. Wang, M. S. Tan, J. T. Yu and L. Tan, Role of pro-inflammatory cytokines released from microglia in Alzheimer's disease, *Ann. Transl. Med.*, 2015, **3**(10), 136.
160. M. E. Bauer and A. L. Teixeira, Inflammation in psychiatric disorders: what comes first?, *Ann. N. Y. Acad. Sci.*, 2019, **1437**(1), 57–67.
161. J. Aber, R. P. Neilson, S. McNulty, J. M. Lenihan, D. Bachelet and R. J. Drapek, Forest processes and global environmental change: predicting the effects of individual and multiple stressors: we review the effects of several rapidly changing environmental drivers on ecosystem function, discuss interactions among them, and summarize predicted changes in productivity, carbon storage, and water balance, *BioScience*, 2001, **51**(9), 735–751.
162. T. J. Schoenfeld and E. Gould, Stress, stress hormones, and adult neurogenesis, *Exp. Neurol.*, 2012, **233**(1), 12–21.
163. B. Gathmann, F. P. Schulte, S. Maderwald, M. Pawlikowski, K. Starcke, L. C. Schäfer, T. Schöler, O. T. Wolf and M. Brand, Stress and decision making: neural correlates of the interaction between stress, executive functions, and decision making under risk, *Exp. Brain Res.*, 2014, **232**(3), 957–973.
164. S. E. Holmes, D. Scheinost, S. J. Finnema, M. Naganawa, M. T. Davis, N. DellaGioia, N. Nabulsi, D. Matuskey, G. A. Angarita, R. H. Pietrzak and R. S. Duman, Lower synaptic density is associated with depression severity and network alterations, *Nat. Commun.*, 2019, **10**(1), 1529.
165. K. Kokubun, K. Nemoto, H. Oka, H. Fukuda, Y. Yamakawa and Y. Watanabe, Association of fatigue and stress with gray matter volume, *Front. Behav. Neurosci.*, 2018, **12**, 154.
166. L. E. Banderet and H. R. Lieberman, Treatment with tyrosine, a neurotransmitter precursor, reduces environmental stress in humans, *Brain Res. Bull.*, 1989, **22**(4), 759–762.

167. C. Hiemke and S. Härtter, Pharmacokinetics of selective serotonin reuptake inhibitors, *Pharmacol. Ther.*, 2000, **85**(1), 11–28.
168. M. Mantzios and K. Giannou, A real-world application of short mindfulness-based practices: a review and reflection of the literature and a practical proposition for an effortless mindful lifestyle, *Am. J. Lifestyle Med.*, 2019, **13**(6), 520–525.
169. M. S. Wheeler, D. B. Arnkoff and C. R. Glass, The neuroscience of mindfulness: How mindfulness alters the brain and facilitates emotion regulation, *Mindfulness*, 2017, **8**(6), 1471–1487.
170. D. Stuckler, S. Basu, M. Suhrcke, A. Coutts and M. McKee, The public health effect of economic crises and alternative policy responses in Europe: an empirical analysis, *Lancet*, 2009, **374**(9686), 315–323.
171. B. G. Bello, S. T. Tula, G. B. Omotoye, A. J. Kess-Momoh and A. I. Daraojimba, Work-life balance and its impact in modern organizations: An HR review, *World J. Adv. Res. Rev.*, 2024, **21**(1), 1162–1173.
172. M. A. Khalid and S. M. Shahid, Environmental Pollution and Climate Change Impacts on Human Health with Particular Reference to Brain: A Review, *Phytopharmaceuticals for Brain Health*, 2017, pp. 39–68.
173. P. K. Kamat, A. Kalani, S. Rai, S. Swarnkar, S. Tota, C. Nath and N. Tyagi, Mechanism of oxidative stress and synapse dysfunction in the pathogenesis of Alzheimer's disease: understanding the therapeutics strategies, *Mol. Neurobiol.*, 2016, **53**(1), 648–661.
174. T. Rajaram and A. Das, Water pollution by industrial effluents in India: discharge scenarios and case for participatory ecosystem specific local regulation, *Futures*, 2008, **40**(1), 56–69.
175. J. Kaushal, M. Khatri and S. K. Arya, A treatise on Organophosphate pesticide pollution: Current strategies and advancements in their environmental degradation and elimination, *Ecotoxicol. Environ. Saf.*, 2021, **207**, 111483.
176. F. Fonnum and E. A. Lock, The contributions of excitotoxicity, glutathione depletion and DNA repair in chemically induced injury to neurones: exemplified with toxic effects on cerebellar granule cells, *J. Neurochem.*, 2004, **88**(3), 513–531.
177. Z. D. Liu and R. C. Hider, Design of iron chelators with therapeutic application, *Coord. Chem. Rev.*, 2002, **232**(1-2), 151–171.
178. A. Neves Carvalho, O. Firuzi, M. Joao Gama, J. van Horssen and L. Saso, Oxidative stress and antioxidants in neurological diseases: is there still hope?, *Curr. Drug Targets*, 2017, **18**(6), 705–718.
179. J. Y. Wang, L. L. Wen, Y. N. Huang, Y. T. Chen and M. C. Ku, Dual effects of antioxidants in neurodegeneration: direct neuroprotection against oxidative stress and indirect protection *via* suppression of gliamediated inflammation, *Curr. Pharm. Des.*, 2006, **12**(27), 3521–3533.
180. A. Bernardo and L. Minghetti, PPAR-γ agonists as regulators of microglial activation and brain inflammation, *Curr. Pharm. Des.*, 2006, **12**(1), 93–109.

181. P. C. Calder, Omega-3 polyunsaturated fatty acids and inflammatory processes: nutrition or pharmacology?, *Br. J. Clin. Pharmacol.*, 2013, **75**(3), 645–662.
182. U. P. Zmrzljak and D. Rozman, Circadian regulation of the hepatic endobiotic and xenobitoic detoxification pathways: the time matters, *Chem. Res. Toxicol.*, 2012, **25**(4), 811–824.
183. T. A. Wadden, M. L. Butryn and C. Wilson, Lifestyle modification for the management of obesity, *Gastroenterology*, 2007, **132**(6), 2226–2238.
184. V. S. Malik, W. C. Willett and F. B. Hu, Global obesity: trends, risk factors and policy implications, *Nat. Rev. Endocrinol.*, 2013, **9**(1), 13–27.
185. H. M. O'Hagan, Chromatin modifications during repair of environmental exposure-induced DNA damage: A potential mechanism for stable epigenetic alterations, *Environ. Mol. Mutagen.*, 2014, **55**(3), 278–291.
186. P. J. Landrigan, D. G. Graham and R. D. Thomas, Strategies for the prevention of environmental neurotoxic illness, in *Neurobehavioral Methods and Effects in Occupational and Environmental Health*, Academic Press, 1994, pp. 691–697.
187. T. Lakes, M. Brückner and A. Krämer, Development of an environmental justice index to determine socio-economic disparities of noise pollution and green space in residential areas in Berlin, *J. Environ. Plann. Manage.*, 2014, **57**(4), 538–556.
188. B. Weiss, Behavioral toxicology and environmental health science: Opportunity and challenge for psychology, *Am. Psychol.*, 1983, **38**(11), 1174.
189. E. Davoud-Monfared, J. K. Mozafari, A. Keramatinia, E. Naimi, P. Amiri, E. Rahimpour, S. Ganji, M. Vejdani and M. Mohseny, A Review on Lifestyle before and after COVID-19 Pandemic: Four Levels of Prevention, *Int. J. Body Mind Cult.*, 2024, **11**(4), 258.
190. J. Rasmussen and I. Suedung, *Proactive risk management in a dynamic society*, Swedish Rescue Services Agency, 2000.
191. E. de Lange, E. J. Milner-Gulland and A. Keane, Improving environmental interventions by understanding information flows, *Trends Ecol. Evol.*, 2019, **34**(11), 1034–1047.
192. V. Krivtsov, Investigations of indirect relationships in ecology and environmental sciences: a review and the implications for comparative theoretical ecosystem analysis, *Ecol. Modell.*, 2004, **174**(1–2), 37–54.
193. N. H. Perlroth and C. W. Branco, Current knowledge of environmental exposure in children during the sensitive developmental periods, *J. Pediatr.*, 2017, **93**, 17–27.
194. G. Perea and A. Araque, GLIA modulates synaptic transmission, *Brain Res. Rev.*, 2010, **63**(1–2), 93–102.
195. W. Jänig and H. J. Häbler, Neurophysiological analysis of target-related sympathetic pathways–from animal to human: similarities and differences, *Acta Physiol. Scand.*, 2003, **177**(3), 255–274.
196. M. Piwecka, N. Rajewsky and A. Rybak-Wolf, Single-cell and spatial transcriptomics: deciphering brain complexity in health and disease, *Nat. Rev. Neurol.*, 2023, **19**(6), 346–362.

197. K. J. Slack, T. J. Williams, J. S. Schneiderman, A. M. Whitmire, J. J. Picano, L. B. Leveton, L. L. Schmidt and C. Shea, *Risk of Adverse cognitive or behavioral conditions and psychiatric disorders: Evidence report*, 2016.
198. R. Altenburger, W. Brack, R. M. Burgess, W. Busch, B. I. Escher, A. Focks, L. Mark Hewitt, B. N. Jacobsen, M. L. de Alda, S. Ait-Aissa and T. Backhaus, Future water quality monitoring: improving the balance between exposure and toxicity assessments of real-world pollutant mixtures, *Environ. Sci. Eur.*, 2019, **31**(1), 1–7.
199. To comprehend the impact of environmental stress on the brain, more accurate models of these coupled stresses.
200. T. T. Huang, E. A. Callahan, E. R. Haines, C. Hooley, D. M. Sorensen, D. W. Lounsbury, N. S. Sabounchi and P. S. Hovmand, Leveraging systems science and design thinking to advance implementation science: moving toward a solution-oriented paradigm, *Front. Public Health*, 2024, **12**, 1368050.
201. L. S. Wakschlag, M. Y. Roberts, R. M. Flynn, J. D. Smith, S. Krogh-Jespersen, A. J. Kaat, L. Gray, J. Walkup, B. S. Marino, E. S. Norton and M. M. Davis, Future directions for early childhood prevention of mental disorders: A road map to mental health, earlier, *J. Clin. Child Adolesc. Psychol.*, 2019, **48**(3), 539–554.
202. A. R. Eiser, *Preserving brain health in a toxic age: new insights from neuroscience, integrative medicine, and public health*, Rowman & Littlefield, 2021.
203. N. A. Farahany, *The battle for your brain: defending the right to think freely in the age of neurotechnology*, St. Martin's Press, 2023.

CHAPTER 10

Future Directions in Neurochemistry and Brain Disorders

PRABHASH NATH TRIPATHI,*[a] MAYANK GANGWAR,*[b] NITISH KUMAR SINGH,[c] VIVEK CHANDRA VERMA,[d] JAIKEE KUMAR SINGH[e] AND JASPREET MARWAHA[f]

[a] Department of Pharmaceutical Sciences, College of Pharmacy, University of Arkansas for Medical Sciences, Little Rock, AR, USA; [b] Clinical Studies and Trials Unit, Division of Development/Indian Council of Medical Research, Department of Health Research (Min. of Health & FW), India; [c] Department of Biosciences, Manipal University Jaipur, Rajasthan, India; [d] Department of Biochemistry, Panjab University, India; [e] Department of Bioinformatics, Faculty of Engineering and Technology, Marwadi University, Rajkot Gujarat, India; [f] Manipal University, Jaipur, Rajasthan, India
*Emails: pntripathi@uams.edu; mayank.gangwar@icmr.gov.in

10.1 Introduction

The 21st century has ushered in a transformative era in neuroscience—an era in which the chemical language of the brain is being decoded with a speed and resolution that would have seemed implausible a generation ago. Neurochemistry, once confined to cataloging classical neurotransmitters and their synaptic choreography, now stands as a richly interdisciplinary science that fuses molecular biology, genetics, systems neuroscience, bioengineering, and computational modeling with clinical translation. The brain, a living network of electrical impulses and chemical gradients,

Chemical Biology No. 26
Chemical Signatures of the Brain: Bridging Biochemistry, Neuroanatomy, and Brain Disorders
Edited by Sachchida Nand Rai and Abhishek Kumar Singh

Published by the Royal Society of Chemistry, www.rsc.org

coordinates cognition, emotion, and behavior through the interplay of neurotransmitters, neuropeptides, neurohormones, ions, lipids, and intracellular signaling cascades that together confer the biochemical identity of the nervous system. Understanding this molecular symphony is no longer a purely academic pursuit: it is a practical necessity for confronting the mounting burden of brain disorders across the lifespan.[1] Demographic shifts and environmental pressures have amplified this urgency. As populations age, neurodegenerative diseases such as Alzheimer's and Parkinson's are rising in prevalence; psychiatric illnesses including depression, bipolar disorder, and schizophrenia continue to exact heavy personal and societal costs; and neurodevelopmental conditions such as autism spectrum disorder (ASD) and attention-deficit/hyperactivity disorder (ADHD) remain areas of significant unmet need. These challenges emerge against a backdrop of lifestyle change, urbanization, and environmental stressors that perturb neurochemical balance.[2] The field is therefore pivoting from descriptive accounts to mechanistic precision: not merely mapping pathways but manipulating them therapeutically, predictively, and preventively. This is both a philosophical and methodological shift transforming neurochemistry from a discipline of observation into an active intervention, capable of forecasting and designing molecular behavior. Historically, the story began with the identification of dopamine, serotonin, acetylcholine, glutamate, and GABA; with experiments that defined their synthesis, storage, release, and reuptake; and with early therapeutics that targeted single receptors or enzymes.[3] Those breakthroughs birthed modern psychopharmacology and still underpin a large fraction of clinical practice. Yet experience has revealed the limits of a "single-molecule–single-effect" paradigm. Many illnesses resist mono-target approaches, relapse rates remain high, and adverse effects are common. In response, neurochemistry has embraced a systems perspective: each neurotransmitter operates within webs of peptides, lipids, and second messengers; feedback and feedforward loops integrate across cell types and brain regions; and cross-talk with endocrine, immune, and metabolic networks constantly reshapes neural function.[4] The future therefore lies in understanding the brain as a dynamic, multiscale chemical ecosystem, one that can be measured, modeled, and steered. The accelerant of this evolution is technology. Neuroimaging has moved beyond hemodynamics to molecular and mesoscopic readouts. Functional magnetic resonance spectroscopy (fMRS) can quantify fluctuations in excitatory and inhibitory metabolites *in vivo*; positron emission tomography (PET) employing receptor-specific ligands and synaptic density tracers visualizes target engagement and synaptic loss; and optical methods, including two-photon and three-photon microscopy, capture activity at cellular and subcellular scales in animal models.[5] When combined with ultra-high-field MRI, layer-resolved fMRI, and cerebral blood volume or metabolic contrasts, these modalities bridge molecular events to network-level phenomena. At the same time, advances in mass spectrometry, particularly LC–MS/MS with ion mobility, data-independent acquisition, and top-down proteomics permit ultra-sensitive, multiplexed measurement of proteins, metabolites, and lipids, revealing disease-specific biochemical fingerprints in cerebrospinal fluid, plasma, and extracellular vesicles.[6] Complementing these are biosensors, genetically encoded fluorescent indicators for neurotransmitters and voltage, electrochemical probes for monoamines and metabolites, and organic transistor platforms that deliver subsecond neurochemical telemetry during behavior. The next horizon is not a single instrument but the convergence of them all into integrated, longitudinal portraits of brain chemistry. Integrating such heterogeneous data requires computational sophistication. Artificial intelligence and machine learning now extract structure from complexity: harmonizing batch effects, aligning multimodal signals, and uncovering latent patterns that predict risk, progression, and treatment response. Models can embed prior knowledge from pathways and protein–protein interactions, infer causal relationships, and simulate the effects of interventions at multiple scales from receptor occupancy to circuit dynamics.[7] This computational turn enables "digital neurochemistry": *in silico* avatars (digital twins) that are parameterized by an individual's omics, imaging, and sensor data, and that can be interrogated to compare therapeutic sequences, doses, and timings before any physical intervention occurs. In short, computation is not a *post-hoc* analytic step but a co-equal experimental partner that helps design assays, optimize trials, and personalize care. Within this framework, biomarkers become the navigational beacons. Traditional diagnosis often relies on clinical phenotypes that emerge late in disease, when neuronal loss and network reorganization are already entrenched. Neurochemical biomarkers measured in CSF, blood, saliva, or even breath offer earlier, more proximal readouts of pathology and plasticity. In Alzheimer's disease, amyloid-β and phosphorylated tau mirror proteinopathy, while neurofilament light reflects axonal injury; in Parkinson's disease, α-synuclein proteoforms and dopamine metabolism illuminate nigrostriatal integrity; in psychiatric disorders, signatures in the kynurenine pathway, redox balance, and inflammatory mediators stratify risk and treatment response.[8] The field is moving from single-analyte tests to panel-based signatures that combine proteins, metabolites, lipids, and EV-derived markers to improve specificity and to map patients onto endotypes, biochemically coherent subgroups that are more likely to respond to a given therapy. Liquid-biopsy approaches, capturing neural cargo trafficked into blood, will enable minimally invasive, repeatable monitoring suitable for prevention and chronic management. Neurotransmitter levels, receptor densities, and enzymatic activities vary across individuals due to genetic polymorphisms, epigenetic regulation, developmental history, and environment. Precision neurochemistry aims to measure this individuality and match interventions accordingly choosing between serotonergic or dopaminergic strategies; identifying who will benefit from glutamatergic modulation or anti-inflammatory approaches; and setting dose and sequence dynamically based on biomarker feedback.[9] Computational models that simulate neurotransmitter dynamics under candidate treatments can test hypotheses virtually, reducing trial-and-error prescribing and accelerating recovery. Over time, such models may serve as clinical copilots, continuously updating risk estimates and treatment recommendations as new data stream in from wearables, clinics, and laboratories. Another profound shift is recognizing that neurons operate in concert with glia. Astrocytes regulate neurotransmitter recycling, ionic homeostasis, and energy distribution; microglias monitor the parenchyma, prune synapses, and orchestrate immune responses; oligodendrocytes and their precursors manage myelin, axonal metabolism, and conduction speed. Disruption of these glial programs—through infection, injury, aging, or metabolic stress—can derail neurochemical signaling and plasticity.[10] Future therapeutics will increasingly focus on glial modulation: re-tuning microglia from pro-inflammatory to pro-resolving states; enhancing astrocytic glutamate clearance and cholesterol transport; and promoting oligodendrocyte differentiation for remyelination. Because glia sit at the interface of the immune, vascular, and nervous systems, interventions may work from the "inside-out" (targeting central pathways) or the "outside-in" (modulating systemic inflammation, gut–brain metabolites, or vascular health) to restore homeostasis.[11] Repair and regeneration, long considered out of reach in the adult brain, are being reimagined as programmable processes. Psychoplastogens, including ketamine-like agents and next-generation serotonergic compounds, can open brief windows of heightened plasticity; when paired with structured behavioral therapy and neuromodulation (rTMS, tDCS, and focused ultrasound), these windows can be steered to rewire maladaptive circuits. At the molecular level, neurotrophic signaling (BDNF/TrkB, GDNF) supports spine growth and synaptic stabilization, while ECM remodeling reduces structural "brakes" that limit adult plasticity.[12] Myelin repair strategies—*via* RXR agonists, thyroid hormone analogs, and anti-inhibitory pathways—seek to restore conduction and metabolic resilience. Gene and RNA therapies add durable levers: vectors that deliver trophic factors, CRISPR-based editors that correct rate-limiting defects and antisense oligonucleotides that rebalance splicing.[13] Crucially, these approaches are being embedded within staged programs: first, reduce injurious inflammation and excitotoxicity; second, open plasticity; third, deliver targeted repair; and finally, consolidate gains through learning aligned to daily life. Daily life is not a backdrop but a set of programmable inputs. Diet supplies neurotransmitter precursors and membrane lipids; microbial metabolites shape BBB integrity, microglial tone, and astrocyte metabolism; exercise elevates BDNF, improves insulin sensitivity, and remodels lipid signaling; sleep and circadian alignment govern synaptic down-scaling, glymphatic clearance, and neuromodulator reset; psychosocial stress and recovery practices adjust autonomic balance and inflammatory tone; and environmental exposures from particulates to heat modulate oxidative stress and vascular function.[14] Lifestyle therefore acts as a scaffold for pharmacology and devices: when nutrition, movement, sleep, light, and stress-recovery are synchronized, plasticity strengthens and inflammation resolves, raising the ceiling for every other intervention. With wearables and periodic biomarker checks, these levers can be personalized and adjusted in feedback loops that are practical at the population scale. Translating discovery into care requires attention to ethics, equity, and implementation. Omics, imaging, and biosensor data can reveal sensitive information; consent must be layered and revisitable, data governance secure, and privacy preserved through federated learning and differential privacy. Algorithms must be audited for bias across sex, ancestry, and age; outputs should be interpretable and uncertainty-aware; and thresholds may require subgroup calibration to avoid systematic over- or undertreatment.[15] Globally, solutions must be affordable and robust: tiered diagnostics that begin with low-cost screens and escalate to confirmatory tests; reagent-sparing assays; device-agnostic software that harmonizes across scanners and vendors; and workforce training that ensures high-quality sampling, QA/QC, and interpretation. Without deliberate design, precision approaches risk widening disparities; with it, they can democratize access to early detection and effective therapy. Taken together, these developments signal a new role for neurochemistry. No longer limited to *post-hoc* explanation, it is becoming an interventional discipline that reads and writes the brain's molecular code. The future promises clinical workflows in which multimodal measurements (SV2A-PET, ultra-high-field MRS, spatial omics, biosensors) flow into AI systems that map patients onto endotypes and personalize care; in which glial states are monitored and modulated to stabilize networks; in which plasticity windows are opened purposefully and filled with rehabilitative content; and in which environment and habit are engineered to support neurochemical wellness.[16] Success will depend on rigorous analytical validation, trustworthy computation, thoughtful ethics, and equitable access—but the trajectory is clear. With chemistry, computation, and compassion aligned, brain disorders can be detected earlier, treated more precisely, and, in some cases, prevented.[17] The molecular dialogue of the brain, once enigmatic, has become a readable and, increasingly, writable script. As tools sharpen and ethical compasses expand, the prospect of a world where mood, cognition, and resilience are proactively maintained moves from aspiration to an attainable goal.[18] This is the promise and responsibility of modern neurochemistry: to illuminate the brain and to empower it, redefining what it means to safeguard human potential in the molecular age (Figure 10.1).

10.1.1 Advancing Neurochemical Research Tools

The remit of neuroimaging has expanded from mapping hemodynamics to directly indexing synaptic density, neurotransmitter pools, laminar microcircuits, and real-time neurochemical flux—often in the same subject, sometimes in the same session. A prominent example is synaptic vesicle glycoprotein 2A (SV2A) positron emission tomography (PET), which quantifies presynaptic terminal density *in vivo* using ligands such as [^{11}C]UCB-J and newer ^{18}F tracers (*e.g.*, SynVesT-1/2).[19] These radioligands have been applied across neurodegeneration and psychiatry to map synaptic loss preceding

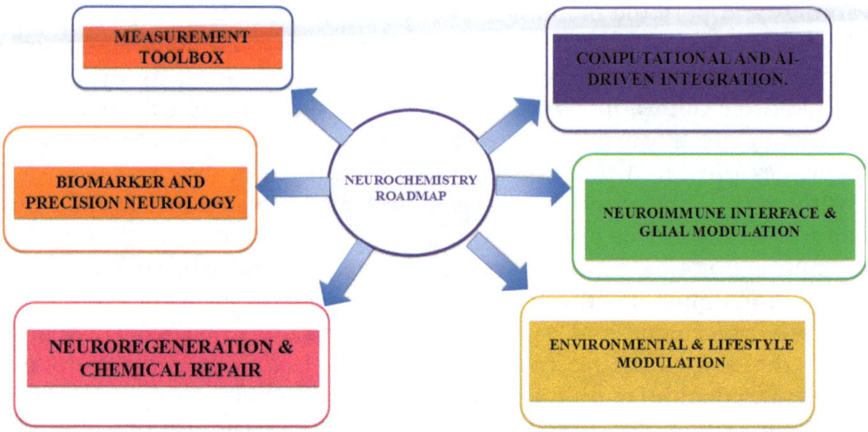

Figure 10.1 Neurochemistry roadmap. This illustration presents a neurochemistry roadmap highlighting six interconnected pillars shaping the future of brain research. It emphasizes the development of advanced measurement tools, progress in biomarkers and precision neurology, and innovations in neuroregeneration and chemical repair. The roadmap also integrates computational and AI-driven approaches, focuses on the neuroimmune interface and glial modulation, and recognizes the importance of environmental and lifestyle factors. Together, these domains outline a comprehensive framework for advancing neurological understanding and therapeutic strategies.

structural atrophy and to probe synaptic plasticity under treatment—shifting PET from "pathology finding" to "plasticity tracking." MRI has similarly moved from millimeter-scale function to mesoscopic and even layer-resolved readouts at ultra-high-field (7 T), enabling separation of feedforward and feedback processing and improving spectroscopic specificity. Laminar fMRI, sometimes with cerebral blood volume (VASO) contrasts or concurrent measures of $CBF/CMRO_2$, now interrogates layer-specific physiology in humans; line-scanning sequences push temporal sampling to the single-digit millisecond regime while retaining submillimeter spatial detail.[20] Together, these developments let investigators examine microcircuit motifs tied to disease (*e.g.*, epileptogenic layers, cortical dysrhythmias) without invasive electrodes. Beyond hemodynamics, molecular MRI methods are starting to "see" chemistry. Chemical-exchange saturation transfer (CEST) has matured to visualize endogenous metabolites; GluCEST maps glutamate distribution and has shown age- and disease-related differences in both animals and humans, with early clinical translation at 5–7 T addressing field-homogeneity and reliability constraints.[21] In parallel, deuterium metabolic imaging (DMI) tracks ^{2}H-labeled substrates (*e.g.*, glucose) noninvasively, offering dynamic 3D readouts of glycolysis and downstream labeling with growing evidence for neurologic applications. These techniques complement proton MRS by providing higher spatial coverage and sensitivity to specific pathways. Crucially, neuroimaging is being fused with optogenetics and optical readouts to assign causality and anchor BOLD signals to defined cell types.[22] Optogenetic fMRI (ofMRI) in rodents and nonhuman primates enables circuit-level perturbation with whole-brain readouts, revealing effective connectivity and validating the hemodynamic signatures of specific neuronal populations. Recent studies report strong correspondence between opto-evoked BOLD maps and tract anatomy and even decode behavioral preferences from ofMRI activation patterns.[23] Hybrid platforms that acquire widefield calcium imaging concurrently with fMRI further link cellular events to mesoscale hemodynamics, enhancing translational bridges from mouse to human. A parallel transformation is occurring in *in vivo* neurochemistry sensors. Genetically encoded reporters—spanning calcium (jGCaMP families), voltage, and a growing catalog of GPCR-activation-based (GRAB) neurotransmitter sensors—now quantify subsecond dopamine, serotonin, norepinephrine, acetylcholine, adenosine, and even neuropeptides in behaving animals.[24] Iterative sensor engineering continues to improve affinity, brightness, and kinetics (for example, next-generation GRAB-NE and new GRAB neuropeptide sensors), making it feasible to align fast neurochemical events with BOLD fluctuations, local field potentials, or behavior. Fast-scan cyclic voltammetry (FSCV) also remains vital, with waveform optimizations increasing sensitivity and selectivity for targets like tryptophan and enabling freely moving recordings that can be co-registered with imaging.[25] On the acquisition side, ultra-high-field MRI is no longer purely research-grade. FDA-cleared 7 T systems are entering clinical workflows for epilepsy, multiple sclerosis, microvascular disease, and movement disorders, where improved SNR and spectral dispersion boost lesion conspicuity and metabolite resolution. Practical implementation now guides document citing, sequence portfolios, and quality-assurance pipelines, signaling steady mainstreaming and paving the way for broader population studies at the mesoscopic scale. Another frontier is functional ultrasound (fUS), which images cerebral blood-volume changes with submillimeter spatial and subsecond temporal resolution. Intra-operative and bedside human applications have demonstrated high-sensitivity mapping, and co-registration with fMRI, and electrocortical stimulation is refining structure–function atlases during awake neurosurgery.[26] Emerging mobile or wearable prototypes and machine-learning-augmented pipelines hint at ecological neuroimaging outside MRI suites, while foundational work explores transcranial feasibility and algorithmic corrections for skull attenuation. Focused ultrasound (FUS) adds a complementary capability: precise, image-guided mechanical energy delivery for both neuromodulation and transient blood–brain barrier (BBB) opening. In conjunction with microbubbles, low-intensity FUS can locally open the BBB to enhance CNS drug uptake; early human studies in Alzheimer's patients report greater plaque reduction within targeted regions when BBB opening is paired with anti-amyloid therapy, motivating larger trials.[27] Methodological roadmaps now outline dosimetry, safety envelopes, and skull-acoustic modeling, while neuromodulatory protocols (tFUS) are being refined with best-practice recommendations from international consortia. Because FUS can also increase bidirectional "traffic" across the BBB, it may augment liquid biopsy by releasing brain-derived biomarkers into blood—potentially coupling therapy and diagnostics in a single session. The next wave is multimodal convergence. PET–MR systems already co-register molecular targets (for example, SV2A for synapses, tau/amyloid for proteinopathy) with high-field structural/functional maps; adding GluCEST or edited MRS layers neurotransmitter pools, while fUS or concurrent electrophysiology supplies vascular and electrical context.[28] ofMRI and widefield optical imaging bridge cellular and systems scales in animals, and advances in line-scanning/laminar fMRI port some of that specificity to humans. Finally, AI pipelines orchestrate these heterogeneous data harmonizing PET kinetic models, laminar BOLD/VASO features, CEST contrasts, and sensor-derived neurochemical time series to produce individualized, mechanistic signatures of disease and treatment response.[29] In aggregate, next-generation *in vivo* mapping is no longer a single technique but an integrated toolbox spanning molecules to microcircuits to whole-brain networks, positioning neuroimaging as a quantitative, interventional science that can track and increasingly, steer neurochemical health across the lifespan.[30]

10.1.1.1 High-resolution Analytical and Omics Platforms

High-resolution analytical chemistry now underpins neurochemistry, delivering molecular readouts that are sensitive, quantitative, and increasingly spatially resolved. At the core is mass spectrometry (MS), where hybrid Orbitrap and time-of-flight (timsTOF) instruments coupled to nano-flow LC provide sub-ppm mass accuracy and deep coverage at low sample input—critical for brain tissue subregions, CSF, and extracellular vesicles. Data-independent acquisition (DIA/SWATH) captures all ionizable peptides per run, while parallel reaction monitoring (PRM) and targeted SRM assays deliver absolute quantitation for translational biomarkers.[31] Ion-mobility separations (FAIMS, trapped ion mobility) add an orthogonal dimension that resolves isobars and conformers, improving precision for neuroactive lipids and post-translationally modified peptides. Proteomics has advanced beyond lists to mechanisms. Top-down MS preserves intact proteoforms, enabling the detection of neurorelevant PTMs (phosphorylation, ubiquitination, acetylation, glycosylation) and truncations implicated in tau, α-synuclein, and TDP-43 pathology.[32] Redox proteomics maps cysteine oxidation states tied to oxidative stress, while cross-linking MS charts native protein–protein interactions in synaptic complexes. Chemoproteomics with activity-based probes and clickable electrophiles profiles enzyme classes (*e.g.*, serine hydrolases, kinases) *in situ* and thermal proteome profiling/CETSA identifies target engagement of CNS drugs directly in brain lysates.[33] Metabolomics and lipidomics decode the neurochemical milieu with complementary platforms. LC–MS/MS, capillary electrophoresis–MS, and high-field NMR quantify amino acids, organic acids, monoamines, and neuromodulators; shotgun and LC-based lipidomics map sphingolipids, ceramides, gangliosides, plasmalogens, and oxysterols central to myelination and neuroinflammation.[34] Ozone-induced dissociation and Paternò–Büchi derivatization localize double bonds in lipid isomers, clarifying the structure–function relationships of neurolipids. Stable-isotope tracing using $^{13}C/^{15}N$-labeled substrates is integrated with metabolic flux analysis to resolve astrocyte–neuron lactate shuttling, glutamate–glutamine cycling, and mitochondrial substrate preferences under conditions of stress, aging, or drug exposure. Single-cell and spatial omics are transforming resolution. SCoPE2/plexDIA workflows and ultra-low-input proteomics quantify thousands of proteins in hundreds of single cells, exposing neuron–glia heterogeneity and disease-associated microglia states. Single-cell metabolomics (nano-LC–MS, CE–MS, LAESI) detects picogram-level metabolites, while spatial transcriptomics (Visium, Slide-seq) integrates with imaging mass cytometry (MIBI) and CODEX to preserve cytoarchitecture.[35] Spatial metabolomics and lipidomics *via* MSI (MALDI, DESI, nano-DESI, AFADESI) generate molecular atlases of cortex, hippocampus, and substantia nigra, registering neurochemical gradients to layers and nuclei. Co-registration with high-field MRI or cleared-tissue light-sheet images bridges molecules to mesoscopic circuitry. Multi-omics integration converts measurements into mechanisms. Matrix factorization (MOFA), DIABLO, and WGCNA uncover cross-omics modules linking proteoforms, metabolites, and lipids to phenotypes such as cognition and inflammation. Bayesian network learning and causal frameworks (including Mendelian randomization when genetics is available) help distinguish drivers from correlates. Knowledge graphs and pathway tools (Reactome, KEGG, and MetaCyc) contextualize features into neurotransmission, redox, and immune signaling. Cloud pipelines (OpenMS, GNPS, and Nextflow) standardize processing; spectral libraries, retention-time indexing, and internal standard panels mitigate batch effects.[36] Rigorous QA/QC—pooled references, drift correction, isotopically labeled standards, and external proficiency testing—ensures reproducibility suited for regulatory-grade biomarker development. Clinically, these platforms enable CSF/plasma biomarker panels (tau/PTMs, neurofilament light, α-synuclein proteoforms), extracellular-vesicle proteomics reflecting brain cell of origin, and neurolipid signatures that track demyelination or microglial activation (Table 10.1).[36,37] Together, high-resolution analytical and omics technologies provide a mechanistic lens on neurochemistry—resolving who (cell type), where (spatial niche), when (dynamic flux), and how (pathway wiring)—and set the stage for precision diagnostics, target discovery, and pharmacodynamic monitoring in brain disorders.[38]

10.1.1.2 Biosensors and Real-time Neurochemical Monitoring

Biosensors have shifted neurochemistry from static snapshots to streaming measurements of the living brain's chemical traffic. At the heart of this shift is the ability to transduce fleeting molecular events—dopamine release,

Table 10.1 The neurochemical toolbox: platforms, samples, and clinical use.[34–38]

Platform	What it measures	Sample	Typical use in brain disorders
LC–MS/MS (discovery DIA/DDA)	Proteins/peptides	CSF, plasma, brain tissue	Find candidate biomarkers and pathways
Targeted proteomics (PRM/SRM)	Specific proteins (absolute)	CSF, plasma	Verify/track biomarker panels over time
Metabolomics (LC/GC–MS)	Small molecules (*e.g.*, kynurenine, redox)	Plasma, tissue	Stratify patients; monitor therapy effects
Lipidomics (LC–MS/MS)	Complex lipids (myelin, sphingolipids)	CSF/plasma, tissue	Track demyelination/ remyelination
Top-down proteomics	Intact proteoforms/ PTMs	CSF, tissue	Tau/α-synuclein proteoform mapping
Ion mobility MS	Isomers/isobars separation	CSF, EVs, tissue	Clarify lipid and PTM isomers
NMR ($^1H/^{13}C$)	Metabolites; flux with tracers	CSF/plasma, slices	Energy metabolism, Glu/Gln cycling
Spatial omics/MSI	Molecules with location	Tissue sections	Map molecular changes to brain regions
EV omics (proteins/lipids/ RNA)	Cell-of-origin cargo	Plasma/CSF	Longitudinal monitoring of synapse/ glia

glutamate spillover, adenosine accumulation, lactate shuttling—into electrical or optical signals with millisecond-to-second precision. Compared with microdialysis, which excels in absolute quantification but averages over minutes and millimeters, modern sensors push temporal resolution to subsecond scales and spatial resolution to tens of micrometers, revealing synaptic and perisynaptic dynamics.[39] This capability allows researchers to align neurochemical pulses with spikes, local field potentials, hemodynamic responses, and behavior, enabling causal inference and closed-loop control that were previously impossible. Electrochemical methods remain the workhorse for monoamines and metabolites. Fast-scan cyclic voltammetry at carbon-fiber microelectrodes applies a rapid triangular waveform to oxidize and reduce target analytes, producing characteristic cyclic fingerprints used to deconvolve dopamine from pH changes and interferents. Improvements in waveform design, chemometric training sets, and electrode surface engineering (carbon nanotube/graphene coatings, nitrogen-doped or boron-doped diamond, and laser-nanostructured carbon) boost sensitivity, reduce fouling, and expand the detectable palette to serotonin, norepinephrine, adenosine, and even tryptophan. Amperometric approaches with enzyme layers (*e.g.*, glutamate oxidase generating H_2O_2) deliver selective detection of nonelectroactive analytes such as glutamate and choline, while permselective overlayers suppress ascorbate and urate.[40] Multiplexed arrays embed several independently addressable sites on flexible shanks, enabling simultaneous mapping across striatal compartments, hippocampal layers, or cortical laminae. Organic and semiconductor transistor platforms are accelerating *in vivo* performance and integration. Organic electrochemical transistors based on PEDOT:PSS translate ionic flux in the brain's extracellular space into amplified electronic signals, supporting low-impedance, low-voltage operation with soft, conformable form factors that better match brain mechanics. Electrochemical aptamer-based sensors and field-effect transistor configurations use conformational changes in nucleic-acid receptors to produce real-time, reversible signals for neuromodulators and peptides in complex biofluids.[41] Aptamers can evolve against targets lacking convenient redox chemistry, expanding the addressable neurochemical space. Anti-fouling strategies—zwitterionic brushes, PEGylation, albumin passivation, and catalytic "self-cleaning" pulses—extend lifetimes from hours to weeks, a prerequisite for longitudinal studies and eventual clinical monitoring. Optical biosensors complement electrochemistry with molecular specificity and genetic targetability. Genetically encoded fluorescent indicators translate ligand binding into changes in brightness or FRET, enabling targeted expression in defined cell types and compartments. GPCR-activation-based sensors (the GRAB family) now exist for dopamine, norepinephrine, serotonin, acetylcholine, adenosine, and GABA, with variants spanning affinities and kinetics to capture both tonic and phasic dynamics.[42] Glutamate sensors (iGluSnFR and fast/hyperaffinity derivatives) quantify synaptic release at single-spine resolution with two-photon microscopy, while red-shifted and near-infrared reporters permit deeper imaging and multiplexing. Voltage indicators and next-generation calcium indicators provide complementary readouts of excitability that can be co-registered with transmitter sensors to parse cause from consequence in circuit computations. Delivering and reading these optical signals in behaving animals relies on fiber photometry, head-mounted miniscopes, and two-photon mesoscopes. Fiber photometry excels at simplicity and stability, reporting population-averaged fluorescence changes from targeted cell types during naturalistic behavior or task performance.[43] Miniscopes and GRIN-lens-based microendoscopes push to cellular resolution, supporting longitudinal tracking of the same neurons across learning or disease progression. Two-photon mesoscopes and three-photon excitation extend depth and field of view, allowing simultaneous monitoring across cortical columns and hippocampal subfields. Combining these tools with optogenetics yields "sense-and-stimulate" experiments, in which detection of a transmitter burst triggers precisely timed inhibition or excitation of upstream circuits—an essential template for closed-loop psychiatry and movement-disorder therapies. A major engineering frontier is multimodality and untethered operation. Hybrid probes unite microelectrodes, micro-LEDs, and microfluidic drug channels on ultrathin polyimide or parylene substrates, enabling simultaneous neurochemical sensing, electrophysiology, optogenetic stimulation, and focal pharmacology in the same tissue volume.[44] System-on-chip front ends integrate low-noise potentiostats with amplifiers and digitizers, shrinking headstages and minimizing motion artifacts. Wireless telemetry *via* Bluetooth Low Energy, NFC, ultrasonic power links, or back-scatter radio eliminates tethers that confound behavior. Battery-free implants, energy harvesting, and adaptive duty cycling address thermal and power budgets, making all-day recordings feasible in freely moving rodents and nonhuman primates, with early steps toward human-compatible form factors. Selectivity, calibration, and drift remain central scientific issues. Electrochemical methods require chemometric models trained on *in situ* datasets to separate analytes from pH and background current; continual recalibration and reference injections help maintain accuracy as tissue encapsulation and electrode aging evolve.[45] Enzymatic sensors can saturate at high substrate levels or lose activity; co-immobilized stabilizers, cross-linked matrices, and microfluidic prefilters mitigate these risks. For optical sensors, expression levels, trafficking, and photobleaching complicate quantitation; ratiometric designs and simultaneous reference channels improve robustness. Across platforms, rigorous ground truth with microdialysis, LC–MS, or stable-isotope challenges is vital for translating relative signals into meaningful concentration and flux estimates. Biocompatibility dictates chronic performance.[45,46] Tissue encapsulation, gliosis, and neuronal dropout can detune sensors and alter the very chemistry under observation. Mechanical mismatch between stiff probes and ultrasoft brain exacerbates damage; softening shanks, hydrogel interfaces, and ultraflexible mesh electronics reduce micromotion and glial scarring. Surface nanotopography and bioactive coatings further mitigate encapsulation. Sterilization protocols must preserve functional layers, and surgical approaches that minimize vascular injury lower inflammatory cascades.[47] Longitudinal histology and two-photon imaging of glial responses are increasingly standard to validate the chronic compatibility of new chemistries and device architectures. Beyond classical transmitters, biosensors are expanding into gaseous and reactive species that shape pathology. Nitric-oxide sensors with metalloporphyrin catalysts, hydrogen-peroxide microelectrodes, and ROS/RNS-responsive fluorescent probes map redox stress during ischemia, neuroinflammation, and neurodegeneration. Lactate and glucose sensors probe astrocyte–neuron metabolic coupling, while K^+ ion-selective electrodes track ionic homeostasis during spreading depolarizations and seizure propagation.[48] Emerging peptide and cytokine sensors seek to connect neuroimmune signaling with behavior on actionable timescales, bridging microglial states to neuromodulatory tone. These developments open a window on pathways that have been historically inferred rather than measured *in vivo*. Clinical translation, once a distant goal, is now tangible. Intraoperative FSCV and enzyme-based probes have recorded human striatal dopamine dynamics during deep brain stimulation lead placement, correlating chemistry with symptom states and stimulation parameters. Sensing-guided DBS—where a biomarker such as adenosine or glutamate triggers adaptive parameter changes—embodies the promise of closed-loop neuromodulation for Parkinson's disease, essential tremor, and potentially depression or OCD. For neuro-oncology and traumatic brain injury, multimodal cortical strips that report K^+, lactate, and glutamate alongside ECoG could guide surgical margins or detect impending edema.[49] Translation hinges on sterilizable, hermetic packaging; ISO-compliant manufacturing; and reproducible calibration against clinical-grade assays, all while satisfying stringent safety, EMI, and cybersecurity requirements for implanted electronics. Real-time neurochemical monitoring is also transforming preclinical models beyond traditional rodents. In nonhuman primates, flexible multisite arrays and wireless photometry enable measurements during complex cognitive tasks, aligning neurochemical fluctuations with saccades, working-memory phases, and social behaviors. In human-derived brain organoids and microphysiological systems, integrated electrochemical and optical sensors map development, network maturation, and drug responses, accelerating discovery while reducing reliance on animal models. Combining sensors with stable-isotope tracing and online mass spectrometry in perfused slices closes the loop between flux, concentration, and enzymatic control, bringing systems biochemistry into intact circuits. The analytical value of these streams is maximized by computational pipelines that fuse modalities and link chemistry to function.[50] Joint models align FSCV and photometry with spikes, pupillometry, and kinematics; state-space methods recover latent neuromodulatory states that drive behavior. Deep-learning architectures classify voltammograms, denoise optical signals, and predict upcoming transmitter release from network activity, enabling proactive interventions. *In silico* twins of an individual's neurochemical network parameterized by sensor data allow simulation of drug dosing or stimulation schedules before application, a step toward precision neuropsychiatry where therapy is titrated to a person's live chemistry rather than retrospective symptoms. Taken together, biosensors and real-time neurochemical monitoring recast the brain as a measurable, controllable chemical system and provide a natural bridge to the computational integration that follows (Table 10.2).

10.1.1.3 Computational and Artificial Intelligence-driven Integration

Computational and AI-driven integration is transforming neurochemistry from a set of siloed assays into a coherent, predictive science in which molecular, cellular, circuit, and behavioral data are fused into actionable models of brain health and disease. At the core is the maturation of machine-learning pipelines capable of harmonizing heterogeneous inputs—LC–MS/MS proteo-metabolomic matrices, lipidomics and chemoproteomics panels, single-cell and spatial omics, PET kinetic parametric maps (*e.g.*, SV2A, tau, amyloid, TSPO), ultra-high-field fMRI/CEST/MRS contrasts, mesoscopic calcium/voltage imaging, functional ultrasound, closed-loop electrochemistry and genetically

Table 10.2 Real-time neurochemical biosensors.[47–50]

Key analyte	Modality	Key strengths	Typical application
Fast-scan cyclic voltammetry (FSCV)	Dopamine, serotonin, adenosine (electroactive)	Very fast; *in vivo* compatible	Link phasic transmitter release to behavior
Enzyme electrodes (amperometry)	Glutamate, choline, lactate, glucose	Selective *via* enzymes	Monitor excitotoxic risk
Genetically encoded fluorescent sensors	DA/5-HT/NE/ACh/GABA/glutamate	Targetable to cell types	Circuit-specific transmitter dynamics
Fiber photometry	Population fluorescence signals	Simple, stable in freely moving animals	Task-locked neurochemical activity
Miniscopes/microendoscopy	Cellular calcium/transmitter signals	Cellular resolution in behaving animals	Track the same neurons across learning
Organic/FET/OECT sensors	Ions/neuromodulators (aptamer-based)	Soft, low-power, amplifying	Toward chronic, flexible implants
Hybrid probes (electrochem + opto + microfluidics)	Chemistry + spikes + stimulation	Multimodal, closed-loop ready	Sense-and-stimulate therapies (*e.g.*, DBS)

encoded neurotransmitter sensors—alongside clinical phenotypes, neuropsychological scores, wearable-derived physiology, and electronic health records.[51] Data engineering is the first frontier: standardized ontologies for metabolites and lipids, robust peak-picking and retention-time alignment, spectral-library curation, batch-effect removal *via* pseudo-reference samples and Bayesian hierarchical normalization, and imaging harmonization with domain-adversarial networks and ComBat variants all reduce technical variance that otherwise overwhelms biological signal. Representation learning then supplies a lingua franca across modalities. Contrastive and multi-view self-supervised methods align molecular profiles with imaging patches and time-series biosensor segments, while graph neural networks embed prior knowledge from pathways and protein–protein interaction maps to constrain learned embeddings within biochemically plausible neighborhoods.[52] Causality is the next essential layer; purely correlational models underperform in translation, so modern pipelines incorporate structure learning (NOTEARS-style and score-based DAG discovery), instrumental-variable strategies leveraging genetics (Mendelian randomization), and dynamic causal models that tie neurochemical states to circuit activity and symptoms. These causal graphs can inform mechanistic simulations, such as predicting the cognitive trajectory if glutamatergic tone is upregulated by 10% or if microglial cytokine signalling is dampened. Such simulations provide a principled scaffold for hypothesis generation and clinical trial design. Such simulations provide a principled scaffold for hypothesis generation and clinical trial design. On top of these foundations, hybrid mechanistic–statistical frameworks merge ordinary differential-equation models of neurotransmitter synthesis, release, reuptake, and receptor occupancy with machine-learned residuals that capture unmodeled biology and patient-specific idiosyncrasies.[53] Physics-informed neural networks and neural stochastic differential equations preserve conservation laws and timescale separations while granting flexibility for real-world data. Digital twins of the brain operationalize this machinery at the individual level: each patient's twin ingests multi-omics baselines, PET/MRI maps, biosensor telemetry, and clinical context to forecast progression (*e.g.*, synaptic density decline, metabolic inflection points), simulate drug exposure–response with pharmacokinetic/pharmacodynamic couplings across the blood–brain barrier, and recommend adaptive dosing or stimulation schedules that are continuously updated *via* Bayesian filtering as new data arrive.[54] In discovery, foundation models pre-trained on massive untargeted metabolomics and proteomics corpora learn fragment- and retention-time-aware embeddings that generalize to low-data, rare-disease cohorts. Generative chemistries (diffusion/transformer architectures) propose BBB-permeable compounds conditioned on target profiles and off-target risk, while multi-objective Bayesian optimization navigates potency, selectivity, solubility, and CNS liabilities *in silico* before synthesis. For imaging, transformer-based reconstructions accelerate high-resolution fMRI and spectroscopic acquisitions; uncertainty-aware deconvolution improves PET kinetic parameter estimation under short scans or low doses; and super-resolution models fuse

PET–MR with CEST or GluCEST to enhance delineation of neurotransmitter pools.[55] Closed-loop experiment design is increasingly guided by reinforcement learning and active learning: agents select the next assay, brain region, or stimulation parameter to maximize expected information gain about a hypothesized pathway, cutting animal usage and time-to-mechanism. In real-time monitoring, state-space models and switching linear dynamical systems reconcile fast biosensor streams with slower hemodynamic and behavioral signals to infer latent neuromodulatory states; Gaussian-process and particle-filter variants deliver calibrated uncertainty, which is critical for safety when models trigger interventions (*e.g.*, adaptive DBS or targeted infusion *via* microfluidics). Interpretability, a traditional weakness of deep networks, is addressed through multi-scale attribution: SHAP and integrated gradients highlight critical metabolites or voxels; concept activation vectors project predictions onto biologically meaningful axes (oxidative stress, glutamatergic drive, microglial activation); and counterfactual explainer modules indicate minimal neurochemical changes that would alter a diagnosis or triage decision, supporting clinician trust and regulatory auditability.[56] Privacy-preserving analytics ensure that sensitive neurochemical and imaging data can be leveraged across institutions *via* federated learning, secure enclaves, and differential privacy, while model cards and data sheets document lineage, limitations, and intended use. Model robustness can be improved through domain generalization using meta-learning across scanners, cohorts, and species while out-of-distribution detection can identify unusual profiles and benchmark tests with hidden pathologies and adversarial perturbations can assess reliability under challenging conditions. Importantly, computational integration does not stop at inference: it structures entire translational programs. Trial-enrichment models rank individuals most likely to progress or to respond to a mechanism-specific therapy, reducing sample sizes and increasing power; synthetic controls built from longitudinal registries and digital-twin trajectories reduce placebo arms where ethical or practical constraints exist; adaptive platform trials use Bayesian bandits to allocate participants to promising arms in near-real time with interim neurochemical endpoints (*e.g.*, SV2A-PET response or CSF proteoform shifts) rather than waiting for late clinical outcomes.[57] In basic science, multi-scale simulations connect enzyme kinetics to synaptic vesicle dynamics to laminar circuit motifs, enabling *ex vivo* and *in vivo* predictions that can be falsified with targeted sensors, optogenetic perturbations, and focal pharmacology. Operationally, modern stacks run on cloud-native, reproducible infrastructure: workflow engines, containerization, and provenance tracking ensure that every figure or biomarker is rebuildable end-to-end; FAIR data principles, extended metabolite/proteoform identifiers, and crosswalks between pathway databases prevent silent semantic drift. Finally, ethical and societal dimensions are embedded by design: bias audits quantify differential performance across sex, ancestry, and age; benefit–risk simulators explore downstream impacts of algorithmic triage; and human-in-the-loop interfaces present uncertainties and mechanistic rationales in clinician-friendly narratives.[58] The result is not merely faster analytics but a qualitatively new capability: the ability to read and write the brain's chemical code with computationally informed precision—forecasting who will decline, which pathway to modulate, when to intervene, and how to titrate therapy in synchrony with a person's evolving neurochemical state.

10.1.2 Biomarker Discovery and Precision Neurochemistry

Biomarker discovery in neurochemistry has moved from single analyte hunting to the construction of multidimensional signatures that reflect pathway activity, disease stage, and therapeutic responsiveness. The central aim is not merely to detect pathology earlier but to define actionable endotypes—biochemically coherent subgroups that guide interventions with higher probability of benefit and lower risk of harm. In this framing, a "biomarker" is less a solitary molecule than a validated signal that survives preanalytical noise, analytical variability, and biological heterogeneity to inform a concrete clinical decision.[59] Specimen strategy is foundational and too often underappreciated. Cerebrospinal fluid (CSF) provides the closest biochemical window to the CNS but requires standardized collection (time-of-day, tube material, first-mL discard, gentle mixing) to reduce artifactual gradients. Blood is attractive for scalability but demands rigorous handling of anticoagulants, hemolysis indices, and freeze–thaw cycles; microsampling (dried blood spots or volumetric absorptive microsampling) is gaining traction for longitudinal, at-home monitoring. Saliva, tears, sweat, and nasal/olfactory mucosa brushings expand noninvasive options and may better capture autonomic or olfactory-circuit pathology.[60] Extracellular vesicles (EVs) isolated from plasma—enriched for neuronal or glial markers using immunoaffinity—offer a compromise between accessibility and cellular specificity, carrying proteoforms, microRNAs, and lipids that mirror CNS states. On the analyte side, neurochemistry-specific classes are diversifying. Beyond canonical proteins, proteoforms (splice variants and post-translationally modified species) often outperform bulk measures; for example, phosphorylation state, truncation, glycosylation, and ubiquitination encode mechanistic information about trafficking, misfolding, or degradation. Ultrasensitive immunoassays (*e.g.*, single-molecule array and proximity extension platforms) have enabled quantification of low-abundance neuroproteins in plasma, while aptamer-based panels widen proteome coverage and discover unexpected axes (complement cascade, coagulation, and angiogenesis) linked to cognitive decline or psychiatric relapse.[61] In lipidomics, sphingolipids, sulfatides, and plasmalogens track myelin integrity and microglial activation; oxysterols and bile acid derivatives connect cholesterol turnover to neuroinflammation. Metabolomics contributes pathway-level readouts—kynurenine/tryptophan balance, one-carbon metabolism, redox couples (GSH/GSSG), and mitochondrial substrate preference—offering direct handles on excitotoxicity, oxidative stress, and bioenergetic failure.[62] Glycomics and autoantibody repertoires add an immune–metabolic dimension, while circulating microRNAs/lncRNAs and RNA-editing footprints provide fast-changing signals suitable for monitoring. Emerging breathomics (volatile organic compounds) and skin or tear chemistries may become practical for ambulatory screening where repeated venipuncture is impractical.[63] Discovery, however, is only the starting line; validation is the race. Analytical validation requires linearity, precision, and accuracy against certified reference materials, matrix-effect assessment, stability studies, and inter-laboratory reproducibility under ISO/CLSI guidance. Preanalytical variables—fasting, circadian phase, menstrual cycle, recent exercise, diet, and medication use—must be explicitly modeled. Clinical validation precedes from well-powered, prospectively collected cohorts with pre-registered analysis plans to avoid overfitting and information leakage.[64] Case–control designs are useful for signal finding but tend to overestimate performance; enriched prospective cohorts and nested case–control within biobanks better reflect real-world prevalence and confounding. Independent replication across ancestries, age groups, and comorbidity patterns is essential, as biomarker distributions can shift with cardiometabolic status, inflammation, and renal function. Crucially, context-of-use drives thresholds and economics: a screening assay in primary care tolerates lower specificity but demands high negative predictive value, whereas a companion diagnostic prior to an expensive biologic must emphasize specificity and clinical utility.[65] To avoid repetition with imaging-centric approaches, consider fluid-plus-function composites. For instance, coupling a plasma proteoform panel with a cognitive stress test or a pupillometry/eye-movement task can clarify whether a biochemical signal has a contemporaneous physiological consequence. Similarly, target-engagement designs in early trials can combine a proximal biochemical marker (*e.g.*, pathway-specific phosphopeptide ratio) with a downstream behavioral or electrophysiological readout to establish the mechanism before committing to large outcomes studies. Pharmacometabolomics—profiling pre-dose and early on-treatment metabolites—helps identify metabotypes that predict efficacy or adverse events and supports adaptive dosing in precision regimens.[66] Because neuropsychiatric syndromes are heterogeneous, endotyping is the linchpin of precision neurochemistry. Rather than mapping biomarkers to DSM/ICD categories, unsupervised or semi-supervised learning can reveal biochemical clusters: inflammatory–metabolic depressions, glutamatergic anxiety endotypes, cholinergic–vascular cognitive impairment, or synaptic failure phenotypes indexed by EV synaptic proteins. Importantly, endotypes must be prospectively stable and treatment-informative; otherwise, they are mere taxonomic curiosities. Stability testing includes bootstrapped clustering consistency, longitudinal membership persistence, and response consistency within endotype-defined strata.[67] When available, instrumental variables (*e.g.*, genetic variants in rate-limiting enzymes) can strengthen causal claims that an endotype reflects a modifiable driver rather than a passenger. Equity and generalizability require design beyond convenience cohorts. Ancestry-aware calibration, sex-stratified reference intervals, and life-stage adjustments (pediatrics, perimenopause, advanced age) prevent systematic misclassification. Environmental exposures—air pollution, pesticides, and heavy metals—are nontrivial neurochemical confounders; incorporating exposomics layers into models improves robustness.[68] For implementation at scale, cost, throughput, and logistics matter: panels using multiplexed immunoassays or targeted LC–MRM transitions on widely available instruments are more deployable than bespoke, discovery-grade workflows. Quality systems must include external proficiency testing and periodic instrument cross-calibration to keep decision limits stable over years. Regulatory and health-technology assessment frameworks shape translation. The biomarker qualification path emphasizes defined context-of-use and evidence that the marker changes decisions and outcomes; payers will ask for clinical utility and cost-effectiveness (*e.g.*, reduced unnecessary imaging or hospitalization, optimized drug utilization).[69] Health-system pilots can demonstrate operational feasibility: turnaround times, reflex testing algorithms, EHR integration with interpretive comments (*e.g.*, "pattern consistent with neuroinflammatory endotype; consider trial X or anti-cytokine strategy Y"), and clinician education to avoid misinterpretation. Theranostics, the union of diagnosis and therapy, is a pragmatic frontier. Companion diagnostics stratify patients to pathway-targeted interventions (for example, an oxidative-stress panel guiding NAD^+/mitochondrial support, or a cholinergic synaptic signature guiding pro-cholinergic agents). Response-adaptive protocols can escalate, switch, or combine treatments based on predefined biochemical milestones (*e.g.*, EV-synaptophysin increase, kynurenine ratio normalization). In refractory illness, N-of-1 designs that cycle micro-interventions while streaming neurochemical measures (from capillary microsamples or wearable biochemical patches now in development) can identify personalized control strategies when group-average evidence fails.[70] The "non-repetition" imperative also highlights adjacent but distinct areas that are often overlooked. Retinal neurochemistry—dopamine metabolites, macular carotenoids, and lipid peroxidation products in the retina or aqueous humor—may reflect central dopaminergic tone and oxidative status *via* the eye–brain axis, offering a uniquely accessible neural tissue. Gut-derived signatures, including short-chain fatty acids, bile acid profiles, and microbial tryptophan catabolites, provide mechanistic links to mood, cognition, and neuroinflammation that could be modulated by diet, pre/probiotics, or postbiotics.[71] Coagulation and endothelial biomarkers (von Willebrand factor multimers, endothelial glycocalyx fragments) capture the vascular–neurochemical interface relevant to small-vessel disease and white-matter injury. Also, autonomic–neurochemical coupling (salivary alpha-amylase with catecholamine panels; heart-rate-variability–aligned sampling) can frame stress-reactivity endotypes with direct therapeutic implications for biofeedback, vagal stimulation, or anti-adrenergic strategies. Finally, precision neurochemistry must be living—capable of learning. Post-deployment, real-world performance monitoring should track drift, subgroup disparities, and outcome alignment, feeding Bayesian updating of thresholds and feature weights. Transparent model cards and patient-facing explanations can reduce overreliance on "black-box" scores. The destination is not a perfect biomarker but a reliable decision ecosystem: standardized sampling; robust, affordable assays; interpretable multi-marker models; and clinical workflows that act on results.[72] When those elements cohere, biomarker-guided care shifts neurochemistry from autopsy-level insight to anticipatory medicine—catching disease before dysfunction, matching therapy to endotype, and adjusting course with biochemical feedback. That, in essence, is the promise of biomarker discovery and precision neurochemistry: to make the brain's chemistry readable in the clinic and writable in the service of recovery.

10.1.3 The Neuroimmune Interface and Glial Modulation

The preceding sections described how biomarkers and precision analytics can partition brain disorders into actionable endotypes; the next step is to understand—and therapeutically steer—the cellular engines that generate those biochemical signatures. At the center of this enterprise is the neuroimmune interface: a continuously active conversation among neurons, microglia, astrocytes, oligodendroglia, vascular and meningeal immune cells, and the peripheral milieu. Far from being mere responders to injury, glia are primary regulators of neurochemical homeostasis—clearing neurotransmitters, sculpting synapses, buffering ions, shaping blood–brain barrier (BBB) permeability, and coordinating metabolic traffic between brain and body.[73] When this choreography is perturbed by aging, genetic risk, infection, vascular stress, or gut–brain signals, glial programs shift toward inflammatory states that distort neurotransmission, weaken myelin, and erode circuit plasticity. These shifts do not occur uniformly; they cluster into pathway-driven endotypes—complement-synaptopathy, purinergic–inflammasome activation, interferon-reactive states—that align with the multidimensional biomarkers outlined earlier and point to glia-directed interventions. This section frames glia not as a homogeneous target but as adaptive networks whose roles depend on context and timing. We begin by surveying microglial, astrocytic, and oligodendroglial contributions to neurochemical balance and their crosstalk with vascular and meningeal immunity. We then examine mechanisms that convert acute, protective responses into chronic, degenerative signaling and propose endotypes that make this complexity clinically tractable. Building on that mechanistic map, we outline a therapeutics playbook—targets, delivery strategies across the BBB, and circuit-level readouts that guard against overcorrection.[74] Finally, we translate these ideas into precision trial design and implementation, so that glial modulation becomes a measurable, staged, and equitable path to true disease modification rather than a broad, nonspecific suppression of inflammation.

10.1.3.1 Microglia, Astrocytes, and Oligodendroglia: Gatekeepers of Neurochemical Homeostasis

Neuroimmune biology reframes the brain as an organ whose health depends on continual dialogue between neurons and glia. Microglia, astrocytes, and oligodendroglial lineage cells act as sentinels, custodians, and repair crews in these chemical commons. Their surveillance is not merely defensive: microglia sculpt circuits through complement-tagged synaptic pruning, regulate extracellular neurotransmitter tone by phagocytosing glutamatergic debris, and release trophic factors that tune plasticity. In healthy states, microglial signaling leans toward homeostatic programs characterized by purinergic sensing (P2Y12), fractalkine–CX3CR1 communication, and "don't-eat-me" cues from neurons.[75] When danger-associated molecular patterns (*e.g.*, ATP, HMGB1) or pathogen-associated motifs engage TLRs and NOD-like receptors, microglia pivot toward inflammatory phenotypes that elevate cytokines (IL-1β, TNF, IL-6), reactive oxygen species, and proteases—molecules that, in excess, degrade synapses and white matter. Astrocytes, the most abundant glia, reconcile immune, metabolic, and neurotransmitter demands. Their endfeet ensheath vessels, shape blood–brain barrier (BBB) integrity and regulate glymphatic clearance—a convective system that removes amyloid, tau, and metabolic byproducts along perivascular routes. Astrocytes maintain neurochemical equilibrium by buffering K^+, recycling glutamate *via* GLT-1/EAAT2, and supplying lactate to neurons for oxidative metabolism. Immune stimuli drive "reactive" astrocyte programs; the field has moved beyond a binary A1/A2 classification toward transcriptomic continua with modules for complement, interferon response, cholesterol transport, and cytoskeletal remodeling.[76] Importantly, astrocytic reactivity is not uniformly deleterious: context and timing determine whether reactivity protects tissue, walls off damage, or sustains chronic dysfunction. Oligodendrocytes (OLs) myelinate axons, accelerating conduction and supporting axonal metabolism; their precursor cells (OPCs) remain distributed throughout adult brain and respond to immune cues. Inflammatory mediators—particularly IFN-γ, TNF, and nitric oxide—impair OL differentiation and myelin maintenance, while microglial-derived growth factors (*e.g.*, IGF-1) and astrocyte-derived lipids foster remyelination.[77] Thus, demyelinating and neurodegenerative disorders often reflect failures in a three-way negotiation among microglia, astrocytes, and OL lineage cells. This triangulated view also reframes neuromodulators. For example, dopamine and noradrenaline alter microglial motility and phagocytic set points; adenosine and ATP control microglia–astrocyte crosstalk; glutamate spillover, if inadequately cleared, drives astrocyte calcium waves that further modulate microglia through ATP release. In short, the neuroimmune interface is a chemically fluent network where "immune" and "neural" signals are inseparable.[78] Two architectural systems knit these players to the periphery. First, the BBB and perivascular macrophages act as a molecular customs checkpoint: peripheral cytokines, lipids, and antibodies influence central glia through endothelial transporters, tight junction tuning, and glymphatic flow. Second, the meningeal immune niche—dura-associated macrophages, border microglia-like cells, T and B lymphocytes—monitors CSF and drains antigens to cervical lymph nodes, creating adaptive-immune memory of CNS events.[79] Together they define "outside-in" axes by which infection, cardiometabolic disease, or gut microbiota composition recalibrate microglial tone and astrocytic reactivity. Understanding these axes at single-cell, spatial, and longitudinal scales sets the stage for targeted glial modulation and precision immunoneurology.

10.1.3.2 *From Acute Defense to Chronic Injury: Neuroinflammation Endotypes and Therapeutic Pathways*

Inflammation is indispensable for acute containment and repair, yet its chronic miscalibration produces a neurochemical milieu that corrodes circuits. Several convergent pathways orchestrate this pivot from help to harm. Pattern-recognition receptors (TLRs, RIG-I-like receptors) and the NLRP3 inflammasome sense cellular distress, licensing caspase-1 to mature IL-1β and IL-18; when activation persists, synapses weaken, glutamate handling falters, and oxidative stress accumulates. Purinergic signaling through P2X7 amplifies inflammasome activity and releases microvesicles rich in proteases and cytokines. Complement cascades (C1q–C3) flag synapses for elimination—adaptive in development but maladaptive when re-engaged in aging or proteinopathies. In parallel, cGAS–STING activation by cytosolic DNA drives type I interferon programs in microglia and astrocytes, shifting energy metabolism toward glycolysis (Warburg-like reprogramming) and away from oxidative phosphorylation—an imprint of "trained immunity."[80] Reactive astrocytes deepen this shift through multiple channels: reduced GLT-1/EAAT2 lowers glutamate uptake, heightening extrasynaptic NMDA activation and calcium load; altered cholesterol synthesis and apolipoprotein trafficking perturb membrane microdomains and receptor function; connexin hemichannels and pannexins leak ATP and glutamate, propagating calcium waves and recruiting microglia. Oligodendroglia, exquisitely sensitive to oxidative and nitrosative stress, succumb to microglia- and astrocyte-derived mediators; demyelination exposes axons, raises energetic demand, and narrows conduction safety margins, rendering circuits brittle.[81] These processes rarely act alone; instead, they assemble into endotypes—coherent, pathway-dominant clusters that cut across diagnoses. Illustrative patterns include a complement-synaptopathy endotype (high C1q/C3 at synapses), an interferon-reactive endotype (strong ISG signatures in microglia and astrocytes), and a purinergic-inflammasome endotype (P2X7/NLRP3 dominance with extracellular-vesicle readouts). Systemic physiology layers further complexity in an outside-in fashion. Metainflammation from obesity and insulin resistance enriches circulating ceramides and pro-inflammatory lipoproteins that signal across, or traverse, the blood–brain barrier; dysbiosis steers microbial tryptophan catabolism toward kynurenine derivatives that bias microglial set points; aging ("inflammaging") elevates baseline IL-6 and TNF, narrows microglial epigenetic plasticity, and degrades glymphatic clearance.[82] Vascular contributions—endothelial senescence, pericyte loss, and basement-membrane thickening—restrict nutrient exchange and perivascular clearance, seeding white-matter injury. Given these diverse drivers, identical neuropathology across patients is improbable, clarifying why single-target anti-inflammatory trials often disappoint. Accordingly, classification should shift from diagnostic labels to pathway activity maps. Longitudinal, multi-omic readouts from CSF/plasma (cytokines, complement breakdown products, lipid mediators), extracellular vesicles (cell-of-origin glial proteoforms), and advanced imaging (TSPO alternatives, SV2A, myelin-sensitive MRI, modern diffusion metrics) can define an individual's active modules. Endotypes must be treatment-informative: complement-high states may warrant C1q/C3 blockade; inflammasome-dominant states may suit NLRP3 inhibitors or P2X7 antagonists; interferon-weighted profiles may benefit from STING modulation or JAK inhibition; and purinergic–excitotoxic coupling may call for adenosine A1/A2A strategies or GLT-1 upregulation.[83] Timing is critical: early microglial activation can be neuroprotective by clearing debris and catalyzing remyelination but becomes destructive if metabolic reprogramming and epigenetic marks "lock in" a pro-inflammatory phenotype. The practical objective is to intercept the right pathway at the right stage, restoring glial–neuronal choreography rather than merely muting immune noise. From this map flows a two-axis strategy for glial modulation: re-tuning state (shifting microglia/astrocytes from inflammatory to homeostatic programs) and restoring function (rebuilding synapses and myelin). For microglia, CSF1R inhibitors can reset populations, but subtler approaches are gaining favor: TREM2 agonists enhance lipid sensing and debris clearance; CX3CR1–fractalkine supports neuron–microglia détente; P2X7/NLRP3 antagonism curbs synaptotoxic cascades; and adenosine A2A antagonism adjusts motility and cytokine release.[84] Complement blockade at C1q or C3 targets aberrant synaptic tagging—particularly compelling where excessive pruning defines the biology. For astrocytes, GLT-1/EAAT2 upregulation (*e.g.*, with β-lactams like ceftriaxone or newer small molecules) restores glutamate clearance; connexin/pannexin modulation reduces pathological hemichannel flux; and tuning cholesterol synthesis/transport (LXR/RXR agonism, ApoE pathway calibration) repairs membrane microdomains crucial for receptor function and myelin lipid supply. In interferon-weighted states, JAK–STAT inhibitors can down-scale astrocytic inflammatory programs while preserving metabolic support. Repairing oligodendroglia demands both immunologic quieting and differentiation cues. RXR agonists, muscarinic antagonists, and thyroid-hormone analogs exhibit remyelinating activity preclinically; agents that neutralize inhibitory cues (*e.g.*, LINGO-1 blockers, despite mixed clinical experience) aim to release OPCs from differentiation blocks. Microglia-derived IGF-1 and TGF-β can be leveraged to support OPC maturation, but inflammatory excess must first be cooled to protect nascent myelin.[85] A sequential regimen—induction to dampen pathogenic inflammation, followed by consolidation to stimulate remyelination—often mirrors successful strategies in other fields. Crossing the BBB remains an engineering bottleneck. Focused ultrasound with microbubbles affords focal, reversible opening for biologics or nanoparticles, enabling region-specific delivery with imaging-based control. Nanocarriers—lipid nanoparticles, polymeric micelles, exosomes—functionalized for transferrin receptors, LRP1, or integrins can bias endothelial transcytosis or lesion homing; pH- and redox-responsive designs release cargo within inflamed niches. A complementary inside-out lever recalibrates systemic inflammation or the gut–brain axis (*e.g.*, bile-acid derivatives, short-chain-fatty-acid mimetics) to shift microglial tone without requiring high CNS exposure. Because every intervention perturbs circuits, pharmacodynamic monitoring is essential to avoid overcorrection. Suppressing microglial surveillance can impair synaptic refinement or antiviral defenses; excessive complement inhibition may preserve dysfunctional synapses that propagate epileptiform activity.[86] Therefore, EV-derived synaptic proteins can index pruning; MR spectroscopy can track glutamate–glutamine balance after GLT-1 interventions; myelin-sensitive MRI and neurofilament light can assay remyelination and axonal integrity; pathway panels (cytokines, complement fragments) confirm on-target effects. Closed-loop neuromodulation adds an orthogonal control: vagus-nerve stimulation engages the cholinergic anti-inflammatory reflex, lowering peripheral cytokines and potentially central glial activation, while transcranial focused ultrasound or targeted electrical stimulation may bias microglial and astrocytic calcium signaling toward pro-resolving states.[87] Ultimately, combination regimens, precisely staged and biomarker-guided, are likely to dominate: a small-molecule inflammasome modulator plus a GLT-1 enhancer; a complement inhibitor followed by a remyelinating agent; or a TREM2 agonist paired with microbiome and lifestyle interventions. Each element is titrated by fluid and imaging readouts that reflect live glial state. This endotype-first, timing-sensitive, and feedback-controlled approach converts neuroinflammation from a monolithic foe into a set of tractable, future-facing therapeutic pathways—aligning glial modulation with the broader aim of precision neurochemistry to achieve true disease modification.

10.1.3.3 Toward Precision Immunoneurology: Biomarkers, Trial Design, and Implementation

Turning concept into clinic demands a precision framework that measures, models, and manages glial biology over time. Measurement starts with harmonized sampling: CSF for proximal signals; plasma for scalable screening; extracellular vesicles immunoenriched for neuronal, astrocytic, or microglial origin; and advanced imaging for spatial context. Panels should cover modules rather than single markers—complement fragments and regulators; inflammasome components (ASC specks, IL-1β/IL-18); interferon-stimulated gene products; purinergic metabolites; lipid mediators (sphingolipids, oxysterols); and synaptic/axonal injury markers. Assays must pass rigorous analytical validation and include ancestry-, sex-, and age-calibrated reference intervals to avoid systematic bias.[88] Wearable and behavioral readouts (sleep fragmentation, autonomic indices) integrate environmental and stress biology that often drives glial set points. Modeling assembles these measurements into endotypes that are stable, interpretable, and actionable. Unsupervised clustering defines candidate endotypes; stability is tested by bootstrap consensus and longitudinal persistence; actionability is proven when endotype membership predicts the response to a targeted therapy.[88,89] Causal anchors—genetic instruments in complement, purinergic, or interferon pathways—help validate that a biomarker module reflects a modifiable driver. Digital twins then simulate intervention sequences allowing trialists to prioritize arms with the greatest expected benefit for specific endotypes. Trial design must break from one-size-fits-all. Enrichment strategies recruit participants with the relevant pathway activity (*e.g.*, high C1q/C3 synaptic deposition by PET/MR or EV markers for complement trials). Early-phase studies emphasize target engagement: within-subject changes in pathway panels, EV-synaptic proteins, MRS-detected glutamate modulation, or myelin imaging, coupled to safety and early clinical signals. Platform trials with adaptive randomization can test multiple, mechanistically distinct glial modulators in parallel, reallocating participants to arms showing biomarker and clinical promise.[89] Synthetic control arms, built from longitudinal registries and digital-twin predictions, reduce placebo burden where ethical considerations loom. Endpoint selection should span tiered horizons: near-term biochemical normalization; intermediate neural repair (synaptic density, remyelination); and longer-term function (cognition, mobility, mood) with patient-reported outcomes. Implementation requires translation into workflows that clinicians can use. Laboratory information systems should return interpretable reports: a traffic-light heat map of modules (complement, inflammasome, interferon, purinergic, astrocytic transport, and myelin) with trend arrows and suggested actions (*e.g.*, "consider complement-pathway agent; confirm stability in 4–6 weeks"). Stewardship programs akin to antimicrobial stewardship can provide consultative guidance for glial-directed therapies, preventing misuse and monitoring adverse effects. Health technology assessments will scrutinize cost-effectiveness; thus, demonstration projects should track reductions in hospitalizations, slower disability accrual, or decreased reliance on broad immunosuppression. Equity mandates inclusive biobanking, community engagement, and subsidized access to assays—without which precision immunoneurology risks widening disparities. Finally, safety and ethics are integral. Over-suppression of immune pathways could invite infection or malignancy; conversely, inadequate control risks accelerating degeneration. Transparent consent should explain pathway targeting, uncertainty, and the plan for continuous monitoring with early stopping rules. As tools mature, glial modulation will join neuromodulation and synaptic plasticity agents in an integrated care stack. The prize is not only symptom control but true disease modification—restoring the glial–neuronal choreography that underwrites neurochemical balance, circuit integrity, and resilient cognition.[90] Through disciplined measurement, mechanistic modeling, and measured intervention, the neuroimmune interface becomes a therapeutic ally rather than an unpredictable foe.

10.1.4 Neuroregeneration, Synaptic Plasticity, and Chemical Repair

Neuroregeneration and synaptic plasticity define the brain's capacity to repair, rewire, and recover function after injury or disease. In the past, repair was framed narrowly as neuronal replacement, but contemporary neurochemistry views repair as a multi-layered program: rescuing stressed neurons and glia; re-establishing synaptic efficacy and network balance; rebuilding myelin and axonal transport; and reprogramming local biochemistry so that circuits can learn again. This perspective unifies cellular biology with chemical therapeutics and device-enabled neuromodulation, emphasizing that regeneration is not a single event, but a staged process guided by molecular cues, activity, and metabolic support.[90] At the molecular level, trophic signaling is the cornerstone of chemical repair. Brain-derived neurotrophic factor (BDNF) acting at TrkB receptors activates PI3K–Akt and MAPK cascades to stabilize synapses, promote dendritic spine growth, and enhance long-term potentiation (LTP). Glial-cell-line-derived neurotrophic factor (GDNF) supports dopaminergic survival and axonal sprouting, while nerve growth factor (NGF) regulates cholinergic tone and attention networks. These pathways intersect with metabolic sensors (mTOR, AMPK) and redox checkpoints (Nrf2), indicating that successful plasticity depends on adequate ATP, NAD^+ balance, and oxidative stress control. Small molecules that boost endogenous trophic tone TrkB agonists or "BDNF mimetics," sigma-1 receptor modulators, and positive allosteric modulators of AMPA receptors can nudge synapses toward a growth-permissive state, particularly when combined with experience-dependent training.[91] A pivotal insight of the last decade is that psychoplastogen agents that rapidly increase structural and functional plasticity can reopen constrained learning windows. Low-dose ketamine and ketamine-like compounds, by disinhibiting pyramidal neurons and activating BDNF–mTOR signaling, promote spine formation within hours, with network-level normalization extending from days to weeks. Classic tryptaminergic psychedelics (*e.g.*, psilocybin) and non-hallucinogenic analogs also induce dendritic arborization and synaptic protein synthesis, potentially *via* 5-HT2A–TrkB crosstalk, glutamate bursts, and downstream mTOR/ERK activation. Importantly, these plasticity surges are content agnostic; without structured behavioral inputs (psychotherapy, motor rehab, cognitive training), they risk reinforcing maladaptive patterns. Precision protocols therefore pair psychoplastogens with task-specific, temporally aligned training to bias plasticity toward therapeutic circuits.[92] Beyond synapses, axonal and myelin repair are vital to regain conduction speed and metabolic resilience. Adult oligodendrocyte precursor cells (OPCs) persist throughout the CNS and can differentiate into myelinating oligodendrocytes when inhibitory cues are removed, and pro-differentiation signals are supplied. Retinoid X receptor (RXR) agonists, thyroid hormone analogs, and muscarinic antagonists have demonstrated remyelination in preclinical models. Parallel strategies aim at release from inhibition: neutralizing myelin-associated inhibitors (Nogo-A, MAG, OMgp) or blocking their downstream RhoA/ROCK pathway can permit axonal sprouting and elongation. Remodeling the extracellular matrix (ECM) can be equally powerful.[93] Perineuronal nets (PNNs) and chondroitin sulfate proteoglycans (CSPGs) restrict structural plasticity; enzymatic digestion (*e.g.*, chondroitinase), sulfation editing, or integrin-activating peptides can soften these molecular "brakes," restoring juvenile-like plasticity while preserving stability through controlled dosing and targeted delivery. Astrocytes and microglia serve as metabolic and synaptic co-therapists in regeneration. Astrocytic lactate shuttling fuels synaptic remodeling, and glutamate transporter (EAAT2/GLT-1) upregulation limits excitotoxicity during the vulnerable phase of spine turnover. Microglia, when biased toward pro-resolving states, clear inhibitory debris, secrete IGF-1 and TGF-β that support OPC maturation, and prune dysfunctional synapses to improve signal-to-noise. These roles argue for two-phase regimens: first, quell destructive neuroinflammatory programs (*e.g.*, P2X7/NLRP3 or complement overactivity), then engage plasticity enhancers; attempting both simultaneously can dilute efficacy and destabilize networks. Modern gene and RNA therapies expand the palette of chemical repair. Adeno-associated virus (AAV) vectors can deliver trophic genes (BDNF, GDNF), synaptic scaffolding proteins, or enzymes that restore neurotransmitter synthesis (*e.g.*, AADC in dopaminergic pathways). CRISPR-based base editors and epigenetic editors enable locus-specific tuning of plasticity genes without double-strand breaks, offering durable, titratable effects. Antisense oligonucleotides (ASOs) and RNAi can downregulate inhibitors of growth (*e.g.*, PTEN in select contexts) or normalize splicing defects. Because these approaches may alter broad networks, activity coupling (timing interventions alongside rehabilitation or neuromodulation) is central to translating molecular changes into coherent behavioral gains. Energy and organellar health are often the rate-limiting steps in regeneration.[94] Mitochondrial biogenesis (PGC-1α), improved mitophagy (PINK1–Parkin), and stabilized axonal transport (dynein/kinesin balance) determine whether new spines and sprouts survive. NAD^+ augmentation (precursors, CD38 inhibition) and Nrf2 activation can enhance antioxidant capacity and synaptogenesis. Lipid metabolism is similarly crucial: neurons rely on astrocyte-derived cholesterol and phospholipids to build membranes; therapies that optimize lipid flux (LXR/RXR agonism, ApoE pathway tuning) support synaptogenesis and remyelination. Here the gut–brain axis contributes short-chain fatty acids and bile acid derivatives modulate microglial tone and astrocyte metabolism, creating a systemic milieu that is conducive to repair. Activity shapes anatomy: thus, device-enabled plasticity has become a core partner to chemical repair. Noninvasive brain stimulation (high-frequency rTMS, intermittent theta burst stimulation) increases cortical excitability and BDNF expression, while transcranial direct current stimulation (tDCS) modulates learning rates and homeostatic plasticity. Closed-loop deep-brain stimulation and focused ultrasound neuromodulation allow circuit-specific timing aligned to behavioral states, reducing noise and stabilizing new synapses. Peripheral vagus nerve stimulation engages neuromodulatory nuclei (LC, NTS) and the cholinergic anti-inflammatory pathway, synchronizing systemic and central conditions for plastic change.[95] The most durable gains arise when precisely timed stimulation converges with plasticity-enhancing pharmacology and task practice, an approach often called "chemically assisted, activity-dependent rehabilitation." Delivery remains a bottleneck, so repair-oriented neurochemistry leans on targeted transport. Focused ultrasound with microbubbles transiently opens the blood–brain barrier to admit biologics and nanoparticles into circumscribed regions; exosomes and lipid nanoparticles decorated with transferrin or RVG peptides exploit receptor-mediated transcytosis; and pH/redox-responsive carriers release drugs preferentially in inflamed microenvironments. For chronic diseases, intermittent, low-burden dosing is ideal to trigger windows of plasticity without constant exposure, *e.g.*, pulsed psychoplastogen sessions coupled to intensive therapy blocks, or periodic trophic gene expression bursts validated by biomarker readouts.[96] Measuring regeneration demands multi-level biomarkers that resolve both structure and function. Synaptic density PET (SV2A ligands) can track synaptogenesis; myelin-sensitive MRI and magnetization transfer index gauge remyelination; diffusion metrics (neurite density, orientation dispersion) reflect microstructural refinement; and MRS glutamate/glutamine ratios capture excitatory tone during plasticity windows. Fluid markers such as extracellular-vesicle-derived synaptic proteins and declining neurofilament light (NfL) indicate circuit stabilization.[97] On the functional side, EEG/MEG gamma power and phase–amplitude coupling mirror local and long-range integration, while task-evoked fMRI or pupillometry quantifies network recruitment and neuromodulatory engagement. Selecting a tight biomarker set aligned to the repair modality (*e.g.*, SV2A-PET plus EV-synaptophysin for synaptogenesis; myelin MRI plus NfL for remyelination) increases translational efficiency and de-risks trials. Safety and ethics are intrinsic to plasticity-first strategies. Excessive or misdirected growth can foster aberrant connectivity, allodynia, or seizures; prolonged disinhibition may destabilize mood or cognition. Hence, guardrails include staged dosing; real-time monitoring of excitatory/inhibitory balance; and rapid "off switches" (*e.g.*, benzodiazepines for runaway excitation, cessation of stimulation, or antisense/CRISPR shutdown constructs). Equally important is access: scalable delivery, behavioral therapy integration, and insurer-recognized biomarkers will determine whether chemical repair remains a boutique intervention or become standard care.[98] In practice, the most promising paradigm is orchestrated repair: (1) map patient-specific barriers (inflammation, ECM constraints, energy failure); (2) relieve dominant brakes (targeted anti-inflammatory or ECM remodeling); (3) open a time-boxed plasticity window (psychoplastogen, BDNF-mimetic, or neuromodulation); and (4) drive learning with structured, goal-relevant training while providing trophic, metabolic, and myelin support. Iterating this cycle with biomarker feedback converts transient molecular surges into durable network changes that restore function.

10.1.5 Environmental and Lifestyle Modulation of Neurochemistry

Neurochemistry is highly responsive to daily inputs, and nutrition provides the biochemical substrate for neurotransmission. Omega-3 polyunsaturated fatty acids (notably DHA) integrate into synaptic membranes to enhance fluidity and receptor mobility and give rise to specialized pro-resolving mediators that temper microglial reactivity. Polyphenols (*e.g.*, cocoa flavanols, tea catechins, berry anthocyanins) and isothiocyanates such as sulforaphane activate Nrf2-dependent antioxidant defenses that protect dendritic spines from redox stress. Micronutrients serve as essential coenzymes: magnesium and B-vitamins (B6, folate, B12) support monoamine synthesis and one-carbon metabolism; iron underpins dopamine biology and myelination; and zinc modulates NMDA receptor function and neurogenesis.[99] In contrast, high-glycemic, ultra-processed dietary patterns increase insulin resistance, ceramides, and advanced glycation end products, which disrupt glutamate homeostasis, stiffen membranes, and bias microglia toward pro-inflammatory set points. The gut acts as a neuromodulatory organ that amplifies nutritional signals. Microbial fermentation of diverse fibers produces short-chain fatty acids (SCFAs) like butyrate and propionate, which tighten blood–brain barrier junctions, influence astrocyte metabolism, and modulate histone acetylation in neural cells. Microbes also transform bile acids into neuroactive derivatives (*via* TGR5/FXR pathways) that affect energy sensing and inflammatory tone. Tryptophan metabolism illustrates this bidirectionality: in eubiosis, flux favors serotonin and indole derivatives that reinforce mucosal integrity and vagal signaling; in dysbiosis, it shifts toward kynurenine and quinolinic acid, tilting circuits toward anxiety, anhedonia, or cognitive fatigue. Precision nutrition operationalizes these principles with lipidomics, metabolomics, and stool profiling to verify biomarker movement—higher DHA-containing phospholipids, lower ceramides, and improved kynurenine/tryptophan ratios—rather than relying on ideology.[100] Timing and context further refine the neurochemical signal. Consistent meal timing entrains peripheral clocks that synchronize with central circadian centers, whereas erratic eating or late-night snacking elevates nocturnal glucose and cortisol and impairs synaptic downscaling. Distributing protein across meals stabilizes precursor availability for neurotransmitter synthesis, and pairing carbohydrates with protein or fat blunts glycemic spikes that otherwise drive adrenergic surges and disturb sleep.[101] Hydration influences cognition *via* osmotic and glial mechanisms; even mild dehydration elevates adenosine and histamine linked to fatigue. Practical steps include daily high-fiber fermented food, two weekly servings of marine omega-3s (or algal alternatives), cooking methods that minimize glycation (moist heat, shorter times), and replacing sugar-sweetened beverages with water or unsweetened tea. With periodic objective checks—omega-3 indices on lipid panels, targeted metabolomics for SCFAs and bile acids, and inflammation markers—nutrition becomes a feedback-guided lever on neurotransmission, glial tone, and membrane biology.[102]

10.1.5.1 Sleep, Circadian Rhythms, and Light: The Daily Maintenance Program

Sleep and circadian alignment are the brain's routine maintenance system, resetting synaptic strength, clearing metabolites, and recalibrating neuromodulators. Slow-wave sleep promotes synaptic downscaling, protecting networks from saturation and energy overload; glymphatic flow intensifies during deep sleep, enhancing convective clearance of lactate, amyloid, and other solutes *via* perivascular channels. REM sleep consolidates emotional and procedural memory, integrates fear extinction, and tunes noradrenergic/serotonergic balance. Chronic restriction or fragmentation elevates extracellular glutamate, disturbs dopaminergic and adenosinergic tone, raises cortisol, and increases amygdala reactivity; the result is a chemistry profile that impairs attention, learning, and mood regulation. Irregular schedules and "social jetlag" desynchronize cellular clocks in astrocytes and microglia, prolonging low-grade inflammation and weakening synaptic plasticity.[103] Light is the master circadian zeitgeber. Bright natural light in the morning anchors the central clock and advances melatonin onset, while evening exposure to strong or blue-enriched light delays circadian phase, compresses slow-wave sleep, and elevates nocturnal arousal signals. Aligning light with biology is a potent neurochemical intervention: 30–60 minutes of outdoor light soon after waking consolidates daytime alertness (*via* orexin and catecholamine tone) and stabilizes nighttime melatonin, whereas dimming household lighting and reducing blue-heavy screens 2–3 hours before bed lowers late-evening cortisol and preserves GABAergic sleep drive. Temperature and sound cues assist; a cooler, quieter bedroom (and a pre-sleep warm shower to induce distal vasodilation and core-temperature drop) facilitates sleep initiation by supporting adenosine-mediated homeostatic pressure. Behavioral design brings the plan to ground. Anchoring consistent sleep/wake times—even on weekends—reduces phase scatter that otherwise accumulates during the workweek.[104] Caffeine is best front-loaded (within 8–10 hours of bedtime) to avoid antagonizing adenosine when sleep pressure is needed; alcohol, despite subjective sedation, fragments slow-wave and REM sleep and raises nocturnal glutamate. Late-night heavy meals increase reflux and sympathetic tone; shifting calories earlier improves insulin dynamics and overnight recovery. For shift workers and students with variable schedules, tactical tools include timed light exposure (portable bright-light devices in the early biological morning), blue-blocking in the late subjective evening, and short strategic naps that restore performance without deep-sleep inertia. Wearables can quantify sleep efficiency, variability, and light exposure; paired with periodic biomarkers (*e.g.*, morning cortisol, evening melatonin, inflammatory panels), data-guided adjustments refine the plan. Crucially, sleep and circadian hygiene amplify other lifestyle levers. Exercise performed in the late afternoon, or early evening improves slow-wave sleep, provided it ends a few hours before bedtime; morning exercise synergizes with light to reinforce circadian amplitude. Nutritional choices—lower evening glycemic load, moderate alcohol, adequate magnesium—stabilize GABA/glutamate balance and reduce nocturnal awakenings.[105] The output of this choreography is measurable: steadier daytime vigilance, faster learning curves, improved emotional regulation, and lower neuroinflammatory markers. In short, consistent sleep and light hygiene are not soft wellness tips; they are high-yield, programmable inputs to neurochemical homeostasis.

10.1.5.2 *Physical Activity, Stress & Recovery, and Environmental Exposures*

Physical activity acts as a systemic neuromodulator. Aerobic and resistance training elevate brain-derived neurotrophic factor (BDNF), promote angiogenesis, and increase insulin sensitivity, collectively enhancing synaptic plasticity and energy availability. Exercise-induced lactate serves as a signaling molecule that supports dendritic remodeling and may upregulate genes for synaptic proteins *via* astrocyte–neuron metabolic coupling.[105] Myokines released by contracting muscle—such as irisin—feed forward to hippocampal neurogenesis and oligodendroglial metabolism. Practically, 150–300 weekly minutes of moderate aerobic activity plus two sessions of resistance work create reliable BDNF and mood gains; even “exercise snacks” (1–3 minutes of brisk stair climbs, or body weight work each hour) interrupt sedentary neurochemistry characterized by lower glymphatic efficiency and endothelial nitric oxide. Aligning workouts with goals matters: skill practice scheduled within several hours after exercise exploits transient “plasticity windows” when catecholamines and BDNF synergize with training to consolidate motor or cognitive gains. Stress and recovery shape the same chemicals from another angle. Chronic sympathetic dominance elevates cortisol and glutamate, suppresses hippocampal neurogenesis, and narrows prefrontal flexibility. Recovery practices are not cosmetic—they are biochemistry in action. Slow diaphragmatic breathing, HRV biofeedback, and mindfulness increase vagal tone, reduce IL-6 and TNF, and stabilize GABAergic inhibition. Nature exposure (green or blue spaces) attenuates amygdala threat reactivity and lowers salivary cortisol; meaningful social contact boosts dopaminergic reward signaling and buffers HPA activation.[105] For individuals with trauma or high baseline arousal, trauma-informed practices and gradual exposure are essential to avoid overshooting arousal windows. Environmental exposures complete the context in which chemistry operates. Fine particulate air pollution (PM2.5), ozone, and diesel exhaust increase oxidative burden, endothelial dysfunction, and BBB permeability; heavy metals and certain pesticides disrupt mitochondrial enzymes and dopamine handling; heat waves amplify dehydration and inflammatory tone. Simple mitigations—HEPA filtration indoors, choosing green routes for commuting, hydrating with electrolytes during heat, and careful handling of household chemicals—translate into measurable improvements in lipidomics, cytokines, and cognitive stamina. Occupational and urban design changes (cleaner transit, shaded walkways, access to parks) turn personal responsibility into shared infrastructure. Digital hygiene is a modern exposure: late-night high-salience screen use pushes dopamine/adenosine balance toward alerting states and displaces slow-wave sleep; time-boxing screens, using grayscale or notification limits in the evening, and prioritizing daytime device interactions reduce circadian disruption. Implementation benefits from closed-loop coaching. Wearables capture steps, heart-rate variability, temperature, and light; smartphones log sedentary time and "stressful" usage patterns; periodic labs (lipids, glucose/insulin, hs-CRP) and, when appropriate, advanced panels (omega-3 index, SCFAs, bile-acid signatures) confirm that choices are moving biology in the intended direction. Programs should be tailored: older adults may emphasize balance and resistance for myelin and vestibular health; individuals with mood disorders might prioritize morning light walks plus breathwork to reset circadian and autonomic tone; people with cardiometabolic risk should focus on post-meal movement to blunt glucose excursions. The throughline is coherence: when movement, recovery, and environment align, neurochemical noise falls, plasticity signals rise, and the brain's ability to learn, regulate mood, and resist degeneration improves in ways that are objective, scalable, and equitable.[105]

10.1.6 Future Outlook

The coming decade will see neurochemistry evolve into an interventional discipline capable of actively reading and writing the brain's molecular code. Multimodal measurement, including SV2A-PET, ultra-high-field fMRI/MRS, spatial omics, and real-time biosensors will provide clinically precise maps of pathway activity, while AI will integrate these data into actionable endotypes that shape individualized treatment.[106] Care will become staged and adaptive, beginning with modulation of harmful inflammation, followed by reopening of plasticity windows, targeted repair of synapses, myelin, and metabolism, and reinforcement of gains through activity and environmental cues aligned with circadian biology. Therapeutic strategies will expand to include psychoplastogens, glial modulators, remyelination agents, gene and RNA editing tools, and closed-loop neuromodulation, all coordinated through digital twins that simulate dosing, sequencing, and risk and update continuously using wearable and fluid telemetry.[107] Rigorous analytical validation, transparent algorithms, bias audits, and privacy-preserving data infrastructures will be essential to ensure reliability and responsible translation, while equitable access supported by tiered diagnostics, low-cost assays, and public-health investment will shape how widely these advances benefit diverse populations.[108]

10.2 Conclusion

The future of neurochemistry is poised to transform brain health by shifting from descriptive observation to precision-guided intervention. As multimodal technologies such as SV2A-PET, ultra-high-field fMRI/MRS, spatial omics, and real-time biosensors converge, they will create a high-resolution, dynamic map of neurochemical activity across individuals. Artificial intelligence will synthesize these complex data streams into mechanistic endotypes that guide targeted and adaptive care. Therapeutic strategies will increasingly operate in coordinated stages—first modulating harmful inflammation, then opening windows of plasticity, followed by synaptic, myelin, and metabolic repair, and finally consolidating gains through behavior, environment, and circadian-aligned routines. Psychoplastogens, glial-state modulators, remyelination agents, gene and RNA editors, and closed-loop neuromodulation will become central tools, optimized through digital twins that simulate intervention timing, sequence, and risk. Ethical implementation will require rigorous validation, transparency, bias auditing, and privacy-preserving data governance. Ensuring equitable access through tiered diagnostics, affordable assays, and population-level support for sleep, light, air quality, and activity will determine who benefits from these advances. As technology, computation, and ethical design evolve together, neurochemistry is set to enable earlier detection, personalized treatment, and meaningful prevention of brain disorders.

Abbreviations

AADC	Aromatic L-amino acid decarboxylase
AAV	Adeno-associated virus
ACh	Acetylcholine
AD	Alzheimer's disease
AIDS	Acquired immunodeficiency syndrome
AIF	Apoptosis-inducing factor
AFADESI	Air flow-assisted desorption electrospray ionization
AMPK	AMP-activated protein kinase
ASD	Autism spectrum disorder
ASO	Antisense oligonucleotide
BBB	Blood–brain barrier
BDNF	Brain-derived neurotrophic factor
BDNF/TrkB	Brain-derived neurotrophic factor/tropomyosin receptor kinase B

BOLD	Blood-oxygen-level dependent
CBF	Cerebral blood flow
CBF/$CMRO_2$	Cerebral blood flow/cerebral metabolic rate of oxygen
[^{11}C]UCB-J	Carbon-11 labeled UCB-J
CD	Cluster of differentiation
CE–MS	Capillary electrophoresis–mass spectrometry
CEST	Chemical exchange saturation transfer
CETSA	Cellular thermal shift assay
$CMRO_2$	Cerebral metabolic rate of oxygen
CODEX	CO-detection by indexing
CNS	Central nervous system
CRISPR	Clustered regularly interspaced short palindromic repeats
CSF	Cerebrospinal fluid
CX3CR1	CX3C chemokine receptor 1
DAG	Directed acyclic graph
DBS	Deep brain stimulation
DESI	Desorption electrospray ionization
DIA	Data-independent acquisition
DDA	Data-independent acquisition/data-dependent acquisition
DIA/SWATH	Data-independent acquisition/sequential window acquisition of all theoretical fragment ions
DIABLO	Data integration analysis for biomarker discovery using latent components
DMI	Deuterium metabolic imaging
DOPA	Dihydroxyphenylalanine
ECM	Extracellular matrix
EEG	Electroencephalography
EHR	Electronic health record
EVs	Extracellular vesicles
FAIMS	High-field asymmetric waveform ion mobility spectrometry
FET/OECT	Field-effect transistor/organic electrochemical transistor
FRET	Förster resonance energy transfer
FSCV	Fast-scan cyclic voltammetry
fMRI	Functional magnetic resonance imaging
fMRS	Functional magnetic resonance spectroscopy
fUS	Functional ultrasound
FUS	Focused ultrasound
GABA	Gamma-aminobutyric acid
GDNF	Glial cell line-derived neurotrophic factor
GLT-1/EAAT2	Glutamate transporter-1/excitatory amino acid transporter-2
GPCR	G-protein-coupled receptor
GRAB	GPCR-activation-based sensor

GRAB-NE	G-protein-coupled receptor activation-based norepinephrine sensor
HS	High sensitivity
HT/NE	Histamine/norepinephrine
IL-1β/IL-6/IL-18	Interleukin-1β/interleukin-6/interleukin-18
ISG	Interferon-stimulated gene
jGCaMP	Genetically encoded calcium indicator (j-series variant)
KEGG	Kyoto encyclopedia of genes and genomes
LAESI	Laser ablation electrospray ionization
LC–MS/MS	Liquid chromatography–tandem mass spectrometry
LTP	Long-term potentiation
LXR/RXR	Liver X receptor/retinoid X receptor
MAG	Myelin-associated glycoprotein
MALDI	Matrix-assisted laser desorption/ionization
MAPK	Mitogen-activated protein kinase
MIBI	Multiplexed ion beam imaging
MOFA	Multi-omics factor analysis
MRS	Magnetic resonance spectroscopy
MSI	Mass spectrometry imaging
mTOR	Mechanistic target of rapamycin
NAD^+	Nicotinamide adenine dinucleotide
nano-DESI	Nanospray desorption electrospray ionization
NE	Norepinephrine
NfL	Neurofilament light chain
NLRP3	NOD-like receptor family pyrin domain-containing 3
NMDA	*N*-Methyl-D-aspartate
NMR	Nuclear magnetic resonance
NOTEARS	Non-combinatorial optimization *via* trace exponential and augmented Lagrangian for structure learning
NPC/OPC	Neural/oligodendrocyte precursor cell
OCD	Obsessive–compulsive disorder
PEDOT:PSS	Poly(3,4-ethylenedioxythiophene):poly(styrene sulfonate)
PET	Positron emission tomography
PET–MR	Positron emission tomography–magnetic resonance imaging
PNN	Perineuronal net
PRM/SRM	Parallel/selected reaction monitoring
PTM	Post-translational modification
QA/QC	Quality assurance/quality control
RNAi	RNA interference
ROS/RNS	Reactive oxygen/nitrogen species
RXR	Retinoid X receptor
SCFA	Short-chain fatty acid
SCoPE2	Single-cell proteomics by mass spectrometry (version 2)
SPECT	Single-photon emission computed tomography
SV2A	Synaptic vesicle glycoprotein 2A

SV2A-PET	Synaptic vesicle glycoprotein 2A positron emission tomography
SWATH	Sequential window acquisition of all theoretical fragment ions
tDCS	Transcranial direct current stimulation
tFUS	Transcranial focused ultrasound
TMS/rTMS	Transcranial magnetic stimulation/repetitive TMS
TSPO	Translocator protein
TrkB	Tropomyosin receptor kinase B
VASO	Vascular space occupancy MRI
VEGF	Vascular endothelial growth factor

References

1. M. S. Bavarsad and L. T. Grinberg, SV2A PET imaging in human neurodegenerative diseases, *Front. Aging Neurosci.*, 2024, **16**, 1380561.
2. L. R. Drake, Y. Wu, M. Naganawa, R. Asch, C. Zheng, S. Najafzadeh, R. Pracitto, M. Lindemann, S. Li, J. Ropchan and D. Labaree, First-in-Human Study of 18F-SynVesT-2: An SV2A PET imaging probe with fast brain kinetics and high specific binding, *J. Nucl. Med.*, 2024, **65**(3), 462–469.
3. D. M. Harrison, P. Sati, E. C. Klawiter, S. Narayanan, F. Bagnato, E. S. Beck, P. Barker, A. Calvi, A. Cagol, M. Donadieu and J. Duyn, The use of 7T MRI in multiple sclerosis: review and consensus statement from the North American Imaging in Multiple Sclerosis Cooperative, *Brain Commun.*, 2024, **6**(5), fcae359.
4. A. T. Cember, R. P. Nanga and R. Reddy, Glutamate-weighted CEST (gluCEST) imaging for mapping neurometabolism: an update on the state of the art and emerging findings from *in vivo* applications, *NMR Biomed.*, 2023, **36**(6), e4780.
5. Y. Yang, B. Li and Y. Li, Genetically encoded sensors for the *in vivo* detection of neurochemical dynamics, *Annu. Rev. Anal. Chem.*, 2024, **17**(1), 367–392.
6. A. R. Rezai, P. F. D'Haese, V. Finomore, J. Carpenter, M. Ranjan, K. Wilhelmsen, R. I. Mehta, P. Wang, U. Najib, C. Vieira Ligo Teixeira and T. Arsiwala, Ultrasound blood–brain barrier opening and aducanumab in Alzheimer's disease, *N. Engl. J. Med.*, 2024, **390**(1), 55–62.
7. A. Gomez-Arboledas, M. M. Acharya and A. J. Tenner, The role of complement in synaptic pruning and neurodegeneration, *ImmunoTargets Ther.*, 2021, 373–386.
8. L. S. Fekonja, R. Schenk, E. Schröder, R. Tomasello, S. Tomšič and T. Picht, The digital twin in neuroscience: from theory to tailored therapy, *Front. Neurosci.*, 2024, **18**, 1454856.
9. V. Balaji, T. A. Song, M. Malekzadeh, P. Heidari and J. Dutta, Artificial intelligence for PET and SPECT image enhancement, *J. Nucl. Med.*, 2024, **65**(1), 4–12.

10. E. S. Finn, L. Huber and P. A. Bandettini, Higher and deeper: Bringing layer fMRI to association cortex, *Prog. Neurobiol.*, 2021, **207**, 101930.
11. M. Piwecka, N. Rajewsky and A. Rybak-Wolf, Single-cell and spatial transcriptomics: deciphering brain complexity in health and disease, *Nat. Rev. Neurol.*, 2023, **19**(6), 346–362.
12. I. W. Tengesdal, M. Banks, C. A. Dinarello and C. Marchetti, Screening NLRP3 drug candidates in clinical development: lessons from existing and emerging technologies, *Front. Immunol.*, 2024, **15**, 1422249.
13. Y. Yu, Z. Wang, Z. Chai, S. Ma, A. Li and Y. Li, Central Nervous System-Derived Extracellular Vesicles as Biomarkers in Alzheimer's Disease, *Int. J. Mol. Sci.*, 2025, **26**(17), 8272.
14. H. Wang, T. Qian, Y. Zhao, Y. Zhuo, C. Wu, T. Osakada, P. Chen, Z. Chen, H. Ren, Y. Yan and L. Geng, A tool kit of highly selective and sensitive genetically encoded neuropeptide sensors, *Science*, 2023, **382**(6672), eabq8173.
15. J. Feng, H. Dong, J. E. Lischinsky, J. Zhou, F. Deng, C. Zhuang, X. Miao, H. Wang, G. Li, R. Cai and H. Xie, Monitoring norepinephrine release *in vivo* using next-generation GRABNE sensors, *Neuron*, 2024, **112**(12), 1930–1942.
16. B. S. Ye, K. W. Chang, S. Kang, S. Jeon and J. W. Chang, Repetitive and extensive focused ultrasound–mediated bilateral frontal blood-brain barrier opening for Alzheimer's disease, *J. Neurosurg.*, 2025, **1**(aop), 1–8.
17. L. Wang, X. Zhang, Z. Yang, B. Wang, H. Gong, K. Zhang, Y. Lin and M. Sun, Extracellular vesicles: biological mechanisms and emerging therapeutic opportunities in neurodegenerative diseases, *Transl. Neurodegener.*, 2024, **13**(1), 60.
18. M. J. Kas, B. W. Penninx, G. M. Knudsen, B. Cuthbert, P. Falkai, G. S. Sachs, K. J. Ressler, E. Bałkowiec-Iskra, F. Butlen-Ducuing, M. Leboyer and H. Marston, Precision psychiatry roadmap: towards a biology-informed framework for mental disorders, *Mol. Psychiatry*, 2025, 1.
19. Y. Chai and R. Y. Zhang, Exploring methodological frontiers in laminar fMRI, *Psychoradiology*, 2024, **4**, kkae027.
20. A. T. Cember, R. P. Nanga and R. Reddy, Glutamate-weighted CEST (gluCEST) imaging for mapping neurometabolism: an update on the state of the art and emerging findings from *in vivo* applications, *NMR Biomed.*, 2023, **36**(6), e4780.
21. I. Polvoy, H. Qin, R. R. Flavell, J. Gordon, P. Viswanath, R. Sriram, M. A. Ohliger and D. M. Wilson, Deuterium metabolic imaging—rediscovery of a spectroscopic tool, *Metabolites*, 2021, **11**(9), 570.
22. S. Soloukey, E. Collée, L. Verhoef, D. D. Satoer, C. M. Dirven, E. M. Bos, J. W. Schouten, B. S. Generowicz, F. Mastik, C. I. De Zeeuw and S. K. Koekkoek, Human brain mapping using co-registered fUS, fMRI and ESM during awake brain surgeries: A proof-of-concept study, *NeuroImage*, 2023, **283**, 120435.
23. Z. Hu, Y. Hu, L. Huang, W. Zhong, J. Zhang, D. Lei, Y. Chen, Y. Ni and Y. Liu, Recent progress in organic electrochemical transistor-structured biosensors, *Biosensors*, 2024, **14**(7), 330.

24. H. A. Clarke, X. Ma, C. J. Shedlock, T. Medina, T. R. Hawkinson, L. Wu, R. A. Ribas, S. Keohane, S. Ravi, J. L. Bizon and S. N. Burke, Spatial mapping of the brain metabolome lipidome and glycome, *Nat. Commun.*, 2025, **16**(1), 4373.
25. L. Krasny and P. H. Huang, Data-independent acquisition mass spectrometry (DIA-MS) for proteomic applications in oncology, *Mol. Omics*, 2021, **17**(1), 29–42.
26. S. J. Finnema, N. B. Nabulsi, T. Eid, K. Detyniecki, S. F. Lin, M. K. Chen, R. Dhaher, D. Matuskey, E. Baum, D. Holden and D. D. Spencer, Imaging synaptic density in the living human brain, *Sci. Transl. Med.*, 2016, **8**(348), 348ra396.
27. A. V. Leopold, D. M. Shcherbakova and V. V. Verkhusha, Fluorescent biosensors for neurotransmission and neuromodulation: engineering and applications, *Front. Cell. Neurosci.*, 2019, **13**, 474.
28. H. Wang, M. Jing and Y. Li, Lighting up the brain: genetically encoded fluorescent sensors for imaging neurotransmitters and neuromodulators, *Curr. Opin. Neurobiol.*, 2018, **50**, 171–178.
29. K. Lebrigand, J. Bergenstråhle, K. Thrane, A. Mollbrink, K. Meletis, P. Barbry, R. Waldmann and J. Lundeberg, The spatial landscape of gene expression isoforms in tissue sections, *Nucleic Acids Res.*, 2023, **51**(8), e47.
30. M. Sans, C. L. Feider and L. S. Eberlin, Advances in mass spectrometry imaging coupled to ion mobility spectrometry for enhanced imaging of biological tissues, *Curr. Opin. Chem. Biol.*, 2018, **42**, 138–146.
31. D. Wang, H. Xiao, X. Lv, H. Chen and F. Wei, Mass spectrometry based on chemical derivatization has brought novel discoveries to lipidomics: A comprehensive review, *Crit. Rev. Anal. Chem.*, 2025, **55**(1), 21–52.
32. R. Aebersold and M. Mann, Mass spectrometry-based proteomics, *Nature*, 2003, **422**(6928), 198–207.
33. G. J. Patti, O. Yanes and G. Siuzdak, Metabolomics: the apogee of the omics trilogy, *Nat. Rev. Mol. Cell Biol.*, 2012, **13**(4), 263–269.
34. P. L. Ståhl, F. Salmén, S. Vickovic, A. Lundmark, J. F. Navarro, J. Magnusson, S. Giacomello, M. Asp, J. O. Westholm, M. Huss and A. Mollbrink, Visualization and analysis of gene expression in tissue sections by spatial transcriptomics, *Science*, 2016, **353**(6294), 78–82.
35. E. Fahy, S. Subramaniam, R. C. Murphy, M. Nishijima, C. R. Raetz, T. Shimizu, F. Spener, G. Van Meer, M. J. Wakelam and E. A. Dennis, Update of the LIPID MAPS comprehensive classification system for lipids1, *J. Lipid Res.*, 2009, **50**, S9–S14.
36. R. M. Caprioli, T. B. Farmer and J. Gile, Molecular imaging of biological samples: localization of peptides and proteins using MALDI-TOF MS, *Anal. Chem.*, 1997, **69**(23), 4751–4760.
37. J. Cox and M. Mann, MaxQuant enables high peptide identification rates, individualized ppb-range mass accuracies and proteome-wide protein quantification, *Nat. Biotechnol.*, 2008, **26**(12), 1367–1372.
38. S. Tyanova, T. Temu, P. Sinitcyn, A. Carlson, M. Y. Hein, T. Geiger, M. Mann and J. Cox, The Perseus computational platform for

comprehensive analysis of (prote) omics data, *Nat. Methods*, 2016, **13**(9), 731–740.
39. T. Patriarchi, J. R. Cho, K. Merten, M. W. Howe, A. Marley, W. H. Xiong, R. W. Folk, G. J. Broussard, R. Liang, M. J. Jang and H. Zhong, Ultrafast neuronal imaging of dopamine dynamics with designed genetically encoded sensors, *Science*, 2018, **360**(6396), eaat4422.
40. F. Sun, J. Zeng, M. Jing, J. Zhou, J. Feng, S. F. Owen, Y. Luo, F. Li, H. Wang, T. Yamaguchi and Z. Yong, A genetically encoded fluorescent sensor enables rapid and specific detection of dopamine in flies, fish, and mice, *Cell*, 2018, **174**(2), 481–496.
41. M. Jing, Y. Li, J. Zeng, P. Huang, M. Skirzewski, O. Kljakic, W. Peng, T. Qian, K. Tan, J. Zou and S. Trinh, An optimized acetylcholine sensor for monitoring *in vivo* cholinergic activity, *Nat. Methods*, 2020, **17**(11), 1139–1146.
42. K. T. Kishida, S. G. Sandberg, T. Lohrenz, Y. G. Comair, I. Sáez, P. E. Phillips and P. R. Montague, Sub-second dopamine detection in human striatum, *PLoS One*, 2011, **6**(8), e23291.
43. D. Bang, K. T. Kishida, T. Lohrenz, J. P. White, A. W. Laxton, S. B. Tatter, S. M. Fleming and P. R. Montague, Sub-second dopamine and serotonin signaling in human striatum during perceptual decision-making, *Neuron*, 2020, **108**(5), 999–1010.
44. B. J. Venton and Q. Cao, Fundamentals of fast-scan cyclic voltammetry for dopamine detection, *Analyst*, 2020, **145**(4), 1158–1168.
45. S. Lucio Boschen, J. Trevathan, S. A. Hara, A. Asp and J. L. Lujan, Defining a path toward the use of Fast-scan cyclic voltammetry in human studies, *Front. Neurosci.*, 2021, **15**, 728092.
46. F. Pisano, M. Pisanello, S. J. Lee, J. Lee, E. Maglie, A. Balena, L. Sileo, B. Spagnolo, M. Bianco, M. Hyun and M. De Vittorio, Depth-resolved fiber photometry with a single tapered optical fiber implant, *Nat. Methods*, 2019, **16**(11), 1185–1192.
47. Z. Wu, Y. Cui, H. Wang, H. Wu, Y. Wan, B. Li, L. Wang, S. Pan, W. Peng, A. Dong and Z. Yuan, Neuronal activity-induced, equilibrative nucleoside transporter-dependent, somatodendritic adenosine release revealed by a GRAB sensor, *Proc. Natl. Acad. Sci. U. S. A.*, 2023, **120**(14), e2212387120.
48. J. S. Marvin, B. G. Borghuis, L. Tian, J. Cichon, M. T. Harnett, J. Akerboom, A. Gordus, S. L. Renninger, T. W. Chen, C. I. Bargmann and M. B. Orger, An optimized fluorescent probe for visualizing glutamate neurotransmission, *Nat. Methods*, 2013, **10**(2), 162–170.
49. N. Arroyo-Currás, J. Somerson, P. A. Vieira, K. L. Ploense, T. E. Kippin and K. W. Plaxco, Real-time measurement of small molecules directly in awake, ambulatory animals, *Proc. Natl. Acad. Sci. U. S. A.*, 2017, **114**(4), 645–650.
50. N. Delmo, B. Mostafiz, A. E. Ross, J. Suni and E. Peltola, Developing an electrochemical sensor for the *in vivo* measurements of dopamine, *Sens. Diagn.*, 2023, **2**(3), 559–581.

51. R. B. Rashid, X. Ji and J. Rivnay, Organic electrochemical transistors in bioelectronic circuits, *Biosens. Bioelectron.*, 2021, **190**, 113461.
52. N. Luo, W. Shi, Z. Yang, M. Song and T. Jiang, Multimodal fusion of brain imaging data: Methods and applications, *Mach. Intell. Res.*, 2024, **21**(1), 136–152.
53. W. Huang and N. Shu, AI-powered integration of multimodal imaging in precision medicine for neuropsychiatric disorders, *Cell Rep. Med.*, 2025, **6**(5), 102132.
54. I. Seth, B. Lim, K. Joseph, D. Gracias, Y. Xie, R. J. Ross and W. M. Rozen, Use of artificial intelligence in breast surgery: a narrative review, *Gland Surg.*, 2024, **13**(3), 395–411.
55. S. Li, H. Hua and S. Chen, Graph neural networks for single-cell omics data: a review of approaches and applications, *Briefings Bioinf.*, 2025, **26**(2), bbaf109.
56. X. Han, B. Wang, C. Situ, Y. Qi, H. Zhu, Y. Li and X. Guo, scapGNN: A graph neural network–based framework for active pathway and gene module inference from single-cell multi-omics data, *PLoS Biol.*, 2023, **21**(11), e3002369.
57. C. Zhao, F. Zhang, W. Lou, X. Wang and J. Yang, A comprehensive review of advances in physics-informed neural networks and their applications in complex fluid dynamics, *Phys. Fluids*, 2024, **36**(10), 101301.
58. N. Rieke, J. Hancox, W. Li, F. Milletari, H. R. Roth, S. Albarqouni, S. Bakas, M. N. Galtier, B. A. Landman, K. Maier-Hein and S. Ourselin, The future of digital health with federated learning, *npj Digit. Med.*, 2020, **3**(1), 119.
59. C. R. Jack Jr, D. A. Bennett, K. Blennow, M. C. Carrillo, B. Dunn, S. B. Haeberlein, D. M. Holtzman, W. Jagust, F. Jessen, J. Karlawish and E. Liu, NIA-AA research framework: toward a biological definition of Alzheimer's disease, *Alzheimers Dement.*, 2018, **14**(4), 535–562.
60. S. Palmqvist, S. Janelidze, Y. T. Quiroz, H. Zetterberg, F. Lopera, E. Stomrud, Y. I. Su, Y. Chen, G. E. Serrano, A. Leuzy and N. Mattsson-Carlgren, Discriminative accuracy of plasma phospho-tau217 for Alzheimer disease *vs* other neurodegenerative disorders, *JAMA*, 2020, **324**(8), 772–781.
61. L. Gaetani, K. Blennow, P. Calabresi, M. Di Filippo, L. Parnetti and H. Zetterberg, Neurofilament light chain as a biomarker in neurological disorders, *J. Neurol. Neurosurg. Psychiatry*, 2019, **90**(8), 870–881.
62. C. Cicognola, S. Janelidze, J. Hertze, H. Zetterberg, K. Blennow, N. Mattsson-Carlgren and O. Hansson, Plasma glial fibrillary acidic protein detects Alzheimer pathology and predicts future conversion to Alzheimer dementia in patients with mild cognitive impairment, *Alzheimers Res. Ther.*, 2021, **13**(1), 68.
63. W. M. Van Der Flier and P. Scheltens, The ATN framework—moving preclinical Alzheimer disease to clinical relevance, *JAMA Neurol.*, 2022, **79**(10), 968–970.

64. O. Hansson, K. Blennow, H. Zetterberg and J. Dage, Blood biomarkers for Alzheimer's disease in clinical practice and trials, *Nat. Aging*, 2023, **3**(5), 506–519.
65. A. Siderowf, L. Concha-Marambio, D. E. Lafontant, C. M. Farris, Y. Ma, P. A. Urenia, H. Nguyen, R. N. Alcalay, L. M. Chahine, T. Foroud and D. Galasko, Assessment of heterogeneity among participants in the Parkinson's Progression Markers Initiative cohort using α-synuclein seed amplification: a cross-sectional study, *Lancet Neurol.*, 2023, **22**(5), 407–417.
66. A. Nakamura, N. Kaneko, V. L. Villemagne, T. Kato, J. Doecke, V. Doré, C. Fowler, Q. X. Li, R. Martins, C. Rowe and T. Tomita, High performance plasma amyloid-β biomarkers for Alzheimer's disease, *Nature*, 2018, **554**(7691), 249–254.
67. E. H. Thijssen, R. La Joie, A. Wolf, A. Strom, P. Wang, L. Iaccarino, V. Bourakova, Y. Cobigo, H. Heuer, S. Spina and L. VandeVrede, Diagnostic value of plasma phosphorylated tau181 in Alzheimer's disease and frontotemporal lobar degeneration, *Nat. Med.*, 2020, **26**(3), 387–397.
68. S. Janelidze, N. Mattsson, S. Palmqvist, R. Smith, T. G. Beach, G. E. Serrano, X. Chai, N. K. Proctor, U. Eichenlaub, H. Zetterberg and K. Blennow, Plasma P-tau181 in Alzheimer's disease: relationship to other biomarkers, differential diagnosis, neuropathology and longitudinal progression to Alzheimer's dementia, *Nat. Med.*, 2020, **26**(3), 379–386.
69. V. Ovod, K. N. Ramsey, K. G. Mawuenyega, J. G. Bollinger, T. Hicks, T. Schneider, M. Sullivan, K. Paumier, D. M. Holtzman, J. C. Morris and T. Benzinger, Amyloid β concentrations and stable isotope labeling kinetics of human plasma specific to central nervous system amyloidosis, *Alzheimers Dement.*, 2017, **13**(8), 841–849.
70. S. E. Schindler, J. G. Bollinger, V. Ovod, K. G. Mawuenyega, Y. Li, B. A. Gordon, D. M. Holtzman, J. C. Morris, T. L. Benzinger, C. Xiong and A. M. Fagan, High-precision plasma β-amyloid 42/40 predicts current and future brain amyloidosis, *Neurology*, 2019, **93**(17), e1647–e1659.
71. N. J. Ashton, T. A. Pascoal, T. K. Karikari, A. L. Benedet, J. Lantero-Rodriguez, G. Brinkmalm, A. Snellman, M. Schöll, C. Troakes, A. Hye and S. Gauthier, Plasma p-tau231: a new biomarker for incipient Alzheimer's disease pathology, *Acta Neuropathol.*, 2021, **141**(5), 709–724.
72. M. Khalil, C. E. Teunissen, M. Otto, F. Piehl, M. P. Sormani, T. Gattringer, C. Barro, L. Kappos, M. Comabella, F. Fazekas and A. Petzold, Neurofilaments as biomarkers in neurological disorders, *Nat. Rev. Neurol.*, 2018, **14**(10), 577–589.
73. B. Stevens, N. J. Allen, L. E. Vazquez, G. R. Howell, K. S. Christopherson, N. Nouri, K. D. Micheva, A. K. Mehalow, A. D. Huberman, B. Stafford and A. Sher, The classical complement cascade mediates CNS synapse elimination, *Cell*, 2007, **131**(6), 1164–1178.
74. S. A. Liddelow, K. A. Guttenplan, L. E. Clarke, F. C. Bennett, C. J. Bohlen, L. Schirmer, M. L. Bennett, A. E. Münch, W. S. Chung, T. C.

Peterson and D. K. Wilton, Neurotoxic reactive astrocytes are induced by activated microglia, *Nature*, 2017, **541**(7638), 481–487.
75. H. Keren-Shaul, A. Spinrad, A. Weiner, O. Matcovitch-Natan, R. Dvir-Szternfeld, T. K. Ulland, E. David, K. Baruch, D. Lara-Astaiso, B. Toth and S. Itzkovitz, A unique microglia type associated with restricting development of Alzheimer's disease, *Cell*, 2017, **169**(7), 1276–1290.
76. R. C. Coll, A. A. Robertson, J. J. Chae, S. C. Higgins, R. Muñoz-Planillo, M. C. Inserra, I. Vetter, L. S. Dungan, B. G. Monks, A. Stutz and D. E. Croker, A small-molecule inhibitor of the NLRP3 inflammasome for the treatment of inflammatory diseases, *Nat. Med.*, 2015, **21**(3), 248–255.
77. M. R. Elmore, A. R. Najafi, M. A. Koike, N. N. Dagher, E. E. Spangenberg, R. A. Rice, M. Kitazawa, B. Matusow, H. Nguyen, B. L. West and K. N. Green, Colony-stimulating factor 1 receptor signaling is necessary for microglia viability, unmasking a microglia progenitor cell in the adult brain, *Neuron*, 2014, **82**(2), 380–397.
78. R. Guerreiro, A. Wojtas, J. Bras, M. Carrasquillo, E. Rogaeva, E. Majounie, C. Cruchaga, C. Sassi, J. S. Kauwe, S. Younkin and L. Hazrati, TREM2 variants in Alzheimer's disease, *N. Engl. J. Med.*, 2013, **368**(2), 117–127.
79. D. Erny, A. L. Hrabě de Angelis, D. Jaitin, P. Wieghofer, O. Staszewski, E. David, H. Keren-Shaul, T. Mahlakoiv, K. Jakobshagen, T. Buch and V. Schwierzeck, Host microbiota constantly control maturation and function of microglia in the CNS, *Nat. Neurosci.*, 2015, **18**(7), 965–977.
80. R. M. Ransohoff, How neuroinflammation contributes to neurodegeneration, *Science*, 2016, **353**(6301), 777–783.
81. C. K. Glass, K. Saijo, B. Winner, M. C. Marchetto and F. H. Gage, Mechanisms underlying inflammation in neurodegeneration, *Cell*, 2010, **140**(6), 918–934.
82. S. Voet, M. Prinz and G. van Loo, Microglia in central nervous system inflammation and multiple sclerosis pathology, *Trends Mol. Med.*, 2019, **25**(2), 112–123.
83. F. L. Heppner, R. M. Ransohoff and B. Becher, Immune attack: the role of inflammation in Alzheimer disease, *Nat. Rev. Neurosci.*, 2015, **16**(6), 358–372.
84. S. Hickman, S. Izzy, P. Sen, L. Morsett and J. El Khoury, Microglia in neurodegeneration, *Nat. Neurosci.*, 2018, **21**(10), 1359–1369.
85. D. V. Hansen, J. E. Hanson and M. Sheng, Microglia in Alzheimer's disease, *J. Cell Biol.*, 2018, **217**(2), 459–472.
86. C. Ising, C. Venegas, S. Zhang, H. Scheiblich, S. V. Schmidt, A. Vieira-Saecker, S. Schwartz, S. Albasset, R. M. McManus, D. Tejera and A. Griep, NLRP3 inflammasome activation drives tau pathology, *Nature*, 2019, **575**(7784), 669–673.
87. Y. He, H. Hara and G. Núñez, Mechanism and regulation of NLRP3 inflammasome activation, *Trends Biochem. Sci.*, 2016, **41**(12), 1012–1021.

88. J. Woodcock and L. M. LaVange, Master protocols to study multiple therapies, multiple diseases, or both, *N. Engl. J. Med.*, 2017, **377**(1), 62–70.
89. D. N. Cagney, J. Sul, R. Y. Huang, K. L. Ligon, P. Y. Wen and B. M. Alexander, The FDA NIH Biomarkers, EndpointS, and other Tools (BEST) resource in neuro-oncology, *Neuro-Oncology*, 2018, **20**(9), 1162–1172.
90. T. Notter, J. M. Coughlin, A. Sawa and U. Meyer, Reconceptualization of translocator protein as a biomarker of neuroinflammation in psychiatry, *Mol. Psychiatry*, 2018, **23**(1), 36–47.
91. S. J. Pittock, A. Berthele, K. Fujihara, H. J. Kim, M. Levy, J. Palace, I. Nakashima, M. Terzi, N. Totolyan, S. Viswanathan and K. C. Wang, Eculizumab in aquaporin-4–positive neuromyelitis optica spectrum disorder, *N. Engl. J. Med.*, 2019, **381**(7), 614–625.
92. C. Ly, A. C. Greb, L. P. Cameron, J. M. Wong, E. V. Barragan, P. C. Wilson, K. F. Burbach, S. S. Zarandi, A. Sood, M. R. Paddy and W. C. Duim, Psychedelics promote structural and functional neural plasticity, *Cell Rep.*, 2018, **23**(11), 3170–3182.
93. Z. He and Y. Jin, Intrinsic control of axon regeneration, *Neuron*, 2016, **90**(3), 437–451.
94. K. K. Park, K. Liu, Y. Hu, P. D. Smith, C. Wang, B. Cai, B. Xu, L. Connolly, I. Kramvis, M. Sahin and Z. He, Promoting axon regeneration in the adult CNS by modulation of the PTEN/mTOR pathway, *Science*, 2008, **322**(5903), 963–966.
95. N. Li, B. Lee, R. J. Liu, M. Banasr, J. M. Dwyer, M. Iwata, X. Y. Li, G. Aghajanian and R. S. Duman, mTOR-dependent synapse formation underlies the rapid antidepressant effects of NMDA antagonists, *Science*, 2010, **329**(5994), 959–964.
96. C. Ly, A. C. Greb, L. P. Cameron, J. M. Wong, E. V. Barragan, P. C. Wilson, K. F. Burbach, S. S. Zarandi, A. Sood, M. R. Paddy and W. C. Duim, Psychedelics promote structural and functional neural plasticity, *Cell Rep.*, 2018, **23**(11), 3170–3182.
97. E. J. Bradbury, L. D. Moon, R. J. Popat, V. R. King, G. S. Bennett, P. N. Patel, J. W. Fawcett and S. B. McMahon, Chondroitinase ABC promotes functional recovery after spinal cord injury, *Nature*, 2002, **416**(6881), 636–640.
98. A. J. Green, J. M. Gelfand, B. A. Cree, C. Bevan, W. J. Boscardin, F. Mei, J. Inman, S. Arnow, M. Devereux, A. Abounasr and H. Nobuta, Clemastine fumarate as a remyelinating therapy for multiple sclerosis (ReBUILD): a randomised, controlled, double-blind, crossover trial, *Lancet*, 2017, **390**(10111), 2481–2489.
99. F. Gómez-Pinilla, Brain foods: the effects of nutrients on brain function, *Nat. Rev. Neurosci.*, 2008, **9**(7), 568–578.
100. C. H. Hillman, K. I. Erickson and A. F. Kramer, Be smart, exercise your heart: exercise effects on brain and cognition, *Nat. Rev. Neurosci.*, 2008, **9**(1), 58–65.

101. B. S. McEwen, Physiology and neurobiology of stress and adaptation: central role of the brain, *Physiol. Rev.*, 2007, **87**(3), 873–904.
102. S. Diekelmann and J. Born, The memory function of sleep, *Nat. Rev. Neurosci.*, 2010, **11**(2), 114–126.
103. L. Xie, H. Kang, Q. Xu, M. J. Chen, Y. Liao, M. Thiyagarajan, J. O'Donnell, D. J. Christensen, C. Nicholson, J. J. Iliff and T. Takano, Sleep drives metabolite clearance from the adult brain, *Science*, 2013, **342**(6156), 373–377.
104. S. Hattar, H. W. Liao, M. Takao, D. M. Berson and K. W. Yau, Melanopsin-containing retinal ganglion cells: architecture, projections, and intrinsic photosensitivity, *Science*, 2002, **295**(5557), 1065–1070.
105. C. A. Czeisler, J. S. Allan, S. H. Strogatz, J. M. Ronda, R. Sánchez, C. D. Ríos, W. O. Freitag, G. S. Richardson and R. E. Kronauer, Bright light resets the human circadian pacemaker independent of the timing of the sleep-wake cycle, *Science*, 1986, **233**(4764), 667–671.
106. A. P. Alivisatos, M. Chun, G. M. Church, R. J. Greenspan, M. L. Roukes and R. Yuste, The brain activity map project and the challenge of functional connectomics, *Neuron*, 2012, **74**(6), 970–974.
107. F. S. Collins and H. Varmus, A new initiative on precision medicine, *N. Engl. J. Med.*, 2015, **372**(9), 793–795.
108. E. J. Topol, High-performance medicine: the convergence of human and artificial intelligence, *Nat. Med.*, 2019, **25**(1), 44–56.

Subject Index

1-methyl-4-phenylpyridinium (MPP^+), 125
2-arachidonoylglycerol (2-AG), 13, 66
4-hydrononeal, 23
4-hydroxynonenal (4-HNE), 146, 147
5-hydroxytryptamine (5-HT), 60
5-hydroxytryptophan (5-HTP), 60
8-hydroxyguanine, 23

β-adrenergic receptors, 63
β-hydroxybutyrate (β-HB), 126
γ-aminobutyric acid (GABA) neurotransmitter system, 56–58

acetylcholine, 49, 64–65
acetylcholinesterase (AChE), 64
acetylcholine system, 65
AChE. *See* acetylcholinesterase (AChE)
AD. *See* Alzheimer's disease (AD)
adaptogens and herbal extracts, 160
ADHD. *See* attention-deficit hyperactivity disorder (ADHD)
aging, oxidative stress and, 150–151
AI. *See* artificial intelligence (AI)
alpha-amino-3-hydroxyl-5-methyl-4-isoxazole-propionic acid-type glutamate receptors, 19
ALS. *See* amyotrophic lateral sclerosis (ALS)
Alzheimer's disease (AD), 93, 115–117, 154–155, 190–191, 217, 221–222, 268
  clinical relevance, 123–124
  ketogenic diet (KD), role in AD etiopathogenesis, 122–123
  medicinal foods, 123
AMPK (AMP-activated protein kinase) and energy sensing, 106
amygdala, 6
amyloid-β (Aβ) oligomers, 23
amyloid precursor protein (APP), 187
amyotrophic lateral sclerosis (ALS), 118–119, 150, 157, 262
anandamide, 13, 66
anesthetics, 57
Angelman and Prader–Willi syndromes, 215
ANLS. *See* astrocyte–neuron lactate shuttle (ANLS)
anti-inflammatory agents, 158
antioxidant-based drugs, 157
antioxidant defense systems, 143
  enzymatic antioxidants, 143
  and Nrf2 pathway, 112–113
  non-enzymatic antioxidants, 143–144
antioxidant response element (ARE), 144
antioxidants, dietary, 160
antisense oligonucleotides (ASOs), 231, 263, 363
Apolipoprotein E (APOE) gene, 263
apoptosis, 19
APP. *See* amyloid precursor protein (APP)
ARE. *See* antioxidant response element (ARE)

aromatic L-amino acid decarboxylase (AADC), 115
artificial intelligence (AI), 260
ASD. *See* autism spectrum disorder (ASD)
ASOs. *See* antisense oligonucleotides (ASOs)
astrocyte–neuron lactate shuttle (ANLS), 105
astrocytes, 4, 356–358
    and their regulatory role, 152–153
astrocytic uptake and metabolism, 113
ATP, 16
attention-deficit hyperactivity disorder (ADHD), 60
autism spectrum disorder (ASD), 204, 214, 215, 219–220
axon, 4

Bacopa monnieri, 160
barbiturates, 57
basal ganglia, 6, 59
BBB. *See* blood–brain barrier (BBB)
BDNF. *See* brain-derived neurotrophic factor (BDNF)
benzodiazepines, 57
biomarker discovery and precision neurochemistry, 353–356
biomarkers. *See also* precision medicine (PM) in neurology
    applications for, in neurological disorders, 277
    types of, in neurology, 261
        genetic biomarkers, 261–264
        metabolomic biomarkers, 267–269
        neuroimaging biomarkers, 269–272
        proteomic biomarkers, 264–267
biomarkers and precision medicine, challenges in implementing, 277
    cost and accessibility of precision medicine in neurology, 280
    ethical and privacy issues with genetic data, 279
    future directions and innovations, 284–286
    integrating multi-omics data for comprehensive disease profiling, 280
    need for global cooperation in neurological research, 280–281
blood–brain barrier (BBB), 4, 5, 21–22, 105
    breakdown of, 154
brain-derived neurotrophic factor (BDNF), 158
brain disorders, lipid dysregulation in, 182–183
brain energy demands and mitochondrial centrality, 103
brain regions, neuroanatomy of, 5–8
brainstem, 4, 7

caffeine, 366
calcium (Ca^{2+}) activity and mitochondrial function, 108
    Ca^{2+} overload, risks of, 109
        Ca^{2+} overload in disease pathology, 110
        moderate mitochondrial Ca^{2+} overload, 110
        protective mechanisms, 110
        severe Ca^{2+} overload and permeability transition, 110
    Ca^{2+} regulation of TCA cycle and respiration, 108–109
calcium, excitotoxicity, and ROS, 148
cAMP. *See* cyclic adenosine monophosphate (cAMP)
carotenoids, 144

catalase (CAT), 107, 143
catecholamine synthesis, 115
    energy and mitochondrial requirements, 115
CB1 receptors, 176
CB2 receptors, 176
cellular and molecular aspects of neurochemistry, 15–17
central nervous system (CNS), 1, 21, 52, 64
cerebellum, 7
cerebral cortex, 6
cerebrospinal fluid (CSF), 353
cerebrum, 4
CEST. *See* chemical-exchange saturation transfer (CEST)
ChAT. *See* choline acetyltransferase (ChAT)
chemical-exchange saturation transfer (CEST), 342
choline acetyltransferase (ChAT), 64
cholinesterase inhibitors, 23
chondroitin sulfate proteoglycans (CSPGs), 363
chromatin remodelling, 208–209, 225
circRNAs. *See* circular RNAs (circRNAs)
circular RNAs (circRNAs), 208, 212–214
CNS. *See* central nervous system (CNS)
cognitive and environmental enrichment, 160–161
corticotropin-releasing hormone receptor 2 (CRFR2), 94
COVID-19 pandemic, 92
CSF. *See* cerebrospinal fluid (CSF)
CSPGs. *See* chondroitin sulfate proteoglycans (CSPGs)
cyclic adenosine monophosphate (cAMP), 63
cytokines and inflammatory mediators, 153

damage-associated molecular patterns (DAMPs), 151
DAMPs. *See* damage-associated molecular patterns (DAMPs)
dementia, 122
depolarization, 9
depolarization-induced suppression of inhibition (DSI), 66
DHA. *See* docosahexaenoic acid (DHA)
disease-linked manifestations of oxidative stress, 149–150
DNA, 149
DNA methylation, 205–206, 223–224
docosahexaenoic acid (DHA), 160, 188
dopamine, 48, 49, 51, 58–60
dopaminergic systems, 12
DSI. *See* depolarization-induced suppression of inhibition (DSI)
dysbiosis, 310

EAATs. *See* excitatory amino-acid transporters (EAATs)
ECM. *See* extracellular matrix (ECM)
ECS. *See* endocannabinoid system (ECS)
eicosanoids, 171
eicosapentaenoic acid (EPA), 160, 188
electrical and sensory stimulation, 159
electron transport chain (ETC), 103, 140
endocannabinoids, 13, 48
endocannabinoid signaling *via* GPCRs, 176
    CB1 receptors, 176
    CB2 receptors, 176
    GPR55 and GPR18, 177
endocannabinoid system (ECS), 188–189
endoplasmic reticulum (ER), 108
endorphins, 13, 49, 66
energetic and mitochondrial integration, 114

energy and mitochondrial requirements, 115
energy metabolism and mitochondrial function, neurochemical pathways in, 102
brain energy demands and mitochondrial centrality, 103
calcium (Ca^{2+}) activity and mitochondrial function, 108
Ca^{2+} overload, risks of, 109–110
Ca^{2+} regulation of TCA cycle and respiration, 108–109
future perspectives, 126–127
mitochondrial basics in neurons
neuronal mitochondrial distribution and trafficking, 104
structure, dynamics, and ATP production, 103–104
mitochondrial dysfunction in brain disorders
epilepsy, 119–120
mood disorders, 119
neurodegenerative diseases (NDDs), 115–119
neurotransmitter metabolism and energy cost, 113
catecholamine synthesis, 115
gamma-amino butyric acid (GABA) shunt and anaplerosis, 114–115
glutamate–glutamine cycle, 113–114
primary energy substrates
astrocyte–neuron lactate shuttle (ANLS) 105
glucose and glycolysis, 104–105
ketone bodies and alternative fuels, 105–106
redox signaling and ROS
antioxidant defenses and Nrf2 pathway, 112–113
sources and physiological roles, 110–112
regulatory pathways
AMPK (AMP-activated protein kinase) and energy sensing, 106
mammalian target of rapamycin (mTOR) and anabolic balance, 108
mitochondrial biogenesis and PGC-1α, 107
sirtuins and NAD^+ signaling, 107
therapeutic interventions
ketogenic approaches and emerging therapies, 122–126
mitochondrial protectors and antioxidants, 120–122
enkephalins, 13
Enterobacteriaceae, 126
environmental and lifestyle modulation of neurochemistry, 365
physical activity, stress and recovery, and environmental exposures, 367–368
sleep, circadian rhythms, and light, 366–367
environmental influence on neurochemistry, 297
dietary components and neurochemical modulation, 307–311
environmental stressors and neurochemical consequences, 312–315
future perspective, 318–319

heavy metals, 301–303
neurotoxicity and contaminants, 303–307
prevention strategies for environmental neurotoxicity, 315–318
enzymatic antioxidants, 143
EPA. *See* eicosapentaenoic acid (EPA)
epigenetic and transcriptomic crosstalk, 223
epigenetic control of transcription, 223–225
non-coding RNAs, 226–227
transcription factors as epigenetic modulators, 225–226
epigenetic and transcriptomic regulations in brain disorders
challenges and future directions, 232
annotation, integration, and interpretation, 233
biological and cellular complexity, 232
future outlook, 233–234
temporal and spatial dynamics of regulation, 232–233
translational and ethical barriers, 233
technological advances and methodologies, 227
epigenomic and transcriptomic profiling, 228
experimental models and computational resources, 229
functional genomics and epigenome editing, 228–229
multi-omics integration, 228
therapeutic implications, 230
epigenetic drugs, 230–231
epigenome editing, 231
neuroprotective and reprogramming strategies, 232
personalized medicine and biomarkers, 231
RNA-based therapeutics, 231
epigenetic drugs, 230–231
epigenetic dysregulation in brain disorders, 214. *See also* transcriptomic alterations in brain disorders
neurodegenerative diseases
Alzheimer's disease (AD), 217
Huntington's disease (HD), 217–218
Parkinson's disease (PD), 217
neurodevelopmental disorders, 214
Angelman and Prader–Willi syndromes, 215
autism spectrum disorder (ASD), 215
epilepsy, 215–216
Rett syndrome, 215
neuropsychiatric disorders, 216
major depressive disorder (MDD), 216–217
schizophrenia, 216
epigenetic mechanisms, 205
chromatin remodelling, 208–209
DNA methylation, 205–206
histone modifications, 206–207
non-coding RNAs, 207–208
epigenetics, 204
epigenome editing, 231
epilepsy, 119–120, 215–216
epinephrine, 12
ER. *See* endoplasmic reticulum (ER)
ER oxidoreductin-1 (ERO1), 142

ETC. *See* electron transport chain (ETC)
EVOO. *See* extra-virgin olive oil (EVOO)
excitatory amino-acid transporters (EAATs), 55, 56
exercise and physical activity, 158
experimental techniques in neurochemistry and neuroanatomy, 26–27
extracellular matrix (ECM), 363
extra-virgin olive oil (EVOO), 123

FAAH. *See* fatty acid amide hydrolase (FAAH)
fast-scan cyclic voltammetry (FSCV), 343
fatty acid amide hydrolase (FAAH), 187
fingolimod, 177
flavin adenine dinucleotide ($FADH_2$), 108
focused ultrasound (FUS), 343
FSCV. *See* fast-scan cyclic voltammetry (FSCV)
functional homeostasis, 7
FUS. *See* focused ultrasound (FUS)
future directions in neurochemistry and brain disorders, 337
  biomarker discovery and precision neurochemistry, 353–356
  environmental and lifestyle modulation of neurochemistry, 365–368
  neurochemical research tools, advancing, 341–353
  neuroimmune interface and glial modulation, 356–362
  neuroregeneration, synaptic plasticity, and chemical repair, 362–365

GAD. *See* glutamic acid decarboxylase (GAD)
GAD65, 57
GAD67, 57
gamma-amino butyric acid (GABA), 2, 12, 48
  GABA transaminase (GABA-T), 114
  shunt and anaplerosis, 114–115
gastrointestinal (GI) peptides, 88–89
genetic biomarkers, 261–264
genomic sequencing and CRISPR technology, advances in, 282–284
Ginkgo biloba, 160
GI peptides. *See* gastrointestinal (GI) peptides
glial cells, 2, 4, 5
glial involvement in oxidative homeostasis, 148
glucose and glycolysis, 104–105
glutamate, 2, 12, 48, 55
glutamate–glutamine cycle, 113
  astrocytic uptake and metabolism, 113
  energetic and mitochondrial integration, 114
  neuronal replenishment, 114
glutamate neurotransmitter system, 52–56
glutamic acid decarboxylase (GAD), 57
glutamine, 55
glutathione (GSH), 143
glutathione peroxidase (GPx), 143
glycine, 66
Golgi apparatus, 13, 85
GPCRs. *See* G-protein-coupled receptors (GPCRs)
GPR18, 177
GPR55, 177
G-protein-coupled receptors (GPCRs), 57, 59, 61, 63, 64, 83, 179
  GPCR–lipid mediator crosstalk and neuroinflammation, 177–178
  and lipid mediators in neurology, 175
    leukotriene GPCRs, 176
    prostaglandin GPCRs, 176
  lysophospholipid GPCRs, 177

GPx. *See* glutathione peroxidase (GPx)
GSH. *See* glutathione (GSH)

HD. *See* Huntington's disease (HD)
heavy metals, 301–303
high-performance liquid
  chromatography (HPLC), 27
histamine, 49, 65, 67
histone deacetylase (HDAC)
  inhibitors, 230
histone modifications, 206–207, 224
HPA axis. *See* hypothalamic–
  pituitary–adrenal (HPA) axis
human brain, schematic
  representation of, 51
human leukocyte antigen (HLA)
  alleles, 263
Huntington's disease (HD), 192,
  217–218, 222
hyperdopaminergic states, 60
hyperphosphorylated tau tangles, 23
hypothalamic–pituitary–adrenal
  (HPA) axis, 63, 123, 298
hypothalamic releasing factors, 86
hypothalamus, 7

induced pluripotent stem cells
  (iPSCs), 229
inducible nitric oxide synthase
  (iNOS), 148
iNOS. *See* inducible nitric oxide
  synthase (iNOS)
ion channel and excitotoxicity
  modulators, 158
ionotropic glutamate receptors
  (iGluRs), 55
ionotropic receptors, 10, 50
iPSCs. *See* induced pluripotent stem
  cells (iPSCs)

KD. *See* ketogenic diet (KD)
ketogenic approaches and emerging
  therapies
    Alzheimer's disease (AD)
      clinical relevance,
        123–124
      ketogenic diet (KD),
        role in AD etiopatho-
        genesis, 122–123
      medicinal foods, 123
    Parkinson's Disease (PD)
      clinical trials, 126
      ketogenic diet (KD),
        role in PD etiopatho-
        genesis, 124–126
ketogenic diet (KD)
    Alzheimer's disease (AD), role
      in etiopathogenesis,
      122–123
    Parkinson's disease (PD), role
      in etiopathogenesis,
      124–126
ketone bodies and alternative fuels,
  105–106

L-3,4-dihydroxyphenylalanine
  (L-DOPA), 58–59, 60, 62
LC. *See* locus coeruleus (LC)
leukotriene GPCRs, 176
ligand-gated ion channels, 15, 61
limbic system, 6
lipid-based biomarkers and
  therapeutics, 183
    endocannabinoid system as a
      potential therapeutic target,
      188–189
    targeting lipid pathways for
      treatment, 186–188
lipid-based drug development,
  challenges in, 192–193
    personalized lipid therapies,
      potential of, 193–194
lipid dysregulation in brain
  disorders, 182–183
lipidomics and brain research,
  advances in, 189
    Alzheimer's disease,
      190–191
    Huntington's disease,
      192
    Parkinson's disease, 191
    schizophrenia, 192

lipid–protein interactions in neural function, 178
intracellular signaling cascades involving lipids, 179–180
synaptic plasticity, impact of lipid modifications on, 180–181
lipids, types of, involved in signaling, 172
lipid signaling, 170–172
lipid signaling pathways in the brain, 172–175
endocannabinoid signaling *via* GPCRs, 176
CB1 receptors, 176
CB2 receptors, 176
GPR55 and GPR18, 177
GPCR–lipid mediator crosstalk and neuroinflammation, 177–178
G-protein-coupled receptors (GPCRs) and lipid mediators in neurology, 175
leukotriene GPCRs, 176
prostaglandin GPCRs, 176
lysophospholipid GPCRs, 177
lncRNAs. *See* long non-coding RNAs (lncRNAs)
locus coeruleus (LC), 62
long non-coding RNAs (lncRNAs), 208, 212
long-term depression (LTD), 10, 51, 56, 66, 93, 181
long-term potentiation (LTP), 10, 51, 55–56, 64, 92, 93, 181
LPA. *See* lysophosphatidic acid (LPA)
LTD. *See* long-term depression (LTD)
LTP. *See* long-term potentiation (LTP)
lysophosphatidic acid (LPA), 177

machine learning, 260, 264
mAChRs. *See* muscarinic acetylcholine receptors (mAChRs)
MAGL. *See* monoacylglycerol lipase (MAGL)
major depressive disorder (MDD), 216–217, 221
malondialdehyde (MDA), 146, 147
mammalian target of rapamycin (mTOR) and anabolic balance, 108
MAMs. *See* mitochondria-associated membranes (MAMs)
MAPKs, 22
MDA. *See* malondialdehyde (MDA)
MDD. *See* major depressive disorder (MDD)
metabolomic biomarkers, 267–269
metabotropic receptors, 10, 50
microglia, 5, 151–152, 356–358
microRNAs (miRNAs), 207, 212
miRNAs. *See* microRNAs (miRNAs)
mitochondria, 15, 16, 147
mitochondria-associated membranes (MAMs), 108
mitochondrial basics in neurons
neuronal mitochondrial distribution and trafficking, 104
structure, dynamics, and ATP production, 103–104
mitochondrial biogenesis and PGC-1α, 107
mitochondrial calcium ions, 108
mitochondrial DNA (mtDNA), 147
mitochondrial dysfunction in brain disorders
epilepsy, 119–120
mood disorders, 119
neurodegenerative diseases (NDDs)
Alzheimer's disease (AD), 115–117
amyotrophic lateral sclerosis (ALS), 118–119
Parkinson's Disease (PD), 117–118

mitochondrial function, Ca^{2+}
activity and, 108
Ca^{2+} overload, risks of, 109
Ca^{2+} overload in disease pathology, 110
moderate mitochondrial Ca^{2+} overload, 110
protective mechanisms, 110
severe Ca^{2+} overload and permeability transition, 110
Ca^{2+} regulation of TCA cycle and respiration, 108–109
mitochondrial protectors and antioxidants, 120–122
mitochondrial ROS production, 140–141
mitochondrial stabilizers, 157
mitogen-activated protein kinases (MAPKs), 153
molecular level, 19
monoacylglycerol lipase (MAGL), 187
monoamine neurotransmitters, 12
monoamines, 2, 48, 49
mood disorders, 119
MS. *See* multiple sclerosis (MS)
mtDNA. *See* mitochondrial DNA (mtDNA)
multiple sclerosis (MS), 155
muscarinic acetylcholine receptors (mAChRs), 64

N-acetylcysteine (NAC), 157
nAChRs. *See* nicotinic acetylcholine receptors (nAChRs)
NADH. *See* nicotinamide adenine dinucleotide (NADH)
NADPH oxidase (NOX), 141, 148
N-arachidonoylethanolamide, 66
ncRNAs. *See* non-coding RNAs (ncRNAs)
NDDs. *See* neurodegenerative diseases (NDDs)
NE. *See* norepinephrine (NE)
necrosis, 19
nervous system, structure of, 3–5, 8
neural plasticity, neurodevelopment and, 17–20
neural proliferation, 18
neural tissue, intrinsic vulnerability of, 146–147
neuroanatomy
of brain regions, 5–8
experimental techniques in, 26–27
future aspects, 27–30
neurochemical research tools, advancing, 341
biosensors and real-time neurochemical monitoring, 345–349
computational and artificial intelligence-driven integration, 349–353
high-resolution analytical and omics platforms, 344–345
neurochemical signals, 3
neurochemistry, 2
cellular and molecular aspects of, 15–17
experimental techniques in, 26–27
future aspects, 27–30
in neurological and psychiatric disorders, 22–25
of neurotransmitters and neuromodulators, 11–14
neurodegenerative diseases (NDDs)
Alzheimer's disease (AD), 115–117, 217, 221–222
amyotrophic lateral sclerosis (ALS), 118–119
Huntington's disease (HD), 217–218, 222
Parkinson's disease (PD), 117–118, 217, 222

neurodevelopmental disorders, 214
Angelman and Prader–Willi syndromes, 215
autism spectrum disorder (ASD), 215, 219–220
epilepsy, 215–216
Rett syndrome, 215, 220
neurodevelopment and neural plasticity, 17–20
neuroimaging biomarkers, 269–272
neuroimmune interface and glial modulation, 356
biomarkers, trial design, and implementation, 360–362
microglia, astrocytes, and oligodendroglia, 356–358
neuroinflammation endotypes and therapeutic pathways, 358–360
neuroinflammation, 151
across neurological disorders, 154–155
astrocytes and their regulatory role, 152–153
blood–brain barrier, breakdown of, 154
cytokines and inflammatory mediators, 153
interface between neuroinflammation and oxidative stress, 154
microglia, 151–152
physiological role of, 151
resolution and regulation of, 155–156
signaling pathways in, 153–154
neurological and psychiatric disorders, neurochemistry in, 22–25
neurological biomarkers, evolution of, in clinical practice, 259–261
neurological disorders, neuroinflammation across, 154–155
neurological homeostasis, neuropeptides and, 93–94
neuromodulators, 2, 11–14
neuronal communication and synaptic transmission, 8–11
neuronal development, regulation by neuropeptides, 92–93
neuronal migration, 18
neuronal mitochondrial distribution and trafficking, 104
neuronal replenishment, 114
neurons, 2, 4, 9, 15, 49
neuropeptide binding receptors, 91
neuropeptides, 2, 48, 49, 65–67, 82
biosynthesis, 85–86
classification, 86
gastrointestinal (GI) peptides, 88–89
hypothalamic releasing factors, 86
pituitary neuropeptides, 86–88
discovery of, 84
functions in brain, 91–95
and neurological homeostasis, 93–94
neuronal development, regulation by neuropeptides, 92–93
and sensory perception, 94
structure, 85
and systemic inflammatory response, 94–95
neuropeptide S, 93
neuropeptide signaling pathways, development of, 89–91
neuropeptide Y, 13
neuroprotection, 21–22, 139
integrated perspective on, 161–162
neuroprotective and reprogramming strategies, 232
neuropsychiatric disorders, 216
major depressive disorder (MDD), 216–217, 221
schizophrenia, 216, 220–221

neuroregeneration, synaptic plasticity, and chemical repair, 362–365
neurotoxicity and contaminants, 303–307
neurotransmission, 48
neurotransmitter, 4
neurotransmitter metabolism and energy cost, 113
  catecholamine synthesis, 115
    energy and mitochondrial requirements, 115
  gamma-amino butyric acid (GABA) shunt and anaplerosis, 114–115
  glutamate–glutamine cycle, 113
    astrocytic uptake and metabolism, 113
    energetic and mitochondrial integration, 114
    neuronal replenishment, 114
neurotransmitter system, 9, 11–14, 47
  acetylcholine (ACh), 64–65
  dopamine (DA), 58–60
  future research, 67–68
  GABA, 56–58
  glutamate, 52–56
  major, 52
  and neuropeptides, 65–67
  neurotransmission, basics of, 49–52
  norepinephrine (NE), 62–64
  serotonin, 60–62
neurovascular unit (NVU), 21
nicotinamide adenine dinucleotide (NADH), 108
nicotinic acetylcholine receptors (nAChRs), 64
nitro-tyrosine, 23
N-methyl-D-aspartate (NMDA), 23, 67
non-coding RNAs (ncRNAs), 207–208, 226–227
  in transcriptomic control, 211
    circular RNAs (circRNAs), 212–214
    long non-coding RNAs (lncRNAs), 212
    microRNAs, 212
non-enzymatic antioxidants, 143–144
norepinephrine (NE), 48, 49, 62–64
Nrf2 activators and redox modulators, 157–158
Nrf2 pathway, transcriptional control *via*, 144–146
nuclear factor erythroid-2-related factor 2 (Nrf2), 144
NVU. *See* neurovascular unit (NVU)

oligodendrocyte precursor cells (OPCs), 362
oligodendrocytes (OLs), 4, 357
oligodendroglia, 356–358
OLs. *See* oligodendrocytes (OLs)
omega-3 fatty acids, 160
omega-3 polyunsaturated fatty acids, 365
OPCs. *See* oligodendrocyte precursor cells (OPCs)
oxidative stress, molecular mechanism of, 140
  antioxidant defense systems, 143
    enzymatic antioxidants, 143
    non-enzymatic antioxidants, 143–144
  imbalance between ROS/RNS and antioxidant defenses, 146
  reactive oxygen species (ROS), sources of, 140
    exogenous sources of ROS, 142–143
    mitochondrial ROS production, 140–141

oxidative stress, molecular
mechanism of (*continued*)
NADPH oxidases (NOX family), 141
other enzymatic and non-enzymatic sources, 142
peroxisomal and endoplasmic reticulum sources, 141–142
reactive nitrogen species (RNS), sources of, 143
transcriptional control *via* Nrf2 pathway, 144–146
oxidative stress and nervous system, 146
and aging, 150–151
calcium, excitotoxicity, and ROS, 148
dietary limitations and future directions, 163–164
disease-linked manifestations of oxidative stress, 149–150
glial involvement in oxidative homeostasis, 148
mitochondria, 147
neural tissue, intrinsic vulnerability of, 146–147
redox-sensitive signaling and gene regulation, 149
targets of oxidative damage, 148–149
oxidative stress and neuroinflammation, therapeutic approaches for, 156
behavioral and environmental approaches, 160
cognitive and environmental enrichment, 160–161
sleep and circadian regulation, 161
social engagement and emotional health, 161
stress reduction and mindfulness practices, 161
herbal and nutraceutical approaches, 159
adaptogens and herbal extracts, 160
dietary antioxidants, 160
omega-3 fatty acids, 160
polyphenols and flavonoids, 159–160
interface between neuroinflammation and oxidative stress, 154
neuroprotection, integrated perspective on, 161–162
non-pharmacological approaches, 158
electrical and sensory stimulation, 159
exercise and physical activity, 158
nutritional interventions, 159
photobiomodulation and hypoxic conditioning, 159
pharmacological approaches, 157
anti-inflammatory agents, 158
antioxidant-based drugs, 157
ion channel and excitotoxicity modulators, 158
mitochondrial stabilizers, 157
Nrf2 activators and redox modulators, 157–158
oxytocin, 13

PAHs. *See* polycyclic aromatic hydrocarbons (PAHs)
PAMPs. *See* pathogen-associated molecular patterns (PAMPs)
parietal lobe, 6
Parkinson's disease (PD), 3, 6, 48, 59, 60, 117–118, 150, 155, 191, 217, 222
  clinical trials, 126
  ketogenic diet (KD), role in PD etiopathogenesis, 124–126
pathogen-associated molecular patterns (PAMPs), 151
pattern recognition receptors (PRRs), 151
PBDEs. *See* polychlorinated diphenyl ethers (PBDEs)
PD. *See* Parkinson's disease (PD)
PDH. *See* pyruvate dehydrogenase (PDH)
PDI. *See* protein disulfide isomerase (PDI)
perineuronal nets (PNNs), 363
peripheral nervous system (PNS), 1, 4
peroxiredoxins (Prx), 143
peroxisomal and endoplasmic reticulum sources, 141–142
persistent organic pollutants (POPs), 303
personalized lipid therapies, potential of, 193–194
personalized medicine and biomarkers, 231
photobiomodulation and hypoxic conditioning, 159
physical activity, stress and recovery, and environmental exposures, 367–368
pituitary neuropeptides, 86–88
PIWI-interacting RNAs (piRNAs), 207
PNNs. *See* perineuronal nets (PNNs)
PNS. *See* peripheral nervous system (PNS)
polychlorinated diphenyl ethers (PBDEs), 305
polycyclic aromatic hydrocarbons (PAHs), 305
polygenic risk scoring (PRS), 264
polyphenols and flavonoids, 159–160
polyunsaturated fatty acids (PUFAs), 147
POMCs. *See* proopiomelanocortins (POMCs)
POPs. *See* persistent organic pollutants (POPs)
Prader–Willi syndromes, 215
precision medicine (PM) in neurology, 272. *See also* biomarkers
  genetic and environmental considerations, 275
    environmental and lifestyle considerations, 276–277
    genetic considerations, 275–276
  personalized treatment plans based on biomarkers, 274–275
  role in neurological diseases, 273–274
Prevotellaceae, 126
proopiomelanocortins (POMCs), 91
prostaglandin GPCRs, 176
protein carbonyls, 23
protein disulfide isomerase (PDI), 142
proteomic biomarkers, 264–267
PRRs. *See* pattern recognition receptors (PRRs)
PRS. *See* polygenic risk scoring (PRS)
Prx. *See* peroxiredoxins (Prx)
PUFAs. *See* polyunsaturated fatty acids (PUFAs)
pyruvate dehydrogenase (PDH), 108

reactive nitrogen species (RNS), 148
  imbalance between ROS/RNS and antioxidant defenses, 146
  sources of, 143

reactive oxygen species (ROS), 148
imbalance between ROS/RNS and antioxidant defenses, 146
sources of, 140
exogenous sources of ROS, 142–143
mitochondrial ROS production, 140–141
NADPH oxidases (NOX family), 141
other enzymatic and non-enzymatic sources, 142
peroxisomal and endoplasmic reticulum sources, 141–142
reactive nitrogen species (RNS), sources of, 143
receptors, classification of, 9–10
receptor tyrosine kinases (RTKs), 179
redox-sensitive signaling and gene regulation, 149
redox signaling and ROS
antioxidant defenses and Nrf2 pathway, 112–113
sources and physiological roles, 110–112
repolarization, 9
Rett syndrome, 204, 215, 220
reward prediction error (RPE), 59
RNA-based therapeutics, 231
RNAi, 363
RNA splicing and isoform diversity, 211
RNS. *See* reactive nitrogen species (RNS)
ROS. *See* reactive oxygen species (ROS)

saltatory conduction, 9
SCFAs. *See* short-chain fatty acids (SCFAs)
schizophrenia, 48, 192, 216, 220–221
Schwann cells, 5
sensory perception, neuropeptides and, 94
serotonin, 48, 49, 61
serotonin neurotransmitter system, 60–62
short-chain fatty acids (SCFAs), 365–366
siRNAs. *See* small interfering RNAs (siRNAs)
SIRT1, 107
sirtuins and NAD^+ signaling, 107
sleep, circadian rhythms, and light, 366–367
sleep and circadian regulation, 161
small interfering RNAs (siRNAs), 207
social engagement and emotional health, 161
SOD. *See* superoxide dismutase (SOD)
sodium channels, 9
sortilin, 91
sphingolipids, 171
sphingosine-1-phosphate (S1P), 177
spinal cord, 2, 4
spinocerebellum, 7
SSADH. *See* succinic semialdehyde dehydrogenase (SSADH)
stress reduction and mindfulness practices, 161
substance P, 13, 49
substantia nigra pars compacta (SNc), 59
succinic semialdehyde dehydrogenase (SSADH), 114
superoxide dismutase (SOD), 143, 147
superoxide dismutase 2 (SOD2), 107
synapses, 17, 47
synaptic plasticity, 17
impact of lipid modifications on, 180–181
synaptic transmission, neuronal communication and, 8–11
systemic inflammatory response, neuropeptides and, 94–95

TBI. *See* traumatic brain injury (TBI)
TCA cycle. *See* tricarboxylic acid (TCA) cycle
TH. *See* tyrosine hydroxylase (TH)
thalamus, 7
TPH. *See* tryptophan hydroxylase (TPH)
transcription factors and co-regulators, 211
transcriptomic alterations in brain disorders, 218. *See also* epigenetic dysregulation in brain disorders
  neurodegenerative diseases
    Alzheimer's disease (AD), 221–222
    Huntington's disease (HD), 222
    Parkinson's disease (PD), 222
  neurodevelopmental disorders
    autism spectrum disorder (ASD), 219–220
    Rett syndrome, 220
  neuropsychiatric disorders
    major depressive disorder (MDD), 221
    schizophrenia, 220–221
transcriptomic regulation in the brain, 209
  non-coding RNAs in transcriptomic control, 211
    circular RNAs (circRNAs), 212
      functions in brain health, 213–214
    long non-coding RNAs (lncRNAs), 212
      functions in brain health, 212
    microRNAs, 212
      functions in brain health, 212
  RNA splicing and isoform diversity, 211
  transcription factors and co-regulators, 211
traumatic brain injury (TBI), 155
tricarboxylic acid (TCA) cycle, 55, 108–109, 114
tryptophan hydroxylase (TPH), 60
tuberoinfundibular pathway, 59
Turkish Chair, 86
tyrosine hydroxylase (TH), 58

uric acid, 144
urocortin 3 (UCN3), 94

vasopressin, 13
ventral tegmental area (VTA), 59
vesicular glutamate transporters (VGLUTs), 55
vesicular monoamine transporter 2 (VMAT2), 59, 61
vestibulocerebellum, 7
VGLUTs. *See* vesicular glutamate transporters (VGLUTs)
vitamin B6, 57
vitamins, 144
VMAT2. *See* vesicular monoamine transporter 2 (VMAT2)
voltage-gated calcium channels, 55
voltage-gated sodium channels, 15
VTA. *See* ventral tegmental area (VTA)

Withania somnifera, 160